STRATIGRAPHY
Principles and Methods

STRATIGRAPHY
Principles and Methods

Robert M. Schoch
Boston University

VNR VAN NOSTRAND REINHOLD
New York

Library of Congress Catalog Card Number 88–26103

ISBN 0–442–28021–1

Printed in the United States of America

Van Nostrand Reinhold
115 Fifth Avenue
New York, New York 10003

Van Nostrand Reinhold International Company Limited
11 New Fetter Lane
London EC4P 4EE, England

Van Nostrand Reinhold
480 La Trobe Street
Melbourne, Victoria 3000, Australia

Macmillan of Canada
Division of Canada Publishing Corporation
164 Commander Boulevard
Agincourt, Ontario M1S 3C7, Canada

16 15 14 13 12 11 10 9 8 7 6 5 4 3 2 1

Library of Congress Cataloging-in-Publication Data

Schoch, Robert M.
  Stratigraphy : principles and methods.

  Includes index.
  1. Geology, Stratigraphic.   I. Title.
QE651.S34   1989   551.7   88–26103
ISBN 0–422–28021–1

PREFACE

The philosophical basis from which a scientist views his subject is just as important as the factual content of his work since interpretation from the facts and observations of his studies must be coloured and moulded by his philosophical outlook. (Blow, 1979, p. 1391)

Certainly, discussions of real scientific results are far more interesting and entertaining than discussions of words, but not necessarily more important. For we have to express our scientific results in words, and if the words are ambiguous and mean different things to you and to me, they will fail to transmit the results between us. (Rodgers, 1959, p. 684)

As currently practiced, stratigraphy is a broad, all-encompassing discipline within the field of geology. No author in a single volume the size of the present book could hope to deal satisfactorily with all the multifarious aspects and ramifications of stratigraphy; thus I have not attempted to do so here. Rather, I have selected certain issues relevant to stratigraphy, for limited discussion. The primary purpose of this book is to provide the reader with a taste or feel for stratigraphy, or more accurately what can be termed stratigraphic philosophy (or stratigraphical philosophy, if you prefer). Central issues of stratigraphic philosophy as I view it are: What is stratigraphy? How do we arrive at stratigraphic knowledge? And how do we express our hypotheses and conclusions through stratigraphic classifications, terminology, and nomenclature?

This book is an introduction to a science in which there are often more questions than answers. Heated discussions of such basic issues as fundamental principles, terminology, definitions, and classification continue to pervade the stratigraphic community. In this book I have attempted to be relatively nondogmatic, but we are all influenced by our own experiences, training, and biases. In particular, as an American writing in English I have perhaps been unduly influenced by what may be termed the "American school" of stratigraphy, the philosophy of which is exemplified by the *North American Stratigraphic Code* (North American Commission on Stratigraphic Nomenclature [NACSN], 1983; reprinted in this book as Appendix 1). Arguably the American school, however, is currently representative of the dominant view of stratigraphic philosophy held worldwide; the first edition of the *International Stratigraphic Guide,* edited by the American Hollis D. Hedberg (1976), espouses the basic premises of the American school of stratigraphy. In the discussions that follow I do not claim to have covered all bases thoroughly (one must always be selective) or to have arrived at definitive conclusions. It must be stressed that this book is intended only to introduce issues with the hope that the reader's curiosity and intellect will be sparked, causing the reader to delve deeper into the literature and science of stratigraphy.

It is assumed that the reader has a basic understanding of physical and historical geology, including some introductory petrology and petrography, and also some notion as to what stratigraphy is all about. This is not a book on sedimentology, geologic field methods, paleogeography, paleoecology, paleoclimatology, historical geology, global plate tectonics, or any of the other numerous disciplines relevant to stratigraphy but not stratigraphy per se. In my mind sedimentology, for instance, is not stratigraphy (as is further discussed elsewhere in this book) despite the fact that in many universities these two subjects are treated as virtually synonymous, at least at the undergraduate level. The reader also should be warned that this is not a compilation of case studies, nor is it a picture book, as so many general works on stratigraphy seem to be these days. Primarily because of space limitations I have been unable to include any formal discussion of quantitative techniques as applied to stratigraphy. For a thorough introduction to the newly emerging field of quantitative stratigraphy the reader is referred to the work of the International Geological Correlation Programme Project No. 148 (Evaluation and Development of Quantitative Stratigraphic Correlation Techniques), cosponsored by the United Nations Educational, Scientific, and Cultural Organization (UNESCO) and the International Union of Geological Sciences (IUGS), as summarized in Cubitt and Reyment (1982) and Gradstein et al. (1985).

The book is divided into seven chapters. Chapters 1 and 2 provide an overview of stratigraphy and the physical constituents (namely, rocks) of the science. Chapter 3 outlines some conceptual foundations and broad principles generally utilized in stratigraphy, and in Chapter 4 various widely adopted conventions of stratigraphic classification and nomenclature receive attention. The final three chapters discuss methods of stratigraphic classification and synthesis, beginning with the actual rocks (lithostratigraphy) in Chapter 5 and becoming increasingly less concrete and more abstract in Chapters 6 (biostratigraphy and magnetostratigraphy) and 7 (chronostratigraphy and geochronology). Some authors consider lithostratigraphy, alternately referred to as physical stratigraphy, descriptive stratigraphy, or prostratigraphy, to be the real core of stratigraphy, whereas others consider lithostratigraphy to be merely a method of classifying rocks before one gets on to the real business of stratigraphy, which is biostratigraphy and/or chronostratigraphy. In a sense chronostratigraphy and geochronology may be seen as extremely abstract fields, as they deal not with material objects per se but with the temporal relations of events recorded in material objects such as rocks. Aside from numerical dating techniques, two of the major ways of deducing the temporal relationships of rocks are through biostratigraphy and magnetostratigraphy; thus these two disciplines might be thought of as transitional between lithostratigraphy and chronostratigraphy/geochronology. Given recent advances in numerical dating and geochronology, it has been suggested that eventually, someday, we may have "a little black box into which we only have to pop our rock specimen for its age to be read automatically on a dial" (Ager, 1981, p. 78). If this were the case, then much of the science of, and controversy over, chronostratigraphy and geochronology discussed herein might become moot. Perhaps only lithostratigraphy would remain a viable and necessary science; if nothing else, lithostratigraphy would be necessary for the location and description of mineral deposits and other economic resources found in the earth.

We are a long way, however, from having numerical dating techniques totally supplant, rather than supplement, conventional stratigraphic methods.

This book has been written in Attleboro, Massachusetts, while I held the position of assistant professor in the Division of Science, College of Basic Studies, Boston University. I thank Brendan Gilbane (dean of the College of Basic Studies) and Charles P. Fogg (chairman of the Division of Science) for their support. Here I must also thank numerous colleagues who have contributed to my studies through correspondence and discussion, and by forwarding their reprints. My debt to such colleagues is manifested in my citations of their papers. I wish to acknowledge in particular the interest and support that Charles S. Hutchinson, Jr. has shown for this project. I especially thank my wife, Cynthia, for her never failing support and help; my young son Nicholas also has been encouraging of my endeavors. I appreciate the professionalism of the personnel at Van Nostrand Reinhold who have been involved in the production and publication of this volume, especially Susan H. Munger, Ella Harwood, Alberta Gordon, and Constance MacDonald. In the end, of course, I am solely responsible for the contents of this book.

Robert M. Schoch
Attleboro, Massachusetts

REFERENCES

Ager, D. V., 1981, *The Nature of the Stratigraphical Record, 2nd edition,* Macmillan, London.
Blow, W. H., 1979, *The Cainozoic Globigerinida,* E. J. Brill, Leiden.
Cubitt, J. M., and R. A. Reyment, eds., 1982, *Quantitative Stratigraphic Correlation,* Wiley, Chichester, England.
Gradstein, F. M., F. P. Agterberg, J. C. Brower, and W. S. Schwarzacher, 1985, *Quantitative Stratigraphy,* D. Reidel, Dordrecht, Netherlands.
Hedberg, H. D., ed., 1976, *International Stratigraphic Guide,* by the International Subcommission on Stratigraphic Classification, Wiley, New York.
North American Commission on Stratigraphic Nomenclature, 1983, North American Stratigraphic Code, *Amer. Assoc. Petrol. Geol. Bull.* 67:841-875.
Rodgers, J., 1959, The meaning of correlation, *Amer. Jour. Sci.* 257:684-691.

CONTENTS

STRATIGRAPHY
Principles and Methods

Chapter 1

THE SCIENCE OF STRATIGRAPHY

AN OVERVIEW OF THE SCOPE OF STRATIGRAPHY

Stratigraphy is fundamental to all geological investigation. Stratigraphic methods, techniques, and principles are applicable to all earth materials and are used in the study of the geometry, structure, sequence, and history of any rock body. Such rock bodies need not even be limited to those on our Earth or to the "rock bodies" produced by nature. Stratigraphic methods can and must be applied to the detailed study of the geology of the Moon and other Earth-like planets (cf. Mutch, 1972; Wilhelms, 1987), and are also of value to the archeologist (Harris, 1979). Within mainstream geology, in the study of fields as diverse as structural geology or paleoecology and paleoenvironmental interpretation (Dodd and Stanton, 1981), a sound stratigraphic framework is necessary. All formally named material-based geologic field units are designated according to stratigraphic principles. Originally geology was virtually synonymous with what is now termed stratigraphy, the only other distinct aspect of geological inquiry being mineralogy (B. Conkin and J. Conkin, 1984). No geologist can be ignorant of stratigraphic concepts, methods, and techniques.

Stratigraphy may be considered the heart of historical geology. It is primarily through the study of the relationships of the rocks on the surface of the earth that we are able to reconstruct the earth's history. Stratigraphy deals with the direct, tangible, physical evidence—the rock record—of bygone eras. In the opening sentence of his classic work, Grabau (1913, p. 1) states: "Stratigraphy in its broadest sense may be defined as the inorganic side of Historical Geology, or the development through the successive geologic ages of the earth's rocky framework or *lithosphere*" (italics in the original). Here Grabau is referring primarily to lithostratigraphy. Of course, stratigraphy today also includes the study of the organic remains, the fossils, found in many rocks; these are the domain of the stratigraphic paleontologist, biostratigrapher, and biochronologist. Through stratigraphic studies one can elucidate the all-important spatial and temporal relations of extinct organisms. Furthermore, organisms often have had a profound influence on the composition and structure of rocks deposited on the earth's surface (Lapo, 1982), and likewise rocks record the environmental conditions with which ancient organisms had to deal. Many workers contend that fossils "are the only direct evidence we have of the history of life on our planet, and in particular of the course of evolution" (Hallam, 1977b, p. v; see also Schoch, 1986b, p. 6; Schoch, 1987) and thus are essential to a full reconstruction of the history of the earth. As is discussed later in this book, the unidirectional nature of evolution often means that fossil organisms become ideal clasts or markers that can be used in correlating rock bodies.

Stratigraphy often is closely associated with sedimentology and the study of sedimentary rocks (e.g., Boggs, 1987; Brenner and McHargue, 1988; Dunbar and Rodgers, 1957; Fritz and Moore, 1988; Krumbein and Sloss, 1963); until about the 1920s there was no real distinction between the disciplines of stratigraphy and sedimentology (Krumbein and Sloss, 1963, p. ix). However, this is no longer the case, and stratigraphy should not be considered synonymous with sedimentary geology, although sedimentological studies often are of great importance to the stratigrapher. Sedimentology in the modern sense is concerned primarily with the study of the description, classification, and interpretation of sediments. Much of modern sedimentology deals specifically with the reconstruction of ancient sedimentary environments. The present book explicitly excludes sedimentological discussions per se.

The term stratigraphy was first coined in 1852 by d'Orbigny as *stratigraphie* (d'Orbigny, 1852; see B. Conkin and J. Conkin, 1984, p. 39; the adjective "stratigraphical" had been used as early as 1817 by William Smith, and the phrases "stratigraphical geology" and "stratigraphic geology" also are sometimes used synonymously with stratigraphy). Stratigraphy comes from the Latin *stratum* and the Greek *graphia,* and literally means the descriptive writing about, or more loosely the study of, strata or stratified rocks (Hedberg, 1976, p. 12; Storey and Patterson, 1959: as has been pointed out by Dunbar and Rodgers, 1957, and Storey and Patterson, 1959, a term such as stratilogy or stratology [*logos*—study of] might be more appropriate etymologically than the term stratigraphy, as a label corresponding to the actual nature of the science). A geologic stratum refers to a layer of rock, a tabular body of rock, or a planar unit of rock (Dunbar and Rodgers, 1957, p. xi; Hedberg, 1976, p. 12). The most typical and common stratified rocks are of sedimentary origin (thus the traditional close link between sedimentology and stratigraphy) and were originally deposited as relatively thin (compared to the areal dimensions of the rock body) sheets on some surface of accumulation (Dunbar and Rodgers, 1957). However, igneous and metamorphic rocks may also show layering or stratification (for instance, original sedimentary bedding in a metamorphic rock; lava flows and ash falls among igneous rocks) and thus fall within the purview of traditional or classical stratigraphy. Such stratified rocks, be they of sedimentary or igneous origin, normally represent surface accumulations that were deposited in layers and conform to the principle of superposition (see below).

Stratigraphy is primarily concerned with the observation, description, and interpretation of stratified rocks. One of the prime concerns of stratigraphy is determination of the succession, age relations, geographic distribution, and temporal correlations of rock strata. Indeed, some authors regard stratigraphy as pertaining solely to the temporal relationships of strata: "the relationship between rock and time is stratigraphy" (Storey and Patterson, 1959, p. 711). "*Stratigraphy* studies the beds of the earth's crust, the rocks, from the point of view of their chronological succession and their geographic distribution" (Gignoux, 1955, p. 1, italics in the original). In the latter quotation Gignoux essentially equates stratigraphy with historical geology, especially as it relates to the reconstruction of changing paleogeographies through time. However, as commonly used stratigraphy now encompasses not only the temporal relations of rocks, or their temporal relations and geographic distributions, but all aspects of rocks in a four-dimensional space–time continuum. "Stra-

tigraphy is the study of the spatial distribution, chronological relations, and forma-
tive processes of layered rocks" (Wilhelms, 1987, p. vii).

Given most stratigraphers' interest in the temporal relationships of rocks, the discipline of geochronology, often more or less intimately associated with stratigraphy,
now often is subsumed under the discipline of stratigraphy in the broad sense (cf.
Harland, 1978). Definable as "the science of dating and determining the time sequence of events in the history of the Earth" (Hedberg, 1976, p. 15), geochronology
includes within its domain (in some authors' minds is synonymous with) geochronometry, the subdiscipline concerned with the quantitative measurement of geologic
time in terms of some standard unit such as years (for further discussion of these
topics, see Chapter 7).

Traditionally many workers have strictly limited the domain of stratigraphy to
the study of surface-accumulated rocks, as indicated by the following quotation:
"The distinction between surface-deposited (stratigraphic) layers and post-depositionally emplaced rocks is fundamental; post-depositionally emplaced bodies, regardless of whether they may be petrologically classed as igneous, metamorphic,
or 'sedimentary' (sandstone dikes and salt intrusions), should not be regarded as
stratigraphic units" (Wheeler, 1959, pp. 694-695). However, it is now increasingly
common to find the term stratigraphy used in a wider context. Interpreted in a broad
sense, stratigraphy can include the study of the composition, properties, geometry,
sequence, and history of virtually all stratified (and associated nonstratified) rocks
and earth materials. In this context, the International Subcommission on Stratigraphic Classification (ISSC) of the International Union of Geological Sciences
Commission on Stratigraphy has defined stratigraphy "simply as the *science of rock
strata*" (Hedberg, 1976, p. 12, italics in the original). The ISSC went on to state:

> As such, stratigraphy is concerned not only with the original succession and age
> relations of rock strata, but also with their form, distribution, lithologic composi-
> tion, fossil content, geophysical and geochemical properties—indeed, with all
> characters, properties, and attributes of rocks as *strata,* and their interpretation
> in terms of environment or mode of origin, and of geologic history. All classes
> of rocks—igneous and metamorphic as well as sedimentary, unconsolidated as
> well as consolidated—fall within the general scope of stratigraphy and strati-
> graphic classification. Some nonstratiform rock bodies are considered under stra-
> tigraphy because of their association with or close relation to rock strata. (Hed-
> berg, 1976, p. 12, italics in the original)

Also interpreting stratigraphy broadly, the North American Commission on Stratigraphic Nomenclature (NACSN) has published the following statement:

> *Stratigraphic procedures* and principles, although developed initially to bring or-
> der to strata and events recorded therein, are applicable to all earth materials, not
> solely to strata. They promote systematic and rigorous study of the composition,
> geometry, sequence, history, and genesis of rocks and unconsolidated materials.
> They provide the framework within which time and space relations among rock
> bodies that constitute the Earth are ordered systematically. Stratigraphic proce-
> dures are used not only to reconstruct the history of the Earth and extra-terrestrial
> bodies, but also to define the distribution and geometry of some commodities

needed by society. *Stratigraphic classification* systematically arranges and partitions bodies of rock or unconsolidated materials of the Earth's crust into units based on their inherent properties or attributes. (NACSN, 1983, p. 847, italics in the original; for the convenience of the reader the 1983 *North American Stratigraphic Code* is reprinted in this volume as Appendix 1.)

Another classic definition and elaboration of the concept of stratigraphy reads as follows:

Stratigraphy is the branch of geologic science that has to do with the definition and description of major and minor natural divisions of rocks, mainly sedimentary, and interpretation of their significance in geologic history. It involves determination of the sequence of rocks, both locally and in the general time scale, tracing their areal distribution, observation of lateral and vertical variations in their characters, correlation of equivalent but possibly widely dissimilar units, and, finally, study of conditions or geologic events that are involved in their genesis. There are no other branches of geology to which stratigraphy is not more or less intimately related, both depending on them and serving them. It is not well to assert that stratigraphy, rather than any other member of the geologic body, represents the heart, but it is evident to everyone that the subject of geologic history, which is the fundamental objective of stratigraphic research, is vital to all divisions of earth science. (Moore, 1941, p. 179)

Here it should be noted in particular that Moore explicitly states that stratigraphy is concerned with "natural divisions or rocks" and the correlation of such units. Whether such natural divisions occur, what their nature might be if they do exist, and how they are to be "correlated" are all points of continuing research and debate that are discussed in this book. At the moment suffice it to say that it has been proposed that some problems in stratigraphy, such as the definition of boundaries of major stratigraphic units, are questions that can be simply and satisfactorily answered by a vote (Cowie et al., 1986). If this is the case, one may conclude that stratigraphic units defined in such a manner are merely convenient artificial divisions of rocks. Carried to an extreme, such views might regard much of stratigraphy as a type of fictionism or nominalism: statements and hypotheses, such as the division of rocks into stratigraphic units, could supply a coherence and order to sets of data, yet do not reflect real or natural entities and processes of nature (see Schoch, 1986b, p. 15).

SUBDIVISIONS OF THE DISCIPLINE
OF STRATIGRAPHY AND TYPES
OF STRATIGRAPHIC UNITS

Stratigraphy *sensu lato* may be viewed as composed of two basic subdivisions: descriptive stratigraphy and interpretive stratigraphy. Descriptive stratigraphy, also sometimes referred to as lithostratigraphy (*sensu lato*), prostratigraphy (see Schindewolf, 1970a,b), or physical stratigraphy, has been defined as follows by Harland (1978, p. 29): "rock stratigraphy which gives the recorded data of stratigraphy, (e.g. in measured stratigraphic sections with fossil lists and other information, and in large scale maps and so divides the earth's crust into convenient, named units to which all data and all interpretations may be referred)." Here it is important to note that not only purely lithologic or petrologic characters of rocks are referred to, but all primary, so-called objective, characters or attributes of rocks such as faunal and floral content, magnetic properties, mineral content, isotopic ratio content, and so on. (Lithostratigraphy as commonly used in the United States is often restricted to the purely lithologic characters of rocks; see NACSN, 1983.) Interpretive stratigraphy, consisting of such activities and subdisciplines as correlation of disparate stratigraphic units and rock bodies, regional stratigraphic syntheses, geochronology, paleoenvironmental studies, and so on, leading to the general reconstruction of geologic history, is based upon the data of descriptive stratigraphy. Of course, the observations and data collection activities of descriptive stratigraphy are never made in a total theoretical vacuum, and there usually is constant interplay between the two subdivisions. In some cases it may be difficult to distinguish between the two.

Many workers have traditionally recognized a "holy trinity" of subdivisions of stratigraphic science (or at least stratigraphic units); however, there has not been universal agreement as to the components of the holy trinity. Furthermore, recently far more than three categories of stratigraphy have been formally recognized (NACSN, 1983). Ager (1984) and the *International Stratigraphic Guide* (Hedberg, 1976; but see ISSC, 1979, 1987a,b,c, 1988) consider the holy trinity of stratigraphy to be composed of (1) lithostratigraphy, (2) biostratigraphy, and (3) chronostratigraphy. Owen (1987; see also Schenck and Muller, 1941, and NACSN, 1983) regards the holy trinity of stratigraphic units to be (1) lithostratigraphic or material rock units, (2) geochronologic or abstract time units, and (3) chronostratigraphic or hybrid time–rock units. Owen (1987) considers biostratigraphic units to be possibly either synonymous with, or a subcategory of, chronostratigraphic units; in this context, Owen (1987) seems to hark back to the old belief that fossils are a primary means of determining the temporal relationships of rocks (see Chapter 7). In contrast to Owen (1987), neither the *North American Stratigraphic Code* (NACSN, 1983) nor the *International Stratigraphic Guide* (Hedberg, 1976) recognizes geochronologic units as stratigraphic units although they are intimately related to stratigraphic units.

The *North American Stratigraphic Code* (NACSN, 1983; reprinted here as Appendix 1) distinguishes between two major types of stratigraphic units or categories: those based on content or physical limits (for example, lithostratigraphic and biostratigraphic units) and those based on, or related to, concepts of geologic age or time (for example, chronostratigraphic and geochronologic units). Within the latter category, one can distinguish between stratigraphic units that are based on some material category (namely, a body of rock) and those units that are not considered strictly stratigraphic as they are based on an abstract concept (namely, time) and do not have a direct material referent. Somewhat similarly, American stratigraphers in particular (American Commission on Stratigraphic Nomenclature, 1957; Wheeler, 1959) have drawn a distinction between stratigraphic units that are considered predominantly "objective units" because they are based primarily on observations with little interpretation (for example, lithostratigraphic units) and stratigraphic units that are considered dominantly "subjective units" based largely on interpretations (particularly temporal interpretations) of observed features of rocks (for example, chronostratigraphic units).

The stratigraphic literature contains an enormous amount of discussion on both particular stratigraphic units and their boundaries (for instance, the question of defining the boundary between the Silurian and Devonian systems) and the general nature and definition of various terms and types of stratigraphic, and related, units. Indeed, some critics would suggest that much of the supposed "science" of stratigraphy consists of endless bickering over definitional, terminological, and nomenclatural matters. The present author would suggest, however, that terminological and classificatory matters can determine the categories with which we think and work, and thus materially affect the conclusions drawn from our science. As Blow (1979, p. 1392) wrote: "Stratigraphic terminology studies are not just an 'academic exercise.' Their results affect the basic philosophy and interpretive work of all conscientious stratigraphers." Likewise Rodgers (1959, p. 684) stated: "Certainly, discussions of real scientific results are far more interesting and entertaining than discussions of words, but not necessarily more important. For we have to express our scientific results in words, and if the words are ambiguous and mean different things to you and to me, they will fail to transmit the results between us." And in a similar vein Harland (1978, p. 9, italics in the original) noted the importance to stratigraphy of distinguishing "between matters of *convention* (e.g. what name or term to use for an object or concept) and matters of *scientific principle* (e.g. what questions are capable of scientific demonstration). It is difficult to separate convention and principle because when we use language for scientific communication it can be translated and deliberately constructed for the purpose; but at the same time, it influences our understanding of the matters of scientific substance." All too often common stratigraphic codes and guides (e.g., Hedberg, 1976; NACSN, 1983) are not restricted to simple nomenclatural issues (what name to apply to a certain concept or rock unit, given that the concept or unit has been adopted by the researcher) or matters of convention, but subtly or overtly legislate the basic units, concepts, and principles to be used in the science.

Much of the present book is concerned with the various types of units that have been utilized in stratigraphy, their definitions, natures, hierarchies (ranks), and relationships. Before proceeding further it is desirable here briefly to describe in a bit more detail the three major categories of stratigraphic (and related) units, along with examples of some of the commonly used units within each category, that have become ingrained in the tradition of American (and Western more generally; see Hedberg, 1976; NACSN, 1983) stratigraphy: (1) material categories based on physical content or physical limits, that is, stratigraphic units composed of specified types of rocks per se; (2) material categories based on geologic age, that is, stratigraphic units composed of rocks formed during a specific interval of time; and (3) nonmaterial temporal units, that is, units that are intervals of geologic time. As mentioned above, the third category of units is considered by many not to be strictly stratigraphic units, although in many cases defined by stratigraphic units.

Material units based on physical content or physical limits are the units of descriptive stratigraphy or lithostratigraphy in the broad sense. The *North American Stratigraphic Code* distinguishes a number of different kinds of units within this category, such as lithostratigraphic units (distinguished on the basis of inherent lithologic or petrographic properties) and biostratigraphic units (distinguished on the basis of fossil content). The basic working unit that falls within this category remains the "formation": essentially a locally or regionally mappable body of rock that is identified and distinguished by certain lithologic attributes. Stratigraphically adjacent formations may be combined into a "group," or a formation may be subdivided into "members" (see Chapter 5).

The most common units that fall into the category of material units related to geologic age are chronostratigraphic (variously termed chronostratic) units. Loosely speaking (see more detailed discussion in Chapter 7), a chronostratigraphic unit is the material stratigraphic unit composed of all rocks, and only those rocks, formed during a specified interval of geologic time. The basic chronostratigraphic unit is the "system"; the rocks formed during approximately the last 600 million years are most frequently divided, on a global basis, into 12 systems (discussed at the end of this chapter; see also Appendix 3). Adjacent systems can be grouped into "erathems," and any system can be subdivided into "series." A major undertaking of one branch of stratigraphy since at least the middle of the nineteenth century has been to devise a universally agreed-upon global chronostratigraphic scale (which, together with the geochronologic units described below, would form the basis of a standard geologic time scale).

The most commonly used temporal and nonmaterial units are geochronologic units. The fundamental or basic geochronologic unit is the period, which is essentially the interval of geologic time corresponding to the rocks of a chronostratigraphic system. Several adjacent periods can be grouped into an era (corresponding to a chronostratigraphic erathem), or a period can be divided into epochs (corresponding to the chronostratigraphic series). Thus the chronostratigraphic and corresponding geochronologic units parallel each other and are intimately associated both theoretically and practically. (Indeed, which set of units, if either, logically or epistemologically precedes the other is subject to debate; both usually are based on the same reference points in rocks, that is, golden spikes in type sections—see Chapter 4; Harland, 1978; Harland et al., 1982.) The commonly accepted hierarchy of chronostratigraphic units and corresponding geochronologic units is as follows:

Chronostratigraphic	*Geochronologic*
Eonothem	*Eon*
Erathem	*Era*
System	*Period*
Series	*Epoch*
Stage	*Age*

The term chronomere has been applied to any chronostratigraphically (chronostratically) defined interval of time, and the term stratomere has been used to describe the rocks formed during a particular chronomere (George et al., 1967; Harland, 1978). Thus particular geochronologic units (such as eras, periods, epochs, and so on) represent particular chronomeres, and the corresponding chronostratigraphic units (erathems, systems, series, and so on) represent stratomeres.

HISTORICAL PERSPECTIVE ON THE ORIGINS OF STRATIGRAPHY

Some Founding Heroes of Stratigraphy

Here are briefly outlined the major contributions of a few of the seventeenth- to early nineteenth-century founding heroes of the science of stratigraphy (from Steno to Lyell). It is important to have at least a modest historical perspective on a discipline, and such a historical sketch also provides a convenient means of introducing some of the basic themes and principles of stratigraphy. As with any brief historical sketch, this review is admittedly idiosyncratic, biased, and a bit Whiggish, but perhaps all historical accounts are to some extent. For more detailed notes on the early history of stratigraphy, along with reprints of many important original documents, the reader is referred to the excellent volume edited by B. Conkin and J. Conkin (1984); for details on the development of stratigraphy in America during the nineteenth and early twentieth centuries the reader is referred to Merrill (1922, 1924) and Moore (1941). The historical development of particular stratigraphic concepts (such as facies, stages, or types of unconformities) is treated under their respective sections.

Prior to the development of modern stratigraphy there were various isolated speculations on the nature of rocks, minerals, and fossils. From antiquity to the Renaissance, fossils (in the modern sense of the term, referring to the remains or traces of ancient organisms—originally "fossil" could refer to rocks, minerals, or any other objects dug from the earth) were variously regarded as "sports of nature" that only fortuitously resembled organisms, or (as in modern science) as the remains of once living plants and animals (Desmond, 1975; Rudwick, 1976). It was widely held that minerals, ores, and coal grow within the earth; periodically mines would be closed down in order that ores could grow and replace the ores that had already been extracted (Mason, 1962). In western Europe during the Middle Ages and well into the eighteenth century much or all of the earth's surficial geology was generally accounted for by the Noachian deluge described in the seventh chapter of Genesis. A notable exception, Leonardo da Vinci (1452–1519) suggested that fossil shells found high in mountains could not be the result of a universal deluge because, among other reasons:

> the Flood could not have carried them there, because things which are heavier than water do not float high in the water, and the aforesaid things could not be at such heights unless they had been carried there floating on the waves, and that is impossible on account of their weight. (Translated by E. McCurdy and reprinted in Mather and Mason, 1939, p. 5)

Da Vinci suggested that layers of fossil shells in rocks were formed when shelled organisms upon a shore or the bottom of the sea were covered by earth or mud, the mud subsequently petrified, and finally the floor of the sea was uplifted (see Kohlberger and Schoch, 1985; Mather and Mason, 1939).

Up to the seventeenth and eighteenth centuries, all of time was believed to be encompassed within approximately six to ten thousand years. Ancient historical records, other than the Judeo–Christian Bible, extended back to only about 2000 B.C., taking the Biblical chronologies into account, this date could be pushed back to about 4000 B.C., when God is said to have created the earth in a single day (Boorstin, 1983; Toulmin, 1962–63). If God created the world in a single day, then all rocks are of basically the same age or, at most, of two ages: those rocks created at the original creation of the earth and those rocks redeposited in the diluvium. Essential to the development of stratigraphy, and a major scientific revolution in its own right, was the discovery that the earth is very old (millions or billions of years old), and that rocks can be differentiated on the basis of time. All rocks are not the same age; until this was firmly realized, the significance of stratigraphic superposition (certain rock layers lie above other layers) could not be realized (Weller, 1960).

Steno. The traditional founding father of stratigraphy, Nicolaus Steno (1638–86: also known as Niels Stenson, Nicoli Stenonis, or Nils Steensen; see Fritz and Moore, 1988; Harris, 1979), is often credited with the first explicit formulation of the principle of superposition, "the basis of all stratigraphy" (Weller, 1960, p. 20). Born in Copenhagen, trained in medicine and well known as an anatomist as well as a geologist, Steno made his geological observations primarily in the Tuscany region of Italy (Boorstin, 1983; Debus, 1968). Early in his geological career Steno correctly interpreted what were called glossopterae or tonguestones as fossil shark teeth (Fritz and Moore, 1988). In the *Prodromus* (1669, *De solido intra solidium naturaliter contento dissertationis prodromus,* Printing Shop under the Sign of the Star, Florence [Prologue to a dissertation on how a solid body is enclosed by the processes of nature within another solid body: Boorstin, 1983; B. Conkin and J. Conkin, 1984; Hancock, 1977]) Steno recognized the concept of strata and argued that "the strata of the earth are due to the deposits of a fluid" (Steno, 1669, in Mather and Mason, 1939, p. 37). Furthermore, Steno recognized that particles would settle out of a fluid suspension according to their relative weights, the denser and heavier ones first. This differential settling of particles would form horizontal layering (stratification).

For Steno all rocks were not of the same age; at least two distinct classes, temporal groups, or cycles of strata (rock) formation existed: (1) strata in which the constituent particles are homogeneous and relatively fine were "produced at the time of creation from a fluid which at that time covered all things" (Steno, 1669, in Mather and Mason, 1939, p. 37), and (2) strata that contain within themselves fragments of other strata, or organic remains, "must not be reckoned among the strata which settled down from the first fluid at the time of the creation" (Steno, 1669, in Mather and Mason, 1939, p. 38). Steno (1669) also argued that where and how a stratum formed (for example, in the sea or as a result of river flooding) could be deduced from an analysis of the particulate matter constituting the rock body. In some respects, this aspect of Steno's work could be viewed as a precursor to modern sedimentological and paleoecological analyses.

Steno (1669) next proceeded to formulate what have become known as his three principles (see Dott and Batten, 1976, p. 17): (1) the principle of superposition—"in a sequence of layered rocks, any layer is older than the layer next above" (Dunbar and Rodgers, 1957, p. 110); (2) the principle of original horizontality—strata originally formed, under the influence of gravitational settling of particles, horizontal to the surface of the earth; and (3) the principle of original lateral continuity—when originally formed, strata were laterally continuous unless they terminated against another solid substance. Concerning these fundamental principles, it is worth quoting Steno (1669) at length:

Concerning the position of strata, the following can be considered as certain:
1. At the time when a given stratum was being formed, there was beneath it another substance which prevented the further descent of the comminuted matter; and so at the time when the lowest stratum was being formed either another solid substance was beneath it, or if some fluid existed there, then it was not only of a different character from the upper fluid, but also heavier than the solid sediment of the upper fluid. [Principle of superposition]
2. At the time when one of the upper strata was being formed, the lower stratum had already gained the consistency of a solid. [Principle of superposition]
3. At the time when any given stratum was being formed it was either encom-

passed on its sides by another solid substance, or it covered the entire spherical surface of the earth. Hence it follows that in whatever place the bared sides of the strata are seen, either a continuation of the same strata must be sought, or another solid substance must be found which kept the matter of the strata from dispersion. [Principle of original lateral continuity]

4. At the time when any given stratum was being formed, all the matter resting upon it was fluid, and, therefore, none of the upper strata existed. [Principle of superposition]

As regards form, it is certain that at the time when any given stratum was being produced its lower surface, as also its lateral surfaces, corresponded to the surfaces of the lower substance and lateral substances, but that the upper surface was parallel to the horizon, so far as possible; and that all strata, therefore, except the lowest, were bounded by two planes parallel to the horizon. Hence it follows that strata either perpendicular to the horizon or inclined toward it, were at one time parallel to the horizon. [Principle of original horizontality] (Steno, 1669, as translated into English by John Garrett Winter [1916] and reprinted in Mather and Mason, 1939, pp. 38–39)

In the *Prodromus* Steno included an epitome of the geologic history of Tuscany (and by extrapolation, the whole world: Gould, 1987, p. 52), illustrated by a series of diagrammatic geologic cross sections (Fig. 1.1; perhaps the first such sections ever published: Weller, 1960, p. 21). According to Steno, the geologic history of Tuscany passed through two cycles (Gould, 1987). Originally sediments forming strata were laid down over the entire land (25 of Fig. 1.1). Cavities, caves, grottoes, or vacuities were excavated below the surface by the action of fire and water (24 of Fig. 1.1), and the upper strata collapsed into the cavities, forming the mountains and valleys (23 of Fig. 1.1); thus ended the first cycle. The sea subsequently invaded the area and deposited new strata in the valleys (22 of Fig. 1.1). Cavities then formed in the younger strata (21 of Fig. 1.1), and the younger strata collapsed to form smaller hills and valleys (20 of Fig. 1.1). Steno's use of the concept of subterranean collapse to explain the present topography of the surface of the earth may have been influenced by his explorations of the grottoes at Lake Garda and Lake Como (see Boorstin, 1983, p. 456).

The influence that Steno's work had on the actual development of stratigraphy is questionable. The *Prodromus* was translated into English only a few years after its publication (B. Conkin and J. Conkin, 1984); however, it was meant to be merely the introduction to a larger work that was never completed (Boorstin, 1983). In discussing the modern development of biostratigraphy by William Smith (1769–1839), Hancock (1977, p. 3) felt compelled to write: "It is true that the law of superposition is contained in the seventeenth-century ramblings of Steno (1669), but to the end of his life Smith had probably never heard of Steno." Zittel (1901) suggested that Steno's work was nearly unknown before the 1830s. Mason (1962), in his popular book *A History of the Sciences,* which includes a chapter on the development of geology, fails even to mention Steno.

Woodward, Strachey, and Hooke. Working in England, John Woodward (1665–1728) corresponded with naturalists around the world gathering data on geology and paleontology while also building up a personal collection of fossils. Woodward (1695, parts of which are reprinted in Mather and Mason, 1939, and B. Conkin and

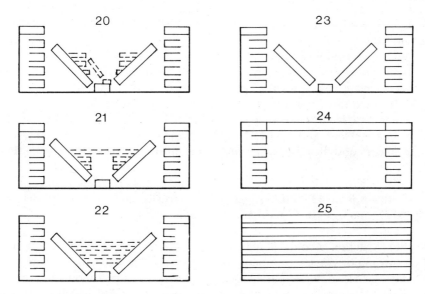

Figure 1.1. Steno's diagram [redrawn] illustrating the geologic history of Tuscany. (After Mather and Mason, 1939, p. 43; see also Davies in Smith, 1981a, p. 14; Gould, 1987, p. 55; and Weller, 1960, p. 21.) Steno's original description of the figure is as follows:

"The *six* last figures do *both* shew, how from the present face of *Etruria* we may collect the six distinct faces of the same Country, above discoursed of, *and* serve also for the more easy understanding of the particulars, we have deliver'd concerning the *Beds* of the Earth. The *pricked* lines represent the *Sandy* beds of the Earth, so nominated from their *main* matter, there being mix't with them divers both Clayie and stony beds. The other lines represent the *Stony* beds, likewise so called *a potiori,* seeing there are Beds found in them that are of a softer substance. . . .

"The 25th figure exhibits the perpendicular Plane of *Etruria,* at the time when the Stony Beds were yet entire, & parallel to the horizon.

"The 24th shews the vast cavities, eaten out by the force of Fire and Water, without any breach in the upper Beds.

"The 23th represents, how Mountains & Vally's came to be made by the ruine of the superior Beds.

"The 22th, that by the Sea new Beds were made in the said Valleys.

"The 21th, that of the new Beds the lower ones were consummed, the uppermost remaining untouch't.

"The 20th, that by the breach of the superior sandy Beds *there* were produced Hillocks and Vallys." (Steno, 1669, as translated into English by Henry Oldenburg [published in London, 1671–73] and reprinted in Mather and Mason, 1939, pp. 43–44, italics in the original.)

J. Conkin, 1984) determined that throughout the world terrestrial rocks are arranged in layers or strata "divided by parallel fissures" (Mather and Mason, 1939, p. 49), and that such strata often contain great numbers of shells and other marine organisms. Woodward suggested that the strata presently observable on the surface of the earth were a result of the Noachian flood, which had churned up and destroyed the surface of the original earth, held pieces of the original rocky material forming the surface of the earth in suspension, and then created the present strata by the settling of the suspended sediments from the waters. Woodward held that fossils were the remains of once living organisms that had been destroyed in the flood. According to Woodward, as the flood waters receded, the heaviest substances (for example, heavy minerals, metals, large fossil bones) settled out first, followed by progressively lighter materials (Mason, 1962; this is reminiscent of some modern-day creationist accounts of why different fossils occur at different levels in the composite global stratigraphic column: Morris, 1974, 1977; Schadewald, 1983; Whitcomb and Morris, 1961).

At about the same time that Woodard was carrying on his studies, John Strachey (1671–1743) was describing the strata of the coal fields of Somersetshire, and used fossils to correlate coal veins that had been offset by a fault (B. Conkin and J. Conkin, 1984). In his 1719 cross section (reprinted in Mather and Mason, 1939, p. 54) Strachey illustrated an angular unconformity between the Paleozoic coal measures of Somersetshire and the overlying Mesozoic strata. It is unclear, however, if Strachey understood the significance of the unconformity he was illustrating (Eyles, 1969; Weller, 1960).

In about 1688 (but not published until 1705) the English physicist and mathematician Robert Hooke (1635–1703) suggested that fossils could be used to determine geological chronology (B. Conkin and J. Conkin, 1984).

Moro and Arduino. Around 1740 Antonio Lazzaro Moro (1687–1764) distinguished two major classes of mountains and rocks. The older mountains consist exclusively of massive, nonstratified rocks, whereas the younger mountains are composed of either stratified rocks or stratified rocks overlying nonstratified rocks (Weller, 1960). Moro suggested that the stratification in certain rocks was due to volcanic eruptions of liquid rock that spread horizontally over the surface of the earth in a particular area and could enclose, as fossils, contemporaneous organisms. Mason (1962) traces the Neptunist versus Vulcanist controversy of the later eighteenth and earliest nineteenth century to the opposite views of Woodward and Moro concerning the origin of stratification. Neptunists (for instance, Woodward and Werner, see below) stressed the role of water in the formation of rocks and strata, whereas the Vulcanists (for example, Moro, and some would include Hutton in this camp) stressed the role of heat. In reality, of course, the debate was not as clearcut at the time as it has sometimes been pictured in hindsight.

Building on the work of Moro, Giovanni Arduino (1713–95), the director of mines in Tuscany and a professor of mineralogy at Padua, suggested a fourfold classification of mountains and their associated rocks (Weller, 1960). Primary mountains are composed of rocks that often contain metal ores but lack fossils. Secondary mountains are composed of well-lithified, bedded rocks containing marine organisms. Tertiary mountains are low mountains or hills that consist of predominantly unconsolidated sediments, such as gravels, sands, and clays, and may be associated with volcanically produced rocks. Finally, the most recently formed strata consist of materials eroded from the above three types of mountains and deposited loosely on the surface of the earth.

Lehmann and Füchsel. In Northern Europe during the middle of the eighteenth century the German geologists Johann Gottlob Lehmann (1719–67: Weller, 1960) and Georg Christian Füchsel (1722–73: Debus, 1968) were undertaking studies similar to those in Italy; and Lehmann in particular developed a classification of rocks and mountains paralleling that of Arduino (Krumbein and Sloss, 1963; Weller, 1960). Lehmann classified rocks and mountains into three basic categories:

1. Urgebirge are primitive mountains and rocks, predominantly crystalline (believed to have been formed by chemical precipitation), lacking fossils but containing masses of metallic ores (perhaps formed secondarily), believed to have been formed at the time of creation. These rocks form the cores of the great mountain chains.

2. Flötzgebirge are layered mountains composed of stratified rocks formed from sediments eroded from the primary mountains and deposited on the sides of primary mountains and in basins between them. These rocks contain fossils, and Lehmann suggested that they may have been formed during the Noachian flood. These rocks are often found on the flanks of large mountains.

3. Aufgeschwemmptgebirge are alluvial, loosely consolidated or unconsolidated sands and gravels formed most recently within restricted geographic areas. Recent volcanic rocks may be associated with these rocks.

Lehmann has been credited with the earliest published description and figuring of an actual geological succession, circa 1756, that of Permian rocks in the Harz Mountains (B. Conkin and J. Conkin, 1984, p. 10). Füchsel, circa 1762, "recognized that rock strata might be grouped into a formation that represented an epoch in the history of the earth" (B. Conkin and J. Conkin, 1984, p. 10).

Buffon. Georges-Louis Leclerc, Comte de Buffon (1707–88), a great popularizer of science and perhaps one of the most influential naturalists (at least among the laymen) of the eighteenth century, proposed an all-encompassing speculative theory of the earth that helped pave the way for the reconstruction of earth history by stratigraphers in terms of a series of distinct periods of great temporal duration. Buffon's theory of the earth is also in some ways typical of the numerous speculative "systems" devised to explain the earth and its geology during the seventeenth and eighteenth centuries. Buffon (see Buffon, 1797 [an English edition of a work originally published around 1749]; Buffon as reprinted in Mather and Mason, 1939; Boorstin, 1983) broke with the tradition of a relatively young earth. Whereas James Ussher (1581–1656) concluded on the basis of biblical chronology that the earth was created by God in 4004 B.C., Buffon suggested in print that the earth is approximately 75,000 years old (in private he may have suggested that the earth is much older: Bowler, 1984). Buffon arrived at this age by suggesting that the earth originated in a molten state when a comet knocked a fragment off the sun. He then calculated how long it would take a liquid body of the size and composition of the earth to cool to its present state. Buffon further postulated seven great epochs encompassing all of the earth's history: (1) the earth and planets take their form; (2) matter consolidates and forms the interior rock of the earth and the "vitreous masses" on the surface; (3) waters cover the continents; (4) waters withdraw from the continents, and volcanoes become active; (5) elephants and other southern animals inhabit the northern lands; (6) the continents separate; and (7) man appears.

Werner. Variously considered the "greatest figure in geology toward the close of the eighteenth century" (Weller, 1960) and the archvillain of modern geology, a man who "retarded geologic progress" (B. Conkin and J. Conkin, 1984, p. 11; Geikie, 1897), Abraham Gottlob Werner (1750–1817) was assuredly an extremely charismatic and influential teacher in his time. Located at the Freiberg School of Mines (Germany) from 1775 to 1817, Werner attracted students from throughout Europe. His theories built upon the work of Lehmann and epitomized the Neptunist school of geology (Mason, 1962).

Werner believed that originally the earth was covered entirely by a primeval ocean from which originated almost all rocks. Initially primitive rocks devoid of fossils, such as granite, crystallized out of the ocean; these represented the Urgebirge series. Next a transitional (Übergangsgebirge) series of rocks, such as graywacke, mica, slate, some limestone and diabase, containing a few fossils, precipitated from the ocean. Third, stratified, sedimentary, fossiliferous rocks (Flötzgebirge) and some basalt formed as the universal ocean began to recede. Fourth, derivative rocks of poorly consolidated sands, gravel, and clays derived from erosion of the previously formed rocks were deposited as the waters withdrew from the continents (Aufgesch-wemmptgebirge). Lastly, according to Werner, obvious volcanic rocks are all of relatively recent origin, locally restricted, and due to the natural burning of coal seams (Krumbein and Sloss, 1963; Mason, 1962; Weller, 1960).

Werner suggested that the mountains were original topographic features of the primitive earth, perhaps formed as the first granite crystallized from the universal ocean. The mountains were not uplifted, and strata currently observed to dip outward from the flanks of mountains were formed in their present positions. Occasionally such strata might slide downward before consolidation, becoming contorted or broken (Weller, 1960). Material deposited in the ocean basins would form strata parallel to the surfaces upon which they were deposited.

Adopting the Wernerian system, particularly for the older categories of rock, one could suggest that distinctive sets of rocks had been deposited successively and synchronously from the universal primeval ocean and were originally (before dissection by erosion or other modification) worldwide in extent. If this were true, logically it would be simple to correlate rocks worldwide simply by their characteristic compositions and physical properties. There would only be one global stratigraphic column, and stratigraphy would be a job for simpletons. In its fully developed form, however, Werner's system was not quite so simplistic. Observed differences in the local geology of various regions could be explained on the basis of uneven original topographies and the irregular retreat of the primeval waters from various geographic regions (Bowler, 1984). Still, based on Werner's theories, gross similarities in the sequences of rocks from different localities might be expected.

It has been suggested that Werner's work helped give the term "formation" its modern lithological and petrographical significance, rather than a biostratigraphical or temporal significance (B. Conkin and J. Conkin, 1984; Schneer, 1969; Weller, 1960). Also, Werner may have helped to inspire others to trace particular strata over wide geographic areas (B. Conkin and J. Conkin, 1984).

Hutton. "Prior to Hutton, geology did not exist" (McIntyre, 1963, p. 2). This statement, even if somewhat exaggerated, illustrates the hold Hutton and his legend have on modern geology. In 1785 James Hutton (1727–97) read a paper to the Royal Society of Edinburgh: "Concerning the system of the Earth, its duration, and stability" (Hutton, 1785). This paper was subsequently published in full in the first volume of the *Transactions of the Royal Society of Edinburgh* (Hutton, 1788), and in 1795 an expanded version was published as a two-volume work (Hutton, 1795). Hutton is often hailed as the founder of modern geology (at least among English-speaking peoples) and is credited with the discovery of three basic concepts: the vastness of geologic time ("deep time" of McPhee, 1980, and Gould, 1987), unconformities (in particular, angular unconformities), and uniformitarianism (although at least some aspects of this concept were not original with Hutton [see, for instance, McIntyre, 1963, and Chapter 3; the term itself was first coined by the philosopher and mathematician William Whewell in 1832 in a review of the second volume of Lyell [1830–33]: see Gould, 1984, and Simpson, 1970).

Although trained as a physician (he never practiced), Hutton became involved in the agrarian and industrial advances of the time in Scotland. He established the first factory in Britain to produce sal-ammoniac (ammonium chloride: Mason, 1962; McIntyre, 1963) and pursued a career as a "gentleman farmer." It was in this capacity that he became very interested in soils, soil erosion, and soil recycling. Hutton came to view the earth as a grand machine, the purpose of which was to produce a habitable world for man and other living forms (Hutton, 1788; see Gould, 1987; Hubbert, 1967). He developed the concept of rock cycling: solid rock is weathered to produce soils, the soil is eroded and transported by running water to the oceans where it is deposited in stratified layers, the sediments are then consolidated into rocks, and finally the rocks are elevated above sea level to be weathered and begin the cycle once again. Hutton suggested that sediments consolidated by fusion as a result of heat and pressure, and that the uplift of sediments took place cataclysmically. As part of his theory, Hutton suggested that granite is of magmatic origin.

To a certain extent Hutton advocated studying rocks themselves and interpreting them in terms of operations and processes that can be observed occurring in the world today (for example, erosion, sediment transport; thus Hutton is traditionally credited with developing the concept of uniformitarianism). However, according to Hutton, some processes we may logically deduce to occur that we actually cannot observe (for instance, fusion of sediments to form rock) or have not observed (for instance, cataclysmic uplift of rocks from the deep oceans). Or as Newell (1967, pp. 357–358) has succinctly summarized Hutton's methodology: "interpretations of earth history are logical inferences reinforced wherever appropriate by analogies between past and present conditions." Hutton rejected divine intervention or supernaturalism as secondary or efficient causes (divine wisdom may have set the earth machine in place originally, and the machine had a definite purpose or final cause—to form a habitable abode for life): "Therefore, there is no occasion for having recourse to any unnatural supposition of evil, to any destructive accident in nature, or to the agency of any preternatural cause, in explaining that which actually appears" (Hutton, 1788, p. 285; quoted in Hubbert, 1967, p. 6).

By examining weathering rates of rock and the lack of significant changes in coastlines since Roman times, Hutton concluded that time is indefinitely long (Hubbert, 1967, p. 6). In keeping with his theory of cataclysmic uplift followed by erosion and deposition of new sediments, Hutton predicted the existence of granitic intrusions in sedimentary rocks and angular unconformities, both of which he later discovered and recognized (see Gould, 1987, p. 70; Lyell, 1857, p. 60). Based on this form of reasoning, Hutton speculated that the earth has gone through a number of cycles, but is essentially ageless (infinitely old). One of Hutton's most famous statements reads: "But if the succession of worlds is established in the system of nature, it is in vain to look for any thing higher in the origin of the earth. The result, therefore, of our present inquiry is, that we find no vestige of a beginning,—no prospect of an end" (Hutton, 1788, p. 304; quoted in Hubbert, 1967, pp. 5-6). Hutton believed that with the exception of man (who is of a relatively recent origin), the same species of organisms have existed somewhere on the earth throughout geologic time. Hutton's cyclical theory of the earth was essentially ahistorical and nondirectional (cf. Gould, 1987).

Traditionally Hutton's original writings have been regarded as obtuse, rambling, and unreadable. Many of Hutton's ideas were anticipated with remarkable precision by Toulmin (1783: see Simpson, 1970; also McIntyre, 1963, who suggests that Hutton must have read Toulmin's work before writing his own papers). After his death,

Hutton's theory was popularized by his friend John Playfair (1748–1819) in a highly readable style, in the book *Illustrations of the Huttonian Theory of the Earth* (1802). As Gould (1987) has aptly discussed, over the years myths have developed around Hutton and his work. In many texts Hutton is viewed as a strict empiricist, an inductivist, a Lyellian uniformitarianist, an observer of nature who drew conclusions only after gathering all of the relevant facts. Hutton came to represent the tradition of fieldwork in geology (cf. Geikie, 1897). Gould (1987) suggests that the empiricist myths concerning Hutton may have originated with Playfair's interpretation of Hutton's theory. Playfair greatly emphasized the mechanics of Hutton's theory of cycles over Hutton's concept of final cause and purpose to the world. Furthermore, whereas Hutton was essentially ahistorical in his original theory, Playfair was interested in unique events arranged in time—the heart of the tradition of historical geology. Unlike Hutton, Playfair interpreted field evidence in explicitly historical terms. Few people read Hutton's work in the original; they read Playfair (1802) and assumed that Playfair had not altered Hutton's meaning or intent.

Smith. William Smith (1769–1839: sometimes nicknamed "Strata": Stanley, 1986) is often credited with founding stratigraphy in the modern sense (Cox, 1948; Hancock, 1977). Born in Oxfordshire, England, Smith had relatively little formal education but trained himself to be a surveyor. All of his life he was a practicing surveyor and civil engineer, working on canals, mines, and stone-quarrying operations. From an early age Smith collected fossils; in his work with coal mines in the district around Bath he noted that not only the coal seams but also the rock formations overlying them succeeded one another in a regular succession. Smith's first great contribution to geology was to demonstrate the constancy of the succession of geological formations over relatively wide geographic areas. Any particular formation (stratum or group of strata) had a lateral (geographic) continuity and could be mapped. It occurred to Smith, who was familiar with agricultural soil maps, that the underlying bedrock might be mapped in a similar fashion. He probably attempted to construct his first geological maps in 1793 or 1794 (Cox, 1948, p. 3). By 1799 Smith was exhibiting such early maps to the Bath Agricultural Society.

In his work Smith sometimes found that different formations in the succession had very similar lithologies. However, sometime before 1796 Smith discovered that strata of similar lithology but in different positions in the succession could be distinguished from one another by their fossil content. Extrapolating from this concept, one can recognize the relative placement of rocks within the stratigraphic column by their fossil content alone; fossil species succeed one another in a regular progression. In 1799 Smith dictated to his friends the Reverends Benjamin Richardson and Joseph Townsend a list of the succession of strata around Bath, along with their thicknesses, localities where they could be studied, and their characteristic fossils (Richardson supplied names for the fossils: Arkell, 1933; Cox, 1948, Hancock, 1977; Phillips, 1844). An outline of this succession first appeared in print in *The History of Bath* (Warner, 1801).

Smith also discovered two other basic principles. First, he determined that a single formation that is lithologically homogeneous sometimes can be subdivided on the basis of the fossil distribution within it (Hancock, 1977). Second, Smith realized the importance of looking beyond the superficial similarities of fossils from similar facies but of different ages. To quote Smith on this point (1817, p. 22; quoted from Hancock, 1977, p. 5): "The upper part of this thick Stratum contains large incurved

oysters or Gryphaea, so much resembling others I have collected from remote parts
. . . as to be distinguished with difficulty; but this is only one of the many instances
of the general resemblances of organized Fossils, where the Strata are similar.''
However, at least as late as 1817 the idea had not yet been developed that different
facies of the same age can be temporally correlated by identical fossils in common
(Hancock, 1977).

In his work as a surveyor and civil engineer Smith traveled throughout England,
continuously taking notes on the geology and collecting specimens. Smith wished to
publish a large treatise recording his discoveries, but his proposed opus magnum
never appeared because of his busy schedule, the fact that writing was difficult for
him, and the problem of finding a publisher to finance the undertaking. However,
Smith did publish numerous shorter works, including a number of charts, geologic
sections, and geologic maps. His most famous publication probably was his great
(in printed form it measures approximately 6 by 8½ feet [1.8 by 2.55 meters]) geologic map entitled ''A Delineation of the Strata of England and Wales with part of
Scotland . . . '' (1815).

Throughout his career Smith made a point of demonstrating the wide practical
applications of his new science of stratigraphy, such as to agriculture and forestry,
mining, canal and road building, problems of erosion and topography, procuring
raw materials for industry, land drainage, irrigation, and the sinking of water wells
(see Smith, 1815, reprinted in Mather and Mason, 1939). According to Smith, each
rock formation has its own characteristic flora and fauna, and human settlements
have been determined to a large extent by the underlying geology (Cox, 1948).

Geological Society of London. Extremely important for the development of modern stratigraphy was the founding of the Geological Society of London in 1807. In
the seventeenth and eighteenth centuries there was a tradition of system building
in geology. Woodward, Burnet (see Bowler, 1984; Gould, 1987), Buffon, Hutton,
Lamarck (see Lamarck, 1984), Cuvier, and others had their own theories of the
earth, variously addressing such grand aspects as how the earth originated, operates,
and will end. These systems were not always based on a high quotient of detailed
observations, although in some cases they did include basic field evidence. The primary concern of some of these theories was the ultimate cause and purpose of the
earth, not the detailed mechanics of earth history; they were not necessarily meant
to be strictly empirical formulations. It would be unfair to judge them by modern
standards.

In the late eighteenth and early nineteenth centuries a new empiricism arose
among geologists, reflected in the attitude of the founding members of the Geological Society of London. In founding the society the members ''called for a morato-
rium on theorizing'' (B. Conkin and J. Conkin, 1984, p. 1) and requested that geologists gather raw data—field evidence. Interpretation of the data was to be
discouraged until such time as a sufficient body of data was gathered to warrant
interpretive analysis. Many members of the society went about gathering field evidence under the auspices of what Gould (1987, p. 152) has termed the ''stratigraphic
research program.'' Raw stratigraphic data were collected and interpreted only to
the point of determining the actual sequence of geological and paleontological
events through time in particular geographic areas.

The earliest members of the Geological Society of London were primarily ''mineralogists of an older school'' (Cox, 1948, p. 5), and William Smith could not gain

recognition or support from the society when he could have used it most. Parkinson (1811, reprinted in B. Conkin and J. Conkin, 1984), in the first volume of the *Transactions of the Geological Society of London,* used Smith's concepts of the continuity of strata and the constancy of superposition of strata, along with Smith's method of recognizing strata by their contained fossils. Parkinson (1811, p. 47) gave specific credit to Smith for these discoveries, stating that these concepts have been "long since recommended by Mr. W. Smith." In 1831 the Geological Society of London recognized Smith as the father of English geology and awarded its first Wollaston Medal to him.

Cuvier and Brongniart. In France in the early nineteenth century the great naturalist and paleontologist Georges Cuvier (1769–1832) and the mineralogist Alexandre Brongniart (1770–1847) were studying the Tertiary strata of the Paris Basin using techniques similar to those of Smith and the English school. Cuvier and Brongniart (e.g., 1808, 1822; Cuvier, 1813—Brongniart was primarily responsible for the stratigraphic work in these papers) recognized a succession of alternating marine and freshwater assemblages of fossils and their enclosed rocks that could be interpreted as representing a series of "revolutions" in the history of life, at least in the district of the Paris Basin. Cuvier and Brongniart (1808, p. 434; translated and reprinted in B. Conkin and J. Conkin, 1984) were also perhaps the first explicitly to describe and recognize a disconformity (although they did not name it as such).

It is sometimes suggested that Cuvier and Brongniart should be given equal credit with Smith for discovering the basic principles of biostratigraphy, but Hancock (1977) in particular has argued otherwise. The majority of Cuvier and Brongniart's work was restricted to the local area of the Paris Basin, and Cuvier's primary concern was not with correlating strata but with reconstructing a general history of life. Brongniart, on the other hand, did realize that fossils could be used in correlating rocks separated by significant geographic distances and of differing lithologies, but he also acknowledged the work of William Smith and the English school of geology upon which he was building (Hancock, 1977).

Lyell. Charles Lyell (1797–1875) was born to a well-off Scottish family in Forfarshire (now Angus) and attended Oxford University; he was trained as a lawyer (barrister) and practiced law for several years (Debus, 1968, Hubbert, 1967). At Oxford, Lyell attended lectures by William Buckland (1784–1856), one of the best-known geologists of the time, and soon came to devote all of his time to geology.

At first Lyell's geological studies followed the catastrophic-diluvial vein of the work of Buckland and other contemporary British geologists (Hubbert, 1967); but as he continued his studies, he came to believe that the history of the surface of the earth can be explained in terms of processes (and phenomena known to be formed by such processes) that are presently operating and observable. Lyell's basic concept, dubbed uniformitarianism by Whewell (1832), is often summarized in textbooks using the old cliché that "the present is the key to the past" (Dott and Batten, 1976; Boggs, 1987: this dictum is attributed to Sir Archibald Geikie [see McIntyre, 1963]). Actually, Lyell's concept of uniformitarianism is a complex set of theories, assumptions, and methodological principles, which are discussed in more detail in Chapter 3.

Lyell found that his ideas were similar to parts of the Huttonian theory of the earth (Hutton, 1788, 1795; Playfair, 1802), and, in part, he expanded upon Hutton's (and Playfair's) work. Lyell envisioned a cyclical, steady-state world that has always operated in the same way. In his view, the earth has not changed in any essentials through time; there are no secular trends in earth history. The causes and processes that are operating now have always operated at about the same rates and intensities; there are no processes that operated in the past that are not presently operating. Likewise, there are no processes operating in the present that did not exist in the past. According to Lyell, the earth always supports approximately the same proportions of continent and ocean, there has always been approximately the same number of species on the earth, the proportions of the various types (groups) of living organisms on the earth are approximately equivalent through time (although he came to modify this view late in his life after the publication of Charles Darwin's work on evolution [Darwin, 1859]), and all the basic kinds of rocks (for example, igneous, metamorphic, sedimentary) can form during any period of geologic history because rocks and minerals are physical objects formed according to chemical laws that operate through all of time.

Unlike Hutton, however, Lyell did not take a totally ahistorical position. Lyell was a historian of earth history who was interested in the ordering of events through time even if he did not accept a progressionist history of the earth. In Lyell's steady-state but cyclical earth, each cycle differed in historical details from all other cycles even if the cycles were equivalent and operating according to the same causes and processes.

Lyell brought together his views in his great work, *Principles of Geology, Being an Attempt to Explain the Former Changes of the Earth's Surface by Reference to Causes Now in Operation* (in three volumes, 1830–33). The *Principles* was extremely successful and influential. Lyell produced eleven editions of the *Principles* between 1830 and 1872, continually modifying, revising, and changing various aspects of the work. The twelfth edition was publsished after his death in 1875.

Of special importance to stratigraphy, Lyell devised an ingenious way to date (temporally order) rocks based on their fossil content; he applied this methodology to dating the rocks of the Tertiary in particular. In the early nineteenth century most stratigraphers were progressionists, believing that fossil forms underwent increasing perfection through time. Thus, at least theoretically, strata could be dated and ordered on the basis of the relative development or perfection of the contained organic beings (Gould, 1987). In practical terms, the more similar the fossils of a rock body were to living forms, the younger the rock body must be. Lyell, however, was a nonprogressionist and thus could not order organisms by the usual criteria of advancement. On the other hand, he did not deny that any particular period in geologic time is distinguished by its own characteristic species (each species is historically unique), an idea that held the key to Lyell's method of ordering strata by their fossil content.

In discussing the use of fossils to date rocks, Lyell states:

First, the same fossils may be traced over wide regions, if we examine strata in the direction of their planes, although by no means for indefinite distances.

Secondly, while the same fossils prevail in a particular set of strata for hundreds of miles in a horizontal direction, we seldom meet with the same remains for many fathoms, and very rarely for several hundred yards, in a vertical line, or a line transverse to the strata. This fact has now been verified in almost all parts of the globe, and has led to a conviction, that at successive periods of the past, the same area of land and water has been inhabited by species of animals and plants even more distinct than those which now people the antipodes, or which now exist in the arctic, temperate, and tropical zones. It appears, that from the remotest periods there has been ever a coming in of new organic forms, and an extinction of those which pre-existed on the earth; some species having endured for a longer, others for a shorter, time; while none have ever reappeared after once dying out. (Lyell, 1857, p. 98)

According to Lyell, although the earth always contains approximately the same number of species, the totality of species is constantly changing (but not progressing) through time. As a species goes extinct, it is replaced by another equivalent species (this process of extinction and replacement may be a purely random process). Any period of geologic time is characterized by a particular combination of unique and recognizable species. Moreover, if we begin with any particular geologic period (call it period S) and then compare the species composition of that period to other geologic periods that recede from it in time (becoming either older or younger; for example, assume that we have periods P, Q, R, S, T, U, V, oldest to youngest), the further removed in time we get from period S, the fewer the number of species we will find in common between period S and another period. Periods R and T may have a large number of species in common with period S because they are directly adjacent to S in time, periods Q and U should have a moderate number of species in common with period S, and periods P and V, being the furthest removed in time from period S, will share the least number of species with period S.

Lyell applied the statistical method of determining the relative ages of rocks, outlined in the last paragraph, to the classification of the Tertiary in particular. Using fossil mollusks (as identified and compiled by the French paleontologist Gérard Paul Deshayes [1795–1875]), Lyell was able to subdivide the Tertiary into four broad time periods (epochs):

At my request he [Deshayes] drew up, in a tabular form, lists of all the shells known to him to occur both in some tertiary formation and in a living state, for the express purpose of ascertaining the proportional number of fossil species identical with the recent which characterized successive groups; and this table, planned by us in common, was published by me in 1833 [*Principles of Geology*, vol. 3, pp. 45–61, reprinted in B. Conkin and J. Conkin, 1984]. The number of tertiary fossil shells examined by M. Deshayes was about 3000; and the recent species with which they had been compared about 5000. The result then arrived at was, that in the lower tertiary strata, or those of London and Paris, there were about 3½ per cent. of species identical with recent; in the middle tertiary of the Loire and Gironde about 17 per cent.; and in the upper tertiary or Subappennine beds, from 35 to 50 per cent. In formations still more modern, some of which I had particularly studied in Sicily, where they attain a vast thickness and elevation above the sea, the number of species identical with those now living was believed to be from 90 to 95 per cent. For the sake of clearness and brevity, I proposed to give short technical names to these four groups, or to the periods to which they respectively belonged. I called the first or oldest of them Eocene, the second Miocene, the third Older Pliocene, and the last or fourth Newer Pliocene [renamed Pleistocene by Lyell in 1839, see below]. (Lyell, 1857, p. 115)

The Development of a Standardized Stratigraphic
Nomenclature for Rocks and Time

Today we have a universally acknowledged, even if not agreed upon, set of names for the major rock groupings and temporal divisions of earth history—for example, the erathem/era and system/period names (see Appendix 3). Most of the common major (period and higher-level) named subdivisions of the composite stratigraphic column were developed, primarily in Europe, between the late eighteenth and first half of the nineteenth century (the majority being named between 1822 and 1854). These "groups," "systems," or "series" of strata were generally considered to be fundamental rock units (natural groupings) that represented fundamental temporal units or episodes in the history of the earth and life on earth. Discussing his classification of fossiliferous strata in Western Europe, Lyell (1857, p. 103) stated:

It is not pretended that the three principal sections in the above table, called primary [Paleozoic], secondary [Mesozoic], and tertiary [Cenozoic excluding the Recent: note that these three terms as used by Lyell do not directly correspond to the same terms as used by Arduino in 1759-60, see below], are of equivalent importance, or that the eighteen subordinate groups [the forerunners of the presently accepted systems/periods] comprise monuments relating to equal portions of past time, or of the earth's history. But we can assert that they each relate to successive periods, during which certain animals and plants, for the most part peculiar to their respective eras, have flourished, and during which different kinds of sediment were deposited in the space now occupied by Europe.

In most cases the systems and so on were originally based on distinct suites of rocks and contained fossils, rather than based on the time when the rocks were formed (Wilmarth, 1925). Often they were originally designated by names referring to their fundamental lithology (such as the White Chalk in the Upper Cretaceous or the Oolitic System, which is now called the Jurassic), or by names based on the locality where they were first recognized or are typically exposed (such as the Devonian System of Devonshire). As originally recognized, systems were generally bounded by major lithologic and paleontologic discontinuities, that is, unconformities (Krumbein and Sloss, 1963). It became a standard practice to correlate and identify systems globally on the basis of the bounding unconformities; this methodology was intimately associated with the belief that the surface of the earth has been subjected to repeated globally synchronous (at least on the scale of geologic time) events (catastrophes?) that caused physical (preserved in the lithological record) and biological (preserved in the paleontological record) changes on the earth.

In the remainder of this section I briefly review the historical origins of the erathem/era and system/period terms (concentrating on the Phanerozoic terms) currently in common use (see Appendix 3). I also include in this review the series/epoch terms of the Cenozoic because in the past they have variously been considered system/period or series/epoch terms. The original authorship and reference (to the best of my knowledge) for each term is enclosed in parentheses after the term in question. When originally defined, the following terms did not necessarily imply the rank with which they are now usually associated; here I associate them with the standard rock terms (eonothem, erathem, system, series) although they all can be used also with the standard time (geochronologic) terms (eon, era, period, epoch; see above and Appendix 1 for explanations of these terms). As seen below, some of these terms were originally defined as time terms, and have been primarily used as such. (Indeed, at present there may be no known terrestrial rocks that are referable to the Priscoan Eonothem although they are certainly known from the Moon.) Most of the original definitions and references relating to these terms are reprinted in Wilmarth (1925); for reprints of other important related papers, see B. Conkin and J. Conkin (1984) and Mather and Mason (1939). Much useful information concerning the origins of these terms is also found particularly in Berry (1968), Gignoux (1955), and Harland et al. (1982).

It is important to keep in mind that the traditional stratigraphic units discussed below were in many cases originally local units based on now classic sections. However, as such units and names are adopted for global use, they may be redefined, often on the basis of localities far from the eponymous locality (Cowie et al., 1986; Harland et al., 1982; see Chapter 4). At present such refinement of the global stratigraphic scale has only just begun. Thus the exact interval currently specified by an old name may no longer correspond precisely to the original interval in the original type or typical locality. Furthermore, in many cases when the following terms were originally coined, their limits (boundaries) were defined only very roughly (at least by today's standards).

Another point that may be obvious, but should be explicitly stated, is that the stratigraphic nomenclature recounted in part below is meant to be applicable only to terrestrial (Earth) rocks. A completely separate and distinct stratigraphic nomenclature (although based on the same basic stratigraphic principles) has been developed for the Moon (Wilhelms, 1987; Appendix 4; see also Chapter 5). As the geology of other planets and planet-like objects is explored in detail by future workers, similar, but to some extent distinct, stratigraphic scales and nomenclature assuredly will be developed. Finally, it should be noted that the terrestrial stratigraphic column is being continually modified and refined, particularly at the lower end of the scale. An immediate goal of many workers is to standardize the nomenclature of the global stratigraphic column down to the level of the stage/age, at least for the Phanerozoic (see especially Bassett, 1985; Cowie, 1986; Cowie et al., 1986; Harland et al., 1982; Holland, 1984).

A note on the spelling of names: For many names and terms in stratigraphy there are competing variations in spelling and typographical convention. As originally used many names first appeared in languages other than English and may have included accents and so on. Even for names rooted in the English language there may be various forms, such as the American Paleozoic (without a diphthong) versus the British Palaeozoic (with a diphthong) or the American Cenozoic versus the British Cainozoic. The recent trend, helping to promote international usage and understanding, has been to use the simplest form of a name whenever a choice is presented, to drop accents, and to avoid diphthongs; in general, streamlined American usage is being adopted worldwide (Harland et al., 1982).

Cryptozoic Eonothem (G. H. Chadwick, 1930, Subdivision of geologic time, *Geol. Soc. Amer. Bull.* 41:47–48): Meaning hidden life, introduced for all of Precambrian time (that is, rocks and time prior to the beginning of the Cambrian, = Proterozoic and Archeozoic, or Archean, of various authors).

Subdivisions of the Precambrian/Cryptozoic: The subdivision of rocks and time before the Cambrian has a long and confusing (and ongoing) history, dating back to the nineteenth century (see, e.g., Harland et al., 1982; Wilmarth, 1925). At present many authorities divide Precambrian time into two eons (e.g., Palmer, 1983, reprinted as Appendix 3). The Archean (= Archaean, also sometimes referred to as the Archeozoic, Archaeozoic, or Azoic) extends from the age of the oldest known terrestrial rocks (circa 3.8 x 10^9 years old: Drury, 1981) to approximately 2500 Ma (million years ago). The Proterozoic is the interval of time from the end of the Archean to the beginning of the Cambrian. The Archean/Archeozoic has also been used as the interval of time that terminates at 2500 Ma, but has no initial boundary (that is, extends back infinitely in time). Harland et al. (1982; see also Harland, 1975, 1978) suggested that the initial Archean boundary be defined arbitrarily at 4000 Ma, and that pre-Archean time be named Priscoan, suggestions adopted by Haq and Van Eysinga (1987).

Phanerozoic Eonothem (G. H. Chadwick, 1930, Subdivision of geologic time, *Geol. Soc. Amer. Bull.* 41:47–48): Meaning evident life, introduced for Paleozoic, Mesozoic, and Cenozoic time.

Paleozoic (= Palaeozoic) Erathem (A. Sedgwick, 1838, *Geol. Soc. London Proc.,* 2(58):684–685; see also J. Phillips, 1840, *Penny Cyclopaedia* 17:153–154, and J. Phillips, 1841, *Palaeozoic fossils of Cornwall, Devon, and West Somerset,* Great Britain Geol. Surv. Mem., p. 160): Sedgwick (1838) originally defined the Paleozoic to include only [using the modern terms] the Cambrian, Ordovician, and Silurian. Phillips (1840) included the Devonian ("Old Red Sandstone") in the Paleozoic and suggested that the Carboniferous might also be included in this era. In 1841 Phillips included the Carboniferous and the Permian ("Magnesian limestone") in the Paleozoic. Loosely translated, Paleozoic means "ancient life"; for many years the oldest known fossils came from the Cambrian.

Cambrian (= Cumbrian) System (A. Sedgwick, 1835, *Edinburgh New Philos. Jour.* 19:390, abstract; see also A. Sedgwick and R. I. Murchison, 1835, On the Silurian and Cambrian Systems, exhibiting the order in which the older sedimentary strata succeed each other in England and Wales, *British Assoc. Adv. Sci. Rept., 5th Meeting,* pp. 59–61): Sedgwick originally based his Cambrian on the "greywacke, or slate series, of the north of England and Wales." The lower part of Sedgwick's Cambrian was nonfossiliferous and is now excluded from the Cambrian; the upper part of his Cambrian has been transferred to the Ordovician. The term Cambrian derives from Cumbria, an ancient British kingdom that occupied what is presently northwest England.

Ordovician System (C. Lapworth, 1879, *Geol. Mag., London,* new ser., 6:12–14): Sedgwick and Murchison became embroiled in a heated argument over the boundary between their Cambrian and Silurian Systems; the disputed strata were generally labeled Lower Silurian. In 1879 Lapworth proposed the new name Ordovician (named after the ancient tribe of Ordovices of Wales) for the intermediate rocks.

Silurian System (R. I. Murchison, 1835, *London and Edinburgh Philos. Mag. and Jour. Sci.,* 3rd ser., 7:46–52): Murchison named the Silurian on the basis of deposits in Wales lying below the Old Red Sandstone [Devonian] and above the "slaty rocks." The name derives from the kingdom of the Silures, an ancient Welsh tribe.

Devonian System (A. Sedgwick and R. I. Murchison, 1839, *Geol. Soc. London Proc.* 3(63):121–123 (abstract); see also A. Sedgwick and R. I. Murchison, 1840,

Geol. Soc. London Trans., 2nd ser., 5(3):701–702): Sedgwick and Murchison based the Devonian on the sparsely fossiliferous upper part of the Old Graywacke of Devonshire that was found to be younger than the Silurian; correlative with this is the Old Red Sandstone.

Carboniferous System (W. D. Conybeare and W. Phillips, 1822, *Outlines of the Geology of England and Wales,* pp. vii, 278, 320–364; see also R. Kirwan, 1779, *Geological Essays,* pp. 290–291, in which he uses the term "carboniferous soils"): The Carboniferous (named after the coal measures, the word Carboniferous means coal-bearing) was originally designated to include (from oldest to youngest) the Old Red Sandstone, the Carboniferous or Mountain Limestone, the Millstone grit and shale, and the Coal Measures; the typical locality was exposures in the Pennine Range of England. Subsequently the Old Red Sandstone was excluded from the Carboniferous and included in the Devonian (see above).

Mississippian System or Subsystem (A. Winchell, 1869 [or 1870], *Am. Philos. Soc. Proc.* 11:79 ["Mississippi group"]; see also H. S. Williams, 1891, *U. S. Geol. Surv. Bull.* 80:135 ["Mississippian series"]): This term was originally proposed by Winchell as "a geographical designation for the Carboniferous Limestones of the United States which are so largely developed in the valley of the Mississippi River." Williams defined it as a "series" of the Carboniferous, stratigraphically occupying the interval between the Devonian and the Coal Measures [Pennsylvanian]. Mississippian is purely an American term, the approximate equivalent being the Lower Carboniferous of Europe.

Pennsylvanian System or Subsystem (H. S. Williams, 1891, *Arkansas Geol. Surv. Ann. Rept.* 4:xiii; see also H. S. Williams, 1891, *U.S. Geol. Surv. Bull.* 80:83–108): The original use of the term Pennsylvanian was as a synonym (in a table) of the Carboniferous Coal Measures, as typically developed in Pennsylvania. Like the Mississippian, the Pennsylvanian is exclusively an American term, the approximate equivalent being the Upper Carboniferous of Europe.

Permian System (R. I. Murchison, 1841, *Philos. Mag.,* 3rd ser., 19:419; see also R. I. Murchison, E. de Verneuil, and A. von Keyserling, 1845, *Geology of Russia in Europe and the Ural Mountains,* pp. 7–8, 140–141): Murchison named the Permian on the basis of exposures in the province of Perm, adjacent to the Ural Mountains, in Russia. The Lower New Red Sandstone and the Magnesian Limestone are synonymous with, or included within, the Permian.

Mesozoic Erathem (J. Phillips, 1840, *Penny Cyclopaedia* 17:153–154; see also J. Phillips, 1841, *Palaeozoic fossils of Cornwall, Devon, and West Somerset,* Great Brit. Geol. Surv. Mem., p. 160): Meaning middle life, the Mesozoic was originally defined to include the upper part of the Permian and Triassic (together forming the "New Red Sandstone"), Jurassic ("Oolitic"), and Cretaceous.

Triassic System (F. von Alberti, 1834, *Beitrag zu einer Monographie des Bunten Sandsteins, Muschelkalks, und Keupers und die Verbindung dieser Gebilde zu einer Formation,* pp. 323–324): Von Alberti united the Bunter sandstone (Buntsandstein) group, Muschelk group, and the Keuper group (from oldest to youngest) of Germany into a single three-part (Trias) "formation." The Upper New Red Sandstone of England is part of the Triassic.

Jurassic System (A. von Humboldt, 1799, *Ueber die unterifdischen Gasarten,* published by Wilhelm von Humboldt: Wilmarth, 1925, was unable to locate a copy of this work; see A. von Humboldt, 1858, *Kosmos, Stuttgart* 4:632; K. A. von Zittel, 1901, *History of Geology and Paleontology,* pp. 497–502): Von Humboldt named the Jurassic on the basis of rocks exposed in the Jura Mountains of France and Switzerland. Originally his term referred to only a small portion of the present Jurassic. The Jurassic deposits are well exposed in England where they have been the subject of much classic work and have been traditionally called the Lias and Oolite (Arkell, 1933).

Cretaceous System (J. J. d'Omalius d'Halloy, 1822, Observations sur un essai de carte géologique de la France, des Pays-Bas et des contrées voisines, *Annales des Mines* 7:373–374): As d'Halloy defined it, the Cretaceous terrane referred to the rocks between the Jurassic and the Tertiary and corresponded in essentials to present usage of the term. The term Cretaceous is derived from the Latin for chalk.

Cenozoic (= Cainozoic, = Kainozoic [apparently the original spelling]) *Erathem* (J. Phillips, 1840, *Penny Cyclopaedia* 17:153–154; see also J. Phillips, 1841, *Palaeozoic Fossils of Cornwall, Devon, and West Somerset,* Great Brit. Geol. Surv. Mem., p. 160): Cenozoic means recent life. Most authorities currently include the Pleistocene and the Recent within the Cenozoic, but in the past many workers have excluded either the Recent or both the Pleistocene and Recent from this era.

Tertiary System (G. Arduino, 1760 [or 1759?], [Letter by Arduino to Antonio Valisuieri, professor of natural history, University of Padua], *Nuova raccolta di opuscoli scientifici e filologici del padre abate Angiolo Calogierà, Venice,* 6:142–143): For Arduino's original concept of the Tertiary, see the above section on "Moro and Arduino." At present the Tertiary is commonly used to refer to the Paleocene through Pliocene. This is the oldest term in stratigraphic nomenclature that is still in common use; however, it is slowly being replaced by the terms Paleogene and Neogene (Harland et al., 1982).

Paleogene System (M. Hoernes, 1856, *Die fossilen Mollusken des Tertiärbeckens von Wien: I, Gastropoden,* Geol. Reichsanst., Abhandl., 3:736 pages, 52 plates; M. Hoernes, 1870, *Die fossilen Mollusken des Tertiärbeckens von Wein: II. Bivalven,* Geol. Reichsanst., Abhandl., 4:479 pages, 85 plates): As discussed by Harland et al. (1982) and Papp (1979), Moritz Hoernes originally introduced the term Paleogene as essentially equal to Lyell's Eocene, and the term Neogene for the Miocene and Pliocene. The term Paleogene is now generally used to refer to the Paleocene, Eocene, and Oligocene; the Neogene is composed of the Miocene and Pliocene.

Paleocene Series (W. P. Schimper, 1874, *Traité de paléontologie végétale,* 3:680–682; see also W. D. Matthew, 1920, Status and limits of the Paleocene, *Geol. Soc. Amer. Bull.* 31:221): Schimper originally described the Paleocene Epoch (meaning ancient recent) on the basis of distinct flora elements that he regarded as intermediate between Cretaceous and Eocene forms. In the early twentieth century the term was revived and adopted by vertebrate paleontologists in particular, in recognition of the distinctive mammalian fauna of the so-called Basal Eocene. As presently used, the Paleocene is the basal series/epoch of the Cenozoic.

Eocene Series (C. Lyell, 1833, *Principles of Geology,* 3:52–55, 57–58): For an account of how Lyell originally defined the Eocene, see the section above on "Lyell." The term means dawn recent.

Oligocene Series (E. Beyrich, 1854, Uber die Stellung der hessischen Tertiär-bildungen, *K. Preuss. Akad. Wiss. Berlin Monatsber.,* November 1854:664–666): Using Lyell's subdivisions of the Tertiary (Eocene, Miocene, and Pliocene), by the middle of the nineteenth century some stratigraphers found that there were some contemporaneous deposits that were classified variously as upper Eocene or lower Miocene by different workers (Wilmarth, 1925); in order to resolve this situation

Beyrich proposed the name Oligocene (meaning small or little recent) for the strata and time under consideration.

Neogene System See "**Paleogene System**" above.

Miocene Series (C. Lyell, 1833, *Principles of Geology,* 3:52–55, 57–58): For an account of how Lyell originally defined the Miocene, see the section above on "Lyell." The term means less recent.

Pliocene Series (C. Lyell, 1833, *Principles of Geology,* 3:52–55, 57–58): For an account of how Lyell originally defined the Pliocene, see the section above on "Lyell." The term means more recent.

Quaternary System (J. Desnoyers, 1829, Observations sur un ensemble de dépôts marins plus récens que les terrains tertiaires du basin de la Seine, et constituant une formation géologique distincte; précédées d'un apercu de la non simultanéité des bassins tertiaires, *Annales Sci. Nat.* 16:171–214, 402–491; see also H. Roboul, 1833, *Géologie de la période quaternaire,* Paris, pp. 1–5; A. Morlot, 1856, Notice sur le quaternaire en Suisse, *Soc. Vaudoise Sci. Nat. Bull.* 4:41–45): This period derives its name in relation to the tripartite division of pre-Quaternary time (Primary, Secondary, and Tertiary) by many geologists in the later eighteenth century. It usually refers to the Pleistocene (glacial period) and the Recent, although it has also been used to refer only to the Pleistocene.

Pleistocene Series (C. Lyell, 1839, *Elements of Geology,* French translation, Paris, appendix pp. 616–621; see also C. Lyell, 1839, *Charlesworth's Mag. Nat. Hist.* 3:323, footnote; E. Forbes, 1846, *On the connexion between the distribution of the existing fauna and flora of the British Isles, and the geological changes which have affected their area, especially during the epoch of the Northern Drift,* Great Brit. Geol. Surv. Mem. 1:402–403: C. Lyell, 1873, *Antiquity of Man,* 4th edition, pp. 3–4): In 1839 Lyell designated his Newer Pliocene the Pleistocene, meaning the most recent (referring to the fossils contained in rocks of this age). Forbes (1846) used Pleistocene to refer to the "Glacial Epoch," which, according to Lyell (1873), Lyell had previously designated Post-Pliocene (see notes under "Recent," below). In 1873 Lyell adopted Forbes's usage and considered Pleistocene to be synonymous with what he had earlier designated Post-Pliocene. Reading Lyell (1857), however, one notes that it appears that the modern usage of the term Pleistocene generally refers to Lyell's (1857) "Newer Pliocene or Pleistocene" and "Post-Pliocene."

Recent Series (C. Lyell, 1833, *Principles of Geology,* 3:52–53): As originally used by Lyell (1833), the term Recent referred to all post-Pliocene (post-Tertiary) time; any rocks that had formed since the appearance of man were considered Recent. By 1857, however, Lyell had restricted the scope of the Recent when he distinguished the following groups of strata, from youngest to oldest (Lyell, 1857, p. 104): (1) Recent, (2) Post-Pliocene, (3) Newer Pliocene, or Pleistocene. The term Recent is now generally considered to refer to time since the end of the Pleistocene and is synonymous with Holocene, a name agreed upon at the 1885 International Geological Congress for the post-Pleistocene (Weller, 1960).

Chapter 2

ROCKS: THE MATERIAL BASIS
OF STRATIGRAPHY

Stratigraphy in the broad sense may be considered to be the study of geologic history, events reconstructed from the past and placed in an abstract temporal framework. Ultimately and proximately, however, stratigraphy is dependent upon actual material referents—bodies of rock materials and their spatial relationships—in order to advance its cause and obtain one of its prime goals, the accurate interpretation of the geology of the particular planet under consideration (or comparable body, such as the earth's moon). Consequently, stratigraphers must be familiar with the description and classification of rocks and their modes of genesis. Because traditionally the prime target of stratigraphy has been, and still is, the interpretation of sequences of sedimentary rocks on the planet Earth (the classically layered rocks), stratigraphy often has been considered intimately connected with sedimentology, sedimentary petrology, and petrography (see, for instance, the popular stratigraphy textbooks by Boggs, 1987; Dunbar and Rodgers, 1957; Krumbein and Sloss, 1963; and Matthews, 1984, all of which deal in large part with sedimentology). Indeed, some works on stratigraphy are in fact devoted almost exclusively to considerations of sedimentology (e.g., Fritz and Moore, 1988). Yet, when one views the products of national and international commissions or other groups concerned with stratigraphy and its clarification (e.g., Hedberg, 1976; NACSN, 1983; Owen, 1987; Schoch, 1988, and references cited in these works), one does not find these groups producing codes or guides to sedimentology or the names to be applied to rock types in petrography, but codes and recommendations applied to rock bodies (masses) on a larger scale. The excellent little book on stratigraphy by Donovan (1966, *Stratigraphy: An Introduction to Principles*) contains no explicit discussion of petrology or petrography per se, but assumes that the reader approaching stratigraphy already has this necessary background.

One might consider stratigraphy to build upon petrology and petrography just as mineralogy or crystallography must necessarily build upon some knowledge of chemistry; this does not mean that chemistry per se is just a subdivision of mineralogy, or vice versa. Likewise, to give a more extreme analogy, the social sciences (for example, cultural anthropology, sociology, psychology, history of man) deal with the human species, and although the general biology of this organism is certainly an important consideration to some aspects of social sciences, few would consider the study of the detailed anatomy, physiology, and functional morphology of *Homo sapiens* to be a subdivision of the social sciences in general. Ultimately, of course, the boundaries between all human disciplines are somewhat arbitrary (all the aspects of the world being to some degree interrelated); but my contention here is that stratigraphy is (or should be) a discipline distinct from sedimentology, petrology, and petrography.

Weller (1960, p. 49) pointed out some of the distinctions between stratigraphy and sedimentary petrology in the following manner: "Stratigraphers and sedimentary petrologists observe the same rocks and sediments but they see them from somewhat different points of view. Stratigraphers are interested mainly in rock masses; their attention is devoted largely to observing gross characters, structures, and relations in the field [and, I would add, the temporal or chronological relationships of rocks]. Sedimentary petrologists are much more concerned with details; mainly they study those features that require laboratory observations. Insofar as the descriptive phases are concerned, stratigraphy and sedimentary petrology are separate branches of geology." Sedimentary petrologists often are primarily interested in reconstructing the specific environment of deposition—the particular mode of genesis—of a rock or sequences of rocks. Stratigraphers, on the other hand, are generally concerned with placing the units or layers of local sequences of rocks within broad (ultimately global) temporal and spatial relationships. Since Weller wrote the above lines, stratigraphers have come to rely increasingly on laboratory techniques performed on rock samples (such as techniques involved with numerical/isotopic dating, stable isotope stratigraphy, and magnetostratigraphy, to name a few); but the distinction between stratigraphers and petrologists or petrographists remains.

In his next paragraph, Weller (1960, p. 49) went on to point out that even though stratigraphy and such disciplines as sedimentology are, or can be considered (as by this author), distinct, there is certainly overlap between them when defined broadly (after all, both study the same material objects), and it is undeniable that they are of benefit to one another.

> Much benefit is to be gained by close coöperation by workers in these fields, and satisfactory solutions to many problems cannot be obtained by either type of study unsupplemented by the other. Stratigraphy particularly stands to gain by coöperative action because petrologic studies are likely to be valuable in confirming or denying conclusions based on more general stratigraphic observations. Such studies may also suggest other explanations and direct attention to features previously overlooked or neglected because their significance was not appreciated. On the other hand, the results of petrologic studies may gain greater meaning than they would otherwise possess by orientation in the more general stratigraphic picture. (Weller, 1960, p. 49)

Certainly studies of the genesis of rock bodies, beds, or layers are of particular importance in ascribing certain stratigraphic interpretations; for instance, specific features may be considered useful in chronocorrelation or event stratigraphy (see Chapter 7) only if specific modes of genesis are ascribed to them. In such contexts not only sedimentologic or petrologic considerations may be of vital importance, but the results of studies of biologic, paleontologic, geochemical, climatic, and other attributes may be just as important. Stratigraphy is, in many ways, by nature an integrative and interdisciplinary science.

With all this in mind, I consider it beyond the scope of this book to discuss sedimentology or petrology in any detail. However, in the section below a very brief introduction to the classification of the materials of physical stratigraphic units (that is, rocks) will be presented, along with an equally brief introduction to their mode of genesis and basic description.

COMPOSITION AND CLASSIFICATION OF ROCKS

Rocks are generally considered to be "naturally occurring aggregates of minerals or mineraloids" (Ehlers and Blatt, 1982, p. 2). Here a mineral can be considered "a naturally formed chemical element or compound having a definite range in chemical composition, and usually [some would insist always] a characteristic crystal form" (AGI, 1976, p. 282); common examples include quartz, feldspars, micas, clays, amphibole, pyroxene, and garnet. Mineraloids are entities such as opal, glass, coal, and so on, that are naturally occurring and form constituents of rocks but are not of definite chemical composition or crystalline. A rock may be composed solely of a single mineral or mineraloid, that is, be composed of an aggregate of identical crystals or particles; but characteristically rocks consist of several or numerous minerals and/or mineraloids.

Traditionally all rocks have been divided into three categories, igneous, sedimentary, and metamorphic, defined as follows by the American Geological Institute (quoted in Ehlers and Blatt, 1982, pp. 2–3):

Igneous rock: A rock that solidified from molten or partly molten material; that is, from a magma.

Sedimentary rock: A rock resulting from the consolidation of loose sediment or chemical precipitation from solution at or near the earth's surface; or an organic rock consisting of the secretions or remains of plants and animals.

Metamorphic rock: Any rock derived from pre-existing rocks by mineralogical, chemical, or structural changes, especially in the solid state, in response to marked changes in temperature, pressure, and chemical environment at depth in the earth's crust; that is, below the zones of weathering and cementation.

However, as Ehlers and Blatt (1982) are quick to point out, there are some borderline or transitional types of rocks that do not fit neatly into one category or the other, but bridge two categories. Based on historical precedent, such rock types usually are relegated to a single category. For example, a strong argument could be made that stratified volcanic tuffs (formed by the settling of rock fragments that have been explosively ejected from a volcano) are sedimentary in nature, yet they are usually classified as igneous. An example of a rock transitional between the igneous and metamorphic categories is serpentinite (Ehlers and Blatt, 1982). Such a rock ultimately originates as a molten rock that cools and crystallizes as an aggregate of olivine and pyroxene (clearly an igneous rock), but during the cooling process water vapor may react with the olivine and pyroxene to form serpentine, so that the rock could be classified as metamorphic. Somewhat similarly, given any sedimentary rock buried at depth in the earth's crust, there may be some discussion as to how much change (for example, diagenesis—compaction, cementation, recrystallization, and perhaps replacement) it can undergo before it is perhaps considered to be a low-grade metamorphic rock. "One person's hard shale [a sedimentary rock] is another person's slate [a metamorphic rock]; a metaquartzite [metamorphic] to one investigator may still be a quartz-cemented quartz sandstone [a sedimentary rock] to another" (Ehlers and Blatt, 1982, p. 250).

In general, if only through practice and historical precedent, the three major classes of rocks are evident and useful to most geologists. It should be explicitly pointed out, however, that this threefold classification is genetically based. Whether a rock is considered to be igneous, metamorphic, or sedimentary is not a function of its mineralogy, texture, structure, and so on, per se, but reflects either an observation or a hypothesis as to the conditions under which it originated. Of course, the inherent characteristics of rocks are what are used in inferring mode of origin; but ultimately it could be argued that one can classify only a particular rock as sedimentary, igneous, or metamorphic, as a result of observing its formation in nature, by analogizing with rocks of known genesis, or by running laboratory experiments in order to determine the parameters within which a certain rock type must have formed.

The general characteristics commonly used to distinguish igneous, sedimentary, and metamorphic rocks from one another can be found in many basic petrology and petrography texts, such as Bayly (1968), Ehlers and Blatt (1982), and Williams, Turner, and Gilbert (1954). As has been discussed previously, the vast majority of stratigraphic work has traditionally dealt with (and still deals with) stratified (that is, layered) rocks, and the vast majority of stratified rocks exposed on the surface of the earth are sedimentary—the primary exceptions being extrusive igneous rocks, which may show layering approaching true stratification (actually, intrusive rocks may also show layering or zoning of mineralogy, depending upon the manner in which the magma crystallized), and metamorphic rocks that were derived from preexisting stratified rocks and still exhibit some of their original stratification. In recent years the concept of stratigraphy generally has been broadened to include the spatial and temporal relationships of all rock bodies, whether stratified or not (see Chapter 1).

Given the basic threefold classification of rocks described above, most stratigraphers would further subdivide each of these broad categories of rocks into a few basic divisions (such subdivisions may vary from investigator to investigator; one example is given by Weller, 1960). Metamorphic rocks, rarely dealt with by the average stratigrapher in any detail, may simply be divided into those that are presumed to have been formed from preexisting stratified rocks and still exhibit some of their original or primary stratification, and those that do not exhibit any original stratification (the original rock material having been of either a sedimentary or an igneous [or a previous metamorphic] nature). Metamorphic rocks exhbiting original stratification then will be treated accordingly by the typical stratigrapher—for instance, perhaps, as sedimentary rocks. Igneous rocks generally are divided into intrusives (molten rock that penetrated other country rock at depth and solidified before it reached the surface) and extrusives (magmas that poured out at the surface, such as lava flows, or were ejected at the surface of the earth). Extrusive rocks are commonly layered or stratified to some extent, and may be divided into various lava flows (such as basaltic or rhyolitic flows) and pyroclastic deposits (such as deposits of volcanic ash, tuff, or agglomerate). Within any of the broad categories mentioned above, igneous and metamorphic rocks may be further subdivided on the basis of their mineralogy, texture, and structure. In the case of metamorphic rocks, they also may be classified according to the degree or grade of metamorphism they exhibit; for example, the series slate, phyllite, schist, and gneiss (listed from low to high levels of metamorphism, respectively).

Of course, the vast majority of rocks dealt with by stratigraphers are sedimentary rocks, which may be roughly classified in various ways. They may be classified according to area and mode of origination, such as terrestrial versus aqueous (marine and freshwater), or placed into the categories water-laid, wind-laid (eolian), and glacially deposited. Very elaborate classifications of sedimentary rocks can be devised that correspond to the depositional environments in which they were formed (see below). Alternately, sedimentary rocks may be classified according to their mineralogy, grain size, and texture. On this basis it is common to subdivide the vast majority of sedimentary rocks into two broad groups, the detrital (noncarbonate) series and the limestone (carbonate) series.

The detrital series consists of various rocks formed primarily from detrital grains where the average size of the constituent grains determines the basic category of rock. Common terms for rocks of the detrital series, corresponding to increasing grain size, are shale, siltstone (shales and siltstones may together be considered mudstones), sandstone, grit, and conglomerate. Following the classic Udden-Wentworth Size Scale, each of these rock types can be defined on the basis of grain size, as shown in Table 2.1. Of course, many rocks contain grains that fall into two or more of the named size categories shown in Table 2.1; thus coarser rocks such as conglomerates and sandstones may contain a fine clay matrix between the larger particles. What is commonly termed a graywacke (or greywacke), generally considered a type of sandstone, is composed of detrital quartz, feldspar, and rock fragments set in a clayey matrix; some of the detrital fragments in such a rock may be larger than what is technically sand size.

The detrital constituents of such rocks commonly will be composed of one or more of the following: rock fragments, mineral grains (such as quartz, feldspar, or mica fragments) that were eroded and weathered out of preexisting rocks, and clays. Here it is important to note that as used by geologists in a loose sense, clay can have any one of several meanings (Weller, 1960, p. 54). In a nontechnical sense the term clay may be applied to any relatively fine-grained sediment or rock that is relatively plastic when wet. As shown in Table 2.1, clay may be defined simply as the fraction of the sediment of a rock with grains measuring less than a certain diameter, such as 1/256 mm on the Udden-Wentworth Size Scale; this is the use of the term clay in a sedimentologic sense. In a mineralogic sense the term clay is applied to a group of hydrous aluminum silicate minerals, such as illite, montmorillonite, and kaolinite. Secondary cement, such as silica or calcite, may be deposited within the interstices between the primary grains of a detrital rock.

Table 2.1 Classification of Detrital Sedimentary Rocks According to the Udden-Wentworth Size Scale.*

GRAIN SIZE IN MM.	SEDIMENT	ROCK TYPE	
Greater than 256	Boulders	Conglomerate	
64–256	Cobbles	Conglomerate	
4–64	Pebbles	Conglomerate	
2–4	Granules	Grit	
1/16 [.0625]–2	Sand	Sandstone	
1/256 [.004]–1/16	Silt [Mud]	Siltstone	Mudstone
Less than 1/256	Clay [Mud]	Shale	

*After Weller, 1960, p. 52, Table 3; see also Ehlers and Blatt, 1982, p. 325, Table 13-1.

The limestone series is the class of calcium carbonate–based (primarily calcite [or aragonite], the second most common carbonate mineral being the magnesium-bearing dolomite) rocks. Limestones may be composed of detrital carbonate grains and thus parallel the detrital series reviewed above, or the lime may be carried in solution and precipitated elsewhere; calcite is readily soluble and therefore a mobile mineral. Limestones commonly exhibit various degrees of recrystallization. Rocks of the limestone series may include various amounts of noncarbonate detritus. Indeed, some rocks may be composed of approximately equal amounts of carbonate material and noncarbonate detritus, for instance, rocks commonly referred to as marls (an ambiguous term applied to a number of different rock types, most of which are formed from the subequal intermixing of clays and fine particles of calcite or dolomite). In such cases there may be no easy way to categorize particular rocks as either of the detrital series or of the limestone series.

Besides the basic categories of detrital and limestone series, various other minor categories of sedimentary rocks may be recognized. Such special categories may include the carbonaceous and bituminous rocks (such as coal and peat), cherts, evaporites (for example, rocks composed predominantly of gypsum, anhydrite, or halite), rock phosphates, sedimentary iron ores, and so on.

In a scheme somewhat similar to that of the detrital series and the limestone series used for the categorization of the majority of sedimentary rocks, some workers consider that greater than 95% of all sedimentary rocks can be classified as either mudrocks, sandstones, or carbonate rocks (e.g., Ehlers and Blatt, 1982). According to such schemes, sandstones are composed of fragmental sediment that ranges between approximately 2 mm and 0.062 mm in diameter, and mudstones are composed of fragmental sediment smaller than about 0.062 mm (see Table 2.1). Carbonate rocks are composed primarily of the carbonate minerals calcite, aragonite (both $CaCO_3$), and/or dolomite ($CaMg(CO_3)_2$). As is to be expected, in some cases certain rocks do not fit neatly into any of these categories; Ehlers and Blatt (1982) cite the example of coquina, which is a rock composed of sand-size fragments of fossil shells that could legitimately be considered either a sandstone or a carbonate. In fact, however, coquinas are generally considered limestones, and perhaps most limestones are coquinas in the sense that they are composed of small grains of organic (shelly carbonate) remains. In other words, generally any rock with a fairly high percentage of carbonate may be considered a carbonate rock.

Given the basic broad categories of rocks reviewed above, what are their relative abundances on the surface of the earth? What types of rocks is the typical stratigrapher most likely to encounter? In a recent summary of such estimates, Ehlers and Blatt (1982) suggest that approximately 66% of the surface area of the continents is covered by sedimentary rocks, with the remaining 34% being areas of exposure of igneous and metamorphic rocks. The authors estimate that the "bulk of the 34% is probably igneous" (Ehlers and Blatt, 1982, p. 5). These same authors also suggest that the average sediment thickness on continental blocks is approximately 1.8 km, and in the ocean basins the average sediment thickness is about 0.3 km. Of course, in certain settings sediments may accumulate to thicknesses of several kilometers. Also, among sedimentary rocks, the amount of outcrop area generally decreases with increasing age; thus there are very few Cambrian exposures and many Tertiary exposures even though the entire Cenozoic, including the Tertiary, has been thus far (the Cenozoic has not yet ended) of a shorter duration than the Cambrian Period.

Ehlers and Blatt (1982) state that fully half of all exposed sediments are of Cretaceous age or younger.

As noted above, using the classification scheme of mudrocks, sandstones, or carbonate rocks, better than 95% of all sedimentary rocks fit into one of these categories. Again according to Ehlers and Blatt (1982), approximately 65% of all known
sedimentary rocks are mudrocks, 20 to 25% are sandstones, 10 to 15% are carbonate rocks, and less than 5% are other types of sedimentary rocks (such as conglomerates, coals, evaporites, cherts, phosphate rocks, and so on).

THE GENESIS OF ROCKS EXPOSED
AT THE SURFACE OF THE EARTH

A primary concern of much of petrology, and particularly sedimentology and facies
analysis, is determination of the genesis of the rocks that make up the physical
materials of stratigraphy. As stated above, it is far beyond the scope of this book
to discuss in detail the origin and genesis of various rock types, but a few comments
are warranted on the subject.

In the discussion of the basic classification of rock types, it was already noted that
the generally accepted tripartite division of all rocks into the categories sedimentary,
igneous, and metamorphic is explicitly genetic. Within these broad categories, however, rocks may be classified on the basis of such features as texture or mineralogic
content, which will necessarily reflect to some degree the genesis of the particular
rock under consideration, but may not uniquely pinpoint its genesis. For instance,
virtually identical sandstones found in the stratigraphic record may have formed
under extremely different conditions (perhaps under freshwater fluvial conditions
as opposed to shallow-water marine conditions).

Sedimentary Rocks

Sedimentologists have proposed numerous differing classifications of depositional
environments under which sediments may be deposited (ultimately, in some instances, to become preserved as part of the stratigraphic record). A typical example
of such a classificatory scheme is outlined in Fig. 2.1 and illustrated diagramatically
in Fig. 2.2 (both figures are from Lewis, 1984). Various sedimentary environments
may also be placed within a plate tectonic framework (Fig. 2.3; also from Lewis,
1984). In attempting to determine the depositional setting of any particular rock
body or group of associated rock bodies, various methods may be used. A basic
approach is to study modern depositional environments in detail, noting the types
of sediments involved, their means of transport to (and within) the site of deposition, and the resulting structures formed upon deposition of the sediment, and then
analogize to ancient sedimentary sequences. One may encounter difficulty in this
approach: in the past there may have been rocks formed under regimes for which
there are no current analogues, yet the rocks formed under past conditions may be
superficially similar to those forming today under very different conditions; investigators must be constantly wary of drawing firm conclusions from ambiguous evidence. In some instances it may be useful to interpret an ancient sedimentary sequence in terms of a unique set of particular attributes and features, observed today
in several differing sedimentary environments. Typical rock features utilized by sedi-

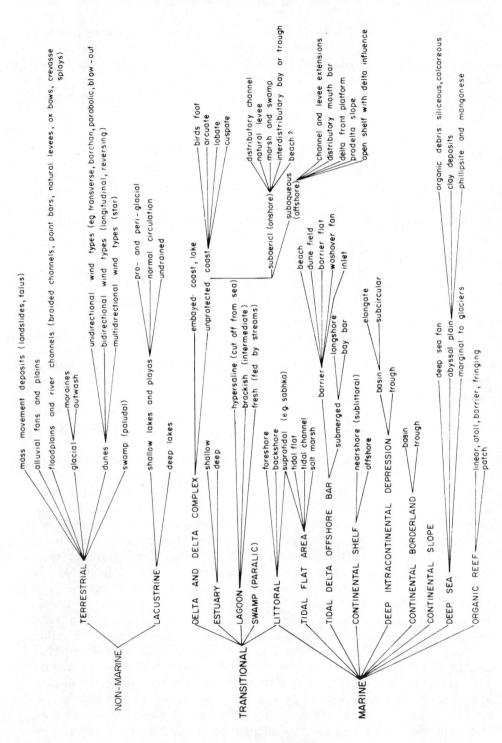

Figure 2.1. One possible classification of depositional environments. (From Lewis, 1984, p. 5.)

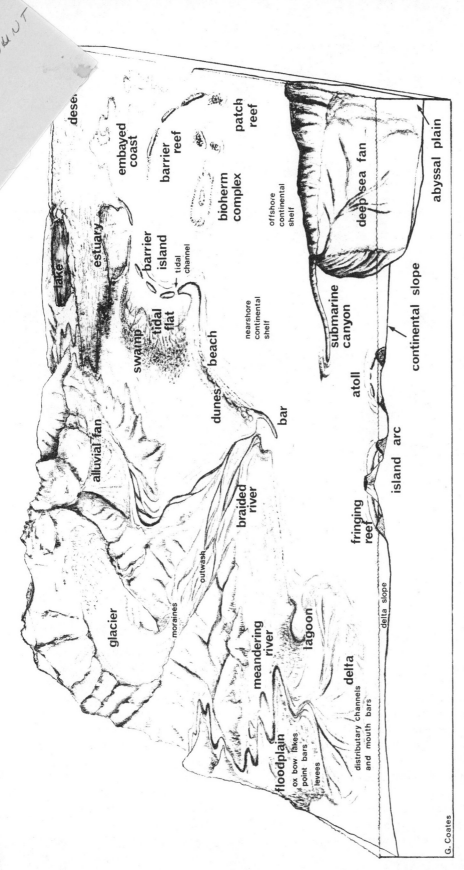

Figure 2.2. "A schematic representation of major sedimentary environments. *Note:* Unrealistic environmental contractions are necessitated by representation of so many settings in one figure." (Sketch by G. Coates; figure and caption reproduced from Lewis, 1984, p. 6.)

35

Rifted Continental Margin

(a) Early phase

(b) After separation

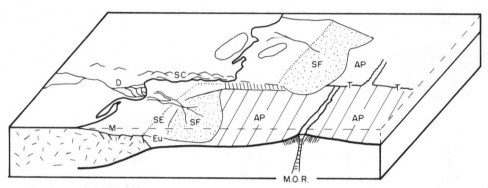

M = Miogeocline		T = Transform
SE = Shelf edge (in modern settings; some question if "shelves" always existed as such		D = Delta
Eu = Eugeocline		SF = Submarine Fan
		SC = Steep Coast, narrow shelf, etc
AP = Abyssal Plain		M.O.R. = Mid Ocean rise / ridge

Figure 2.3. Depositional environments and sedimentary basins placed within plate tectonic settings. (Sketches by J. D. Bradshaw; reproduced from Lewis, 1984, pp. 130–131.)

mentary petrologists in the genetic interpretation of rock bodies include, for example, mineralogy (composition), sediment texture (such as grain size, sorting, roundness, and sphericity of the sedimentary particles), diagenetic effects, and sedimentary structures. The latter category includes various aspects of physical, inorganic bedding and stratification (such as graded bedding, cross-bedding, scour marks, ripple marks, load structures) and biogenic (organic) sedimentary structures formed by organisms (such as trace fossils [ichnofossils, Lebenspuren]). Such sedimentary structures, along with any preserved fossils, may provide important information on the particular conditions (environment) under which the particular sediments were deposited.

When studying sediments and sedimentary rocks in a genetic context, one cannot concentrate only on modes of sediment transport and deposition, but must also consider the nature of the origins of the sedimentary particles making up any partic-

CONVERGENT MARGIN SETTING

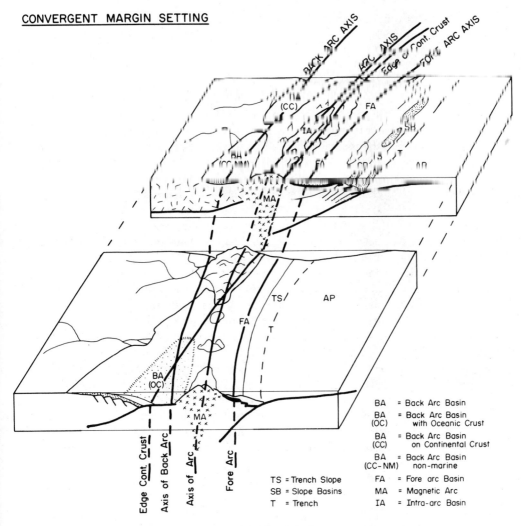

BA	=	Back Arc Basin
BA (OC)	=	Back Arc Basin with Oceanic Crust
BA (CC)	=	Back Arc Basin on Continental Crust
BA (CC-NM)	=	Back Arc Basin non-marine
TS = Trench Slope	FA	= Fore arc Basin
SB = Slope Basins	MA	= Magnetic Arc
T = Trench	IA	= Intra-arc Basin

Figure 2.3. (*Continued*)

ular rock, along with any diagenetic effects the sedimentary grains may have experienced. Ultimately sediment originates on the earth's surface because igneous and metamorphic rocks formed below the surface, and brought to the surface, are unstable at surface conditions (Ehlers and Blatt, 1982; Weller, 1960). At the surface, igneous and metamorphic rocks undergo chemical and mechanical weathering and disintegration; by far the most significant weathering is of a chemical nature. Once a sedimentary rock is formed, it too may be subjected to weathering and erosion on the surface and itself become the source of additional sediment. During weathering, various ions may go into solution and later be deposited as chemical precipitates. Organic constituents also are often typically found as detritus in sedimentary deposits; such organic matter (used in the loose sense) may range from whole organisms to free amino acids. Once deposited, the loose sediment may be subjected to various diagenetic effects such as compaction, cementation, replacement, and mineralogic recrystallization (in some cases diagenesis may border on metamorphism) so that many primary mineralogical and textural features, and sedimentary structures, may be altered.

Igneous Rocks

Igneous rocks are those of a magmatic origin and, as previously noted, can be broadly grouped into intrusive and extrusive igneous rocks. Intrusive igneous rocks have solidified beneath the earth's surface, whereas extrusive igneous rocks have solidified on the surface of the earth. Any igneous rock may incorporate foreign objects, or inclusions, within its body; for example, such inclusions may consist of fragments of the preexisting country rock (known as xenoliths) or older crystals (known as xenocrysts). One may speak of the contact between an igneous rock and the surrounding country rock as being either concordant or discordant. A concordant contact is parallel to the bedding (or possibly other linear features, such as foliation) in the country rock, whereas a discordant contact is not.

The magmatic origin of extrusive igneous rocks can be observed directly wherever there are active volcanoes, so the basic genesis of such rocks has been long known. However, the magmatic origin of intrusive igneous rocks was long doubted and debated. According to the standard interpretations of geologic history, one of James Hutton's contributions was to demonstrate the igneous nature of intrusive granites (see Chapter 1), but it was not until the 1870s that the question of the magmatic origin of intrusive igneous rocks was finally settled by the American geologist Grove Karl Gilbert (1843–1918; see Ehlers and Blatt, 1982). Today it is generally believed that various igneous rocks now incorporated in the crust of the earth and observed at its surface may have originated from melting and emplacement at a variety of depths. Some igneous rocks may have formed early in the earth's history from the melting and differentiation of the mantle, some igneous rocks have originated from either ancient or recent melting at depths of up to 100 km, and still other igneous rocks have formed from melting at the surface (such as melting induced by meteorite impacts). A magma may cool and crystallize at or near its place of origin, or it may migrate considerable distances, either vertically or laterally.

Igneous rocks may have a variety of mineral compositions (see any standard reference that covers igneous petrology/petrography), but commonly such rocks are broadly categorized on the basis of their silica (SiO_2) content. For example (taken from Ehlers and Blatt, 1982), those with greater than 66% SiO_2 content are labeled acidic (such as granite, syenite, quartz diorite), those with 52 to 66% silica are considered intermediate (for instance, granodiorite), those with 45 to 52% silica are basic (such as diorite, gabbro, basalt, andesite), and those containing less than 45% silica are ultrabasic (for example, peridotite, dunite). Intrusive igneous rocks may be emplaced at a variety of depths, from relatively shallow (hypabyssal) to very deep intrusions. The resulting intrusive bodies of rock may also take on a variety of external morphologies and may range in size from meters to thousands of kilometers. A variety of names have been coined to describe the basic shapes of intrusive rock bodies. To give a few common examples, sills are concordant tabular bodies that are typically parallel to any bedding of the country rock. Dikes are tabular rock bodies that are discordant; dikes cut across the bedding of the country rock, perhaps following old joint systems. Laccoliths are mushroom-shaped intrusions, and batholiths are simply large intrusive plutons that have steep sidewalls and often no known base (that is, their vertical thickness is not known). Within any intrusive igneous rock body there may be layering or differentiation. A single igneous complex may have formed as the result of several injections of molten rock, perhaps producing some sort of gross layering. As a molten body of rock cools, heavy, early-formed crystals may settle to the bottom so that the composition of the resulting rock varies in the vertical direction.

Igneous extrusive rocks often are literally stratified (in a conventional stratigraphic sense), are tabular in form, and conform to the principle of superposition. Volcanic dust and ejected debris may be deposited at extreme distances from the source area, and volcanic ash in particular may form thin isochronous surfaces that are extremely useful in chronocorrelation (for example, bentonites; see J. Conkin and B. Conkin, 1983, 1984a,b). In such cases the materials of volcanic origin are essentially sedimentary particles and part of the sedimentary record *sensu stricto*. Lava flows, ash flows, nuée ardentes, and the like are peculiar to igneous extrusive rocks. The consideration of the stratigraphy of these types of rocks is of particular importance for extraterrestrial (for instance, lunar: Mutch, 1972; Wilhelms, 1987) stratigraphy. On the earth certain large areas are also covered by thick sequences of extrusive igneous rocks.

Here we should note that although the vast majority of stratigraphy is concerned primarily with sedimentary rocks, they are not the only rocks to which stratigraphic procedures and principles apply. Although some stratigraphers actively question whether intrusive igneous and high-grade metamorphic rocks are, or should be, part of the domain of stratigraphy, extrusive igneous (including volcanic) rocks are particularly amenable to standard stratigraphic procedures and classification originally developed primarily with sedimentary rocks in mind. Under the *North American Stratigraphic Code* (NACSN, 1983), extrusive igneous, metavolcanic, and metasedimentary rocks are all explicitly classified lithostratigraphically using the same system as for sedimentary rocks.

Molton extrusions (lavas) may erupt at the surface along narrow fissures, sometimes giving rise to lave plateaus, or extrusions may erupt from centralized vents commonly labeled volcanoes. In some cases, a chain or line of volcanoes may form along what is essentially a fissure. As described by Mutch (1972), volcanoes can be classified into four broad categories, although any particular volcano may well exhibit features of more than one category. Shield volcanoes are broad and gently sloping, formed by the successive outpourings of floods of liquid lava. Strato-volcanoes are formed when episodes of lava flooding alternate with the explosive ejection of particulate material; the rocks composing and surrounding strato-volcanoes consist of sequences of stratified and interlayered lava flows and pyroclastic sediments. Cinder cones are formed where particulate material is ejected without associated liquid lavas. Finally, highly viscous magmas may be emplaced at the surface of the earth as relatively high and steep extrusive domes.

The geomorphological form of a volcano, and the types of rocks deposited, are determined in good part by the type of magma extruded. Basic magmas tend to be relatively hotter, less viscous, and more fluid, and therefore may form large shield volcanoes or great plateaus composed of basalts (see below). Basic magmas that explosively lose volatiles may form cinder cones and strato-volcanoes. In contrast, more acidic magmas tend to be cooler and more viscous, and thus do not flow as far as basic magmas. Acid magmas may tend to form localized domes. If a viscous magma has a high volatile content, violent explosions may occur, spreading volcanic ash and debris over large areas (Mutch, 1972; Rittman, 1962). A highly heated and gas-charged mass of lava and ash particles may be ejected essentially horizontally from a vent and continue downhill as an avalanche, forming a nuée ardente that may be deposited to form a welded tuff or an ignimbrite.

In certain regions large bodies of extrusive igneous rocks are extremely conspicuous and stratigraphically important. The Deccan traps or lavas (Wadia, 1953; see also Mutch, 1972), of Cretaceous to Eocene age, cover a wide geographic area in western India and the surrounding region. The Deccan traps are basically a flood basalt, composed of a series of successive flow sheets, averaging 600 meters thick, but ranging up to 3000 meters thick in some areas. Individual flows making up the unit are approximately 5 to 30 meters thick. In the case of the Deccan traps, the lavas are believed to have flowed from vents and fissures, forming a relatively monotonous sequence of horizontal flows. The rocks are extremely uniform petrographically, and have been stratigraphically subdivided primarily on the basis of thin fossiliferous beds that separate some of the flows.

Metamorphic Rocks

As noted above, metamorphic rocks constitute a class of rocks derived from preexisting rocks because of conditions (essentially of temperature and pressure) that have caused changes in the texture and mineralogy of the rock. The limits of what constitutes metamorphism are not clearly defined, but it is generally accepted that if such textural and/or mineralogical changes in a rock "occur at pressures and temperatures above those of diagenesis and below those of melting, they are referred to as metamorphism" (Ehlers and Blatt, 1982, p. 511). At the lower end of the scale (where diagenesis grades into metamorphism), incipient metamorphism is usually promoted primarily by increased temperatures. Some workers define low-level metamorphism by the first appearance of some mineral that does not normally form on the surface; examples of such minerals include albite, epidote, lawsonite, muscovite, and pyrophyllite (Ehlers and Blatt, 1982). Depending on the specific geochemical and mineralogical environment, such minerals may form in the subsurface at temperatures within the range of 150 to 350°C. At the upper end of the scale, metamorphics are replaced by igneous rocks when melting occurs; depending on rock types and lithostatic pressures, melting may begin in the range of 650 to 800°C. Rocks known as migmatites exhibit a combination of metamorphic and igneous features and occupy the boundary between metamorphic and igneous rocks. A rock is referred to as "polymetamorphic" if it is believed to have been subjected to more than one metamorphic event during its history (that is, a metamorphic rock that has been metamorphosed).

Low-grade metamorphic rocks may retain structures and features of the original parent rock, such as bedding or even in some rare instances traces of fossils (given that the parent rock was sedimentary in nature). As metamorphism increases, such original features are progressively obliterated, and new features, characteristic of metamorphic rocks, develop. As a result of the pressures and unequal stresses to which metamorphic rocks are commonly subject, parallel features develop in such rocks, for instance, foliation (including schistosity and slaty cleavage; essentially the planar and parallel arrangement of minerals and/or textural features) and lineation (essentially linear fabrics developed in metamorphic rocks). Various types of foliation in particular have at times been naively mistaken for traces of original bedding.

In fact, foliations may develop at any angle to original features, such as primary bedding.

One can distinguish several basic modes of origin and occurrence of metamorphic rocks, such as regional metamorphism, contact metamorphism, burial metamorphism, and cataclastic metamorphism (Ehlers and Blatt, 1982). The most common form in this context is regional metamorphism, where metamorphism has occurred over a widespread area. Regional metamorphism is particularly associated with orogenic belts and is the result of the high temperatures and pressures that normally accompany mountain-building processes. Intensity of metamorphism, and therefore the metamorphic grade of the resulting rocks, may vary systematically relative to a geographic or structural axis or axes. Metamorphic grades of the rocks are often indicated by the occurrences of various mineral zones (such as chlorite, biotite, garnet, staurolite, kyanite, and sillimanite zones, from low-grade to high-grade metamorphism, respectively) that can be mapped, and the boundaries between such zones are often referred to as isograds. Ultimately regional metamorphism and related igneous and orogenic activity should be relatable to a general global tectonic framework.

Contact metamorphism essentially refers to baking or other alteration of the country rock that is adjacent to an igneous intrusion. The scale of contact metamorphism will vary according to the size, temperature, and so on, of the igneous body involved, but typically zones of contact metamorphism around an igneous body may range in thickness from less than a meter to several kilometers. Thick regions of contact metamorphism may exhibit an internal zonation of mineralogy and texture away from the igneous intrusion. Simple baking of country rock, without compositional changes, may result in what are termed hornfels. Alternately, fluids may emanate from the intrusion into the country rock, or may simply be mobilized in the country rock as a result of the intrusion; these processes are referred to as metasomatism and may result in the formation of skarns or tactites.

The phenomenon that has been termed burial metamorphism by some investigators grades into what others refer to as high-grade diagenesis. Deeply buried sediments and/or volcanic rocks may experience high temperatures and pressures that promote mineralogical changes (for example, the formation of zeolites) while at the same time the original structure of the rocks is preserved. Cataclastic metamorphism refers to the local fracturing and crushing of rocks in fault zones and other structural situations. Fault breccias and mylonites may be the result of faulting at relatively shallow or deeper levels, respectively.

BASIC DATA GATHERING IN THE FIELD

The crux of much of stratigraphy is the spatial relationships of rocks over geographic areas. In order to identify, map, correlate, and otherwise interpret rocks, observations must be made, and basic data are gathered in the field. The stratigrapher must search for areas where the rocks or unit(s) of interest are exposed on the surface (that is, outcrops must be located), or the rocks may be sampled and data gathered by drilling or by remote methods such as seismography (see following sections). The mainstay of much stratigraphy traditionally has been the observation and interpretation of surface exposures in the field; and even as other techniques come to play increasingly important roles, traditional field work will remain important—direct observation and sampling of rocks have yet to be replaced.

The scale and detail with which observations are made will depend on the nature of the particular project, and the nature of a single project may change as it progresses. Such factors as time and resources available, basic goals, amount of work done in the area previously, and so on, are all of extreme importance. Reconnaissance work in an unexplored territory inevitably will be different from a project that involves detailed analysis and refinement of the geology and stratigraphy of a classic region (perhaps an area where stratigraphers have worked since the late eighteenth and early nineteenth centuries, such as parts of England and the continent). However, for any stratigraphic study it is standard practice when in the field to note, at any particular exposure, certain characteristics of the rocks, as described below (after Weller, 1960; see also such works as Compton, 1962, 1985; Kottlowski, 1965; Krumbein and Sloss, 1963; Lahee, 1923; Lewis, 1984; and Prothero, in press):

1. The general lithologic characters, and sequences and relationships (for example, interbedding, mixing, or gradation) of lighologies should be noted. Mineral compositions should be determined as accurately as possible. Common descriptive terms are normally best used, at least initially; interpretations can be postulated later, or noted but kept distinct from primary observations. Degree of induration should be recorded.

2. Coloration of the various lithologies observed should be carefully recorded; common and fairly unambiguous terms are best used. The standard *Rock Color Chart* (Goddard et al., 1951; see also Lewis, 1984) or similar devices and systems may be used for this purpose. When colors are recorded, it should be stated whether the rock is wet or dry; if possible separate colors for the rock when both wet and dry should be given.

3. Bedding characteristics (parallel, cross bedded, wavy bedding, and so on) and bedding scale (finely laminated, thin bedded, thick bedded, massive—it is usually least ambiguous simply to state [using a linear scale, commonly metric] the thickness of the bedding) should be noted.

4. Textural features, such as grain size and distribution, sorting, roundness or angularity, and crystallinity (such as of limestones or igneous rocks) should be carefully recorded.

5. Features such as the massiveness of beds, ripple marks, concretions, and so on, should be noted.

6. All weathering characteristics of the rocks should be noted. Leaching, alterations in color, detailed weathered surface forms, and general geomorphologic expression of the rocks should be recorded. Whenever possible, fresh rock should be carefully compared to typical weathered rock commonly found on the surface, and any similarities or differences noted. Characteristic soils, flora, and fauna developed on particular rocks or stratigraphic units also should be noted; such information can be particularly useful in areal mapping of stratigraphic units.

7. Any and all fossils should be noted and their occurrences in the sections or exposures recorded as precisely as possible. As accurately as possible (but be wary of overidentifying) the fossils should be identified, and the absolute numbers and relative abundances of different species (or other low-level taxonomic groups) at various stratigraphic levels should be recorded. The condition (fresh, abraded, replaced, and so on) and position (for example, whether they are in life position or have been noticeably transported since death) of fossils should be noted.

8. Thicknesses of lithologies and stratigraphic units (even if only informal or provisional) should be measured or estimated whenever possible.

9. Strikes and dips of bedding planes and any structural features such as joints, cleavage, fractures, faulting, folding, or other deformation should be noted.

In the field, all significant exposures should be accurately located, preferably both on a good topographic map and also via directions or instructions on how to get back to the particular exposure. In pursuing fieldwork it is generally useful to record as much as possible, not only concerning the geology per se, but also any unusual conditions or events, weather patterns, names of local contacts and guides, places to eat or stay, and so on. If in doubt, record. One can always ignore extraneous information or notes later; one does not want to be in the position of having lost potentially valuable information.

It is also advisable to record not only with words, but with sketches and photographs. If practical, samples also may be gathered. Samples may be collected on a casual basis to represent the basic lithologies and fossil contents encountered, or samples of a particular size may be collected systematically throughout the sections encountered, perhaps with the intention of later rigorously analyzing their contents and subjecting the results to various sorts of statistical manipulation.

Two basic techniques of the field stratigrapher, and of field geologists in general, are the measuring of stratigraphic sections and the construction and compilation of geologic maps (in their most basic forms, maps that record the geology exposed at the surface of the earth, such as rock types, over a geographic area; see Figs. 2.4 through 2.6). Space dictates that these topics cannot be elaborated upon here. Techniques used in geologic mapping and in the measuring of stratigraphic sections are throughly introduced in Compton (1962, 1985); Kottlowski (1965) is devoted exclusively to the subject of measuring stratigraphic sections in the field.

GEOPHYSICAL WELL LOGS

In many areas adequate exposures of rocks on the surface are lacking; so to elucidate the stratigraphy of a region one may have to resort to data gathered from the subsurface. These data may take the form of direct samples (cores) drilled from the rock (in which case one is essentially creating artificial exposures), or may be procured in a much less direct form, such as recordings of particular seismic properties of the rocks (seismic stratigraphy: see following section). In many cases the ideal might be viewed as the direct recovery of continuous cores mechanically drilled from wells, but in most cases, perhaps for monetary reasons, this is impractical. Moreover, as Rider (1986) points out, if cores are taken from a deep well, their position within the vertical dimension of the well may be ambiguous. One usually determines the depth from which a continuous core is taken by adding together all the lengths of drill cable used, but mistakes often can occur. Even if a complete series of contiguous cores is attempted, there may be recovery problems that throw off the vertical positioning of cores. Consequently, Rider suggests that, given a typical well or borehole of several thousand meters' depth, the calculated drill depths of a core may be as much as 5 to 15 meters different from those of well logs of the same sequence.

In drilling a well or borehole, what is commonly known as a "mud log" is maintained as standard practice. Important information recorded in this log includes the drilling rate over various parts of the well, the depths at which the bit is changed, problems faced, information gathered about the lithogies encountered as well as any other information about the rocks (such as gas content), and any other events in the drilling of the well. The drilling rate in particular (typically given in terms of

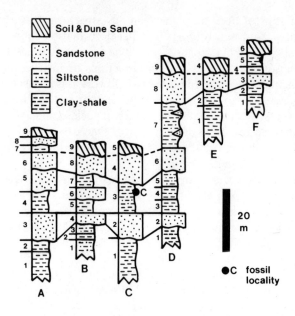

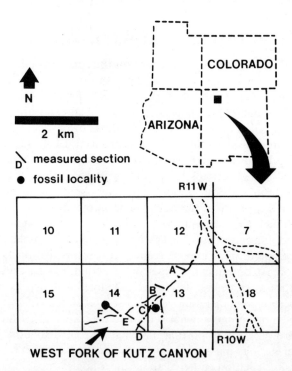

Figure 2.4. Example of measured stratigraphic sections with proposed lithocorrelations (top of figure); location of the measured sections in the west fork of Kutz Canyon, San Juan Basin, New Mexico, is shown below. (From Schoch and Lucas, 1981, pp. 3 and 15.)

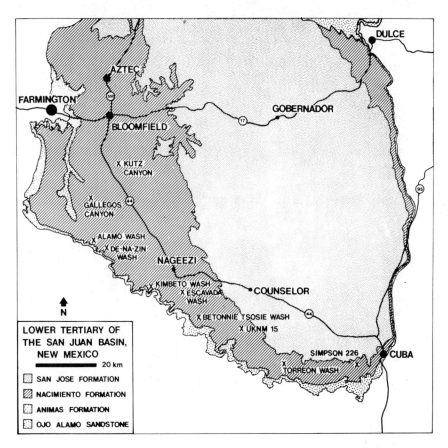

Figure 2.5. Example of a geologic map: geologic map of the Paleocene and Eocene strata of the San Juan Basin, New Mexico. (From Schoch, 1986a, p. 125.)

minutes per meter) may shed important information on the boundaries between major changes in lithology; for example, the drilling rate may suddenly increase in going from a sandstone to a shale.

A common means of drilling wells employs a lubricating mud, which is forced down into the hole to the drilling area and then returned to the surface, carrying away the rock cuttings. The cuttings or chips can be recovered to some extent, so that a direct record of the lithologies encountered can be made. However, such cuttings often record only very imprecisely the formations and lithologies found in the drilling. Samples are not taken continuously, but usually on the order of every 20 to 30 minutes; the closeness of stratigraphic spacing of such samples will depend on the drilling rate (the rock may be sampled every 25 meters or every 2 meters). Depending on the rock type, samples might be completely pulverized and inseparable from the drilling mud once they are returned to the surface. Or, because of differing densities of the lithologies in the section, various rock chips might be forced to the surface of the well at different rates, thus mixing with one another or perhaps being recorded out of sequence. Furthermore, the drilling mud and any samples it contains are not returned to the surface simultaneously with the drilling of the particular lithology. There is always a lag time between the penetration by

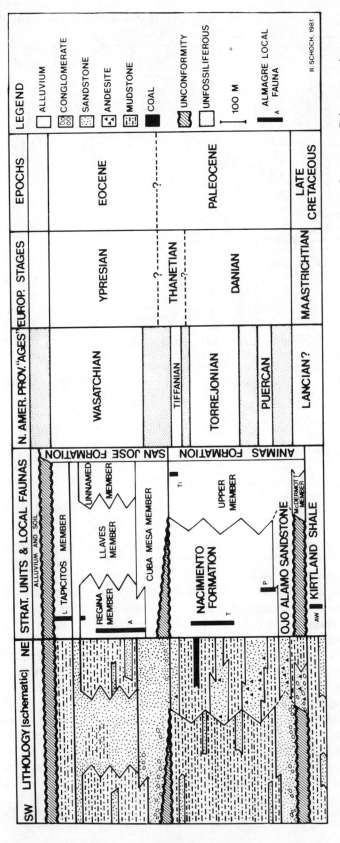

Figure 2.6. Example of a composite stratigraphic section for a local sedimentary basin: diagrammatic summary of the latest Cretaceous, Paleocene, and Eocene strata of the San Juan Basin, New Mexico. (From Schoch, 1986a, p. 126.)

the drill of a certain rock and the return to the surface of the mud and samples; this lag time depends on many factors—the rate of circulation of the mud, the drilling rate, the current depth of the well, and so on. In order to place any cutting samples on the log of the well with any degree of accuracy (that is, in order to determine the depth from which they originated) one must estimate the lag time for any particular sample. For a well several thousand meters deep, the lag time for some samples may be well over an hour. Obviously the reliability and usefulness of cutting samples are limited.

After a well is completed, it is possible to sample the sides of the well or borehole directly by taking sidewall cores; but such sampling is limited to certain predetermined points along the wall of the borehole—it is not practical to sample the wall continuously. Sidewall cores tend to be of very small dimensions (on the order of a few centimeters at most in the longest dimension) and may not even be representative of the characteristic lithology of the interval sampled (for example, in a rock that is predominantly sandstone with some shale lenses, the sidewall core may by chance sample a shale lens). In taking a small sidewall core, certain characteristics of the rock being sampled may also be physically destroyed.

For these types of considerations, once a well is drilled, data on the rocks must be gathered through various secondary means. A basic tool for gathering information used in subsurface stratigraphy is the geophysical well log. As described by North (1985), Rider (1986), and many other authors, a geophysical well log is simply a continuous (or pseudocontinuous, the parameter being recorded at relatively closely spaced intervals) recording of some geophysical parameter along a borehole (Fig. 2.7). Such a log also may variously be called a wireline geophysical well log, an electrical log (because the first such logs recorded the electrical properties of rocks), or simply a well log or log. However, one must be careful not to confuse the last two terms (especially if taken out of context) with other types of logs, such as logs of lithologies (for instance, based on cuttings sampled during drilling operations or based on continuous cores brought to the surface during mechanical coring) or logs of organic or paleontological content. As is discussed below, common geophysical well log types include temperature logs, sonic logs, resistivity and conductivity logs, gamma ray logs, neutron logs, and caliper logs (to name but a few). The discussion in the remainder of this section is meant to provide only a general introduction to some of the major geophysical well logs currently in use. For more detailed information the reader is referred to such sources as Asquith and Gibson (1982), Brenner and McHargue (1988), Cant (1984), Lewis (1984), Merkel (1979), North (1985), and Rider (1986), references cited in these works, and the information and manuals published by the commercial logging industry (for example, Dresser Atlas, 1982, 1983, and Schlumberger, 1972, 1974, 1986). Perhaps the best single source devoted solely to the geological interpretation of well logs is Rider (1986); I would encourage the interested reader to peruse this work in particular.

The usual way to record a geophysical well log is to lower the recording equipment (the logging instrument, often called a sonde) down to the bottom of the borehole (which is done after the drilling equipment has been removed and the borehole has been stabilized, if necessary; in some cases wells may have to be cased in order to avoid collapse, but this will restrict the number and kind of logs that can be run), and then to pull it back up to the surface while simultaneously recording the geophysical parameter(s) of interest. Logging tools used to make the recordings are normally multifunctional (they can record numerous different parameters simulta-

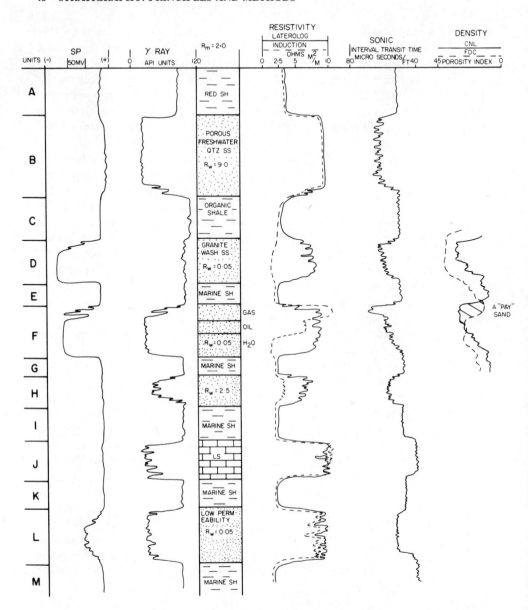

Figure 2.7. Examples of idealized well log responses. (From Lewis, 1984, p. 199.)

neously), of very thin diameters (on the order of 7.5 to 10 cm) such that they easily fit into the boreholes, and at times very long (up to 18 m according to Rider, 1986). The sampling rate for current logging equipment is usually between once every 3 cm and once every 15 cm per log (Rider, 1986); this information is recorded digitally and stored on magnetic tape, and also is processed by a computer on the surface. After the equipment is lowered to the bottom of the hole, it is pulled up by its cable at speeds generally in the range of 0.3 to 1.8 km/hour; thus, a 2000-meter hole may typically require on the order of 2 hours for a single ascent of a particular piece of equipment. Despite the multifunctional nature of modern equipment, gathering a complete set of possible logs from a single hole may require that several descents and ascents be made, using different pieces of equipment. Also, because of the length of any single piece of equipment the recording devices for various logs will not be at the same depths simultaneously. On the surface such discrepancies must be compensated for (normally by use of a computer).

A basic type of well log is the caliper log. A simple mechanical caliper tool measures the diameter of the hole along a vertical profile. Dual-caliper tools also are commonly in use to measure borehole shape, and recently devices for representing the three-dimensional geometry of boreholes have been developed (see Brenner and McHargue, 1988; North, 1985; Rider, 1986, and references cited therein). Caliper logs are extremely important as an indication of the nature of the borehole and its walls, which may in turn seriously affect the quality of various geophysical logs that one wishes to run on the well. Some logging equipment designed primarily for other purposes is automatically fitted with a caliper, used to ensure that the actual measuring device is applied to the borehole wall. In other cases, depending on the nature of the borehole wall (for example, if serious caving is present), corrections may have to be applied to the geophysical logs made, or it may not be worthwhile to run certain logs.

A simple caliper log may record that the hole at any particular point is larger than the bit size of the drill used to excavate the hole, that the hole is smaller than the bit size, or that it is the same size as the drill bit (in which case it is said to be "on-gauge"). Holes that are significantly larger than the bit size may be said to have "caved" or "washed out"; the walls may have caved in or been eroded away by the lubricating mud. Such responses are typical of poorly consolidated shales. On-gauge holes are typical of relatively dense or firmly consolidated lithologies such as massive limestones or calcareous shales (Brenner and McHargue, 1988; Lewis, 1984; Rider, 1986). In some cases the borehole diameter may be smaller than that of the drill bit size; this may be indicative of what has been termed "mud-cake build-up," or it may indicate sloughing of the rock. In either case, such narrow zones may form tight spots, past which it may be difficult to run some of the equipment. Sloughing may be caused by such minerals as smectite, which, when penetrated by the drill, takes up water from the drilling mud and swells or expands. True mud-cake build-up may form along the borehole walls where beds of relatively high permeability are encountered (Rider, 1986).

In order to interpret various other types of logs accurately (for instance, the electrical resistivity of rocks is affected by temperature), and as an adjunct to geochemical, diagenetic, and rock maturity studies (particularly of hydrocarbons and organics), it often is important to record borehole temperatures (Merkel, 1979). As a general rule, temperature increases with depth as one approaches the center of the earth; that is, heat (thermal energy) appears to flow from the interior of the earth to the surface, and we can speak of a temperature or geothermal gradient for the earth (which may be conventionally expressed in degrees centigrade per kilometer). However, in any particular section (borehole) there will not necessarily be a strict linear increase in temperature with depth because the rocks encountered may not be homogeneous in their thermal properties. Various lithologies have characteristic thermal conductivities and thus transmit heat with various efficiencies. Shale and coal, for example, have relatively low conductivities and thus may actually act as thermal insulators, whereas sand and salt bodies have relatively high thermal conductivities (Rider, 1986). The general net result is that when rocks are characterized by high thermal conductivities, the local geothermal gradient will be relatively low (the rate of temperature increase with depth will be relatively low as the heat is efficiently conducted to the surface and escapes), whereas when rocks of low thermal conductivity are encountered, the geothermal gradient will increase (temperature will increase quickly with increasing depth).

The maximum temperature encountered at the bottom of a well or borehole can be measured, and from this an overall geothermal gradient can be calculated, or a thermometer tool can be used to produce continuous temperature readings along the length of the hole. In either case, however, if temperatures are measured either during the drilling process or shortly thereafter, they will not record the true temperatures of the rocks but rather the temperature of the borehole mud. Such temperature readings are consistently lower than the true temperatures of the rocks because the circulating mud draws heat from the rocks. If the rocks are left to sit, eventually equilibrium will be reached between any mud in the borehole or well and the surrounding rocks, and the true rock temperatures can be taken. The time required to attain such equilibrium, however, can be on the order of four months; so various mathematical models have been devised to approximate "true" (equilibrium value) temperatures based on the lower temperatures usually recorded during or shortly after drilling (see references cited in Rider, 1986).

Resistivity and conductivity logs measure the resistance of the rock bodies (formations) under consideration to the passage of an electrical current and their ability to conduct an electrical current, respectively (Asquith and Gibson, 1982; Lewis, 1984; Merkel, 1979; Schlumberger, 1972, 1974, 1986). Resistivity and conductivity are reciprocals of one another, and whether resistivity or conductivity is actually measured along a particular borehole, it is common practice to express the results in terms of resistivity.

The resistivity of rocks is primarily a function of the resistivity of the rock materials and the fluids in the pores of rock bodies. As a first approximation, one may consider most rock materials themselves to be essentially insulators, but any rock body normally includes spaces, voids, or pores between the grains of the actual rock material, which are normally filled with some liquid, such as water, oil, or natural gas. Most formational water within rocks contains dissolved salts (typically being much saltier than average seawater) that can conduct an electric current. As a general rule, the higher the concentration of salts, the higher the conductivity (and, of course, the lower the resistivity). However, if the same rock type or formation is hydrocarbon-bearing, then the conductivity will be very low and the resistivity extremely high (North, 1985; Rider, 1986).

The conductivity or resistivity of a certain rock type is not simply a function of the conductivity of its contained fluids; the geometry of the pores, particularly the connections between the pores, also plays an extremely important role in determining the conductivity/resistivity of a particular formation. A rock with large pores that are directly connected with one another through large openings will provide a relatively easy path for the movement of electric currents and thus will add to the conductivity of the formation. The size, spacing, and interconnections between the fluid-filled pores or voids in any particular rock as they relate to conductive properties constitute what is commonly known as the Formation Resistivity Factor. Even if a rock itself is composed solely of nonconductive material, the geometry of the voids in the rock passively affects the conductivity of the unit.

Within any particular rock body the Formation Resistivity Factor in a particular area may be directly related to the porosity of the rock; however, as noted above, this factor is also dependent on how the pores or spaces are connected to one another. The Formation Resistivity Factor is independent of the particular fluid that may happen to fill the pores of the rock; the relation of this factor to the overall resistivity of a particular rock in a given area is:

$$R_o = F \times R_w$$

where R_o = the overall rock resistivity (that which may be measured directly in logging), R_w = the resistivity of the fluid filling the pores' spaces, and F = the Formation Resistivity Factor.

Not all rock materials play a totally passive role in determining the electrical conductivity/resistivity of a rock body. Clay minerals (for instance, in shales), depending on the particular mineral species involved, may have the ability to conduct electricity along their surfaces independent of any fluids contained in the rock. Such factors will of course influence the Formation Resistivity Factor and can create very complex situations, as in the case of shaly sands.

When a borehole is drilled, a lubricating mud is used; and this mud (with its own characteristic resistivity—the type of mud or muds used in drilling must be recorded) invades the rock and any fluid that the rock originally contains, naturally changing the resistivity of the rock immediately adjacent to the borehole. The influence of the invading mud diminishes laterally away from the borehole until a point is reached where virgin rock again is met. Obviously, one must take into account such considerations when running and interpreting resistivity logs. The effect of the invading drilling mud sometimes can be compensated for mathematically, graphically, or by the use of logging devices that reach the uninvaded zone of virgin rock.

In taking resistivity/conductivity measurements, two basic methods can be used: a current can be passed from one electrode to another through the rock, and the potential drop in current will measure the resistivity directly; or a current can be induced in the rocks by means of a transmitting coil on the measuring tool and detected by a receiving coil on the tool—the ability of the rocks to carry the current being a measure of their conductivity. With various sophisticated resistivity and induction tools, it now often is possible to measure the resistivity of a rock or formation anywhere laterally from the area of the borehole wall to the virgin rock. Induction logs tend to penetrate the farthest from the borehole, but they also give the coarsest resolution.

Resistivity logs of various resolutions have a number of potential applications, especially when used in conjunction with other types of logs (as is true for virtually all well logs; none is commonly used in isolation). In hydrocarbon exploration, resistivity logs can be used to identify and quantify oil reservoirs. For the stratigrapher, in particular, resistivity logs can be useful in the determination or interpretation of rock porosities, compactions, general textures, lithologies, facies, and bedding characteristics. Resistivity logs may be used directly in suggesting subsurface correlations of well sections.

When a series of differing rock types (formations) in a borehole are electrically connected by a conductive fluid (such as the drilling mud), electrical disequilibrium can be created; and due in large part to the electrochemical effects generated by salinity (ion concentration) differences between the borehole fluid and the formation fluids, spontaneous currents may be generated (see Asquith and Gibson, 1982;

Lewis, 1984; Merkel, 1979; Rider, 1986; Schlumberger, 1972). The self-potential or spontaneous potential log (the SP log) is simply a recording of the natural potential differences between a reference electrode on the surface of the hole and an electrode that is lowered into the borehole.

Spontaneous potential logs indicate changes in potential, not absolute values, and are recorded in terms of plus or minus millivolts. SP logs are used primarily to aid in the determination of the resistivity of the water or fluid in a rock, and in the determination of permeabilities. Spontaneous potentials are due in large measure to the free movement of ions and thus are naturally related to the permeabilities of the rocks involved. Generally any SP log deflections toward the negative (minus millivolts) are assumed to indicate that the rock corresponding to the deflection is a permeable bed. In some situations SP logs have been particularly useful in identifying shales, because of their generally low permeability.

The majority of rocks emit natural radioactivity to some degree. This natural radiation is due almost exclusively to three families of radioactive elements: the uranium–radium family, the thorium family, and the potassium isotope ^{40}K. As a generalization, each of these elements (uranium, thorium, potassium) contributes approximately the same magnitude of radioactivity in a "typical" rock body. Although potassium is by far the most abundant of the elements listed above, its contribution to natural radioactivity is relatively small per quantity of potassium, whereas a very minute amount of uranium, for instance, makes a large contribution to the general radioactivity of a rock. As a rule of thumb, igneous and metamorphic rocks tend to be more radioactive than most sedimentary rocks, and among sedimentary rocks shales often emit the strongest natural radiation.

The gamma ray log measures and records the overall radiation emitted from the rocks along a borehole wall; this is the sum of the radiation due to the uranium, thorium, and potassium. A spectral gamma ray log records the radioactivity of the three elements separately (Merkel, 1979). The devices used to record the natural radiation are based on the fact that these radioactive elements spontaneously emit gamma rays (high energy photons), which can be detected with a scintillation counter or a similar device. The simple gamma ray tool merely records the number of gamma ray emissions per unit time, which is proportional to the amount of radioactive elements in the particular rock under consideration. The spectral gamma ray tool follows the same principle, but this more sensitive tool distinguishes between energy levels of the emitted gamma rays, which are indicative of the original elements from which they were derived (i.e., uranium, thorium, or potassium). Gamma ray logs are recorded in terms of the "API unit" (American Petroleum Institute unit), which is derived from a standard reference well at the University of Houston, in Texas (see Rider, 1986).

In running and interpreting gamma ray logs, one must take several considerations into account. Gamma rays emitted from a source do not maintain their energy indefinitely; on the contrary, they rapidly lose their energy (degrade) and finally are no longer detectable. The denser the material through which the rays must pass, the more rapid this degradation. The gamma rays also are simultaneously scattered as they interact with material through which they pass, such as the rock body. (These combined effects of degradation and scattering are referred to as Compton scattering.) Thus, as the gamma rays emerge from the wall of the borehole, they may have a variety of energy values, depending in part on how far within the rock they originated and the density of the rock. The vast majority of natural radioactivity detected by the logging tool probably originates less than a few tens of centimeters from the borehole wall. Before being detected by the gamma ray tool, the gamma rays commonly must also pass through some drilling mud, which itself will add to the Compton scattering and may lower the apparent radiation. One must also be aware, however, that the drilling mud may contain natural radioactivity that also will be detected by the gamma ray tool. In the latter case, if the amount of drilling mud is relatively constant throughout the borehole, this may simply have the effect of increasing the baseline of radioactivity along the borehole while the relative difference in radioactivity from one formation to another remains the same.

A more serious problem can occur when caving or other relatively large irregularities are found along the length of the borehole. In such instances the irregularities may be filled by a significantly increased volume of drilling mud relative to the remainder of the borehole, separating the actual gamma ray detector from the borehole wall. This increased amount of drilling mud will tend to increase the Compton scattering of the photons being emitted from the rock, changing the gamma ray log values. Also, if the mud is characterized by a significant degree of radioactivity, this added radioactivity will not be evenly distributed along the length of the borehole, but rather will be highest where the drilling mud is thickest (for example, in areas of caving of the drilled rock) and thus will compromise the results of the log.

The speed with which the gamma ray or spectral gamma ray tool is brought up through the borehole while making measurements can dramatically influence the results. The tool must count photons (emitted randomly by the source elements) over a unit time, and the more time it has to count photons at any one spot, the more accurate the results will be. However, the gamma tool usually moves constantly along the borehole and thus detects emissions from a vertical interval of rock along the borehole in a unit time. As a result, the unit of time used to count emissions must be made relatively small to avoid mixing lithologies in a single count—although full and accurate counts then are compromised. In order to achieve more accurate counts, the unit of time can be enlarged, but with the increased chance that the constantly moving tool will encounter two or more different lithologies. In that case, the shapes of beds, formations, and contacts between them will be blurred and distorted. In general, the more slowly the tool is pulled up the borehole, the better the results will be, although the expense necessarily will increase as well.

Traditionally, simple gamma ray logs have been used to identify shales in sedimentary sequences (Asquith and Gibson, 1982), but this is now seen as an oversimplification; not all shales are necessarily radioactive, and not all radioactivity is concentrated in shales. In stratigraphy the simple gamma ray log and the spectral gamma ray log record important compositional aspects of the rocks and thus can be extremely useful in attempting to determine lithologies, facies, and mineral contents of rock bodies. Proper interpretation of the spectral gamma ray logs in particular is dependent on knowledge of the geochemistry of potassium, uranium, and thorium. In many cases these logs continue to be useful in identifying shales and shale or clay components, as radioactive potassium and thorium in particular are often found concentrated in shales and clay minerals. Potassium alone is found in a variety of minerals, such as clays, micas, and feldspars, and may also be concentrated in evaporite deposits. Uranium often is adsorbed by, or found associated with, concentrations of organic matter. Uranium peaks on the spectral gamma log also are typically found associated with condensed sequences and unconformities. Thorium is very stable chemically, does not pass into solution easily, and is often found as a residual component in weathered areas or in the form of detrital thorium-bearing mineral grains in sediments. Thorium appears to have a greater affinity for minerals typically found in the terrestrial realm than those typical of the marine realm (Rider, 1986).

As Rider (1986; see also Asquith and Gibson, 1982; Schlumberger, 1974) has pointed out, gamma ray logs often are used for correlation purposes. They lend themselves to this use for a number of reasons. Gamma ray logs are run on most boreholes; so the data are readily available. As discussed above, gamma ray logs generally are considered to give at least a gross indication of lithology, or lithologic characters—so correlation on the basis of these logs can be viewed as a type of direct lithocorrelation. The radioactivity producing such logs is not affected by depth, pressure, borehole temperature, or other variables; thus it is relatively repeatable (although because of the random nature of radioactivity any two runs along the same borehole will not be exactly the same in fine details). For rocks that have a large shale component, the radioactivity often is extremely variable throughout the shales vertically but relatively consistent horizontally throughout one bed, layer, or formation. This phenomenon may be the result of certain depositional environments that produced such shales, which were laterally persistent over short intervals of time, but changed through time (such changes being now recorded in the sequences of stratigraphic columns). The net result is that the matching of gamma ray logs may be extremely useful in proposing correlations, although one must be wary of overcorrelating. Depending on various parameters such as the diameter of the borehole, the drilling mud used, the measuring device used, and so on, the baselines and sensitivities of various gamma ray logs from different wells or boreholes may vary greatly. This inconsistency, along with the fact that the finest peaks on a typical gamma ray log are due to statistical variation (and thus are not useful in correlation), must be kept in mind when one proposes gamma ray log-based correlations.

A sonic or acoustic log (Lewis, 1984; North, 1985) measures and records the capacity of the rocks along a borehole or well to transmit sound waves. Sonic log values normally are given in terms of interval transit time (for example, microseconds per unit of linear measurement, commonly the foot [= approximately 30 cm]); this sonic transit time is the reciprocal of the velocity. The actual sonic logging tool simply measures the length of time it takes a sound pulse emitted at one end of the device to travel by compressional waves (P-waves) through the rock to a receiver or receivers elsewhere on the tool. The path taken by the sound waves through the rocks is very close to the borehole wall; consequently, caving or rugosity along the surface of the borehole may compromise the sonic log. The walls of boreholes tend to deteriorate with time, so it is best to run sonic logs very soon after drilling.

Sonic logs can be used to calculate porosities of rocks, but according to Rider (1986) porosities calculated using neutron or density logs (see below) are generally more reliable than these values. Perhaps the most important current use of sonic logs is their application to seismic work. Both sonic logs and seismic sections (see section below) are based upon acoustic velocities in rocks. A time–depth curve—the time it takes a sound wave to reach a certain depth in the particular section penetrated by the borehole—can be derived from the sonic log. This curve can be independently calibrated or checked by lowering a geophone into the borehole to precise levels (usually of presumed stratigraphic significance) and firing shots on the surface; thus a second time–depth curve is developed. Given the time–depth curve, a sonic log then can be replotted (via computer) against a linear travel time scale (rendering it comparable to seismic sections in which the vertical dimension is time rather than depth) as opposed to the original linear depth scale. When the information from a sonic log is combined with the information from a density log run along the same borehole, then the acoustic impedances of the various rock layers found in the section can be calculated. From this information, with the aid of a computer, a synthetic seismic log or trace can be constructed for the borehole or well (see Rider, 1986; Sheriff and Geldart, 1982, 1983; and references cited in these works). Direct comparisons between the sonic log plotted against travel time, the synthetic seismic trace, and an actual seismic section can be extremely important in interpreting seismic work.

Sonic logs also can be used directly in stratigraphic work. For most major lithologic types there is a great amount of overlap in the acoustic velocities, yet sonic logs may be of some use in distinguishing lithologies from one another. For example, dolomites, anhydrites, and hard rocks (igneous and metamorphic rocks) tend to be characterized by relatively high velocities, whereas coals and some shales have characteristic low velocities. Even if in many cases a particular velocity is not adequate for identifying a precise lithology apart from other evidence, the sonic log is extremely sensitive to lithologic changes. Rider (1986) compares the sonic attributes of rocks in wells and boreholes to color: neither is diagnostic of particular lighologies, but both color and sonic attributes may be characteristic of a particular formation or unit, and changes in either may indicate subtle, but significant, changes in lithology. For this reason the matching of sonic logs may be very useful in lithocorrelations. What otherwise appears to be a fairly homogeneous lithologic unit may even show small, correlatable variations on the sonic log. Such variations may be due to very fine textural or mineralogical variations in the rock body. Sonic log trends may even reflect fining-up or coarsening-up sequences in sedimentary sections.

Sonic velocities may reflect compaction. In general, as a rock becomes more compact, the velocities increase. Particularly in thick shale sequences there may be a general trend of increasing velocity with depth. Such general compaction trends, and any shifts or abrupt changes in them, have been utilized in attempts to identify unconformities, estimate the amount of erosion at a particular unconformity, or even to estimate the relative amount of uplift a particular section underwent in the past.

The density log records the overall (average) or bulk density of the rocks encountered in a borehole or well—that is, the density of the actual rock material (matrix) and any fluid that it might contain. The density is calculated in standard fashion by bombarding the rock with gamma rays and then measuring their attenuation (cf. Compton scattering, discussed above) via a detector (North, 1985). In relatively dense rocks the scattering and attenuation of gamma rays is greater than in less dense formations; so differing numbers of gamma rays are picked up by the detector, depending on the rock's formations. Unfortunately, the density tool apparently penetrates only very shallowly from the wall of the borehole into the rock (perhaps only on the order of 10 cm; see Rider, 1986), and this, of course, can compromise its usefulness. Density effects of the drilling mud per se may be automatically compensated for by the tool, but caving, rough spots, or mud-caking may all influence the density measurements. Furthermore, because of its shallow penetration, the device may be measuring the zone around the borehole invaded by the drilling fluid.

One of the principal uses of the density log is to calculate the porosities of the rocks encountered. Knowing the bulk density of the rock, the lithology and grain density of the rock (derived from other methods or even hand samples), and the composition and density of the formational fluid (again either derived or estimated from other sources), one can then calculate the porosity of the rock. As discussed above, combined with the sonic log the density log is also useful in calculating the acoustic impedance of rock units along a well or borehole. In combination with other logs, density logs are used in attempts to interpret lighologies, and they are also important in some studies of compaction and diagenesis. Certain relatively homogeneous and pure rock or mineral types may be provisionally identified by their characteristic densities, such as coal (low density), pyrite (high density), and specific evaporite deposits of intermediate densities (such as halite, anhydrite, and gypsum).

When a rock is bombarded with fast neutrons, these neutrons travel through the rock colliding with various nuclei, lose their energy, and are finally captured by the nucleus of some atom, which then emits gamma radiation (Schlumberger, 1972). The most significant energy loss of the neutrons is experienced when they collide with nuclei that are of essentially equivalent mass to the neutron, that is, hydrogen nuclei. Consequently, the rate of energy diffusion and eventual capture of the neutrons is in large part a function of the number of hydrogen atoms contained in the rock being tested. Most of the hydrogen in a typical rock is in the form of water (H_2O), either as a fluid in pore spaces or as water bound up within or between crystals of the rock matrix. The neutron log is a record of how the rocks in a well or borehole react to neutron bombardment in this respect.

The neutron logging tool consists of a fast neutron source (such as plutonium–beryllium or americium–beryllium) and detectors. Like that of the density tool, the depth of penetration of the neutron tool into the wall of the borehole is relatively shallow, although it may be as much as 60 cm with some tools in some formations; again, such factors as caving, rugosity of the borehole wall, mud-caking, and influences of the drilling mud may need to be taken into consideration.

A standard use of the neutron log is to calculate porosity (Asquith and Gibson, 1982; Merkel, 1979), and neutron logs are commonly plotted in terms of neutron porosity units. Here the idea is that in certain water-bearing formations the only hydrogen present will be the hydrogen in the formational fluid, so that the reading of the neutron tool is directly related to the volume of the water-filled pore space, which is a measure of porosity. In some pure limestones containing formational water the neutron porosity is equal to the true porosity. However, depending on the nature of the rock materials (matrix) and their specific effects on the neutrons, compensations or corrections may need to be made. The readings of the neutron tool also can be described in terms of a hydrogen index, where the reading of a pure matrix (totally free of water and its hydrogen) would be zero, and the reading of pure water would be one.

Oil has essentially the same hydrogen index as water (Rider, 1986), but gas that may fill pore spaces in rocks is less dense and therefore has a low hydrogen index, which will result in a lowered neutron porosity reading. As bound hydrogen occurs in clays that compose shales, such rocks will exhibit heightened neutron porosities. Although both neutron and density logs reflect porosity, they approach the subject in different ways; in isolation neither is particularly useful in identifying specific lithologies. However, when used in combination, neutron and density logs can be extremely effective indicators of lithologies.

The dipmeter log is intended to record direction and degree of apparent dip of the strata encountered in a well or borehole (North, 1985). Dip is calculated by recording microresistivity logs from various sides of the borehole (for instance, four electrodes can be set at 90 degrees from one another) and then comparing the detailed microresistivity curves from the various faces of the borehole. It is assumed that identical beds or layers will have equal resistivities across the diameter of the borehole, so that by matching up resistivities the strike and dip of particular beds can be determined. Such dipmeter logs have been used to help identify particular types of sedimentary sequences, upon which paleoenvironmental interpretations can then be imposed, and have also been used in deformational and tectonic studies.

Recently Rider (1986) has suggested that the true interpretation of dipmeter logs is ambiguous. Rider believes that there is not enough field evidence for what the actual meaning of dipmeter logs might be, stating (Rider, 1986, p. 166): "The geologist has no field model of dip changes every 10 cm vertically [the approximate spacing between dip readings given on current dipmeter logs], although he should have. The only way in which the dipmeter may be reliably interpreted is by the use of field examples."

Once a series of well logs have been run on a borehole, they may be used in a number of different ways. They may be manually interpreted in terms of lithologies, facies, or depositional environments—which appears to me to be both an art and a science. For lithologic interpretations it is common to compare the values of all the various logs recorded for a certain depth and from this deduce the possible lithology. As noted above in discussions of the various individual well logs, although any particular value on a single well log may not be indicative or diagnostic of a unique lithology, it may suggest certain lithologies or exclude others. When used simultaneously, the information from numerous different types of logs may appear to fix a lithology fairly accurately, but any conclusions should also be compatible with the larger picture, such as the presumed sequence of lithologies (does the sequence make sense? does it fit into a regional picture?—although negative answers to these questions do not preclude that the interpretation is correct) and any direct evidence of the lithology (such as cuttings or corings). Manual interpretations of series of well logs may be greatly aided by sophisticated statistical and computer analyses, especially because most logs now are originally recorded digitally and stored by computer means (such as on magnetic tapes).

Well logs also are finding wide application in the interpretation of sedimentological conditions, facies, or environments of deposition. For this work, it is frequently the shape of a particular log or series of logs—that is, the vertical trends in logs— that is regarded to be of prime importance. For example, rocks that are composed predominantly of sands may be characterized by the shape of their self-potential or gamma ray logs, which in turn are interpreted as indicating vertical trends in grain size that may be diagnostic of particular facies or depositional environments and useful in the reconstruction of the local geologic history; for instance, progradation or marine transgression may be identified in this manner.

Well logs are very valuable in basic stratigraphic studies *sensu stricto*. As is discussed above, particular types of well logs have proved extremely useful in basic correlation (that is, correlation or matching of similar points and sequences from one borehole to another). Without being given other information, such "correlations" are essentially matchings or lithocorrelations. Indeed, suites of well logs may be used to identify and even define basic lithostratigraphic units. As Rider (1986) points out (see also Chapter 4 on type sections), stratotypes of certain wholly subsurface lithostratigraphic units may be wells or boreholes, and the formation may in practice be characterized or defined primarily by well log attributes.

Given adequate interpretation, true chronocorrelations may be proposed on the basis of well logs. (Similarly, adequate interpretation also must be applied to any surface data in order to propose chronocorrelations.) One must always be wary of matching similar (but not necessarily correlative) facies and then interpreting such matches as litho- or chronocorrelations. Horizons that are presumed to be of time significance might be identifiable in well logs, such as volcanic beds or other marker horizons. Also useful in postulating correlations may be unconformities that, in some cases, are readily indentifiable on well logs.

Currently some stratigraphers consider that one of the most promising applications of well logs is as an adjunct or supplement to seismic stratigraphic studies. As various authors have stressed (see, for instance, papers in Payton, 1977), the interpretation of seismic stratigraphic sections requires the integration of seismic data with all other available information, much of which is derived from well logs. By such techniques as generating synthetic seismic traces, particular well logs can be matched directly to seismic sections; seismic horizons can be identified on the well logs, and horizons or units that are correlated among well logs should in turn be identifiable on the seismic sections. Major reflections on seismic sections usually correspond to major changes in lithology (often marking unconformities); the lithologies and their changes can be documented on the well logs. The depositional sequences identified in seismic sections may be, or should be, readily identifiable on well logs.

SEISMIC SECTIONS

Seismological studies date back to the first half of the nineteenth century, and even earlier (Sheriff and Geldart, 1982), but the beginning of the application of seismic methods to stratigraphic work (specifically petroleum exploration) dates from around the time of World War I (1910s). In using seismology to detect subsurface layers and rock structures, two basic methodologies have been applied: refraction and reflection seismic methods (Boggs, 1987; Sheriff and Geldart, 1982). Early stratigraphic exploration used primarily refraction methods, but these were soon supplanted in most cases by reflection methods, which now form the primary basis of seismic stratigraphic research (see Figs. 2.8 through 2.10). In the remainder of this section I will briefly review the principles underlying refraction and reflection seismic methods, and the generation of modern seismic stratigraphic sections (the actual techniques used have changed greatly over the years, and are continually being improved; furthermore, at any one time numerous techniques are in use simultaneously). The following discussion is based on such works as Anstey (1982), Bally (1983, 1987), Berg and Woolverton (1985), Boggs (1987—a general textbook overview), Brown and Fisher (1980), Davis (1984), Neidell (1979), North (1985), Payton (1977), Sheriff (1980), and Sheriff and Geldart (1982, 1983). Here it is possible only to give the briefest overview of the generation and interpretation of seismic sections; for further information the reader should consult any of the excellent works cited above.

Seismic waves (sometimes colloquially referred to as energy waves, acoustic waves, or simply sound waves) are usually classified into two basic forms: compres-

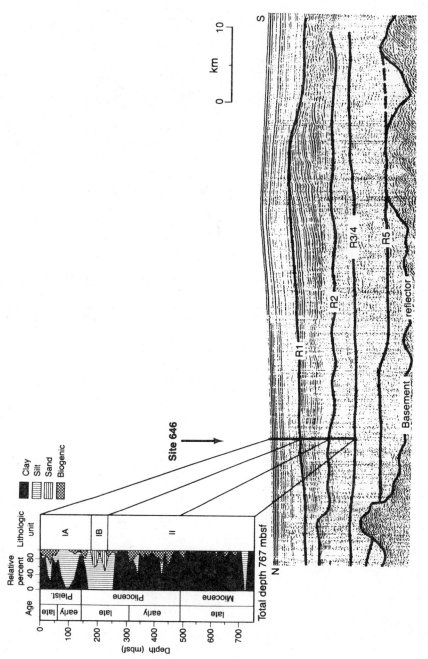

Figure 2.8. Example of a seismic section showing proposed correlation of ages and lithologies with reflector horizons on a seismic profile (seismic section taken in the Labrador Sea). (From Firth et al., 1987, p. 5, reprinted by permission of the Cushman Foundation for Foraminiferal Research.)

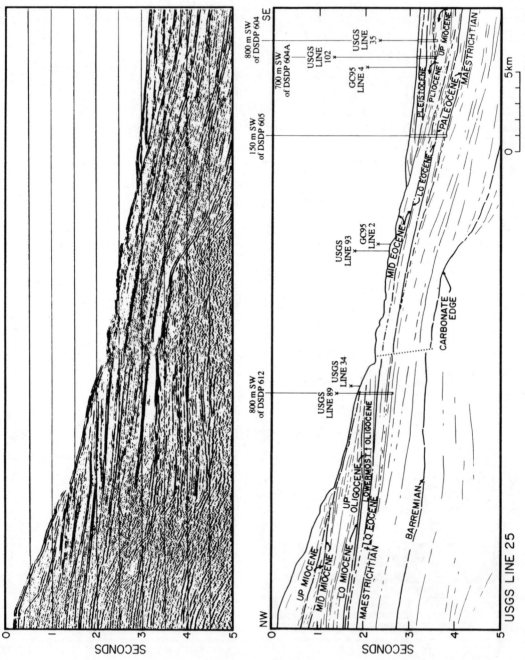

Figure 2.9. Example of a seismic section and its interpretation. Top, original seismic section (taken from offshore New Jersey); bottom, interpretation of the seismic section along with the location of other seismic lines and relevant Deep Sea Drilling Project cores. (From Mountain, 1987, p. 50, reprinted by permission of the Cushman Foundation for Foraminiferal Research.)

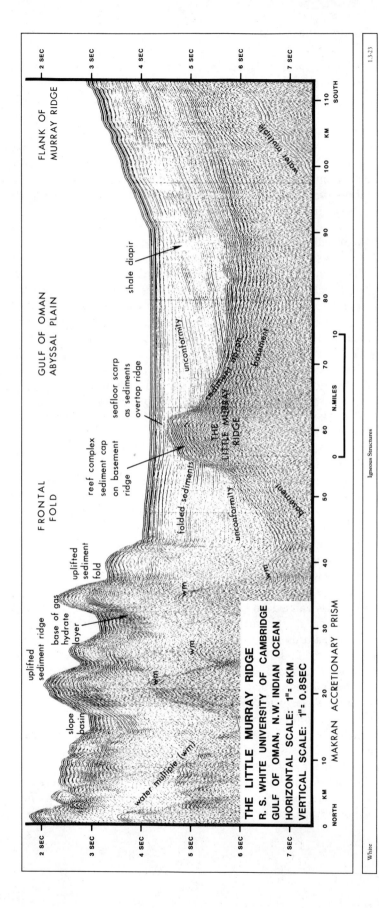

Figure 2.10. Example of an interpreted seismic section, The Little Murray Ridge, Gulf of Oman, N.W. Indian Ocean. (From White, 1983, p. 1.3–23, reprinted by permission of the American Association of Petroleum Geologists.)

1.3-23

sional or P-waves, and transverse, shear, or S-waves. Compressional or P-waves involve series of compressions and rarefications of the rock serving as the transmitting medium; that is, the transmitting medium rapidly changes its volume and shape as it transmits a P-wave (North, 1985). Compressional waves are the dominant waves used in exploration seismology, such as the generation of seismic stratigraphic sections (Sheriff and Geldart, 1982). Shear, transverse, or S-waves involve displacements of the transmitting medium in a direction perpendicular to the direction of propagation; the medium changes shape but not its volume. As seismic waves pass through the materials that make up the earth, they travel with specific velocities, which are a function of the elastic properties (particularly the rigidity, density, and resistance to volume changes) of the particular rocks through which the waves pass. For any single type of sedimentary rock (for instance, sandstone with 30% porosity) velocity tends to increase with depth. Comparing different broad categories of rocks, the following can be arranged in order of increasing average seismic wave velocities (based on North, 1985, his Fig. 21.10, p. 418): alluvium and clay, sandstone and shale, limestone and dolomite, granites and metamorphic rocks, salt and anhydrite. One must keep in mind, however, that there can be a considerable range in overlap of velocities associated with rocks in these various classes. At contacts between rocks with differing seismic properties (such as differing densities or rigidities)—that is, at velocity interfaces—seismic waves will be both refracted and reflected; it is these refractions and reflections that are utilized in refraction and reflection seismic methods, respectively. Seismic waves may be generated naturally, such as by earthquakes, or they may be generated artificially—the latter being the type used in developing seismic sections.

In both refraction and reflection methods, seismic waves are generated at the surface, the waves travel into the subsurface, and eventually some of the waves return to the surface. The time it takes the waves to return and the properties of the waves as they return shed information on the underlying rock structures and stratigraphy. In order to artificially produce seismic waves on land, an explosive charge (for example, dynamite), the shot, traditionally is exploded in a vertical hole near the surface (perhaps at a depth of 6 to 60 meters: Sheriff and Geldart, 1982; North, 1985). Mechanical thumpers have also been used to generate the waves; in this method a heavy weight simply is dropped to the ground from the back of a truck in order to produce the seismic waves. More recently vibratory methods have been used: a heavy mass carried under a specially designed truck is hydraulically pressed against the surface of the earth to the point where it lifts the truck off the ground. The whole apparatus then is mechanically vibrated, with the result that seismic waves are generated and directed into the earth (North, 1985). At sea, explosives have been thrown overboard and detonated to produce the wave sources, but more recently such techniques as the release of compressed air have come into more general use to generate waves when running seismic sections beneath the oceans.

The seismic waves, transmitted through the rocks by temporary displacements of the particles of which the rocks are composed, travel as spherical wavefronts from the particular point source of wave generation. After traveling through the subsurface, reflecting or refracting, returning waves are detected by devices known as geophones on land and hydrophones at sea. The information from any one geophone (or set of geophones acting as one) may be recorded on a continuously moving strip of paper or a magnetic tape, or by some similar means, where a long dimension represents time (since the "shot"), and the dimension perpendicular to the first dimension represents the amplitude of pulses of returning seismic waves.

The refraction method is based on the principle that when seismic waves, that are generated near the earth's surface and penetrate the rocks vertically, intercept a velocity interface (perhaps, for example, the contact between two layers of differing lithologies), some of the waves will be refracted into the underlying layer and travel horizontally (parallel to the surface) along the layer at a faster rate than waves traveling more shallowly (based on the principle that seismic velocities tend to increase with depth). As the waves move through the rock layer under consideration, they will generate disturbances along the velocity interface that will be manifested as seismic waves traveling back to the surface. If a geophone is placed an adequate distance from the shot point, then the first waves to be detected will be those that are the result of penetration to a deep, faster-velocity layer where such penetrating waves traveled horizontally at a rate considerably faster than the velocity of the seismic waves traveling through shallower beds. Typically, in refraction seismic methods geophones are placed laterally from the shot point several times the vertical distance from the surface to the rock layer of interest. If one knows or estimates the velocities by which seismic waves are transmitted from the surface to the layer under consideration and the velocity of waves through the particular layer, one then can use the total of the travel time of the wave from the shot point to the layer, along the layer, and back to the surface, where it is recorded by the detector (geophone) to calculate the depth to the layer. As Sheriff and Geldart (1982, p. 3) point out, the applications of refraction methods are considerably more limited than those of reflection methods (discussed below). With refraction methods, only a layer, bed, or stratum that has a seismic velocity significantly higher than that of any of the beds above it can be mapped—an example might be a saltdome. Refraction methods usually involve much greater distances between the shot point and the geophones; therefore very strong sources of artificial seismic waves may be required. Large blasts of dynamite or other explosives necessary for refraction methods may be expensive, dangerous, damaging to the natural environment, and objectionable to the general public in the areas being explored.

In reflection seismic methods, the waves reflected directly back from subsurface interfaces to the surface are recorded and used in seismic analyses. Multiple arrays of geophones, connected to amplifiers, are arranged in a pattern extending outward from the shot point, and each geophone (or set of closely spaced geophones acting as a single composite geophone) records the reflections from the shot at a different locality. At sea the seismic wave source and hydrophones can be towed behind a moving ship. The data from the various geophones can then be processed and integrated with the result that two- or three-dimensional visual displays of the reflecting interfaces of the subsurface are generated.

In processing seismic data, numerous factors must be taken into account. Because of the spherical nature of the wavefronts from a point source, and the varying distances of the geophones from the shot point, uncorrected recordings of reflections from various geophones along a single line on the ground will not correlate along straight lines, but along curved paths. This curvature must be removed in order to produce a seismic recording that approaches a conventional stratigraphic section. Noise also must be eliminated or otherwise handled. Seismic noise may originate from extraneous sources (such as nearby motor vehicle traffic) or be due to waves traveling along the surface to the geophone without having penetrated to and been reflected by a subsurface interface, or seismic waves may be distorted irregularly by zones of weathered rock near the surface. Sometimes the same interface may show multiple reflections from a single shot. Strongly dipping reflecting horizons may appear to be in the wrong positions and, based on a knowledge of velocities in the rocks, have to be shifted. To improve the precision and detail of a seismic section along a single linear transect, several shots can be recorded with the geophones and shot point in slightly different positions relative to one another.

Typically the data are computer-processed, and seismic sections (seismic profiles), analogous to two-dimensional conventional stratigraphic cross sections, are generated and displayed via photographic or dry-paper records. Differing intensities of light and dark (conventionally white and black, but colors also can be incorporated into seismic displays) may be used to represent the differences in wave amplitude from reflecting surfaces. Single vertical seismic traces still may be discernible, with the peaks (for instance) emphasized by being filled in with black and the troughs left blank. The horizontal axis of such seismic sections will be linear distance along the surface. The vertical axis, however, commonly is not depth to the reflecting surfaces but two-way travel time from the surface to the reflector and back. Given velocity data, sections showing time can be converted to sections showing depth; however, such processing is further removed from the actual data. Unless all the rocks in a particular section happen to be characterized by identical seismic velocities (which would be an extremely rare circumstance), there will not be a simple linear relationship between the travel time of the seismic waves and the depths to the reflectors being recorded. In fact, the velocities of the rocks above a particular layer or stratum of interest may vary laterally such that in a seismic time section (where the vertical dimension represents time) a particular reflector (representing a certain rock layer or stratum) will appear to be higher in the section where it is below rocks of relatively high velocity, and lower in the section where it is below rocks of relatively low velocity. Without critically interpreting the seismic time section, one may casually interpret the apparent offset between different parts of the same reflector as some type of structural feature, such as a fault. A properly corrected depth section, however, may show that the different parts of the reflector occur at the same absolute level in the section and are probably representative of a continuous, unfaulted layer or bed.

Discussion of the general analysis and interpretation of seismic sections is beyond the scope of this book, and the reader is referred to the references cited above (recent summaries of the seismic analysis of sedimentary facies regimes are provided by Boggs, 1987, and Davis, 1984); certain aspects of the interpretation of seismic sections that relate particularly to stratigraphy are considered elsewhere in this book (see Chapter 7). Here it should be stressed, however, that great care must be taken in postulating interpretations of seismic sections. In isolation, seismic sections can be extremely ambiguous, and even misleading, to the untrained eye; seismic data should be integrated with all other available data, such as surface outcrop data and well log data. Part of the problem is that seismic sections can appear deceptively simple when in reality they are not. Reflector horizons on seismic sections do not necessarily correspond to lighologic or formational boundaries that are of prime importance to the stratigrapher; on the other hand, large-scale structures may be revealed in seismic sections that are not readily apparent on the surface or from the data of only a few wells or boreholes. One must also keep the matter of scale in mind when working with seismic sections. On depth-corrected sections the vertical scale is commonly on the order of thousands of meters so that what appears to be a thin reflector horizon on the section may represent a layer that is tens or hundreds of meters in thickness. The details most familiar to the field stratigrapher (concentrating on outcrop sections) certainly will not appear on a seismic section.

Chapter 3

SOME CONCEPTUAL FOUNDATIONS, BASES, AND UNDERPINNINGS OF STRATIGRAPHY

HISTORY

In many ways stratigraphers are historians of the earth. Historical geology, to which stratigraphy is central, consists primarily of an inquiry into events placed in a temporal sequence—history. Thus it is appropriate that we look briefly at the concept and meaning of history and the historical method.

A first question is: Is there a history to reconstruct, or is the world as a whole essentially ahistorical? This question usually is seen as trivial in the context of modern geology (yes, there is a history to reconstruct), but this has not always been the case. Well into the seventeenth century the earth was generally viewed as less than 10,000 years old, and, as Gould (1987) has demonstrated, Hutton (1788, 1795) hypothesized an infinitely old earth but denied any importance to history per se (see Chapter 1); Hutton actually adopted an antihistorical point of view. Even now certain fields of endeavor, such as some aspects of chemistry, physics, philosophy, and mathematics, are essentially ahistorical (see Toulmin, 1962–63).

In order to have history, at least history in a modern Western sense, we need to have time (cf. Kitts, 1966; Toulmin, 1962–63). Time is often conceived of as a dimension (a fourth dimension), and by analogy it is seen as linear and unidirectional. "Time's Arrow" has only one direction (McLaren, 1978). As Harland (1978, p. 10) notes, temporal and spatial terms often are used interchangeably; for example, we may speak of a "point" or "position" in time when referring to an "instant" or "moment" in time, and likewise we may speak of some phenomenon (such as pebbles occurring in a stratigraphic section) as being "frequent" when we really mean "abundant." What time is, especially on a historical or geohistorical scale, is unclear. This is so, at least in part, because geohistorical time cannot be experienced directly by our senses and thus may not necessarily be merely an extension of our "common-sense time" (Kitts, 1966). Historical and geohistorical time may be viewed more as a human mental construct than as a true phenomenon of nature, although intimately associated with natural phenomena and processes. In line with this train of thought, Kitts (1966, p. 127) suggested that: "Historic time has its own special properties which are imposed upon it by the assumptions and procedures in terms of which we construct it." Likewise, Harland (1978, p. 14) stated: "The concept of time would be impossible without knowledge of natural processes. We interpret each experience in a space–time framework. The framework depends on phenomena for identification while phenomena depend on the framework for comprehension. While comprehension of nature depends on a space–time frame to display it, the frame has no separate existence. Moreover, our only interest in the frame is the history it displays."

There are different ways to reconstruct history once it is established or suggested that there is a history to reconstruct. For the last several thousand years we may have first-hand human testimony, historical records, but these are far from complete and soon fail us totally as we travel back in the past. As Gould (1987) has discussed in connection with the work of the early geologists Burnet, Hutton, and Lyell, one can recognize that history has occurred and then reconstruct it from the "imperfections" seen in the modern world. If the world were ahistorical, any object or thing should have the attributes it has for some practical, pragmatic, or essential reason (unless gods or devils are playing tricks on us). A high degree of optimality of design would be expected to exist in the world. However, objects often carry with them traits, attributes, imperfections that seem to be explained only by the hypothesis that they are merely the product of history—historical artifacts. At the extreme of the spectrum, if something is in a state of total ruin, this implies that the thing is a product of history, for "a ruin can only be a wreck of something once whole" (Gould, 1987, p. 43; of course this immediately raises the question of what is a "ruin" and whether we can consistently recognize ruins). In biology and paleontology the historical "imperfections" (or more simply, without implying a judgment value, "traits") in organisms, passed on from generation to generation, from species to species through time, can be analyzed and are the primary data used in reconstructing the evolutionary history of organisms (see Schoch, 1986b). Likewise, analogous methods can be used to reconstruct the history of the inanimate world. In other words, we must use the evidence at hand (what we observe in the world around us today) to reconstruct history, and history is a construction or hypothesis (see Hubbert's concept of history, quoted below).

A way of inferring history (once it is recognized or assumed that there is a history to infer) is to assume that a known (presently operating and observable) process is responsible for the event or observation we wish to explain. We can then extrapolate from the known present to the unknown past—a form of uniformitarianism (Gould, 1987; see section below). Another methodology, once it is established or assumed that there is a history to infer, is to take a sequence of configurations or phenomena that appear to be related causally (or were produced by similar causes) and order them into a gradational series of stages (transformation series, morphocline, or order: Harland, 1978; see Schoch, 1986b), the whole of which is subsequently interpreted as representing temporal history. Initially such a sequence or ordering of configurations may lack direction of aging or polarity; that is, which end of the sequence comes first (is oldest) and which last (is youngest) may not be known. Such polarity of a temporal sequence may be established in any of several ways. One end of the sequence may be rooted in the present (Recent) and thus, from this perspective, establish the direction of oldest to youngest. It has also been suggested that, given that all other considerations are accounted for, both entropy (disorder) and complexity (or information content) tend to increase with time (Harland, 1978;

Layzer, 1975). Another way of stating the latter hypothesis is that in general as time passes, more (rather than less) history (historical information) is recorded by the configurations of phenomena or material objects. This rule may not literally hold true for infinite expanses of time, but does appear to be generally applicable to time on the order of magnitude of the temporal duration of our planet.

Finally, we may discover simple "laws of nature" that can be used directly and unambiguously (at least in the simplest cases) to order phenomena in a temporal context. Examples of such laws might be Steno's principle of superposition or the rules of radioactive decay of unstable isotopes. With such generalizations we may feel empowered to read history directly from nature.

What is the purpose or role of the historian? Here it is worth quoting the classic empiricist concept of Buffon (1797, pp. 4–5): "It is an historian's business to describe, not invent; that no suppositions should be admitted upon subjects that depend upon facts and observation; that his imagination ought only to be exercised for the purpose of combining observations, rendering facts more general, and forming one connected whole, so as to present to the mind a distinct arrangement of clear ideas and probable conjectures." In other words, one begins with observations, that is, "facts" (some would call them the lowest-level hypotheses), and then slowly combines them to form a more general edifice. In the same sentence, Buffon states that the historian should not invent or hypothesize ("supposition"), but then allows that imagination may be used in generalizing from the observations at hand. This points up the obvious, and as yet unsolved, general problem of how do we separate description from interpretation in the historical sciences. For example, in the basic description of a stratigraphic section, how often are terms that imply a particular genesis for a rock used? Are the best historians, and scientists, those with the best imaginations?

What is history? What is the meaning of history? These are questions that historians of the human race have grappled with in particular. According to Partin (1969, p. 435), in reference to the meaning of human history, the ancient definition of history is "philosophy teaching by example, and also by warning." Such a concept hardly seems appropriate to geologic history (or, I believe, to any history) unless we interpret historical events as signs from the gods or perhaps strictly as a catalog (an incomplete one, at that) of possible configurations of phenomena. Also, it appears that this definition may treat history as something known, rather than a thing constructed (or reconstructed).

Benson (1984, p. 40) has suggested that "history is the science—or, if you will, the art—of learning from unique events." This is superficially similar to, but profoundly different from, the concept of history recounted by Partin. We learn from events, we discover generalities, we impart meaning to unique events, but the events do not actively instruct or teach us. It can perhaps be suggested that only by reducing the uniqueness of particular events can historical sciences proceed. We classify unique events into groups that are supposed to have properties in common and draw generalizations from them (ultimately this may lead to ahistorical concepts; cf. Hutton's cyclical concept of events on the surface of the earth). We reconstruct events of the past and infer processes to explain them (or vice versa); usually in science both steps of the process are based ultimately on observations of the present. Hubbert (1967, p. 30) has thus suggested: "History, human or geological, represents our hypothesis, couched in terms of past events, devised to explain our present-day observations.''

CONVENTION AND SCIENTIFIC PRINCIPLE

A continuing problem that has plagued much of stratigraphic inquiry, both on applied and theoretical levels, is the question of the distinction between matters of *convention* and matters of *scientific principle* (to use Harland's, 1978, terminology; see quotation in Chapter 1). Matters of convention may be viewed as being concerned not only with what name or term to apply to a particular object or concept (nomenclature), but also with such issues as where to draw an artificial or arbitrary boundary (for instance, between two rock units or rock types that seemingly grade into one another). Matters of scientific principle concern questions where reason, logic, and scientific analysis can be applied to arrive at a "correct," "probable," or "preferable" answer. Unfortunately, the distinction between matters of convention and matters of scientific principle is not always clear-cut. The placement of the Silurian–Devonian boundary, for instance, may be a matter of convention for some workers but a matter of scientific principle for others, depending on numerous factors such as one's philosophical background and scientific training. Some questions may even be viewed as simultaneously partly matters of convention and partly matters of scientific principle (this may be the case in certain exercises of lithocorrelation). There is not always universal agreement as to the appropriate scientific questions that should be asked. Furthermore, seemingly semantic arguments over terminology and nomenclature may mask more fundamental differences concerning scientific principles and methodologies. Even the categorization of various questions asked in stratigraphy, and science more generally, into either matters of convention or scientific principle imposes a certain way of thinking on the subject at hand.

Other distinctions made in stratigraphy, more or less analogous to the distinction between matters of convention and scientific principle, are the distinctions between *normative* and *positive* aspects of science, and *artificial* (= *quiet*) and *natural* entities and phenomena (such as stratigraphic units and boundaries: Harland, 1978; McLaren, 1978). The idea is that normative, artificial, or conventional questions concern simple matters of definition, terminology, and convention and thus can be simply settled by a vote or by a dictatorial fiat, whereas questions of positive science, scientific principle, or natural boundaries and relations cannot. The latter set of questions must be solved by scientific inquiry, demonstration, and analysis.

The distinction between formal and informal usage in stratigraphic classification and nomenclature also may be analogous to matters of convention versus scientific principle, or normative versus positive aspects of science (Harland, 1978). "Formal nomenclature is established or accepted according to some distinct procedure. Its advantage is that the names are unique, unambiguous, and not subject to frequent change. Informal usage has no such constraint and is more suited to positive scientific development because it is more flexible and experimental. The difference is important for us in so far as any development of formal language needs some international discussion" (Harland, 1978, p. 10). Harland's last point. although well taken, is debatable. Certainly when a formal language is developed, it will be maximally useful to those adopting it. Realistically, if a formal language is to be adopted globally, it will probably require some international discussion, but this is not a necessity for formal language a priori. Once an apparatus of conventions for generating different types of formal names, for instance, were established, it conceivably could run autonomously with little or no discussion.

UNIFORMITARIANISM

Uniformitarianism, sometimes referred to as the principle of uniformity (Hubbert, 1967), is often viewed as the closest thing geologists have to a fundamental "law" within their discipline. Uniformitarianism has been considered the basic principle underlying historical geology (including stratigraphy), as well as being fundamental to all historical science in general. The concept of uniformitarianism has been viewed as geology's major contribution to science and philosophy in general (Simpson, 1963, 1970). However, since the time of Hutton, uniformitarianism (in its various guises) has come under attack. It has been dismissed as encompassing nothing more than principles common to all science, not unique or original to geology. It also has been asserted that at least some of the principles traditionally associated under the rubric of uniformitarianism are blantantly false. Much of this confusion is due to the fact that the single term uniformitarianism has been used to refer to many different concepts and principles over the last 150 years. Hubbert (1967, p. 4) offers four common, but not necessarily equivalent, answers to the question: what is the principle of uniformity?

(1) The present is the key to the past.
(2) Former changes of the earth's surface may be explained by reference to causes now in operation.
(3) The history of the earth may be deciphered in terms of present observations, on the assumption that physical and chemical laws are invariant with time.
(4) Not only are physical laws uniform, that is invariant with time, but the events of the geologic past have proceeded at an approximately uniform rate, and have involved the same processes as those which occur at present.

Despite the antecedents of Hutton and earlier workers (see Chapter 1), Charles Lyell, more than any other single man, was responsible for formulating the traditional geological concept of uniformitarianism. As Gould (1965, 1987), Hooykaas (1963, 1970), Porter (1976), and Rudwick (1976) (among others) have pointed out, the concept of uniformitarianism, as employed by Lyell (1830–33 and later editions) and many authors since Lyell's time apparently unites two categories of assertions concerning the way science should proceed and the way the world operates. In brief, these two categories concern: (1) methodological principles and assumptions that, it is often asserted (but see Simpson, 1970), are generally considered common to all science (at least in a Western tradition) and therefore rarely questioned within a traditional scientific context (this is referred to by Gould [1965, 1987] as "methodological uniformitarianism," and is referred to as "actualism" by Hooykaas [1963]); and (2) substantive beliefs, claims, and assumptions about the actual operations of the world (for example, the assertion that rates of global tectonic uplift are, and have always been, relatively slow, steady, and gradual: these substantive beliefs are labeled "substantive uniformitarianism" by Gould [1965, 1987] but are simply called "uniformitarianism" by Hooykaas [1963])—beliefs that need not necessarily follow from methodological principles and hence may be highly questionable.

Within this framework Rudwick (1976; see also Gould, 1987) has distinguished four basic aspects of Lyell's (1830–33) concept of uniformitarianism: (1) methodological claims about the uniformity of natural laws and (2) natural processes, and substantive claims concerning (3) rates of processes (Lyell advocated gradualism) and the (4) overall state of the world through time (Lyell advocated nonprogressionism, or a dynamic steady state world). Simpson (1970) suggests that most issues involved in the discussion of uniformitarianism do indeed fall into two general classes or categories, roughly corresponding to the two categories recognized by the authors cited above, but these categories as labeled are somewhat too restrictive, rigid, and extreme as compared to the actual views of Hutton, Lyell, and other workers. The deep issues, according to Simpson (1970), involve: (1) inherent properties of the universe, that which is *immanent* to the universe—"immanence"; and (2) "configuration"—the study of the actual *configurations* that have arisen through time in historical sequences (for example, the actual, detailed history of the earth—such configurations must be in accordance with the immanent properties of the universe, but the same immanent properties may allow multiple configurations in history). The perceived duality between immanence and configuration relates, in part, to Simpson's (1963) distinction between the historical and nonhistorical sciences. Methodology, for Simpson (1970, p. 60), involves the relationship between issues of immanence and configuration—although it could be argued that as a prerequisite to any methodology some assumptions must be made as to the immanent or inherent properties of the universe. Gould (1965, 1987), Hooykaas (1963, 1970), Hubbert (1967), Rudwick (1976), and Simpson (1963, 1970) have raised a number of specific issues within the general topic of uniformitarianism, some of which merit review here; the impetus for the present discussion is Simpson (1970).

Perhaps the most profound aspect of uniformitarianism, although it is now universally taken for granted in the sciences, is the concept of naturalism. Methodologically, procedurally, and heuristically, science and scientific explanation do not invoke the intervention of gods, the supernatural, the nonnatural, the preternatural, or the noumenal. Although Hutton believed in a divine first cause, he accepted only rational, naturalistic second causes. In his science, particularly in considering the earth's history and workings, Hutton did not consider any causes outside of what appears to be, or could be posited to be (even if unobserved at the moment) normal, regular, or ordinary processes of nature. That is, in his science Hutton dealt with only second causes that were considered to be rational and comprehensible without invoking preternatural intervention. This is a position still popular today. For virtually all scientists naturalism is axiomatic for their work as scientists, although they may hold varying opinions as to first causes and providence. Some contemporary scientists hold that the ultimate first cause is outside the scope of inquiry of science, although others appear, to some observers, to be actively engaged in attempting to pursue at least aspects of "first causes" or "origins" (for example, the origin of man, the beginning of the universe). Hutton himself was providentialistic and somewhat teleological in that he held that the reason behind the earth as we see it today is to provide a fit abode for man and other life forms. Even today certain scientists may hold similar beliefs as long as, like Hutton, they do not allow such beliefs to enter into their science. Scanning through a reference such as Marquis (1985), one may note that a considerable number of persons involved in science or science-related professions also list religious affiliations for themselves.

Simpson (1970) suggests that Hutton's greatest contribution was to ban any sort of supernaturalism or preternaturalism from the study of the history of the earth. This concept may not have been original to Hutton, but he stressed it in his writings. Simpson (1970) further views naturalism as a necessary prerequisite for uniformitarianism (especially as espoused by Lyell and developed by succeeding investigators), but not as special or peculiar to uniformitarianism; naturalism underpins all modern science.

Perhaps closer to the core of uniformitarianism as used in the geological community is the concept of actualism. Actualism refers to the notion or principle that we should attempt to explain the past in terms of causes and processes now existing or known in the present time (Gould, 1987; Simpson, 1970). Or, to put it another way, it is postulated that processes that exist now also existed in the past. Simpson (1970) expresses this in terms of inherent characteristics and properties (presumably expressed in the laws and processes of nature) of the universe: the immanent properties of the present universe also existed in the past. In many ways the concept of actualism is synonymous with the aphorism "the present is the key to the past."

Simpson (1970) notes that along with the concept of actualism it is usually assumed that the immanent characteristics or properties of the universe are presently observable, or at least potentially observable. It is also usually assumed or implied that not only did the immanent characteristics seen in the present universe exist in the past, but all of the immanent characteristics of the universe of the past still exist in the present universe. Simpson (1970) coined the term "preteritism" for the latter concept. As Simpson notes, these two principles complement one another, but are not equivalent. Loosely, however, they are usually both encompassed in the term "actualism."

Gould (1965, 1987) uses the term actualism to refer specifically to the uniformity or invariance of processes through time, which Gould identifies as equivalent to the principle of simplicity (economy, or Occam's Razor); the uniformity of natural law in space and time is singled out by Gould (1987, pp. 119–120) as essentially the principle of induction. Together these two principles form what Gould calls methodological uniformitarianism. According to Gould (1987, p. 120), "the first two uniformities [those mentioned above in this paragraph] are geology's versions of fundamental principles—induction and simplicity—embraced by all practicing scientists both today and in Lyell's time." The implication is that these aspects of uniformitarianism are not particularly interesting per se, or at least not unique to geology; they are merely necessary if geology is a science. Simpson (1970) has countered by suggesting that the concept of actualism is not equivalent to either the principle of induction or the principle of simplicity.

Briefly stated, the principle of simplicity suggests that the simplest hypothesis that fully and adequately explains the data (as perceived and interpreted at the particular time) is to be preferred (Schoch, 1986b). Simpson (1970, p. 44) expresses this concept as follows: "no additional properties [of matter and energy] should be postulated unnecessarily"; here Simpson is referring explicitly only to the present time. The extension of this general principle through time, postulating that the properties of matter and energy have not changed with time, is for Simpson not simplicity but actualism—that is, simplicity is atemporal, whereas actualism involves a temporal dimension. Gould (1987, p. 120) defines the principle of simplicity as "don't invent extra, fancy, or unknown causes, however plausible in logic, if available causes suffice.'' This immediately opens up the question of when do we know, or judge, that available causes suffice. How complicated a scenario do we hypothesize using available causes before we postulate a new, simpler cause? Do we accept an ether, action at a distance, propagation of a wave through a vacuum? Moreover, even as Gould phrases the principle of simplicity (above), no processes are "available" in the past until we invoke the concept of actualism to suggest that presently observed processes (the only processes we can ever experience first-hand) were "available" in the past.

As for the concept that actualism is in part merely a restatement of the principle of induction, Simpson (1970) asserts that uniformity in nature (actualism in a loose sense) is rather a necessary prerequisite for inductivism (especially as applied to cases that involve any temporal disjuncture). To take individual cases and successfully generalize into the future or back toward the past requires that there be some uniformity in nature through time—the concept of actualism. Any belief or faith in the validity of inductivism stems from experiences that nature does indeed seem to be uniform. However, one can certainly accept the uniformity of natural law and process through time—one can accept actualistic tenets—without being an inductivist. How much science is carried on in an inductive mode is questionable; hypothetico-deductive methods of scientific reasoning and investigation are perhaps primary (Hempel, 1965, 1966; Popper, 1959, 1968, 1984; Schoch, 1986b).

Although the principle of simplicity, or some other criterion (see Schoch, 1986b), may be necessary in order to do science (one must be able to choose between competing hypotheses), it is not self-evident that actualism per se is necessary to carry on a scientific research program. Perhaps rather than postulating that the immanent properties and characteristics of the universe are unchanging with time, one could postulate that the immanent properties of the universe themselves are changing over time, but changing in a systematic and predictable way. Perhaps what is important to science is simply the ability to predict. Observing the present only, one might find it impossible to distinguish between actualistic and nonactualistic universes (but rational ones in either case—that is, the principle of simplicity, or some similar principle, would hold). It is only with a temporal perspective, with history, that such considerations become important. In the last two hundred years actualistic systems have generally proved themselves superior to nonactualistic systems in their heuristic and predictive values in historical geology; therefore, they generally have been preferred.

BEDDING, STRATIFICATION, AND LAYERING OF ROCKS

The aqueous rocks, sometimes called sedimentary, or fossiliferous, cover a larger part of the earth's surface than any others. These rocks are *stratified,* or divided into distinct layers, or strata. The term *stratum* means simply a bed, or any thing spread out or *strewed* over a surface; and we infer that these strata have been generally spread out by the action of water, from what we daily see taking place near the mouths of rivers, or on the land during temporary inundations. For, whenever a running stream charged with mud or sand, has its velocity checked, as when it enters a lake or sea, or overflows a plain, the sediment, previously held in suspension by the motion of the water sinks, by its own gravity, to the bottom.

In this manner layers of mud and sand are thrown down one upon another. (Lyell, 1857, pp. 2-3, italics in the original)

Sedimentary rocks are commonly arranged in beds or layers, that is, they are stratified. Indeed, the term stratigraphy is derived from this characteristic of most sedimentary rocks. Stratification is apparently due to many different causes (briefly discussed below). Although the topic is extremely fundamental to the study of sedimentary rocks, we probably still can agree with Shrock (1948) and Weller (1960) that causal processes responsible for the most common forms of stratification, namely, types of parallel bedding, have received relatively little attention.

Before we proceed further, it should be noted that the terms bed, layer, stratum, and lamina have been used in various manners by different authors. Shrock (1948, pp. 5-6) used the terms bed and layer "interchangeably in referring to any tabular body of rock lying in a position essentially parallel to the surface or surfaces on or against which it was formed, whether these be a surface of weathering and erosion, planes of stratification, or inclined fractures. The body need not have been formed in a horizontal position, although generally its original position will have closely approached horizontality. Used in this broad sense, the terms apply to stratified and massive sedimentary rocks and their metamorphic derivatives, clastic dikes and the fillings of crevices, and tabular masses of igneous rocks such as flows, sills, sheets, and dikes." Shrock (1948) restricted the application of the words stratum and lamina to sedimentary and metamorphic (presumably metamorphosed sediments) rocks. A stratum for Shrock (1948) is any single layer or bed, no matter what thickness; a lamina is a layer less than about 10 to 12 mm thick.

Various other authors have utilized different definitions of bedding and lamination. Matthews (1984), for instance, following the usage of McKee and Wier (1953), uses the term lamina to refer to a stratum less than a centimeter thick and the term bed for any stratum thicker than one centimeter: very thin beds are 1 to 5 cm thick, thin beds are 5 to 60 cm thick, thick beds are 60 to 120 cm thick, and very thick beds are greater than 120 cm thick. In contrast, Ehlers and Blatt (1982), following the usage of Ingram (1954), use the following terms: thinly laminated (bedding thinner than 0.3 cm), thickly laminated (0.3-1 cm), very thinly bedded (1-3 cm), thinly bedded (3-10 cm), medium bedded (10-30 cm), thickly bedded (30-100 cm), and very thickly bedded (thicker than 1 meter). Ehlers and Blatt (1982, p. 331) define the concept of "bedding" itself as "the subplanar discontinuity that separates adjacent layers of rock."

Stratification is generally produced by some complex interaction of physical, chemical (including diagenetic), or biological processes and conditions (see Weller, 1960, and Ehlers and Blatt, 1982, for generalized discussions of stratification; also Ricken, 1986, on diagenetic bedding). For convenience, however, we can briefly treat each of these factors separately.

As suggested in the quotation from Lyell cited above, stratification may result from the nonuniform deposition of sediments. This may involve changes in the nature of the material that is being deposited and/or changes in the rate of deposition. Variations in sediment flow into a basin of deposition may be correlated with such factors as changing climatic conditions (involving different types of erosion and different competencies of sediment transport), changing rocks outcropping at a source area or changing source areas, and changing topographic conditions. In many cases on a short time scale changes in the transporting medium of the sediment may be the most important factor in determining bedding. A stream may vary in its discharge and power and alternately lay down coarser or finer sediments. Likewise, changes in the course of a stream could create a definite stratification. Changes in depth of water in a lake or sea can alter sedimentation patterns and produce stratification. Or, to use an example cited by Weller (1960), Pleistocene glacial deposits may be stratified because of the alteration of ice-borne till and wind-blown loess.

Weller (1960) suggests that once sediments are initially deposited they may be physically disturbed, and these disturbances may produce bedding. The sedimentary particles may be stirred up (for instance by wave action) and then resettle in beds in essentially the same area. Unconsolidated sediments may be quickly eroded and moved to a different area where they are then redeposited in beds. Post-depositional compaction and diagenesis may form or enhance bedding. Diagenesis may accentuate what was originally a gradational boundary between two layers or rock types (Ehlers and Blatt, 1982). Ricken (1986) has demonstrated that in marl-limestone sequences rhythmic bedding, or diagenetic bedding or stratification, may develop during diagenesis. Ricken (1986, p. 2) suggests that "the processes which produce diagenetic bedding always generate sequences with more pronounced rhythmicity than that found in the primary sediment." One must be wary of falsely interpreting such diagenetic bedding strictly in terms of cyclic depositional processes, such as rhythmic climatic oscillations (for example, Milankovitch cycles: Fischer, 1980; Gilbert, 1895; Milankovitch, 1930).

Stratification, or presumed stratification or bedding, may also form in rocks through various chemical conditions. Chemically precipitated sediments, such as salt deposits (for example, gypsum, anhydrite, halite) may exhibit stratification. Changing chemical environments, for instance, alteration between relatively oxidizing and reducing environments, may promote bedding or at least colored layers, as in iron-based pigmentation. Likewise, terrestrial weathering profiles may appear in the stratigraphic record as layering or bedding. Concretions, nodules, styolites, or solution features may develop parallel to bedding, or even produce a pseudostratification.

Of course, much stratification in the sedimentary record is biologically produced and controlled. Many sedimentary rocks are composed partly or wholly of clasts that were produced by biological organisms. Such organisms can be extremely important in controlling geochemical conditions that allow the precipitation or dissolution of certain minerals (the calcite group in particular); also, organisms may bioturbate and otherwise disturb layers of sediment once they are laid down.

While on the topic of bedding, stratification, and layering in sedimentary rocks, we should note that many authors would suggest that the vast majority of bedding planes or surfaces represent diastems or unconformities (see next section), periods of time not represented by rock. As Ager (1981, p. 29) puts it: "But what are all those bedding planes? What is any bedding plane if it is not a mini-unconformity? If we really had continuous sedimentation then there would surely be no bedding planes at all." Many areas may be characterized by extremely episodic sedimentation patterns (Dott, 1983), and in general the stratigraphic record is more "gap" than material record (Ager, 1981; Sadler, 1981). To quote Ager's (1981, p. 35) delightful prose once again: "Perhaps the best way to convey this attitude is to remember a child's definition of a net as a lot of holes tied together with string. The stratigraphical record is a lot of holes tied together with sediment. It is as though one has a newspaper delivered only for the football results on Saturdays and assumes that nothing at all happened on the other days of the week. To change my metaphor yet again, I would compare the stratigraphical record with music. Just as the intervals between the notes in music are every bit as important as the notes themselves, so the bedding planes are as important as the beds."

Here it should also be pointed out that not all sedimentary bedding is strictly horizontal and parallel (the reader unfamiliar with basic sedimentary structures should consult any of numerous sedimentology or petrology texts, such as Blatt, Middleton, and Murray, 1972, 1980; Boggs, 1987; R. A. Davis, 1983; Ehlers and Blatt, 1982; Fritz and Moore, 1988; Lindholm, 1987). Sedimentary particles may be stacked or layered at an angle to the major bedding planes of a deposit—for example, the imbrication of pebbles and cobbles in conglomerates (where the gravel, pebbles, or cobbles dip upstream), or the cross bedding of sand and coarse silt (where the cross beds dip downstream). Cross bedding may be produced in a variety of ways, such as by the avalanching of sedimentary materials down the downstream slopes of ripples and dunes or by the deposition of particles on other types of sloping surfaces (for instance, on the face of a beach).

As was mentioned in Chapter 2, extrusive igneous rocks may show definite stratification or layering, for example, corresponding to a series of laval flows. Intrusive igneous rocks may likewise exhibit layering, for instance, corresponding to the differential crystallization and settling of various materials. Metamorphic rocks, if formed from preexisting sedimentary rocks, may contain relict bedding or stratification.

UNCONFORMITIES

The concept of unconformities is fundamental to stratigraphy, and while there is a general consensus as to the nature and classification of unconformities, there is still disagreement on the particulars.

As Newell (1967) has noted, the term conformable has been used to describe sequences of parallel strata (in a structural sense with no implications of lack of temporal gaps), but the term has also been used to refer to sequences in which it is believed that there was a continuity of sedimentaton, deposition, and preservation of rock. However, Newell asserts that strict "continuity of sedimentation" (a phrase that he never defines) rarely, if ever, occurs, and furthermore is probably unrecognizable even in sections where it does occur. "Considering the oscillatory nature of sedimentary processes, the probability of strict continuity of sedimentation and environmental facies anywhere for long intervals, even in the deep ocean basins, seems to be vanishingly small. *It is virtually impossible to demonstrate that a given sequence is free from stratigraphic breaks,* although many experienced stratigraphers speak confidently of continuous sequences 'especially in the basins'" (Newell, 1967, p. 349, italics in the original). The very fact that rocks are stratified may suggest that they do not represent strictly continuous sedimentation. Layering and parallel stratification may be a result of the nature of the deposited sediment fluctuating over time (changes in sedimentary regimen over time) even though the deposition of sediment is continuous. Or, perhaps more commonly, layering, stratification, laminae, bedding, and so on, may be the result of shorter or longer periods of temporary nondeposition (interruptions in deposition), perhaps even accompanied by some erosion (see above on the origin of stratification). In order to avoid misunderstandings that can result from the different meanings and implications attached to the term conformable, Newell (1967) suggested that the term concordant be applied to the description of parallel sequence of strata, and the term discordant be applied to angular contacts.

There is general agreement that an unconformity, as the term is now commonly used in America (see below), simply refers to any significant break in time within a stratigraphic column, any important stratigraphic discontinuity (e.g., Boggs, 1987; Weller, 1960), or an interval representing missing time and leaving no tangible (material) stratigraphic record (Visher, 1984). Of course, this immediately leaves open the question of what constitutes a significant break in time, an important stratigraphic discontinuity, or an "interval" of missing time; many workers have recognized degrees of unconformity (major unconformities, minor unconformities). Mitchum, Vail, and Thompson (1977, p. 56) recognize a "significant hiatus," which defines an unconformity, when "at least a correlatable part of a geochronologic unit is not represented by strata." Similarly, Boggs (1987, p. 525) suggests that "unconformable strata are strata, in a vertical sequence, that do not succeed underlying rocks in immediate order of age or do not fit together with them as part of a continuous whole. The contacts between such strata are called unconformities."

The implication of Boggs's (1987) definition appears to be that if we have a sequence where Miocene Formation B overlies Eocene Formation A, there is an unconformity between Formations A and B even if there is no clear physical evidence of a significant time break between the formations other than their dating (perhaps they are dated micropaleontologically). However, if the age of Formation B is later reevaluated and considered to be Oligocene, is there still an unconformity between A and B? Now the formations succeed each other "in immediate order of age" and assuming they do fit together as "a continuous whole," perhaps we must then consider them as conformable. Finally, if with more refined dating techniques it is found that the age of Formation A is Early Eocene and the age of Formation B is Late Oligocene, must we now postulate or label an unconformity between the two formations once again? The point of this hypothetical example is to make clear from the start that not all unconformities are totally objective entities of nature. Unconformities can vary in degree (length of geologic time missing between preserved strata) as well as in kind (for example, disconformities versus angular unconformities, see below). What constitutes an unconformity may depend on the scale at which we are analyzing a stratigraphic section or sequence, and may vary from worker to worker. Furthermore, in some cases whether an unconformity is recognized or not (or at least, how significant it is considered) may simply be a function of the arbitrary units of geologic time being applied to the section.

Returning once again to the hypothetical example given above, if in 1980 Formation A were considered to be Eocene in age and Formation B were considered to be Miocene in age, most workers would immediately recognize some kind of unconformity between the two formations because the Oligocene is obviously unrepresented. However, if the same formations were assigned the same ages (Eocene and Miocene, respectively), in 1850 Formation B would have been viewed as following Formation A immediately in age because in 1850 the concept of the Oligocene had not yet been described. Of course, if one assumes that all major units of geologic time and rock strata are separated by major unconformities (diastrophic revolutions or catastrophisms—see Chapter 1), then simply by virtue of the fact that two formations are assigned to different epochs (even if the epochs are contiguous), one could argue that there must be an unconformity between them.

If an unconformity is simply a surface (a two-dimensional plane, but not necessarily a flat plane—an unconformity can be an irregular surface) between unconformable strata representing a significant, substantial, or important stratigraphic discontinuity (and therefore time break) between rocks, then an unconformity logically could represent either a surface of nondeposition, a surface of erosion, or a combination of both. An unconformity has been defined as simply a surface of nondeposition or erosion (e.g., Krumbein and Sloss, 1963). An unconformity also has been defined as "a surface of erosion that separates younger strata from older rocks" (American Geological Institute, 1976, p. 448). If taken literally, this last definition would apparently exclude surfaces of nondeposition that lacked erosion from the concept of an unconformity (although in practical terms it may be difficult or impossible to demonstrate that no erosion occurred on a surface of nondeposition). Originally, however, the term unconformity referred solely to what in America is now termed an angular unconformity. Angular unconformities invariably include some erosion between the unconformable rock bodies. Referring explicitly to Hutton's original concept of angular unconformities, Gould (1987, p. 62) has defined the term as follows: "An unconformity is a fossil surface of erosion, a gap in time separating two episodes in the formation of rocks. Unconformities are direct evidence that the history of our earth includes several cycles of deposition and uplift."

An angular unconformity usually is defined as an unconformity between two sets of strata in which the bedding planes of the two sets are not parallel to one another; the two sets of strata are angularly discordant (Fig. 3.1). Often the underlying (older) strata are at a sharp angle to the overlying (younger) strata. The younger strata are usually essentially horizontal (at least when originally deposited on the older strata), whereas the older strata may have been steeply tilted, perhaps folded or crumpled, and were eroded before deposition of the younger (overlying) strata. Given the correct conceptual framework, angular unconformities are the most obvious and conspicuous breaks in the stratigraphic record.

Although Strachey (1719) clearly illustrated an angular unconformity, it was Hutton who first recognized the significance of unconformities. Based on his cyclical theory of the world machine, Hutton predicted that angular unconformities should exist. In 1787 he observed his first angular unconformity at Loch Ranza, Isle of Arran, Scotland (B. Conkin and J. Conkin, 1984; Gould, 1987). Later the same year he observed a second unconformity at Jedburgh, Scotland (illustrated in Gould, 1987, p. 60), and in 1788 he observed his most famous unconformity at Siccar Point, Berwickeshire, Scotland (illustrated as the frontispiece of Lyell, 1857, and in Dunbar and Rodgers, 1957, p. 96). At Siccar Point the Devonian Old Red Sandstone lies at a very low angle upon the nearly vertical upturned edges of Silurian strata ("Primary schistus" of Hutton, 1795).

Apparently Hutton never actually used the words unconformity or unconformable; Dunbar and Rodgers (1957, p. 116) suggest that the term unconformable was first introduced into English from the German by Bakewell (1815). Throughout the nineteenth century the terms unconformable and unconformity were used to describe only what is here termed angular unconformity; the terms are still commonly used with this meaning in Great Britain (Roberts, 1982). However, as geology progressed, it was realized that there are many significant breaks in the stratigraphic

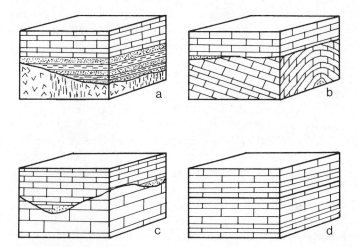

Figure 3.1. Schematic diagrams representing four basic types of unconformities: (a) nonconformity, (b) angular unconformity, (c) disconformity, and (d) paraconformity. (After Dunbar and Rodgers, 1957, p. 117.)

record that are not marked by angular discordance between the strata concerned. By the beginning of the twentieth century the term unconformable was being extended, at least in North America, to include the latter concept. In 1913 Grabau wrote: "Strata separated by an unrepresented time interval are generally spoken of as unconformably related. Two types of such unconformable relation may be recognized, the *stratic* where the stratification of the formation on both sides of the plane of nonconformity [i.e., unconformity] is parallel or nearly so, and the *structural,* where the two sets of strata are inclined at a greater or less angle with reference to each other" (Grabau, 1913, p. 821, italics in the original). Grabau then coined the term disconformity "for the first type [of unconformity], in which no folding of the older set of strata is involved . . . with the corresponding limitation of the term *unconformity* to the second type, or that in which folding plus erosion of the first set of strata precedes the formation of the second set" (Grabau, 1913, p. 821, italics in the original).

Many workers in North America, however, continued to use the term unconformity in the generic sense to encompass the concepts of both angular unconformity and disconformity. According to Dunbar and Rodgers (1957, p. 117), Pirsson (in Pirsson and Schuchert, 1915) used the term unconformity in a generic sense, accepted Grabau's term disconformity, and coined the term nonconformity to encompass both the concept of angular unconformity between layered strata and the concept of an unconformity where stratified rocks lie directly on a nonstratified rock body (such as an igneous or a highly metamorphosed body).

Types of Unconformities

As generally applied in North America today, the term unconformity is used broadly to apply to several different types of stratigraphic contacts. Here are briefly reviewed the basic classes of unconformities along with a few related terms and concepts.

Angular unconformity: This term refers to an angular discordance separating two units of stratified rocks and is discussed and defined above (refer also to Fig. 3.1). Angular unconformities usually are thought of not as simply geometric configurations per se, but also as the result of large-scale diastrophism. Certain structures produced by submarine slides, cross bedding, and so on, are not generally considered to be angular unconformities (Newell, 1967).

Disconformity: As originally defined by Grabau (see above), the term disconformity referred to any unconformity among stratified rock bodies where the bedding of the rocks involved is essentially parallel above and below the surface of unconformity (as opposed to angular unconformities, where this is not the case). Dunbar and Rodgers (1957, p. 119) proposed that the term disconformity be restricted to cases "in which the two units of stratified rocks [above and below the surface of unconformity] are parallel but the surface of unconformity is an old erosion surface of appreciable relief" (see Fig. 3.1). Newell (1967) elaborated upon Dunbar and Rodgers's (1957) redefinition of the term:

> Disconformities are lithostratigraphic erosional breaks in parallel sequences. They may, or may not, coincide with paleontologic discontinuities. Disconformities are based on physical evidence of hiatus, usually displaying more or less conspicuous features generally attributed, rightly or wrongly, to subaerial erosion and weathering at a stratigraphic interruption. They are frequently identified by such features as superjacent conglomerate, subjacent leaching, channeling, joint fillings, mud cracks, and oxidized zones. (Newell, 1967, p. 352)

This concept of disconformity has not been universally accepted: Weller (1960) objected to the restriction, Krumbein and Sloss (1963) and Matthews (1984) continued to use the term in Grabau's (1913) original sense, and the American Geological Institute (1976, p. 123) defined the term disconformity as an "unconformity between parallel strat[a], *e.g.,* with strata below not dipping at an angle to those above." J. Conkin and B. Conkin (1973) considered paraconformities and paracontinuities to be types of disconformities (see below). On the other hand, Boggs (1987), Roberts (1982), and Visher (1984) use the term essentially in Dunbar and Rodgers's sense. Roberts (1982, p. 218), however, treats disconformities as relatively local and small-scale unconformities; as used by Roberts, in a disconformity "the beds above and below the erosion surface still remain conformable to one another on a large scale."

The terms *para-unconformity* and *parunconformity* (Crosby, 1912; not to be confused with paraconformity, see below) are essentially synonyms of Grabau's (1913) disconformity.

Nonconformity: Dunbar and Rodgers (1957, p. 119) also restricted the term nonconformity (see above), using it to refer only to an unconformity consisting of stratified rocks lying unconformably on older nonstratified rocks (usually plutonic igneous rocks or massive metamorphic rocks; see Fig. 3.1). A distinct erosional surface or features may have developed on the nonstratified rocks before they were covered by sediments. This concept of nonconformity now has been generally accepted.

Paraconformity: Dunbar and Rodgers (1957, p. 119) introduced the term paraconformity for an unconformity "in which the beds are parallel and the contact is a simple bedding plane." As used by Dunbar and Rodgers, the concept of a paraconformity may refer to an obscure, questionable, uncertain, or presumed unconformity. These authors further elaborate on this concept as follows: "There is a special need for a term for this nonevident type of unconformity in which the contact is a simple bedding plane, since its recognition is commonly subjective, especially during reconnaissance study or where fossils are rare or lacking. Important breaks of this sort have often been recognized by one stratigrapher only to be denied by another. Commonly the hiatus is inferred because of an abrupt faunal change, but such changes may be susceptible of more than one interpretation. They may be due to loss of part of the evolutionary record, for example, or to merely change in the environment without loss of time [theoretically, of course, these cases could be distinguished from one another]. It will make for objective reporting, therefore, if such inferred breaks in the sequence are identified by a distinct name" (Dunbar and Rodgers, 1957, pp. 119–120).

Newell (1967) further defined (and in part redefined) the term paraconformity:

Paraconformities are biostratigraphic discontinuities within sequences of parallel strata, based and evaluated solely on paleontologic evidence. They usually coincide with a conspicuous bedding plane, but they may fall between conspicuous lithologic surfaces. Hence, their significance is necessarily controversial in early phases of study and after extensive studies they may or may not be judged to coincide with disconformities. (Newell, 1967, p. 352)

Newell (1967) noted that with use of this definition, paraconformities may be laterally coextensive with—may laterally grade into—disconformities (as Newell used this term, see above).

In coining the term paraconformity, Dunbar and Rodgers (1957) appear to have included several different concepts under one name: (1) a true (that is, unequivocally recognized and agreed upon) unconformity where the surface of unconformity, as identified, is parallel to the bedding planes of the stratified units both above and below the unconformity; (2) an obscure unconformity where it is known that two stratigraphically adjacent units are unconformable with one another (they may differ widely in age and therefore an unconformity must exist between them), but where the precise surface of unconformity is very difficult to recognize ("Such phenomena occur [for example] when a deep residual soil is developed on an erosional surface and is subsequently partially reworked and redeposited with the basal beds of the succeeding unit" [Krumbein and Sloss, 1963, p. 305]); (3) an uncertain, possible, subjective, or disputed "unconformity."

Roberts (1982, p. 218) uses the term paraconformity in a restricted and slightly modified sense. According to Roberts, periods of nondeposition, resulting in breaks in the stratigraphic record on the order of magnitude of one or more biostratigraphic zones (now missing in the section), "are termed non-sequences or paraconformities if they are simply the result of non-deposition" (if an erosion surface is formed one is dealing with a disconformity). According to Roberts, paraconformities can be recognized, whereas diastems (see below) cannot.

Dunbar and Rodgers (1957, p. 121, their Figure 61) illustrated what they considered to be an example of a paraconformity. The caption to their photograph reads, in part: "Beargrass Quarry near Louisville, Kentucky, showing paraconformity between the Louisville limestone (Middle Silurian) and the Jeffersonville limestone (lower Middle Devonian)." Apparently on this basis J. Conkin and B. Conkin (1973) designated the unconformity between the Middle Silurian and Middle Devonian limestones in the Louisville, Kentucky area to be the "type paraconformity." J. Conkin and B. Conkin (1973, p. 13) then attempted to define the nature of a paraconformity by analyzing the "type paraconformity," with the following results: "the type paraconformity . . . is defined by a faunal discontinuity of large magnitude while the physical evidence exhibited along the surface of paraconformity is (at most places in the Louisville area), upon cursory inspection, merely suggestive of a minor break (diastem) in the record. The physical surface of this paraconformity does, however, exhibit upon careful observation, a relief of a foot or so at particular localities . . ." Ten years later, Conkin, Layton, and Conkin (1983, pp. 112, 114) revised this description, stating:

> The surface of paraconformity exhibits a physical stratigraphic rock gap as great as 20 feet . . . where the upper part of the Middle Devonian (Eifleian) *Brevispirifer gregarius* Zone of the Jeffersonville Limestone lies on the Louisville Limestone. Even in the type area, a physical strata gap of 4.10 to 9.80 feet exists. The basal beds of the Jeffersonville Limestone (Schoharie of New York) bear silicified stromatoporoid fragments reworked from the underlying Louisville Limestone and usually bear some rounded, frosted quartz sand grains, both expected above a physical surface of unconformity, but inconsistent with Dunbar and Rodgers'[s] (1957) definition of a paraconformity. The paraconformity should be restricted to a disconformity which exhibits a large faunal break; however, a paraconformity may, or may not, exhibit [a] physical surface of channeling from a few inches, to a few feet, to several feet in the immediate area of outcrop.

In this statement Conkin, Layton, and Conkin (1983) have essentially synonymized the concept of paraconformity with the concept of disconformity, or at least rendered any distinctions between the concepts virtually meaningless. However, the whole process of designating a "type" for a conceptual idea, such as a paraconformity, and then adjusting the original author's intention and meaning on that basis, is questionable (unless, perhaps, the author explicitly tied the original concept to a "type," and not just an example, in the first place). In zoology, for instance, material types are used to stabilize nomenclature, not to legislate concepts. In the world of ideas the type of a word or concept is the author's original definition in words and original use of the term in written or verbal context. Thus the type definition of paraconformity is Dunbar and Rodgers's (1957, quoted above) explanation of the term; at most Conkin, Layton, and Conkin (1983) have demonstrated that Dunbar and Rodgers picked a poor example to illustrate their conception of a paraconformity. Furthermore, as Newell (1967) noted early on, Dunbar and Rodgers's (1957) conception of a paraconformity may be laterally coextensive with an erosional disconformity.

Paracontinuity: In conjunction with their assertion that diastrophism and unconformities are critically important in chronostratigraphy (see discussion of these subjects in Chapter 7), J. Conkin and B. Conkin (see especially 1973, 1984a,b, 1985) have introduced and named their concept of a paracontinuity, which, using their terminology, is a type of disconformity.

> The term paracontinuity is derived from para (nearly or almost) and continuity (a continuum); the implication is one of an apparently continuous sequence of sedimentation but one which is broken by a slight, though significant, faunal

discontinuity and slight, though clearly discernible, channeling along the surface of disconformity. The paracontinuity is, then, a kind of large scale diastem [see below] which is geographically widespread and which exhibits a physical and faunal discontiniuty. (J. Conkin and B. Conkin, 1973, p. 12)

A paracontinuity displays itself as a slight but significant physical discontinuum coincident with a slight but significant biologic discontinuum that thus allows for [chronostratigraphic] boundary determination on both physical and biological bases. (B. Conkin and J. Conkin, 1984, pp. 4-5)

In outcrop there is always a transgressive detrital unit (varying in thickness from "a single grain to a few feet") immediately above the physical surface of the paracontinuity (J. Conkin and B. Conkin, 1985). The "temporal gap along paracontinuities in general may range from the magnitude of a bed . . . to a member, formation, group, and stage" (J. Conkin and B. Conkin, 1985, p. 491). These authors defined a "type" paracontinuity between the Upper Devonian and Lower Mississippian in Illinois (see J. Conkin and B. Conkin, 1973, 1985, for details).

In 1985, J. Conkin and B. Conkin revised their concept of paracontinuities to encompass:

three orders of paracontinuities based on the hierarchical ranks of the chronostratigraphic units they separate and the temporal gap magnitudes existent along the surfaces of paracontinuities. . . . A first order paracontinuity is herein defined as a chronostratigraphic discontinuity separating world-wide chronostratigraphic units of hierarchical ranks of eonothem, erathem, series, and stage. . . . We propose the term second order paracontinuity for those paracontinuities which separate chronostratigraphic units of hierarchical rank below stage. . . . The third order paracontinuity differs from both first and second order paracontinuities in that it exhibits a larger time gap. It is, in fact, the kind of unconformity which is generally understood to be a disconformity in Grabau's (1905 and 1913) sense. . . . The boundary defined by a third order paracontinuity is not a chronostratigraphic boundary. . . . If the temporal gap along the surface of a third order paracontinuity increases to that of a series or more, the disconformity becomes a paraconformity (Dunbar and Rodgers, 1957). (J. Conkin and B. Conkin, 1985, p. 491)

It appears that for J. Conkin and B. Conkin virtually all chronostratigraphic boundaries are, by definition, paracontinuities (although not all paracontinuities are chronostratigraphic boundaries). As J. Conkin and B. Conkin (1985) are aware, some "paracontinuities" that separate stages or higher units also simultaneously separate units below the level of the stage and thus are both first and second order paracontinuities—which seriously limits the usefulness of their classification. J. Conkin and B. Conkin's concept of paraconformity is further hampered by associating it with hierarchical ranks that are arbitrarily assigned and reassigned to stratigraphic sections. Furthermore, as is evident from the quotations given above, J. Conkin and B. Conkin's modification of the concept of paracontinuity renders the term essentially synonymous with any nonangular unconformity between stratified rock bodies (= disconformity in the sense of Grabau, 1913).

Use of the term paracontinuity in the literature has been restricted to that of J. Conkin and B. Conkin and their associates (see B. Conkin and J. Conkin, 1984; J. Conkin, 1985, 1986; J. Conkin and B. Conkin, 1973, 1975, 1979, 1983, 1984a, b, 1985; J. Conkin et al., 1980, 1981, 1983).

Diastem: The term diastem is generally applied to very minor (temporally) pauses or breaks in deposition with little or no erosion before deposition is resumed. Weller (1960, p. 392) states that "any minor stratigraphic discontinuity not considered important enough to be an unconformity may be termed a diastem."

Barrell (1917) first coined the term diastem for the innumerable small stratigraphic gaps (temporal breaks) that are often so obscure as to be essentially unrecognizable, but which he reasoned must exist between many bedding planes. Barrell (1917) originally applied the concept to marine strata, but it has since been applied to rocks formed in any environment. According to Roberts (1982), diastems are genuinely unrecognizable (which does not mean they do not exist); once they are recognized by some physical feature, they are usually classified as paraconformities or disconformities.

Hiatus: The term hiatus, as commonly used by stratigraphers, refers to geologic time that is not represented by strata at a certain point (such as at an unconformity or between particular beds) in a stratigraphic section or sequence (e.g., Boggs, 1987; Mitchum, Vail, and Thompson, 1977; Weller, 1960). When used in this manner, the term hiatus may be essentially synonymous with the term lacuna. Hiatus is also used at times in a more restricted sense to refer only to the time that was never represented by the deposition of rock in a particular stratigraphic section or sequence (see below under "Lacuna").

Lacuna: The term lacuna, as commonly used by stratigraphers, refers to a gap or break in the stratigraphic record. The lacuna may refer to the part of the stratigraphic record destroyed at an erosional surface (previously deposited strata are removed; = erosional vacuity) and time that was never represented by a rock record (= hiatus in the restricted sense; see American Geological Institute, 1976; Krumbein and Sloss, 1963, p. 360).

Relations of Strata to Surfaces of Unconformity

Mitchum, Vail, and Thompson (1977) have distinguished and classified some basic relationships that may exist between the strata either above or below an unconformity and the surface of unconformity itself. In their publication these authors use the following terminology to refer to the relations of strata to the boundaries of their "depositional sequences" (see Chapters 5 and 7), but in most cases these boundaries are unconformities.

If the strata above or below a surface of unconformity are parallel to the surface of unconformity (whether the surface of unconformity is flat and horizontal, inclined, uneven, or undulating), the strata are said to be concordant to the unconformity (Fig. 3.2). Such concordance is also normally exhibited, of course, among conformable strata. If strata are not in a concordant relation relative to an unconformity, then they are discordant (or exhibit discordance) relative to the unconformity.

Strata, whether immediately above or immediately below a surface of unconformity, may exhibit various forms of lapout. "*Lapout* is the lateral termination of a stratum at its original depositional boundary" (Mitchum, Vail, and Thompson,

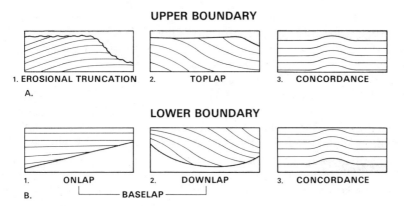

Figure 3.2. The relations of strata to boundaries of depositional sequences, as defined by Mitchum, Vail, and Thompson (1977, p. 58, reprinted by permission of the American Association of Petroleum Geologists):

"*A.* Relations of strata to upper boundary of a sequence. *A1.* Erosional truncation: strata at top of given sequence terminate against upper boundary mainly as result of erosion (e.g., tilted strata terminating against overlying horizontal erosional surface, or horizontal strata terminating against later channel surface). *A2.* Toplap: initially inclined strata at top of given sequence terminate against upper boundary mainly as result of nondeposition (e.g., foreset strata terminating against overlying horizontal surface at base-level equilibrium where no erosion or deposition took place). *A3.* Top-concordance: relation in which strata at top of given sequence do not terminate against upper boundary.

"*B.* Relations of strata to lower boundary surface of a sequence. *B1.* Onlap: at base of sequence initially horizontal strata terminate progressively against initially inclined surface, or initially inclined strata terminate updip progressively against surface of greater inclination. *B2.* Downlap: at base of sequence initially inclined strata terminate downdip progressively against initially horizontal or inclined surface (e.g., initially inclined strata terminating against underlying initially horizontal surface). *B3.* Base-concordance: strata at base of sequence do not terminate against lower boundary."

1977, p. 57, italics in the original). Any stratum has upper and lower boundaries, as well as lateral boundaries; laterally a stratum may taper to a feather edge or end abruptly against another surface. Lapout describes the nature of the lateral termination of strata as they are originally deposited (that is, before erosion, structural deformation, or other post-depositional modification)—for example, how the lateral margins of the strata rest against the surface of the boundary upon which they were deposited or the orientation they took just below their upper boundary.

Baselap refers to the lapout of strata immediately above the surface of unconformity and is classified in two basic categories by Mitchum, Vail, and Thompson (1977). Onlap is the situation in which horizontal (at least at their time of deposition) strata lap out against an inclined surface (Fig. 3.2), or in which initially inclined strata are deposited against a surface with an inclination that is greater than that of the strata being deposited. If strata initially deposited at an angle terminate downdip (at their bases or lower boundaries) against either a horizontal or an inclined surface, the situation is referred to as downlap (Fig. 3.2). Distinguishing downlap from onlap may, in some situations (for example, areas that have been structurally disturbed), require that the orientation of the initial surface on which the strata were deposited be reconstructed. If it is impossible accurately to reconstruct the orientation of this surface, it may be impossible to distinguish between the two types of baselap described above. The terms proximal and distal can be combined with the terms for various types of baselap to signify the regional position of an incident of baselap relative to the position of the source of the sediment supplying the strata forming above the surface of unconformity. Thus a stratum formed by clastic terrigenous debris feeding into an ocean basin from a continent may be marked by proximal onlap landward (toward the source of sediment supply) and distal downlap seaward (away from the source of sediment supply).

Mitchum, Vail, and Thompson (1977) suggest that in most cases baselap relationships indicate that a nondepositional hiatus is present between the surface of unconformity and the strata immediately above, rather than an erosional hiatus.

Mitchum, Vail, and Thompson (1977) note that diagrammatic illustrations in two dimensions (such as in Fig. 3.2) are generally assumed to be illustrative of relationships between strata and unconformities as seen in cross-sectional exposures perpendicular to the strike of the strata under consideration. Cross sections parallel to strike, or at some angle to strike other than 90 degrees, may show apparent (and non-true) relationships of strata and surfaces of unconformity. In the case of simple onlap, a section taken parallel to the strike of the strata above the unconformity might exhibit parallel traces of strata that appear to be in conformable relationship to the surface of unconformity.

Lapout just below the surface of an unconformity (at the upper boundary of one of Mitchum, Vail, and Thompson's [1977] depositional sequences) is referred to as toplap. Toplap may occur in strata initially deposited at an angle, such as foreset beds, where the strata taper updip to terminate against what will become in section a surface of unconformity or boundary between two sequences of strata (Fig. 3.2). Toplap relations may mimic true classical angular unconformity (formed by a profound structural episode followed by erosion) or erosional truncation. As Mitchum, Vail, and Thompson (1977) note, toplap usually may be taken as evidence for a nondepositional hiatus, perhaps with some minor erosion. Toplap relationships usually are the result of a situation where the depositional base level does not permit the strata being formed to extend farther updip. Toplap relations often develop in deltaic and other shallow marine deposits, but also can be found in deep-marine deposits.

The strata immediately below a surface of unconformity may be truncated by erosional processes, especially if the strata have been structurally tilted. However, strata also may be truncated simply by the cutting of a channel or valley. Mitchum, Vail, and Thompson (1977) refer to this category of relations between strata and the surface of an unconformity as erosional truncation, which includes as a subcategory the relationship observed between strata and the surface of unconformity seen in many classic angular unconformities. Structural termination (Mitchum, Vail, and Thompson, 1977) is the situation in which strata (as presently observed) are terminated laterally against a fault, an igneous intrusion, or some other structural disruption. Such structural termination usually is secondary (formed after deposition of the strata under consideration) and does not necessarily represent a hiatus or an unconformity. However, in some cases structural termination, and the resulting discordant relations of strata and surfaces that may be produced, can be misidentified as an unconformity.

Overstep

In cases where the bedding planes of the rocks above and below an angular unconformity are at a high angle to each other, the angular unconformity often will be evident at a single outcrop. In many situations, however, the bedding planes of the rocks above and below the unconformity will appear to be parallel in a single out-crop or within a limited area, but on a regional scale some degree of angularity will be detectable between the rocks above and those below the unconformity. In cases of locally or regionally evident angular unconformity, the surface of the unconformity (for instance, an old erosional surface upon which strata were subsequently deposited) is not parallel to the underlying rocks but cuts across their bedding and structures. The manner in which such an unconformity (and the rock sequence lying above the unconformity) cuts across the underlying rocks commonly is referred to in the British literature as overstep and in the North American literature as overlap (Roberts, 1982). The rocks lying immediately above the unconformity overstep or overlap the upturned (either very slightly upturned or dramatically upturned) and eroded underlying beds, and the rocks below the surface of unconformity are said to be overstepped or overlapped.

Roberts (1982; see also ISSC, 1987a,c) notes that the simplest angular unconformities, observed only on a regional scale, are formed when the underlying rocks are tilted only minutely (perhaps by slight tilting or flexuring over a large region) before the formation of the overlying rocks. The strata above such an unconformity will gradually, over large geographic distances, overstep (overlap) or cut across and contact various formations of the underlying sequence. Thus in Fig. 3.3 the difference in dip in the rocks above and below the unconformity, and thus the angular unconformity itself, may not be evident at any one local outcrop or exposure. However, in some places Formation A rests on Formation W, whereas in other exposures it rests on Formation X, Y, or Z. As Roberts (1982, p. 222) points out, if the difference in dip between the strata above and below the unconformity were only 0.5 degree, a formation (such as W, X, Y, or Z in Fig. 3.3) that was 100 meters thick would be overstepped (overlapped) by the overlying formation (Formation A in Fig. 3.3) over a distance of less than 10 km.

Angular unconformities can be recognized on geologic maps by virtue of the fact that the base of the sequence of rocks immediately overlying the unconformity will truncate or overstep (overlap) the sequence of formations underlying the unconformity. Of course, such a pattern may not always be evident for any number of reasons. A very subtle or slight angular unconformity may require that a large geographic area be mapped before it is detectable. Topographic relief (perhaps preserving outliers and eroding away other rock bodies) in the area under consideration may mask, obscure, or confuse the interpretation of the area. If the rocks overlying the unconformity are themselves folded, faulted, or otherwise structurally disturbed, the recognition of an unconformity may be rendered more difficult. Likewise, the bases of the definitions of the stratigraphic units used and the manner in which they are mapped may render an unconformity more or less difficult to recog-

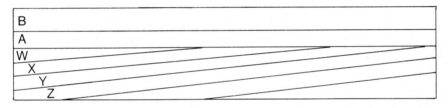

Figure 3.3. Stratigraphic cross section demonstrating the concept of overstep (vertical scale is greatly exaggerated; B, A, W, X, Y, and Z represent formations or other stratigraphic units; see text for further explanation).

nize. For example, thinner and more finely divided stratigraphic units that are carefully mapped may render a slight angular unconformity identifiable, whereas mapping of the same area with only a few coarser and thicker units might not reveal the unconformity.

Related to the concept of overstep (overlap) per se, in some cases the rocks underlying an unconformity may genuinely lack any angular discordance relative to the rocks overlying the unconformity. Still, an unconformity may be detectable if the strata overlying the unconformity "overstep" ("overlap") or cross-cut faults or igneous intrusions that are within the underlying rocks but not in the overlying rocks.

Overlap

The term overlap is commonly used in the British literature, and the term onlap in the North American literature, to describe the manner in which the strata overlying an unconformity are deposited relative to the unconformity and to one another (Roberts, 1982). In some cases the surface of unconformity is not parallel to the bedding planes of the overlying strata; this may occur, for instance, when the overlying beds (assumed to have been deposited horizontally) are formed on a surface that is inclined relative to the horizontal (see Fig. 3.4a). Thus in Fig. 3.4a, each succeeding formation (or other stratigraphic unit) above the surface of unconformity extends farther laterally (over a wider area) than the formation below; furthermore, each formation wedges out (its thickness decreases) and comes to a feather's edge along its lateral margin against the surface of the unconformity (that is, the top surface of the rocks underlying the unconformity). In such a situation, referring to the strata of the sequence of rocks above the unconformity, the higher beds stratigraphically are said to overlap (onlap) the beds that are lower stratigraphically. Observable overlap (onlap) normally indicates that the base of the sequence of strata

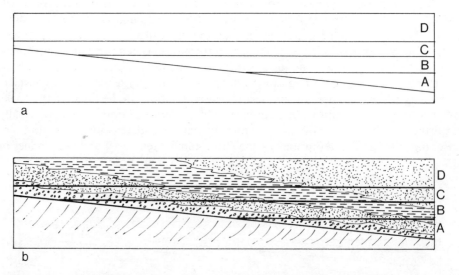

Figure 3.4. Stratigraphic cross sections demonstrating the concept of overlap (onlap). (a) Overlap (onlap) of formations (or other stratigraphic units) A, B, C, and D. (b) Overlap (onlap) of stratigraphic units (e.g., chronostratigraphic units) A, B, C, and D not reflected in the lithologies of the rocks (see text for further explanation).

above the surface of unconformity is diachronous. In cases where the erosion surface marking the surface of an unconformity is perfectly horizontal, there may be no detectable overlap (onlap) of the overlying formations.

The concept of overlap (onlap) is a concept applied primarily to temporally (chronostratigraphically) defined units. A sequence of rock strata subdivided and classified chronostratigraphically may exhibit marked overlap (onlap), whereas the same rocks classified by some other means (for instance, lithostratigraphically) may not reflect the overlap (onlap; see Fig. 3.4b). In other instances, lithostratigraphic units may only imperfectly suggest or reflect the overlap (onlap) pattern.

Overlap (onlap) patterns may be very subtly developed, and only evident on a regional scale, if the surface of unconformity upon which the sediments are deposited is flat and very nearly horizontal. In other cases, however, the surface of unconformity may exhibit considerable topographic relief (as in the case of buried landscapes), and overlap (onlap) patterns of the overlying sediments may be extremely pronounced even in isolated exposures (cf. concept of disconformities, above). In certain cases strata might be deposited with an appreciable dip against a nonhorizontal erosion surface (departing from the principle of original horizontality), but any such initial dip will be less than the slope of the surface of erosion against which they are deposited (Roberts, 1982).

In plan view on a geologic map, overlap (onlap) will be indicated by the fact that among the sequence of rocks overlying the unconformity differing units in different locations will rest upon the surface of unconformity. In plan view, successively higher (stratigraphically) units will appear to pinch out between the boundary of the unconformity (as seen in plan view) and the overlying formations. The same types of factors that may obscure the recognition of overstep relationships (mentioned above) may also obscure overlap relationships among strata. In general, the greater the relief of the old erosional surface that forms the surface of unconformity, the smaller the map area needed to unequivocally identify overlap (onlap) relationships.

Use of Unconformities in Dating Geological Events

During the time interval represented by an unconformity (that is, the time interval corresponding to the gap in the stratigraphic record at the unconformity) many important geological events may have taken place in the area under consideration—for instance, structural events including folding, faulting, and uplift; igneous intrusions; or various metamorphic events. Such features may be datable to the time interval represented by the unconformity. For instance, if the strata lying below an unconformity are folded and/or faulted, but the strata above the unconformity are not affected by the folding or faulting (the strata above the unconformity overstep the strata below the unconformity), then these structural events must have taken place after the rocks immediately below the unconformity were deposited, but before the strata above the unconformity were formed. Furthermore, it is generally accepted that in most cases strata now seen to be folded were once buried with large accumulations of sediment before such folding took place. Then, after folding took place, there must have been uplift and erosion to form the surface of unconformity. Therefore, the folding predates the uplift and erosion that actually formed the surface of unconformity; it might be said that the folding predates the unconformity.

It may not always be a simple matter to determine whether the rocks below an unconformity were affected by a particular episode of folding or faulting that did not affect the rocks overlying the unconformity. In many cases, multiple episodes of folding or faulting have affected the same region at different times. The trend of one episode of folding or faulting may be at an angle to the trend of another episode, or folding and faulting may undergo rejuvenation in a particular area modifying and accentuating structural features already present in the rocks rather than creating totally new features. Any sequence of rocks observed today represents a cumulative record of all of the episodes of deformation that have occurred in its locality since the time when the rocks were deposited. Thus the rocks may be folded both above and below an unconformity, and the axes of the folds may trend in the same directions, yet the rocks below the unconformity may have been folded initially by a structural episode of folding that predated the unconformity. The first episode of folding only was modified by a second episode of folding that postdated the unconformity and the sequence of rocks above the unconformity; and the rocks immediately above the unconformity were folded only by the latter episode of folding. Casual inspection of the rocks, however, may fail to reveal that two distinct episodes of folding occurred, one predating the unconformity (or rather one that took place during the time interval now unrepresented by strata in the section) and the other postdating the unconformity and deposition of sediments above the surface of unconformity.

Following the same basic principles outlined above for dating episodes of folding and/or faulting, igneous intrusions (and accompanying thermal metamorphism) and regional metamorphism often can be dated relative to unconformities. Igneous intrusions and episodes of metamorphism always must be younger than the country rocks that they introduce or affect. If the strata above an unconformity overstep intrusive igneous bodies or are not affected by a certain episode of metamorphism, then the igneous intrusion or episode of metamorphism must predate the oldest strata above the surface of unconformity. However, one must be cautious in drawing conclusions as to the age of intrusive bodies. For instance, it cannot necessarily be asserted positively that an intrusion predates an unconformity solely because all of the intrusive rocks lie stratigraphically below the unconformity. Unless the strata immediately above the unconformity are observed to overstep the intrusive body (indicating that the intrusive rocks were exposed at the surface and eroded to some degree before deposition of the overlying strata), the possibility may remain that the igneous rocks were intruded at a date after the deposition of strata above the unconformity. Of course, circumstantial evidence may strongly support a particular date for the intrusion, even if it does not date it definitively.

As Roberts (1982) has pointed out, it may be possible to place an upper limit on the date of an intrusion if the intrusive rock is of a distinctive type such that detrital fragments of the rock formed by subsequent weathering and erosion can be identified in overlying sedimentary layers. In such a case the intrusive body must predate the rocks, including weathered detritus from the intrusive. Also, in rare instances it may be possible to date an igneous intrusion if the intrusion can be associated with volcanic activity that produced rocks at the surface that subsequently became incorporated into the stratigraphic sequence. In such a case the intrusion is of either approximately or exactly the same age as the volcanic rocks.

In cases where rocks affected by regional metamorphism found only below an unconformity are overstepped by nonmetamorphosed strata immediately above the unconformity, it can safely be concluded that the episode of regional metamorphism occurred after the deposition of the rocks exhibiting metamorphism and before the deposition of the rocks above the unconformity. In fact, it is usually assumed that most regional metamorphism occurs at great depths, and so it is usually assumed that a large accumulation of rocks once must have been present above the rocks now metamorphosed. Subsequent to the episode of metamorphism there must have been uplift and erosion, forming the surface of unconformity, prior to the deposition of the strata immediately above the surface of unconformity (and now in contact with the metamorphosed rocks below). For these reasons, in situations where regionally metamorphosed strata underlie an unconformity and strata that have not been affected by metamorphism, it is common to assume that the unconformity represents a large gap in the stratigraphic record corresponding to both an appreciable amount of time and a thick sequence of sediments that was once deposited on top of the remaining metamorphosed rocks, but was subsequently eroded away (Roberts, 1982). In many cases higher-grade regional metamorphism may be accompanied by folding and other structural deformation of the rocks, and thus the episode of metamorphism and folding can be dated relative to the unconformity, as outlined above for dating episodes of folding and faulting.

Roberts (1982, p. 242) suggests that it may be possible for regional metamorphism to occur at depth in a certain locality while rocks continue to be deposited on the surface, but that "there is rarely any evidence to support such an interpretation." Lacking other information to the contrary, and given that in many cases the rocks may have undergone relatively high-grade metamorphism associated with folding, "it is usually assumed that sedimentary rocks undergo regional metamorphism after deposition has come to an end so that it may be regarded without any qualification as a later event" (Roberts, 1982, p. 242).

BASIC PRINCIPLES OF STRATIGRAPHY

Various "principles" or "laws" have been proposed to be fundamental to the practice of stratigraphy as a scientific discipline. These include Steno's principle of superposition (see Chapter 1), Smith's concept of faunal or biotic succession (generally this concept has been tied to the notion of unidirectional and nonreversible organic evolution, but this is not necessary—it is enough that fossils merely be observed as consistently occurring in the same sequence within stratigraphic sequences; see Chapter 1), and Walther's law (dealing with the successions of facies, see below). Here it should be explicitly noted that these principles or laws often must be applied with some discretion and judgment—and perhaps they are not so inflexible or foolproof as the laws of physics or mathematics are sometimes taken to be. Steno's law of superposition, for instance, should not be taken to imply that "higher" (in terms of topographic elevation) rocks or strata must be younger than topographically lower deposits. For example, rocks may be slumped, overturned, faulted, or otherwise structurally deformed or displaced; younger rocks may fill cavities, caves, or crevasses in older rocks, and for some sorts of deposits (such as river terraces) it may be routine for younger deposits to form at topographically lower elevations (see Fig. 9 of the *North American Stratigraphic Code,* reprinted here as Appendix 1). Likewise Smith's principle of biotic succession does not mean that any "primitive"-looking fossil is necessarily older than any relatively more "advanced"-appearing fossil.

Also extremely important in stratigraphy are concepts used in the interpretation of cross-cutting relationships among strata and emplaced rock bodies (for example, an igneous rock body cutting across a sandstone would be considered younger than the sandstone, for the sandstone must have existed before the igneous rock was intruded into it; likewise, later igneous veins may cut across, and displace, older igneous veins indicating the sequence of intrusions [Ehlers and Blatt, 1982]: see Fig. 3 of the *North American Stratigraphic Code*). The last-mentioned concept has been termed the principle (or law) of cross-cutting relationships (e.g., Fritz and Moore, 1988). A related concept, sometimes termed the law of inclusions, is that a clast (a xenocryst or a xenolith— a foreign crystal or rock fragment, respectively, within an igneous rock) in a particular rock, such as a sandstone pebble in a conglomerate, must have existed prior to (and is therefore older) than the rock or sedimentary layer in which it is found. Following the same line of reasoning, the International Subcommission on Stratigraphic Classification recently restated the principles of cross-cutting relationships and inclusions as follows: "An intrusive body . . . is younger than the rocks it intrudes, and older than those that intrude it. An eroded intrusive or metamorphic rock body is older than the strata that were laid down unconformably upon it, and older, too, than the clastic sedimentary deposits containing the products of its erosion" (ISSC, 1987b, p. 441). These principles may be applicable to nonmatarial features in the stratigraphic record, such as surfaces of bedding or unconformity, as well as material units such as rock bodies.

All the principles just mentioned are used primarily in interpreting the relative temporal relationships of rock bodies by their positional relationships to one another (Kitts, 1966, has termed this "gross ordering principles"). It has been suggested that all of these principles are subsumed under one basic principle of historical geology—what may be termed the principle of positional relationships of rock and mineral bodies (McLaren, 1978). McLaren (1977, pp. 27–28) expressed this concept in writing: "All historical inference in geology comes from the positional relationships of rock and mineral bodies. Stratigraphy is a special case of this general law, and our sole knowledge of the orderly evolution of life as represented by fossils, comes from their mutual relations in stratified bodies. The only proof that one fossil is younger than another lies in the relative position of the two in a sequence of rock." According to McLaren, studies concerned with any features of rocks, such as lithology, magnetism, isotope ratios, and so on, are dependent on this principle. Even work with numerical ages of rocks or minerals, such as those derived by radiometric or isotopic dating, still ultimately depends upon the principle of positional relationships in order to be placed within a geological context. "As with fossils, a mineral or whole rock containing isotopes that can be analyzed by the dating systems must be set into the relative framework by the principle of positional relationships. The mineral or rock must be fixed geologically by its position or any determination of its age in years would be irrelevant" (McLaren, 1978, p. 3).

In contrast to McLaren's views, some authors would distinguish two separate principles applicable to the data and material of stratigraphy (successions of rock types, fossils, apparent isotopic ages, and so on):

Interpretations of earth history depend on two different systems of logic, both of which arrange geologic observations into sequences of events. The first and most widely used is the logic of superposition: the ordering of events iteratively in a system of invariant properties simply by determining the physical relationship of features in the rocks. This is what is meant by the word *stratigraphy*. The second logical system depends on the recognition of an ordinal progresssion which links a series of events in a system of irreversibly varying properties. This provides a theoretical basis outside of the preserved geologic record by which the nature and relation of the events in the progression can be recognized or predicted, and according to which missing parts of the record can be identified.

Geology is an historical philosophy, so the ordinal progressions to which we refer are progressions in time, just as geologic time is perceived by the progress in one or another ordinal series of events. This is what is meant by the word *geochronology*. (Berggren and Van Couvering, 1978, pp. 39–40, italics in the original)

Thus Berggren and Van Couvering would restrict stratigraphy *sensu stricto* to essentially what has sometimes been referred to as descriptive stratigraphy or lithostratigraphy *sensu lato* (somewhat similarly, Odin [1982b] has distinguished between "relative stratigraphy" and "numerical dating" of strata). Geochronology, still intimately related to stratigraphy even in the strictest sense, is based on any system of varying, but irreversible, properties. Thus, for example, positing that biological evolution is a unidirectional, irreversible process, a geochronologic system (biochronology) can be derived from the biostratigraphic record (the physical record of the remains of successions of life forms). Likewise, a radiochronology (Berggren and Van Couvering, 1978) can be developed on the basis of the decay rates of unstable isotopes.

On a large scale, however, it can be argued that the earth's history as a whole has been unidirectional over the last 4500 million years (4.5×10^9 years). Plate tectonics may in some respects be an irreversible, one-way process; the earth appears to be undergoing irreversible geochemical differentiation, and certainly the appearance and development of life has profoundly affected the rocks formed on the surface of this planet (see, for instance, Dineley, 1984; McLaren, 1978; Simpson, 1970). It has been suggested that, at least to a certain extent, the various geological systems may represent unique conditions on the earth that never can be replicated (cf. Ager, 1981).

Other principles that have been suggested as essential or significant to stratigraphy in particular take the form of generalizations applicable to the formation and/or preservation (in the stratigraphic column) of most stratified rock bodies. In this class are included Steno's principles of original horizontality and original lateral continuity (Chapter 1), the baselevel concept of Barrell (1917), and Wheeler's (1959) principle of datum (or datum–surface) variance.

As formulated, Steno's principle of original horizontality states that all sedimentary strata were originally deposited in horizontal layers; if we find them today inclined from the horizontal, then they must have been tilted subsequent to their original formation. As pointed out by numerous authors, however, this is perhaps an overstatement that has been applied to situations where it is not strictly true. In some situations, such as among nonmarine, relatively coarse-grained deposits, deposition can take place on considerable slopes (up to 30 degrees: Fritz and Moore, 1988). In other cases strata may be deposited nearly horizontally, but not exactly so (being perhaps a degree or so from the absolute horizontal). In part, whether strata are considered to be deposited horizontally or not is a question of scale. In situations involving cross bedding or cross lamination, beds or laminae that are deposited obliquely, transversely, or inclined to the main bedding planes (such as foreset beds in deltaic deposits) may appear nonhorizontal when viewed in detail, but from a larger perspective the true overall stratification may be horizontal. It has been suggested that the principle of original horizontality can be reformulated as the concept that all sediments originally will be deposited at some angle less than their angle of repose ("the maximum angle at which sediments can be deposited on a sloping surface" [Ehlers and Blatt, 1982, p. 12]), and in many cases the angle of repose will be nearly horizontal.

Steno's principle of original lateral continuity indicates only that originally any single stratum or sedimentary layer probably was continuous throughout its maximum areal extent (for example, within a depositional basin). This concept is important in attempting to correlate now isolated portions (for instance, rocks on either side of a valley) of what once was presumed to be a single, continuous stratum. Eventually any particular stratum will end, perhaps tapering out to a feather edge, ending abruptly against the side of a basin, or grading laterally into another stratum or facies. The principle of original lateral continuity does not imply that any stratum or layer ever continued indefinitely, eventually to cover the whole surface of the earth.

Prior to Barrell's (1917) work the term baselevel was used primarily to refer to the level (depth) to which rivers could erode their channels (at that time often believed to be approximated by the contemporaneous mean sealevel). Barrell redefined the concept of baselevel as follows (Barrell, 1917, p. 778; also quoted in Wheeler, 1959, p. 701): " . . . the sediments whose interpretation form the basis of earth history have been characteristically deposited with respect to a nearly horizontal controlling surface. This surface of control is baselevel, but for continental and marine deposits the baselevel is determined by different agencies and is a word of more inclusive content than the sense in which it has generally been used by physiographers as a level limiting the depth of fluviatile erosion. Sedimentation as well as erosion is controlled by baselevel, and baselevel, local or regional, is that surface toward which the external forces strive, the surface at which neither erosion nor sedimentation takes place." In a sense Barrell's (1917) concept of baselevel is a refinement of Steno's principle of the original horizontality of strata. As Wheeler (1959) has noted, baselevel may be at different levels at different times at the same point on the surface of the earth, and at different levels simultaneously at different points. Baselevel may be closely tied to sealevel, or it may be completely unrelated to sealevel.

Wheeler's (1959) principle of datum–surface variance encompasses the concept that many rock unit boundaries and other surfaces or horizons cut across, transgress, or are at an angular relationship to one another and/or ideal synchronous time horizons. Originally this principle was applied only to the relationship between many lithostratigraphic unit boundaries and time planes, and was thus termed the principle of temporal transgression by Wheeler and Beasley (1948). In 1959 Wheeler expanded the idea to include the idea that any two stratigraphic surfaces may be nonparallel (for example, a surface defined on the basis of a last occurrence of a particular species and a surface forming the boundary of a lithostratigraphic unit). Wheeler's (1959) principle of datum variance actually appears to be nothing more than a denial of the idea or assumption that all stratigraphic surfaces and all boundaries of stratigraphic units are ideally parallel and in some cases coincident. As Wheeler (1959, p. 704) states, "Except for this factor [the principle of datum variance], the Wernerian concept of universal concentric lithostratigraphic spheres would obtain, and all stratigraphy would be of the 'layer-cake' variety." It is questionable whether one needs to formalize a principle that merely denies a faulty principle.

Although it is not a principle of stratigraphy in the strict sense, when pursuing stratigraphic studies, one must always keep in mind the need to determine the order of succession or sequence in layered rocks. Given a stratigraphic sequence, how do we know which rocks are on the top of the pile, and which are on the bottom? Given a single stratum or bed, how do we know which is the top and which is the bottom? Without such information, the principle of superposition, perhaps the most important principle in stratigraphy, cannot be applied. In relatively undisturbed areas the strata or rocks may be situated in essentially their original attitude with the top layers or beds literally on the top of the sequence; in such instances, determination of the correct stratigraphic sequence (bottom to top) is not a problem. In structurally disturbed or deformed areas, however, stratification, bedding, or layering may be inclined at a high angle to the horizontal, folded, or perhaps vertically positioned or overturned. In such cases it is important, but sometimes difficult, to determine the tops and bottoms of beds, layers, and stratigraphic sequences. Many different methods can be utilized to accomplish this goal. The top of a questionable sequence may be determined by correlating the sequence under consideration to sequences in areas that are not or are only minimally deformed (so that the tops of the sequences are known), or by correlating it to a composite regional sequence whose top is known. Tops of beds and layers may also be determined by careful analysis of internal structures in the rocks themselves—for instance, sedimentary structures such as ripple marks, internal grading and grain-size distribution of sedimentary particles, clast orientation and distribution, trace fossils, and biogenic structures; vesicle and flow structure distributions in lavas: compositional and textural features in various igneous rocks; and relict structures in metamorphic rocks. It is beyond the scope of this book to discuss such material; perhaps the single best work devoted primarily to a discussion of the determination of sequences in layered rocks (including sedimentary, igneous, and metamorphic rocks) is the classic book by Robert R. Shrock (1948) entitled *Sequence in Layered Rocks.*

FACIES

The stratigraphical term facies is derived from the Latin *facia* or *facies,* referring to the external appearance, look, figure, aspect, face, or condition of an object (Hallam, 1981; Walker, 1984a). The term itself was introduced into geology by Nicholaus Steno in 1669 (Teichert, 1958), but a modern stratigraphical usage of the term is generally credited to Amanz Gressly (1838, see Chapter 6). Since Gressly's time the term facies in geology has come to have many varied usages and applications. According to the *International Stratigraphic Guide* (Hedberg, 1976, p. 15, italics in the original), the term facies "in stratigraphy, can mean *aspect, nature,* or *manifestation of character* (usually reflecting conditions of origin) of rock strata or specific constituents of rock strata. It is also used as a substantive for a body of rock strata distinctive in aspect, nature, or character. The general term 'facies' has been greatly overworked. Rock strata may show differences in facies of various sorts so that one may speak of lithofacies, biofacies, mineralogic facies, marine facies, volcanic facies, boreal facies, and so on. If the term is used, it is desirable to make clear the specific kind of facies to which reference is made." Nonstratigraphers have applied the term and general concept of facies to such usages in geology as metamorphic facies, igneous facies, and tectonic facies (tectofacies), whereas biologists, especially ecologists, have written of biological "facies," such as the oyster-bank facies (Eskola, 1915, 1922; Krumbein and Sloss, 1963).

In stratigraphy *sensu stricto* the term facies generally has been employed to refer to one or more of the following concepts (Weller, 1960, p. 505, italics in the original): (1) "the *appearance* of a rock body," (2) "the *composition* or actual nature of a rock body," (3) "the rock body *itself* as identified by its appearance of composition," and (4) "the *environment* that is recorded by a rock body." In connection with the last point, it should be noted that many recent workers concerned with the delimitation and interpretation of facies ultimately are interested in delimiting actual units or bodies of rock that will serve as the basis for reconstructing and interpreting ancient environments (Middleton, 1978; Walker, 1984a,b). To give two typical examples, Matthews (1984, pp. 28–29, italics in the original) states: "The word *facies* is commonly used to denote the product of an environment. Thus, a highly deformed terrain containing many rock types might be mapped as a single *metamorphic facies* if we wish to convey that all of the rocks within that terrain have been subjected to the same temperature–pressure conditions. Similarly, the concept of *sedimentary facies* may group many different lithologies under the concept that all of these lithologies represent various subenvironments of an overall sedimentary environment which can be given a convenient name [an example of such a convenient name given by Matthews is "lagoonal facies"]." Laporte (1979, p. 37, italics in the original) writes that "lateral variations within a sedimentary basin are termed *sedimentary facies.*"

Weller (1960) further notes that the term facies in stratigraphy has been applied to widely divergent scales and levels. A single facies, when considered as a rock body, may encompass several adjacent large rock bodies (perhaps formally or informally considered formations) or only a small part of a thin bed. Differences between facies, or the characteristics of a single facies, may be extremely obvious features of overall gross lithology, or very subtle differences in only a few specific characteristics (for instance, fossil content).

Historical Perspective

Some early geological theorists, such as Abraham Werner (1750–1817; see Chapter 1), had a conception of the material geological record as being composed predominantly of uniform sheets of rock laid atop one another (this concept of the rock record has sometimes been referred to disparagingly as "layer cake" geology). Any one of these uniform rock layers extended far laterally—in fact, it might be hypothesized as of global extent—and was perhaps deposited relatively simultaneously from a large body of water or in a catastrophic process. As Krumbein and Sloss (1963, p. 299) pointed out, such a vision of the geological record does make sense. The sedimentary pile from the late Precambrian to the present at any one spot on the earth very rarely exceeds 8 to 10 km in thickness (stated in terms of about 5 or 6 miles by Krumbein and Sloss) and is commonly much thinner. Individual rock bodies, that is, rock bodies composed of a relatively homogeneous gross lithology, are only very rarely more than a few thousand meters thick, and are usually much thinner. In contrast, however, the lateral (that is, areal or geographic) extent of such rock bodies may be on the order of tens, hundreds, or even over a thousand kilometers. Thus most rock bodies in this sense are, on the scale of the earth, paper-thin and may in a certain region appear to extend endlessly. The most obvious lithologic changes in such rocks occur in the vertical dimension perpendicular to the bedding and the horizontal surface of the earth; for instance, a sandstone may sit above a shale. Rock units typically are defined on the basis of such vertical changes. Any lateral changes in a rock body may not be immediately evident, particularly in isolated outcrops and local sections that record primarily vertical changes, and thus the significance of lateral changes might not be realized (or at least might be deemphasized). This vision of the rock record was further reinforced by the linear nature of most outcrop belts (Krumbein and Sloss, 1963).

The concept that most rock bodies or units extend laterally (until their limits are reached) with very little change was held by various geologists throughout the nineteenth and early twentieth centuries, particularly in North America, and is still often used, at least as an initial hypothesis to be tested, at the present time. In the beginning of this century the American E. O. Ulrich (see below) was a particularly ardent proponent of this viewpoint, and in England in the late nineteenth and early twentieth centuries S. S. Buckman (see Chapter 6) also tended to ignore lateral facies differences (Hallam, 1981). In Europe, however, lateral differences in the composition of rock units received attention early on.

By the late eighteenth century it was realized by some investigators that at the present time many different rock types are being deposited in different areas simultaneously, and when traced laterally rock types and their organic content may vary (note, however, that bedding planes or surfaces may pass continuously from one environment to another despite the changing rock types). Thus the French chemist Antoine-Laurent Lavoisier (1743–94) published in 1789 a diagram showing the relationship of coarse littoral and fine pelagic sediments being deposited simultaneously off the northern French coastline; gravels would be deposited near the shore, whereas finer sediments could be carried into deeper waters. Lavoisier also associated particular faunas with each sediment type and the environment it represented (Prothero, in press). Similarly, Young and Bird (1822; see Hancock, 1977) stated that some types of rock strata are "nearly allied" to one another and may laterally pass into one another within the same beds, such as sandstone grading laterally into sandy shale or oolite into gray limestone. Also, according to Hancock (1977) and Wells (1963), in 1828 Eaton in America recognized that in the Devonian of New York State the red Catskill rocks are a lateral facies of marine gray beds. In the 1830s Sedgwick and Murchison (see Chapter 1) realized that the Old Red Sandstone of South Wales and the marine beds of Devonshire were equivalent in age (both being Devonian), but differed in lithology and paleontology because of contrasting environments of deposition. Similarly, in the 1830s and 1840s the French stratigrapher Constant Prévost (1787–1856) pointed out that temporally equivalent rock bodies may differ in lithologic and paleontologic characters, whereas similar rocks and fossil assemblages may be indicative merely of similar environments of deposition (Hallam, 1981; Hancock, 1977; Krumbein and Sloss, 1963). The individual generally credited with introducing the term and concept of facies in its modern sense is the Swiss geologist Amanz Gressly (see Chapter 6).

In studying the Mesozoic stratigraphy of the Jura Mountains of Switzerland, Gressly (1838) not only was interested in the vertical succession of rocks but also traced each rock unit laterally (along the horizontal dimension, that is, along the strike of the unit). In doing so he encountered lateral changes in the lithology and paleontology of particular units (for example, formations), which he termed facies (see Dunbar and Rodgers, 1957). In his original concept of facies Gressly stressed lateral changes in the appearance of a single stratigraphic unit, but he also wrote of facies, representing similar depositional environments, appearing or recurring in vertical succession. In this context Krumbein and Sloss (1963, p. 318) suggested: "Gressly was trying to point out that rocks of the same lithologic and paleontologic aspect appear in vertical succession without regard to stratigraphic boundaries, but this has been widely, and, the present writers believe, mistakenly interpreted to mean that facies and facies changes refer to vertical as well as lateral differences in character."

Once Gressly used the term facies, however, it came into wide circulation with a number of meanings, and there has been much discussion about the proper or appropriate uses of the term (e.g., see Dunbar and Rodgers, 1957; Krumbein and Sloss, 1963; Middleton, 1978; Moore, 1949; Teichert, 1958; Walker, 1984a; Weller, 1960; and see further discussion below). Much of this discussion has focused on whether the term facies refers only to a set of characteristics of a rock body or to the rock body itself, whether the term should refer only to restricted parts (usually of restricted areal extent) of a particular stratigraphic unit (such as a formation) or should be applied to stratigraphically unconfined rock bodies or appearances (that is, a particular facies might extend vertically through more than one recognized stratigraphic unit), and whether facies terms or designations should be purely descriptive or include environmental or other interpretations (Walker, 1984b; Weller, 1960). There is no universal consensus among practicing stratigraphers as to how any of these issues is to be decided; the reader of any particular work must interpret the way the author uses the term facies.

Along with the various stratigraphic usages of the term facies have developed a series of related terms, a few examples of which follow (see also below). Prefixes (or preceding descriptors) often are added to the term facies to designate that only certain aspects or characteristics of the rocks under consideration are of significance, or are being considered, in a certain context. Thus the terms lithofacies (lithologic facies) and biofacies (biologic facies) are widely used to designate the petrographic and biotic aspects, respectively, of a body of rock or part of a stratigraphic unit. Mojsisovics (1879) distinguished between isopic facies (simiar facies) and heteropic facies (differing facies). Isopic facies bear the same characteristics and are repeated in a vertical sequence, whereas heteropic facies are of the same age but bear differing characteristics and replace each other laterally. Caster (1934, reviewed by Krumbein and Sloss, 1963) proposed that a lithologic assemblage that is representative of a distinct depositional environment be termed a magnafacies; such a magnafacies was essentially a contiguous body of rock and might be designated by a geographically derived name. The rock body representing a magnafacies might be (or probably would be) time-transgressive. A parvafacies (Caster, 1934) was a part of a magnafacies that could be identified between two time-stratigraphic boundaries, that is, within a chronostratigraphic unit. Any particular chronostratigraphic unit should contain numerous different parvafacies, corresponding to different depositional environments present during the particular time interval. As Krumbein and Sloss (1963) note, the parvafacies of Caster (1934) are essentially equivalent to the heteropic facies of Mojsisovics. The concept of the magnafacies is very similar to the concept of the lithosome subsequently introduced by Wheeler and Mallory (1956), where a lithosome is a particular body of rock of uniform lithology that may laterally intertongue with adjacent rock bodies of different lithologies (see also

Krumbein and Sloss, 1963, and Weller, 1960). A single lithosome or a magnafacies may encompass two or more stratigraphic units, such as formations.

Walther's Law

A sort of patron saint of much modern work on facies analysis and the development of facies models, particularly as concerned with environmental interpretation, is the German geologist Johannes Walther (1860–1937). Walther strongly advocated the study of events of the past through the study of modern analogues and phenomena, and he stated (quoted in the preface to Walker, 1984a) that "only the ontological method can save us from stratigraphy." In this statement, by the "ontological method" Walther referred to the study of modern depositional environments in particular and by "save us from stratigraphy" apparently meant to save us from the cataloging and correlation of stratigraphic sequences without imposing environmental interpretations upon them (see preface to Walker, 1984a). Certainly not all stratigraphers would agree, however, that we need to be saved from stratigraphy (or from, for that matter, the implicit assumption that environmental interpretations imposed on stratigraphic sequences are inherently more interesting than other aspects of sequences). Indeed, my view is that the environmental interpretation of stratigraphic layers and sequences is a concern primarily of sedimentology and interpretive petrology, not stratigraphy *sensu stricto*. Such interpretations, of course, are potentially useful inasmuch as they might bear on the delimitation and correlation of stratigraphic units; however, in some cases one perhaps should be wary of classifying and correlating interpretations rather than the actual rocks themselves. It is in part due to these types of considerations that the *Code* and the *Guide* insist that at least certain stratigraphic units be defined on the basis of objective physical criteria, not environmental or genetic interpretations.

Walther defined a facies as "the sum of all the primary characteristics of a sedimentary rock" (quotation from Krumbein and Sloss, 1963, p. 325). Essentially, for Walther (and many other European investigators) a facies description should include all those characteristics, and only those characteristics (secondary alterations of the rocks are excluded), that are potentially useful in positing environmental interpretations of the conditions under which sedimentary rocks were formed. In his research Walther (1893–94) realized that the environments responsible for producing the various facies of any given stratigraphic unit occur in natural associations. Certain depositional environments will form in a laterally contiguous manner (that is, adjacent to one another in a regular order) in a given region, but with time the environmental conditions may change, and environments (and their corresponding facies) may shift or migrate laterally so that in the preserved stratigraphic column one facies may lie above another facies. Thus there may exist a relationship between the sequence of vertically stacked facies seen in a single stratigraphic section and the laterally adjacent facies mapped within a chronostratigraphic unit (for example, mapped between two bedding planes that approximate time surfaces). Walther (1894) wrote (translated by Middleton, 1973, and also quoted in Hallam, 1981, and Boggs, 1987): "The various deposits of the same facies-area and similarly the sum of the rocks of different facies-areas are formed beside each other in space, though in a cross-section we see them lying on top of each other . . . this is a basic statement of far-reaching significance that only those facies and facies-areas can be superimposed primarily which can be observed beside each other at the present time.'' Walther's statement generally has been interpreted to mean that the facies that occur in a conformable vertical sequence (in a local stratigraphic section) probably are representative of the facies that were being deposited in the local region at any one time in laterally adjacent environments. This concept often has been termed Walther's law (or rule) of the correlation (or succession) of facies (in this context Hallam, 1981, suggests that physics has laws, but that geology only has rules).

A key aspect of Walther's law is that it refers explicitly and only to vertical successions of conformable strata. Vertical successions of facies may reproduce a horizontal (lateral) sequence of facies if there are no unconformities in the section, but in many cases such unconformities may be very subtle and perhaps not readily distinguishable; some changes in facies (changes in environments of deposition) may occur very rapidly and leave no stratigraphic record (Prothero, in press). However, in stratigraphic sequences that are believed to contain no major unconformities, Walther's law may provide a powerful tool in interpreting and elucidating the nature and origin of certain facies. Presumably some facies are relatively easily interpreted in terms of modern analogues (see, for instance, Hallam, 1981; Reading, 1986; Walker, 1984a); however, as Walther himself was aware, other ancient facies may have no close modern equivalents. In a conformable sequence Walther's law may help narrow the environmental interpretations that can be imposed upon a particular facies of unknown origin, given that it is laterally conformable with one or more other facies of presumably known environmental origin. Thus a fairly nondescript shale, that might on the basis of its inherent characteristics be considered of either freshwater or marine origin, might be interpreted as of marine origin if it is found in a conformable relationship with a littoral sandstone below and a shallow marine limestone above.

Before leaving Walther, it is interesting to note that until at least the 1960s Walther's law was generally unknown, or at least not articulated, by American stratigraphers (either in the sense of being familar with Walther's work or in recognizing the same phenomena as described by Walther). As competent a stratigrapher as A. B. Shaw (1964) independently derived the same conclusions as had Walther nearly 70 years earlier. As mentioned previously, in the early part of the twentieth century much of American stratigraphy tended to ignore or downplay the importance of facies relationships; so the situation described in the following quotation from the preface to A. B. Shaw's influential book *Time in Stratigraphy* (1964, pp. ix–x) may have been typical: "When this book was still an unfinished draft, one reader mentioned that my thoughts on diachronism and the mirroring of lateral facies by vertical changes in a section were presented by Walther before the end of the nineteenth century. At first, this worried me greatly. If someone had already said what I wanted to say, should I bother to write it down again? Then it occurred to me that in the more than two decades I have studied geology, the name and thoughts of Walther were never brought to my attention [Shaw must never have read Grabau, 1913]. This in itself suggested that either I have been singularly unobservant, or his ideas are not universally familiar." The latter was probably the case, as Shaw (1964, p. 53) states later in his book that he could not recall ever seeing an article in which the principles encompassed in Walther's law and independently discovered by himself were articulated or applied.

Since the 1960s, however, there has been much interest in Walther's law and the whole notion of facies in the stratigraphic record—so much interest that some authors appear to have perhaps even overemphasized the importance of facies, as in the following quotation: "As a result of this lateral variation in environment, we must always expect a particular sedimentary rock in the geologic record to change in facies, sooner or later, as we trace it laterally from one place to another. The facies concept is counter to an earlier, now outdated, notion often referred to as the 'layer-cake concept,' whereby rock strata were visualized as relatively homogeneous layers that recorded uniform environments" (Laporte, 1979, p. 38). But the layers of the "layer-cake" are still there, and in some instances they genuinely do appear to record uniform environments; however, it is now hypothesized that in most cases such layers (strata or stratigraphic units) transgress time. There is no reason necessarily to assume a priori that a particular sedimentary rock (a lithosome—see below) will change in facies when traced laterally, unless by the words "traced laterally" one actually is referring to temporal correlation.

Weller's Contributions to the Concept of Facies

In his book *Stratigraphic Principles and Practice,* J. Marvin Weller (1960) included an extensive discussion and analysis of various twentieth-century interpretations of facies and related concepts, and made some tentative proposals as to possibilities for their hierarchical classification. Even if we do not fully agree with Weller's particular analysis and conclusions, review of his contributions in this area may shed light on the general nature of facies concepts currently in use.

Weller (1960) suggested that the term facies originally was used in geology to refer to the general appearance or aspect of a rock—that is, a facies was an abstract set of characteristics of a rock body, and not the rock itself. Whereas a facies was determined primarily by the lithology and paleontology of a particular rock (cf. E. Haug's [1907] classic definition of a facies as "the sum of the lithologic and paleontologic characteristics of a deposit at a given place" [quoted in Krumbein and Sloss, 1963, p. 319; see also above]), facies is now commonly used as a general term that serves both as an adjective and a noun, and it is not appropriate at this point to try to restrict its meaning to its original usage. "*Facies* can be accepted as referring to rocks or sediments, their aspects, composition, or environments as these are differentiated in any way that seems to be interesting or important" (Weller, 1960, p. 520, italics in the original). Thus basic meanings accorded to the term facies, which should be evident from the context in wich the term is used, include: the generic use of the term to distinguish any types of facies (lithofacies, biofacies, and so on); the general appearance or aspect of a particular body of rock or rocks, or the aspect of a rock or rocks with regard to one or more selected characteristics; or all rocks of some particular kind (some particular composition), such as a (the) black-shale facies composed of all black shales (anywhere on the earth, regardless of their form, age, or geographic location; Krumbein and Sloss, 1963, p. 319, consider this an "indiscriminate application" of the term facies that "contributes nothing to description or analysis")—or the term facies can be applied to a particular body of rock with certain characteristics.

A lithologic facies is distinguished on the basis of any lithologic characters, that is, observable features of a rock, including such aspects as mineral and grain (clast) composition, color, stratification, and so on. Strictly speaking, any fossil components of a rock may be considered a part of the lithologic composition of the rock (Hallam, 1981). Weller (1960) suggests that lithologic facies may be considered or distinguished vertically and/or horizontally as vertical variations in a sequence or section or as lateral variations among rocks, respectively; or lithologic variations may be distinguished without a consideration of stratigraphic relations. In Weller's (1960) suggested use of the term, lithofacies is not necessarily a synonym of the term lithologic facies. According to Weller, a lithofacies refers to a specific rock body. A lithofacies, properly speaking, is a lateral subdivision of a stratigraphic unit (such as a lithostratigraphic unit) that is distinguished from adjacent lateral subdivisions of the unit by lithologic characteristics. Thus a particular formation may be areally or laterally divided into two or more lithofacies, and Weller (1960) suggests that such lithofacies may be given geographic names and be used as recognized stratigraphic subunits of a larger stratigraphic unit. The stratigraphic unit that is divided into lithofacies may, in many cases, be a lithostratigraphic unit, but it may theoretically be any kind of stratigraphic unit, formal or informal, biostratigraphic, chronostratigraphic, unconformity-bounded, allostratigraphic, and so forth.

As used by Weller (1960), the upper and lower boundaries of a lithofacies correspond to the vertical boundaries of a stratigraphic unit. The boundaries between laterally adjacent lithofacies commonly are distinguished in one of three basic ways. Two lithofacies may be separated from one another "qualitatively at more or less indistinct and arbitrary vertical boundaries on the basis of generalized lithologic differences" (Weller, 1960, p. 521). Similarly, adjacent lithofacies may be distinguished from one another statistically, and at some point an arbitrary vertical cutoff plane (cf. Wheeler and Mallory, 1953) between the two laterally adjacent lithofacies established. For example, the boundary between two adjacent lithofacies might be determined on the basis of some statistic involving the relative percentages of shale, sandstone, and limestone in the rocks. Finally Weller (1960) suggests that laterally adjacent lithofacies can be separated by intertonguing contacts or boundaries. Given two lithofacies whose boundary is defined in terms of intertonguing, in the local area of lateral contact between the two lithofacies one might observe in a particular section a vertical sequence composed of alternating manifestations of the two lithofacies.

As should be evident from the above discussion, as Weller (1960) applies the term lithofacies it is more or less a stratigraphic unit; furthermore Weller (1960) considers many stratigraphic units to be representative of facies. "Insofar as a formation is a lithologic unit it is a more or less broadly conceived facies" (Weller, 1960, p. 521). Weller has used the term "stratigraphic facies" to refer to facies that are stratigraphically important, suggesting that the major conventional types of stratigraphic units that are related to vertical sequences (lithostratigraphic, biostratigraphic, and chronostratigraphic [time-stratigraphic] units) are paralleled by equivalent stratigraphic facies (lithologic or lithofacies, biologic or biofacies, and temporal facies) that express lateral relations of rocks. Weller, however, failed to define or elaborate upon the notion of temporal facies; perhaps it is the concept that a time-transgressive stratigraphic unit conceivably could be divided into lateral facies distinguished on the basis of their differing ages. Furthermore it should be pointed out that Weller (1960, pp. 508–509) believed that whereas most lithostratigraphic and biostratigraphic units are "somewhat generalized facies," chronostratigraphic units are not because they "are independent of all material considerations." Yet the rocks representing a chronostratigraphic unit represent various facies. "Both vertical and horizontal variability in sedimentary rocks are aspects of stratigraphy recording differences that are related in time and in space respectively. Both kinds of variability can be expressed either in terms of conventional stratigraphic units or in terms of facies.

The most practical facies also are stratigraphic units and they do not differ importantly from the more familiar rock stratigraphic units except as their mutual relations are emphasized. Thus ordinary stratigraphic units are considered principally in vertical sequence whereas the lateral relations of facies are primarily important" (Weller, 1960, p. 507). In this connection one may note that for most formal stratigraphic units, as outlined in the *North American Stratigraphic Code* (NACSN, 1983) and the *International Stratigraphic Guide* (Hedberg, 1976), the emphasis traditionally has been placed on the definition of upper and lower boundaries and not on the precise definition of the lateral extent of such units; indeed a stratigrapher might extend a particular conventional stratigraphic unit laterally as far as it was deemed appropriate (as long as there was a basic uniformity in defining characteristics and perhaps a continuity of rock—of course, chronostratigraphic units are ideally of global application and have no lateral boundaries). As Weller (1960) defined certain stratigraphic facies (such as lithofacies, see below), the upper and lower boundaries of a particular facies were already set by being coterminous with the vertical boundaries of the stratigraphic unit of which the facies formed a part; in essence a facies might be viewed as a way to deal with lateral boundaries of stratigraphic (rock) bodies.

While Weller preferred to restrict the concept of lithofacies to a lateral subdivision of some stratigraphic unit, he also recognized general stratigraphic facies (and such facies might even be termed lithologic facies or lithofacies) whose boundaries were not necessarily congruent with any of the boundaries of any other (formally or informally) recognized stratigraphic units. Furthermore, a facies might be composed of two or more conventional stratigraphic units. Finally, as is demostrated by the following quotation, Weller found it acceptable to delimit certain boundaries of a facies wholly arbitrarily within a uniform rock body—a notion I would reject.

> Stratigraphic facies need not correspond, however, with any formally recognized stratigraphic unit. For example, a stratigraphic unit may be subdivided into successive vertical parts, each of which is differentiated separately into facies. These parts may be determined by any features that are distinctive. If no such features permit consistent subdivision, the unit may be sliced into horizontal parts [each part being termed a slice; the upper and lower boundaries of a facies within a slice would correspond to the upper and lower boundaries of the slice itself, which are defined totally arbitrarily], each of which constitutes some arbitrary fraction of the whole unit's thickness. More general [facies] units also may be recognized for operational purposes which combine several formations or even transcend systems. Altogether great latitude is possible in the selection of stratigraphic units that are to provide the basis for stratigraphic facies differentiation. The units chosen are likely to be determined equally by the limitations of practical stratigraphy and by the purposes motivating a particular facies study. (Weller, 1960, p. 508)

One might add that choice of a particular facies unit may be only as valid or as useful as any other stratigraphic units upon which it may be based.

Weller (1960) considered biologic facies and biofacies, and tectonic facies and tectofacies, to be roughly analogous to lithologic facies and lithofacies. Indeed, Weller (1960) considered tectofacies to be merely a sort of lithofacies based on tectonic features, or at least the interpretation of such presumed features. Biofacies are distinguished on the basis of biological features of rocks and may grade into lithofacies as some rocks are composed of predominantly organic remains or organismal productions. However, Weller acknolwedges that one may also define or distinguish biologic facies solely on the basis of organismal content without even including the rocks enclosing the organisms within the concept of the biologic facies. Biofacies or biologic facies may be distinguished on the basis of ecological, taxonomic, or other differences between organisms. When recognized on the basis of taxonomic differences, such facies may approach conventional biostratigraphic units, but again it is the lateral relationships between biofacies that are emphasized as opposed to the vertical relationships among conventional biostratigraphic units.

Weller (1960) also distinguished as distinct types of facies, of importance to the stratigrapher, what he termed structural facies, environmental facies, genetic facies, and geographic facies. All of these types of facies are fundamentally interpretations of basic lithologic and/or biologic facies. By structural facies Weller referred primarily to structures, such as reefs and other bioherms, that may be found within the stratigraphic record and may be considered a type of facies because such structures are distinct from the surrounding rock. Within Weller's concept of structural facies might be included buried cinder cones and so on. Geographic facies are facies that are identified with, or associated with, some geographic area, in most instances a paleogeographic subdivision or feature. Thus a facies may be identified on the basis of an ancient seaway, but as Weller points out, such a facies probably would be recognized by a combination of structure and lithology. Environmental facies usually involve the interpretation of lithologic or biologic facies in terms of the particular depositional setting or environment prevailing at the time and place of the formation of the rock record under consideration. Thus a particular sandstone may be distinguished as a lithofacies on the basis of its lithology, or it can be interpreted as representing a shallow-water, near-shore marine environment and thus form the basis of an environmental facies. Genetic facies are based on interpretations as to the mode or process of origin of a particular deposit; thus one might interpret a certain rock body to represent a storm-deposit facies or a turbidity-current facies.

Dealing with lithologic facies concepts, Weller (1960, p. 514) proposed a hierarchical grouping scheme of various facies units. In Weller's terminology a lithotope, the smallest (least inclusive) unit of his lithologic facies hierarchy, is an area characterized by a uniform environment of deposition (Wells, 1947; see also Krumbein and Sloss, 1963). As an area a lithotope is a two-dimensional concept; it lacks a third (vertical) dimension. The body of rock formed in an area of a lithotope is a lithostrome (Wheeler and Mallory, 1956). A lithostrome is the record of a lithotope and is usually manifested as a bed (or beds), layer, or stratum of relatively uniform (including heterogeneously uniform) character. Some authors have used these terms in a different manner, however. To cite one example, Dunbar and Rodgers (1957, p. 137) considered a local assemblage of animals and plants to be a facies of the larger biota of the region, the environment under which the assemblage lived to be its biotope, and the record in the rocks of a particular biotope to be a lithotope. Thus the lithotope of Dunbar and Rodgers (1957) is essentially equivalent to the lithostrome of Weller (1960), and the biotope of Dunbar and Rodgers might be considered a type of lithotope by Weller.

A lithosome, the category next higher (more inclusive) in Weller's (1960) scheme, is a particular body of rock characterized by specific lithologic attributes, and is usually considered the rock record of a particular sedimentary environment in a certain region over a given interval of time (see also Wheeler and Mallory, 1956). A lithosome may be composed of a series of lithostromes. The lateral boundaries of a lithosome usually are considered to intertongue with adjacent facies; the boundaries of a lithosome are not defined arbitrarily or statistically (see above). As generally used, the term lithosome simply is applied to any three-dimensional rock mass or body characterized by a uniform lithology (see Krumbein and Sloss, 1963). Parallel to the terms lithosome and lithostrome, one may refer to biosomes and biostromes for three-dimensional rock bodies and single beds, respectively, characterized by uniform biologic content.

Weller (1960) proposes that the term lithofacies may be utilized as a term for a facies unit potentially composed of several lithosomes. As discussed above, for Weller a lithofacies is a lateral subdivision of a conventional stratigraphic unit; therefore, Weller regards a lithofacies as essentially a stratigraphic unit. Other authors, however, have used the term lithofacies in a more general sense, simply for a particular kind of rock (specified individually or collectively) without specific regard for the relationship of the particular lithofacies under consideration to other lithofacies.

Next in order of increasing generality in Weller's (1960, p. 514) hierarchy of facies concepts is what Weller labeled "Facies 2," defined as a "body of related rocks not defined with respect to others." It appears to me that we might interpret Weller's Facies 2 as being composed of a group or set of contiguous or adjacent (vertically and/or horizontally) lithosomes that in sum form a single body of rock. Analogously, among conventional stratigraphic units a group is composed of a series of contiguous and adjacent formations. Finally, at the top of his facies hierarchy, Weller (1960, p. 514) distinguished a "Facies 1," defined as a "general association of certain related types of rocks." It seems to me that here Weller may have intended to encompass "all rocks of some particular kind without reference to their form, age, or geographic occurrence" (Weller, 1960, p. 520). To use Weller's examples, all red beds can be thought of as representing a red-bed facies, and all black shales represent a black-shale facies. In fact, using this concept of Facies 1, one might even suggest that there is an ideal of a red-bed facies (to use an example) that is approximated in nature by various particular lithosomes and lithofacies. Thus Weller's Facies 1 concept lends itself to the formation of generalized facies models and environmental interpretations intertwined with such models (see Walker, 1984a, and articles therein).

Weller (1960) also proposed several other (primarily nonhierarchical) schemes of classification of facies concepts. Weller noted that facies are often distinguished on the basis of particular elements, characteristics, or aspects that are considered of prime importance. As discussed above, various authors have distinguished facies based on lithologic, biologic, and tectonic aspects (roughly corresponding to lithofacies, biofacies, and tectofacies, respectively); thus in Weller's (1960, p. 515) terminology, facies may be classified by their "facies aspect." Likewise, facies may be distinguished from one another by their three-dimensional forms and their mutual relationships in space; Weller uses these criteria to distinguish facies classes. As an example of facies classes, for lithologic facies one may recognize lithofacies, lithosomes, lithostromes, and lithotopes (see below; here it should be pointed out that when used to distinguish facies classes by Weller, these terms may have slightly different meanings from their connotations when used by Weller in establishing a facies hierarchy [discussed above]).

Treating facies as stratigraphic units (and likewise treating certain conventional stratigraphic units as facies) and utilizing the concepts of facies aspects and facies classes, Weller (1960) devised a classification of facies wherein formations, lithofacies, and lithosomes are all considered comparable units, but differ from one another in their three-dimensional forms and relationships to one another in space. In this classificatory scheme within the facies aspect of lithology, delimiting the general class of rock stratigraphic (lithostratigraphic) units, Weller (1960, p. 516) distinguished formations as "vertically successive stratigraphic units," lithosomes as "laterally intertongued rock bodies," lithofacies as "laterally equivalent, statistically differentiated rock bodies," and lithostromes as "layers of uniform character." Using the facies aspect of biologic content to delimit biostratigraphic units, Weller (1960) recognized the following units corresponding to the four types of lithologic facies units listed above: zones, biosomes, biofacies, and biostromes. In this classification Weller recognized stages as the only named time-rock stratigraphic (that is, chronostratigraphic) units/facies; stages were considered comparable to formations and zones.

Treating facies as interpretive environmental units, Weller (1960, p. 516) distinguished physical, biologic, and structural facies, each designated by the appropriate prefix (litho-, bio-, and tecto-, respectively). An area of uniform sediment deposition or rock formation (reflecting a uniform environment) was dubbed a lithotope, biotope, or tectotope. The various "-topes" are not volumes of rock but two-dimensional surfaces or areas. A body of rock in three dimensions, that is, the physical record of a -tope, was called by Weller (1960) a lithosome, biosome, or tectosome. Finally, "vertically bounded lateral parts (facies) of stratigraphic units" were termed lithofacies, biofacies, and tectofacies by Weller (1960, p. 516).

Still not entirely satisfied with any of the classification schemes for facies described above, Weller (1960, p. 517) proposed that there is a more fundamental division of facies into three types. Weller's "Type I" facies, which he also termed petrographic facies, are facies differentiated on the basis of appearance or composition (that is, petrography—including any fossils that can be regarded as part of the rock constituents), irrespective of the form, boundaries, or relationship of the facies (rock body) to any other facies. Petrographic facies usually are represented by material rock bodies, but they are not necessarily stratigraphic units. Rather, they are generalized facies and even may be considered to be ideal facies models. Within his concept of petrographic facies, Weller distinguishes two "classes": generalized facies consisting of all rocks of a certain kind (matching a certain description) and facies consisting of actual bodies of rock, but of unspecified or indefinite form, extent, and mutual relations. As a facies that falls into this second class of petrographic facies becomes better known (perhaps the particular material rock body is carefully mapped and analyzed stratigraphically), it may become an example of Weller's next fundamental type of facies, stratigraphic facies.

Weller's (1960) "Type II" facies, stratigraphic facies, are considered to be stratigraphic bodies of various forms. Stratigraphic facies are material rock bodies that are distinguished by characteristic compositions and differentiated from one another on the basis of their three-dimensional forms, their mutual boundaries, and their stratigraphic relationships to one another. Within the concept of stratigraphic facies, Weller distinguishes three classes. Weller's first class is composed of units that correspond more or less to conventional stratigraphic units such as formations and zones. These stratigraphic facies are distinguished primarily in vertical successions, the boundaries between adjacent facies are horizontal planes, and the lateral relationships between such facies often are left unspecified. Weller's second class of stratigraphic facies are those that are part of a stratigraphic unit; they laterally intergrade into other parts of the same unit, and the lateral boundaries of such facies are vertical cutoff planes, the position of which is often defined arbitrarily or statistically. Finally, Weller distinguishes a class of stratigraphic facies that again are usually parts of conventional stratigraphic units, but have irregular lateral boundaries that intertongue with laterally adjacent facies. Two such laterally adjacent facies often may be characterized by sharply contrasting characteristics. As Weller (1960) noted, however, the three classes of stratigraphic facies that he distinguished will, in nature, often grade into one another. All conventional stratigraphic units (that is, material units, such as lithostratigraphic units), even if their upper and lower boundaries are emphasized in their definition and delimitation, must end laterally at some point. Some facies may have lateral boundaries that differ in different areas. A single stratigraphic facies may have a lateral boundary that is defined by arbitrary cutoff in one direction or area, as well as a lateral boundary in another direction or area that forms an intertonguing relationship with a second stratigraphic facies.

"Type III" facies of Weller's classification are not material stratigraphic bodies, but observations and interpretations that are concerned predominantly with reconstructing the environments represented or recorded by material rock bodies. The environments under consideration here may be of any nature—lithologic (particularly sedimentologic), biologic, tectonic, and so on. Type III facies appear to approximate the original nineteenth-century concepts of facies as being the aspect of appearance (the face) of a particular rock body rather than the material rock body itself. Weller (1960) did not further subdivide his Type III facies.

In some final comments on facies nomenclature, Weller (1960) suggests that in most cases there is no need for such terms as lithosome and lithostrome—this despite his long discussion of these terms. Instead of speaking of a lithosome, one may refer to an intertongued lithofacies. According to Weller, words such as layer, stratum, bed, or member are adequate as replacements for lithostrome. In concluding his discussion of facies concepts, classification, and nomenclature, Weller (1960, p. 524) notes that, at least as he uses the term, lithofacies has a specific meaning (or meanings, depending on the context) and should not be used as a general synonym for lithologic character.

In contrast to Weller, many investigators have found the concept of the lithosome, in particular, extremely useful. As described by Boggs (1987) and Krumbein and Sloss (1963), one can picture a lithosome as a single continuous body of rock of uniform lithology that would be clearly evident if one could dissolve away, or otherwise remove, all other rock bodies surrounding it (Fig. 3.5). In most cases such a rock body probably would appear as a roughly tabular mass, but the lateral boundaries in particular might be very intricately shaped. In this sense, a lithosome may represent only one part of a heterogeneous formation, or a lithosome, particularly if of complex geometry, may itself be divided into a number of formations or other stratigraphic units.

Many descriptors have been applied to the gross shapes of lithosomes (rock bodies), some of which are purely descriptive (for example, sheet, lens, blanket, wedge, and so on) while others carry genetic implications (for example, reef, channel, bar, and so on). Rock bodies, lithosomes, are most often of generally tabular shape, and Krynine (1948) proposed a useful descriptive classification of the shapes of such

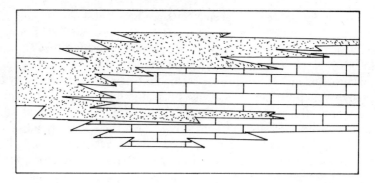

Figure 3.5. Two-dimensional diagrammatic representation of two interfingering lithosomes, a sandstone lithosome (stippled) and a limestone lithosome (rectangles). (After Krumbein and Sloss, 1963, p. 302.)

lithosomes on the basis of their width-to-thickness ratios (see also Krumbein and Sloss, 1963). According to Krynine's classification (Fig. 3.6), a blanket has a width-to-thickness ratio of greater than 1000 to 1, a tabular-shaped lithosome has a corresponding ratio between 50 to 1 and 1000 to 1, a prism has a ratio in the range between 5 to 1 and 50 to 1, and a shoestring has a ratio of less than 5 to 1. Of course, both the width and thickness of a lithosome may vary at different points along its length and width, respectively; the above classificatory scheme is only approximate and may be based on either maximum or average widths and thicknesses.

Given the concept of lithosomes, one also can discuss the vertical and lateral relationships among adjacent lithosomes (Krumbein and Sloss, 1963). The vertical relationships among lithosomes involve the same concepts as among conventional stratigraphic units (conformity or unconformity; see above). As discussed briefly above in terms of facies (particularly lithofacies), laterally lithosomes may end abruptly, pinch-out, intertongue, or grade into a distinct lithosome. A sedimentary lithosome may end abruptly if, for instance, it consists of a rock type that was deposited in a basin with very steep (approaching vertical) confining walls, or if the preserved lithosome is terminated laterally by some means such as a sharp erosional contact, a fault, and so forth. A lithosome composed of igneous rock, such as a granitic pluton, may end abruptly where it contacts the country rock into which it intruded.

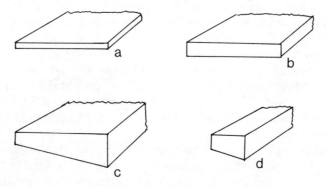

Figure 3.6. The shapes of lithosomes as described by Krynine (1948): (a) blanket, (b) tabular, (c) prism, (d) shoestring. (After Krumbein and Sloss, 1963, p. 303.)

A lithosome or stratigraphic unit is said to pinch-out if, when traced laterally, it progressively thins to a feather edge and thence finally to a point where it is no longer distinguishable (the thickness is zero). In pinching-out, the lithosome simply may thin to zero without being replaced by another lithosome (thus the local stratigraphic sections decrease in thickness laterally); or as one lithosome thins, another vertically adjacent lithosome may thicken (that is, pinch-out in the opposite direction) to replace it. Intertonguing refers to the situation where, along its lateral boundaries, a lithosome splits into many thin units, each of which progressively thins and eventually pinches-out. In the region of intertonguing between two laterally adjacent lithosomes, one will observe in the local stratigraphic sections successive alterations in the vertical dimension between tongues of the two (or more) lithosomes involved. In lateral gradation, a lithosome of a distinctive rock type, or with distinctive characteristics, gradually is replaced by a different rock type or by different distinctive characteristics when traced laterally. To give a typical example, a sandstone may grade laterally into a shale—and this may occur on a scale of hundreds of meters or even several kilometers. Any lateral boundary between two such lithosomes may have to be established arbitrarily or statistically. Krumbein and Sloss (1963, p. 310) note that such lateral gradation between lithosomes may be either mixed or continuous (as is also true of vertical gradational boundaries between adjacent lithosomes or stratigraphic units). In mixed gradation two distinct types of rock or rock clasts (sediment clasts) slowly mix with each other to form a gradation between end members. An example of a mixed gradation would exist where a lithosome of pure shale gradually grades into a lithosome of pure sand by the admixture of increasing amounts of sand grains with the clay grains of the shale. In contrast, continuous gradation involves the progressive and gradual change in a rock type or the sedimentary clasts composing a rock from one end member to the next. An example of such continuous gradation would be the case where a lithosome of pure shale grades into a lithosome of pure sand by a progressive increase in the average grain size of the particles composing the lithosomes (that is, the grains range continuously along a lateral direction from clay size, to silt size, to sand size).

Facies Sequences, Facies Models, and Environmental Stratigraphy

Laporte (1979; see also Newton and Laporte, in press) has defined environmental stratigraphy as the study and reconstruction of ancient environments utilizing techniques and information from sedimentology and paleoecology. Facies as environmental indicators are of prime importance in such endeavors: "The recognition and interpretation of lithofacies and biofacies is the crucial goal of environmental stratigraphy" (Laporte, 1979, p. 38). Major techniques used by such environmental stratigraphers involve the analysis of vertical sequences of facies in the field and laboratory and the construction and refinement of facies models (see papers in Walker, 1984a).

When observed in the field, facies, like any other stratigraphic units, succeed one another in vertical sequences in local sections. After careful observation and thorough familiarization with the rocks in a local area, the field stratigrapher will subdivide the rocks into working facies units (units characterized by distinctive lithologies) and record the pertinent details of the sequences of such units in the various local sections that are available. The degree of preliminary subdivision carried out in the field will depend on various factors, such as objectives of the particular study and length of time available to carry it out; but as Walker (1984b) has stated, if one is in doubt as to how detailed a facies subdivision should be, it is better to over-subdivide. Over-subdivided facies always can be recombined in the laboratory, whereas it usually is impossible to refine a coarse field subdivision in the laboratory. Initial subdivision can be made on the basis of qualitative judgments on the part of the stratigrapher observing the rocks, or in some cases statistical methods can be applied to various quantitative parameters in order to distinguish facies from one another.

In the field it is often the case that a single rock layer, a single lithology, or a presumed single facies may have an ambiguous environmental interpretation. Based on a single rock type, one may not be able to interpret the environment of deposition correctly; thus, virtually identical sandstones may be formed in numerous subenvironments in both the freshwater (for instance, fluvial) and marine realms. In such cases (which probably include the vast marjority actually encountered in nature) Walther's law as applied to the vertical successions of facies is of utmost importance to environmental interpretation. Although we may not be able unambiguously to interpret any single facies, lithology, or stratum, given a genetically related sequence of such facies the interpretation may be much clearer. To quote Walker (1984b, pp. 2–3): "Many facies defined in the field may at first suggest no interpretation at all. The key to interpretation is to analyze all of the facies communally, in context. The sequence in which they occur thus contributes as much information as the facies themselves."

When analyzing sequences of facies in terms of Walther's law, one must pay particular attention to the horizontal contacts between vertically successive facies. As noted above (see section above), the proper application of Walther's law requires that there be no major hiatuses or unconformities among successive sequences, or if there are any unconformities, that these at least be accounted for. Consequently in the field it is important to note the nature of the boundaries separating vertically successive facies; gradational contacts or transitions between facies usually are taken to imply that the two facies represent environments that were laterally adjacent to one another, whereas in the case of sharp contacts (including contacts bearing evidence of erosion) one cannot be certain if the facies represent environments that were laterally adjacent or geographically separated from one another. Sharp breaks between facies sometimes are interpreted as indicating that fundamental changes in the depositional environment took place (see Walker, 1984b, and references cited therein).

In the field, stratigraphic sections may show what appear to be a number of cycles, repetitive sequences, or packages of facies. Environmental stratigraphers often have found it useful to quantitatively document the number of transitions between the various recognized facies in a local section or area (noting also the nature of the boundaries between the facies) and record or summarize such information in weblike facies relationship diagrams that record the interrelationships and transitions among facies (see examples in Walker, 1984a). From such information one can attempt to derive a summary or composite sequence of the typical facies, and the order in which they occur, in a particular stratigraphic section or local area of investigation. The development of accurate summary diagrams of facies transitions is not always an easy matter, however. Each actual sequence may differ from the others in particular details; not all facies transitions may necessarily occur in all of the sequences in the field, and one must be concerned with matters such as the probability that a particular facies always will be followed successively by a second identified facies. One must consider the possibility that some transitions between successive facies are essentially random (see especially Harper, 1984; Walker, 1984b; and articles cited in these papers). Once a summary sequence is hypothesized, then individual sequences or cycles can be compared to the summary sequence.

Summary sequences, and supplementary information, from both ancient and modern examples can be synthesized into what have been termed facies models. Walker (1984b, p. 5) has defined a facies model as "a general summary of a specific sedimentary environment" (it seems that essentially the same definition could also be applied to nonsedimentary environments in which rocks form). Such facies models may be expressed in numerous different ways, such as in the form of idealized sequences of vertically successive facies, idealized stratigraphic cross sections through particular environments, diagrammatic three-dimensional block diagrams showing the typical environment and the various facies (rock types) that might form within it and their mutual relations, or various graphs and equations of parameters important in recognizing various facies. Such facies models may be extremely broad and of large scale, such as a general deltaic model, or they may be very small scale and narrow, such as a facies model for a river-dominated, birdsfoot-type delta (Miall, 1984c).

Walker (1984b) suggests that any good facies model should be a distillation of a number of local examples (the model takes into account the similarities among the cases studied, and local variability is generally eliminated from the final model), and there should be continued two-way feedback between the model and further observations and studies made in the field. Once established, a facies model fulfills four main functions, according to Walker (1984b): (1) it acts as a norm or standard for purposes of comparing actual local examples to it, and also for comparison to other facies models of similar but nonidentical environments; (2) it acts as a framework and guide in which future observations can be integrated (as Walker notes, however, one must not allow a particular preset notion of an environment and facies model to cause one to ignore evidence in the field that is not suggested by the model); (3) the model itself is also a hypothesis from which predictions can be derived, particularly with regard to new local situations (on confronting a new geological terrain, one might hypothesize that the rocks were formed in a deltaic setting and thus use deltaic facies models to make predictions about the nature and relationships of the sequences of rocks, predictions that subsequently can be tested); and (4) the model is a scientific interpretation of the environmental conditions that it is presumed to represent.

CORRELATION IN STRATIGRAPHY

The concept of correlation is fundamental, or even central, to stratigraphy as a science. Without the procedure or process of correlation, there could be no synthesis of stratigraphic deposits that are geographically (or otherwise) separated. If stratigraphers could not correlate stratigraphic units, their discipline might consist of little more than the description of innumerable bodies of rock that bore little or no apparent relationship to one another—classic "stamp-collecting" as used in a pejorative sense. Yet in spite of the overriding importance of correlation to stratigraphy, or perhaps because of its importance, the very meaning of correlation, as applied to stratigraphy, is ambiguous and open to debate. In some respects the manner in which a particular author treats correlation may be indicative of the essence of the author's basic stratigraphic philosophy.

The *International Stratigraphic Guide* (Hedberg, 1976, p. 14) states that "to correlate, in a stratigraphic sense, is to show correspondence in character and in stratigraphic position." The *Guide* does not specify exactly what sorts of entities are to be correlated; presumably stratigraphic uints, horizons, surfaces, unconformities, or simply dimensionless points within rock bodies could be correlated to one another. The *North American Stratigraphic Code* (NACSN, 1983, p. 851, italics in the original) states that "*correlation* is a procedure for demonstrating correspondence between geographically separated parts of a geologic unit. . . . *Correlation* is used in this Code as the demonstration of correspondence between two geologic units both in some defined property and [in] relative stratigraphic position." Note that it is stratigraphic units, either formal or informal, that according to the *Code* are correlated. Taking the *Code* at face value, unless unconformities, horizons, and so on, are considered to be stratigraphic units, such entities are not said to be correlated to one another, although as they pertain to stratigraphic units, they may be involved in establishing correlations. It is unclear from the quotation given above if the *Code* intends differing parts of the same stratigraphic unit to be correlatable to each other, or if correlation is intended to refer primarily to the establishment of some sort of correspondence between different geologic units (such as between different formally named formations in different basins of deposition). Furthermore, the *Code* (NACSN, 1983, p. 851) states that "the term 'correlation' frequently is misused to express the idea that a unit has been identified or recognized." This last statement might be interpreted as implying that when geographically separated parts of the same unit (for example, a formation) are recognized as belonging to the same unit, this is not considered correlation, yet the *Code* (see above) also explicitly defines correlation as "demonstrating correspondence between geographically separated parts of a geologic unit." In practice, many stratigraphers utilize the term correlation simply in the sense of indicating any correspondence or equivalence of stratigraphic units or stratigraphic features (horizons, points in rocks, and so on), whether the same or different stratigraphic units are involved.

After presenting basic definitions of correlation, both the *Code* and the *Guide* proceed to state that there are different types of correlation, depending on the characters, features, or properties used or emphasized in the correlation. Three major examples of types of correlation provided by the guide are: lithologic correlation (lithocorrelation), based on the similarity or correspondence of lithology and lithostratigraphic position; biostratigraphic correlation (biocorrelation), based on similarity or correspondence of fossil content and biostratigraphic position; and chronostratigraphic (temporal) correlation (chronocorrelation; = chronotaxis, Harland, 1978), based on correspondence in age and chronostratigraphic position.

Among practicing stratigraphers, however, there has not been universal agreement with the broad concept of correlation, including under its aegis several very different ideas labeled types of correlation, as espoused by the *Guide* and the *Code*. This is particularly true of the older stratigraphic literature, but the disagreements remain today even if they are now expressed using a different terminology. The occurrence of these disagreements should be understandable when one considers that theoretically lithocorrelation and chronocorrelation, for example, have very different purposes and may produce very different results, yet in practice one strati-

grapher's lithocorrelations or biocorrelations subsequently may be innocently interpreted by the same stratigrapher or another worker as chronocorrelations. Even
today the concept that "correlation is correlation" seems to pervade some stratigraphic work, and subtle distinctions between the information that formed the basis
of a specific correlation and the type of interpretations subsequently placed on the
"correlation" may be extremely elusive.

A few examples of the way that the concept of correlation has been used are as
follows: "Stratigraphic correlation is the demonstration of equivalency of stratigraphic units" (Krumbein and Sloss, 1963, p. 332; this concept of correlation is
extremely inclusive and recognizes different bases for correlation as described in the
Code and the *Guide*). "Correlation is the process of determining mutual relations.
In stratigraphy this term commonly is used with a much restricted meaning and
refers principally to the establishment of equivalent relations with respect to time"
(Weller, 1960, p. 540). "Two units, belonging to different local sections, are said to
be correlative if they are judged to be time equivalents of each other, and *correlation*
is the process by which stratigraphers attempt to determine the mutual time relations
of local sections. Thus correlation is concerned with the synthesis of the data of
established local sections into a composite time scheme applicable to a whole region" (Dunbar and Rodgers, 1957, p. 271, italics in the original). Likewise, Rodgers
(1959, p. 690, italics in the original) has argued that "in stratigraphy the term *correlation* should and in fact ordinarily does mean the attempt to determine time relations among strata, however they may be divided into stratigraphic units" (note that
in Rodgers's conception it is not stratigraphic units but strata that are correlated
[cf. the *Code*—of course, "strata" could be considered the smallest stratigraphic
units]). "*Correlation* in stratigraphy should always imply time-significance" (Storey
and Patterson, 1959, p. 712, italics in the original). "Rock correlations or electric
log correlations or heavy-mineral correlations are all as much forms of correlation
as is any form of time correlation. It would appear to be simpler just to state the
kind of correlation one means, e.g., 'rock correlation' or 'time correlation'"
(A. B. Shaw, 1964, p. 82). "Correlation in a geochronologic context is understood as
time correlation. However, it is worth making a clear distinction between correlation
which is intended to be chronologic (i.e., *chronotaxis* of Weller, 1960) and correlation of other characters (for example the *homotaxis* of Huxley, 1862). . . . The word
correlation [that is, chronocorrelation in the context or Harland's paper] refers to
the activity of correlation with the objective of making some positive chronotaxial
statement. It does not imply any particular degree of success" (Harland, 1978, pp.
16–17, italics in the original). "The process of correlation is the determination of
geometric relationships between rocks, fossils, or sequences of geologic data for
interpretation and inclusion in facies models, paleogeographic reconstructions, or
structural models" (B. R. Shaw, 1982, p. 7).

As should be apparent from the representative quotations given above, there are
at least two major issues in the debate over the concept of correlation. First is the
question of the nature of correlation: does correlation refer primarily or solely to
equivalency in time, or are there many kinds or types of correlation, or is correlation
primarily a working out or determination of the geometric relationships of material
rock bodies (cf. B. R. Shaw and the opinions of some industrial geologists)? Some
stratigraphers do not even accept the concept of chronocorrelation (time correlation) per se, independent of some other form of correlation (such as biocorrelation—of course, there always must be some physical basis for determining time rela-

tions). This viewpoint is expressed in the following statement by O. H. Schindewolf (printed in Hedberg, 1961, p. 31): "I am not able to adopt a distinction between biostratigraphic and chronostratigraphic units. As far as I can see biostratigraphy and chronostratigraphy are the same thing or merely two slightly different versions of the same thing. The methods in both cases are exactly the same." The second issue is: does correlation imply or require the use of stratigraphic units?

Storey and Patterson (1959) suggest that as early as the late 1790s William Smith demonstrated the equivalency or continuity of particular strata laterally, despite lithologic (facies) changes, primarily on the basis of faunal continuity (see Chapter 1). This might be considered the beginning of the concept of correlation in modern stratigraphy. "It was clear to Smith that strata are objective observable time-significant features which continue universally irrespective of lateral lithologic changes within them (intrastratal) except where they are absent through erosion or non-deposition" (Storey and Patterson, 1959, p. 708). Further developments and refinements in correlation, again particularly using paleontological data, were carried out by D'Orbigny, Oppel, and other workers of the early and middle nineteenth century (Storey and Patterson, 1959; see Chapter 6).

In a historical review of the origins and usage of the term correlation in stratigraphy, Rodgers (1959) notes that the actual word "correlation" may have been used as early as 1849 by R. I. Murchison, but it appears to have been uncommon in the literature until about 1880. In the late 1880s and early 1890s the word was used by J. W. Powell (director of the U. S. Geological Survey) in reference to the time relations of formations. In the 1890s to the 1910s the term correlation commonly (and perhaps universally?) had the connotation of time correlation (Rodgers, 1959). Where they mention the concept of correlation, major textbooks of the time (e.g., Chamberlin and Salisbury, 1909; Pirsson and Schuchert, 1915) refer to it in a temporal sense. In the years around 1888–90, according to Rodgers (1959), Powell clearly distinguished between conceptual geologic time units (equivalent to a geochronologic scale of modern usage) and actual rock formations (primarily cartographic units; that is, the formations could be mapped) that were correlated to the geologic time units. However, this distinction was lost in the early twentieth century, particularly because of the influence of E. O. Ulrich. "Ulrich consistently built up the time-stratigraphic framework of his [correlation] charts out of formations as building blocks, and he and his followers tended to consider the formations themselves as time-defined, mutually exclusive units. Thus stratigraphic correlation came to include the interrelation of formations, defined by time, with each other as well as with an abstract time sequence" (Rodgers, 1959, p. 685). Whereas originally one correlated stratigraphic units to an ideal (abstract and conceptual) time sequence, subsequently stratigraphers came to correlate stratigraphic units directly with one another and in the process came to think of the stratigraphic units as themselves representative of (or equivalent to) time units.

Beginning in the 1920s, in large part as a result of the needs of the petroleum industry, stratigraphic work rapidly expanded, and by the 1930s and 1940s stratigraphers again began to distinguish explicitly between stratigraphic units based primarily on lithologies and stratigraphic units based on time (Rodgers, 1959). However, the term correlation was now applied in two different ways: (1) to the determination of lithologic correspondence or equivalency of strata or stratigraphic units to one another, and (2) to the determination of temporal correspondence or equivalency of strata or stratigraphic units either to each other or to an ideal time scale. This, according to Rodgers (1959), is a thumbnail sketch of the origin of the debate of the meaning of correlation in stratigraphy.

Wheeler (1959) points out that by the late 1940s (Hedberg, 1948, 1958), various workers had begun to note that whereas various sources of evidence might be extremely useful in temporal correlations, these sources of evidence do not necessarily guarantee accurate or precise temporal correlations. Thus in many instances fossils may be considered the best indicators of temporal correlation, but biostratigraphic units only approach ideal chronostratigraphic units (isochronous units with synchronous boundaries), and biostratigraphic correlation only approaches ideal temporal correlation. Thus arose a presumed distinction between objective and subjective stratigraphic units. For example, the American Stratigraphic Commission (1957; quoted from Wheeler, 1959, p. 697, with Wheeler's notations in the brackets) considered physically based stratigraphic units, such as biostratigraphic units and lithostratigraphic units, to be "dominantly objective units [in the sense that they are] based largely on observation and measurement [and that chronostratigraphic units belong in a category of] dominantly subjective units based largely on interpretation." (Note the apparent implicit belief that we can arrive at stratigraphic units and classification that are interpretation-free. One can argue first that it is unclear if this ever would be possible, and second that a theory-free classification would be of questionable scientific relevance [cf. Schoch, 1986b, on theory in scientific classification].) Likewise, not all correlation was considered equivalent. Certain types of correlation were relatively "objective," as, for example, when exactly the same fossils were recognized and correlated with one another, or identical lithologies were traced laterally. However, it might be considered somewhat more "subjective" to interpret features of strata in terms of temporal correlation.

In part the above line of thinking can be related to the distinction between normative and positive aspects of stratigraphic classification discussed in a previous section in this chapter. The activity of lithostratigraphic classification and correlation, for example, contains both positive and normative aspects (Harland, 1978). In establishing an initial lithostratigraphic classification, or revising an existing stratigraphic classification, the author(s) is the final authority in making normative judgments, which are correct by definition (until revised). Once a particular lithostratigraphic classification is adopted, however, there may be aspects of positive science in lithocorrelating rocks of other sections to the type sections of the units composing the lithostratigraphic classification. Many would suggest that chronocorrelation, in contrast to lithostratigraphic classification and correlation, has a much higher quotient of uncertainty (or "subjectivity"), in that we never can be absolutely certain that our statements (hypotheses) concerning the temporal correlations of rocks, stratigraphic units, and features are absolutely true. Yet uncertainty is the hallmark of positive science (cf. Harland, 1978; Popper, 1959; Weller, 1960). We may increase our precision and reduce our uncertainty (sometimes to virtual certainty), or quantify the uncertainty with probabilistic statements; however, if we eliminate all uncertainty, we no longer have a science but a dogma.

In order to justify their traditional (some would view it as reactionary) view on the meaning of correlation, Dunbar and Rodgers (1957, p. 271) cite the example of a time-transgressive lithostratigraphic unit that is time-transgressive (for example, the depositional system forming the unit migrated geographically with time) such that in some areas the unit is on the order of one or more ages younger than the same unit in other areas. Dunbar and Rodgers suggest that to state that various parts of such a unit are simply correlative obscures both the temporal and the stratigraphic relationships among the parts of the unit. Even when it is qualified, such as by stating that one is proposing a rock or rock-unit correlation (lithocorrelation in the terminology of the *North American Stratigraphic Code*), Dunbar and Rodgers find the term confusing. These authors suggest that correlation should be restricted to time relations, and that lithocorrelation might be termed physical facies correlation, for instance; if dealing with stratigraphic units that contain the same or similar faunal elements, one should label the relations of the units so derived not as biostratigraphic correlation (biocorrelation) but perhaps as faunal equivalence. Similarly, in discussing the terminology of correlation, Storey and Patterson (1959, p. 712) make the following suggestion: "If time-significance is not intended, then 'lithologic comparison,' or 'tracing' of a particular lithology in outcrop, or in the subsurface are more correct expressions." Of course, if during a study an investigator utilizes lithologic characteristics, biological (faunal) features, or any other aspects of the strata or stratigraphic units involved in an attempt to determine time relations of strata, then one can apply the term correlation to the procedure and results. Proposed correlations need not be true or accurate to be considered correlations; they merely need to be intended to deal with time relations.

Expanding upon these views, Rodgers (1959) discusses various criteria commonly used in working out the continuity, equivalency, or correlations of strata and stratigraphic units (such as lithology and fossil content). Rodgers (1959) concludes that even when one is attempting to determine the "true" (or best estimates of) lithologic relations and lithologic continuity of rocks (that is, lithocorrelation), the best "correlations" are not a result of simply tracing a large shale unit (for example), but are determined by tracing the stratal continuity of individual beds or horizons. A single bed, if traced in detail (perhaps by simply walking it out on the ground) may cut across various lithologic units (formations)—perhaps our hypothetical bed begins in some places within a shale unit, but in other areas within a limestone unit or a sandstone unit. Likewise biocorrelation does not usually consist of merely tracing and correlating various particular environments (for example, the intertidal environment even though the particular species represented change from one locality to another), but rather identical species (or closely similar species) are correlated with one another, and the intertonguing of lithologies is used to determine the relations between differing contemporaneous environments. As any particular study develops and matures, we use lithologic, paleontologic, and all other criteria together, in many cases mutually reinforcing each other, to arrive at a correlation of the strata under consideration, not only among themselves but also with rocks outside the study area. Rodgers's (1959) main point is that ultimately what are generally considered the best correlations are those that approach temporal correlations. "If we choose the stratal continuity of individual beds instead of the more obvious stratal continuity of the shale unit, we do so because we think the beds are more likely to be contemporaneous throughout than the formation. Similarly, if we reject 'facies' fossils and correlate instead by the few that cross facies boundaries, we do so because we think the latter are more significant timewise. Ultimately, in other words, we evaluate our criteria of correlation by their usefulness in demonstrating the true time relations" (Rodgers, 1959, pp. 689–690). In Rodgers's (1959) opinion, the neat theoretical separation of lithocorrelation, biocorrelation, and chronocorrelation (and any other type of correlation) is contradicted in practice in every real example of stratigraphic correlation. I would hasten to add that in many areas initial correlations may be based on only one set (or a few sets) of features, such as field lithology, and thus we may speak of the lithocorrelation of the rocks of an area; but as stratigraphic studies mature in any area, various criteria inevitably are used in order to arrive at more accurate correlations, and whether or not the investigator involved is explicitly attempting chronocorrelation per se, ultimate correlations (taking all the data into account) often will converge on chronocorrelations. Time is a unifying theme in stratigraphy.

The views of Rodgers (1959) and like-minded geologists have not prevailed, however, in the codification of terminology in the *North American Stratigraphic Code* and the *International Stratigraphic Guide* (see above). These documents adopt the term correlation in a broad sense to refer to all classes of correspondence and/or equivalency in a stratigraphic sense. Correlation approximately *sensu* Rodgers (1959) is referred to as chronocorrelation. However, I am not convinced that correlation *sensu* Rodgers (1959) and chronocorrelation are precisely the same thing. Correlation *sensu* Rodgers (1959) is derived from the traditional concept of correlation (see above) where all characteristics of strata (rocks) are taken into account in attempting correlations, and the preferred correlations are chosen on the basis of how well they reflect the temporal relations of the strata (rocks) involved. Chronocorrelation of the *Code* and the *Guide,* at least ideally or conceptually, appears to be a bit more definitive or absolute. "Chronocorrelation demonstrates [the term "expresses" is used in the *Code*] correspondence in age and chronostratigraphic position" (Hedberg, 1976, p. 14). But chronocorrelations never really are known; all we can do is correlate on the basis of lithology, paleontology, isotopic dates, and other evidence that is often less than perfect. Correlations *sensu* Rodgers (1959) may be arrived at, although they may not be definitive and subsequently are refined. Chronocorrelation, by comparison, appears to be more an ideal that we strive for, but may never perfectly reach. Still, for all practical purposes correlation *sensu* Rodgers (1959) and chronocorrelation closely approximate one another. At this point in the development of the science of stratigraphy the terminology for correlation proposed by the *Code* and the *Guide* is defined explicitly, is being generally adopted (e.g., Boggs, 1987), and is probably best accepted for future use. One should not lose sight, however, of the more traditional and restricted view of correlation used in much of the early classic stratigraphic literature.

B. R. Shaw (1982) has suggested that for a process to be considered geologic correlation (presumably more or less synonymous with stratigraphic correlation), two basic qualifications must be fulfilled: "the use of stratigraphic units and the ability to demostrate correspondence of these units" (B. R. Shaw, 1982, p. 7). He places special emphasis on the point that, in his opinion, the concept of correlation in geology (stratigraphy) applies only to "the establishment of the equivalency of stratigraphic units," and "the object of similarity measures must be stratigraphic units" (B. R. Shaw, 1982, p. 8). If objects of correspondence being compared cannot be directly related to some stratigraphic unit, then B. R. Shaw (1982) terms the procedure matching rather than correlation. "For example, a technique termed seismic-stratigraphy is generating interpretations in areas unreachable by existing sampling methods. However, unless the object of correspondence on the seismic trace, the wavelet, can be related directly to a stratigraphic unit, the resulting interpretations are matches, not correlations" (B. R. Shaw, 1982). Yet it is unclear precisely what a stratigraphic unit, according to B. R. Shaw, is. In Shaw's view the *Guide* and the *Code* have been established primarily to define and clarify the concepts of stratigraphic units—but these documents recognize a number of different types of stratigraphic units (Hedberg, 1976; NACSN, 1983), and continue to add to the proliferation of types of stratigraphic units (e.g., ISSC, 1979, 1987a,b). B. R. Shaw (1982) recognizes various types of stratigraphic units, such as lithostratigraphic units and biostratigraphic units, as well as formal and informal stratigraphic units. On a certain level one has to question whether Shaw's distinction between correlation and matching is primarily a matter of how a particular investigator defines stratigraphic units and what types of stratigraphic units are considered valid. Conceivably one can use particular attributes or features to find matches, and then define stratigraphic units on the basis of the same attributes or features in order to turn the matches, in B. R. Shaw's terminology, into correlations. Given the example cited from B. R. Shaw above, if the wavelets cannot be directly related to any previously defined stratigraphic units, then stratigraphic units (perhaps informal stratigraphic units) can be defined on the basis of the wavelets, and the matches will be, by definition, correlations. Or is this a perversion of B. R. Shaw's intention? Perhaps Shaw meant to imply that the features used should be relatable to some previously established, or generally accepted, stratigraphic units in order to qualify as the process of correlation.

Or did B. R. Shaw simply mean to distinguish between correct and incorrect correlations, to distinguish between correlations among "equivalent" (however the term is defined) and nonequivalent stratigraphic units, by labeling the former correlations and the latter matches? In his article, B. R. Shaw (1982, p. 10; see also Boggs, 1987, p. 550; Fig. 3.7 of this book) includes a figure (modified from A. B. Shaw, 1964, pp. 214–215) that shows two different possible correlations between lithostratigraphic units. In one case, similar strata (strata with the same lithologies) are correlated, but the correlations are considered incorrect and therefore termed "matches." In a second case, the same rocks are correlated in a different manner, a manner that is considered correct, so the correlations are termed "correlations." However, B. R. Shaw (1982) has modified A. B. Shaw's (1964) original figure and intention. Reading A. B. Shaw (1964), it is clear that he is distinguishing between different types of correlations, namely, rock correlations and time correlations (Fig. 3.8). Admittedly, however, the issue is somewhat confused because in his text A. B. Shaw (1964) uses the term miscorrelations for time correlations that have been erroneously interpreted as lithologic correlations (thus implying lithic continuity between a unit of rocks in one outcrop and that of another).

Assume that one is given two geographically separated sections (Fig. 3.8a), X and Y (where the two sections lie at some distance from each other perpendicular to the depositional strike of the contained strata); each section contains the sequence of features shale, fossiliferous limestone, shale, unconformity, shale, fossiliferous limestone, and shale. Next assume that two different faunas have been identified, faunas A and B, in the lower and upper limestones of section X, respectively, and furthermore that fauna B has been identified in the upper limestone of section Y. A. B. Shaw (1964) notes that, given this situation, many stratigraphers would readily (or initially) correlate the limestones containing fauna B with one another and from that initial correlation infer that the lower limestones, and intervening shales and unconformities, are also correlative (Fig. 3.8a). However, what does the initial correlation of the limestones containing fauna B really mean? And are the inferred correlations of the lower limestones, shales, and unconformities justified? A. B. Shaw (1964) suggests that many stratigraphers, at least initially, would draw the

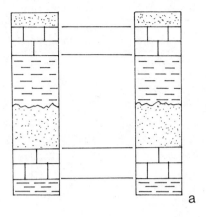

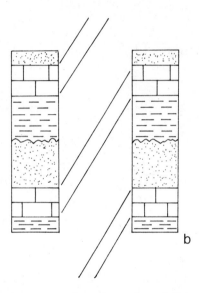

Figure 3.7. Figure illustrating the concepts of matches and correlations as used by B. R. Shaw (1982). Both (a) and (b) represent two stratigraphic sections containing similar rock and fossil subdivisions; (a) indicates matches, whereas (b) indicates correlations. (After B. R. Shaw, 1982, p. 10.)

conclusion that the limestones in the two sections, and also by inference the intervening shales, are time correlatives. Furthermore, many stratigraphers would readily suggest that the limestones and shales in section X are probably laterally continuous with the limestones and shales in section Y; that is, corresponding time and rock correlations are established, given the initial data (Fig. 3.8a).

A. B. Shaw (1964) suggests, however, that such extrapolations and inferences from the initial data (namely, that the upper limestones in both sections contain the same faunal content) may be misguided. A different, and in some cases more likely, interpretation of the two sections is shown in Fig. 3.8b. According to Shaw (1964, p. 215, italics in the original): "Faunal content can be used to prove contemporaneity, but it *cannot* be used to prove lithic continuity. Lithic continuity can be proved only by lateral tracing of the rock bodies." Given the idea that faunal similarity is

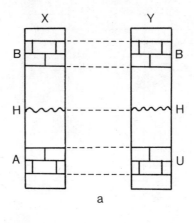

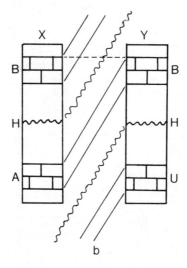

Figure 3.8. Figure illustrating various types of correlations as discussed by A. B. Shaw (1964). In both (a) and (b), sections X and Y represent two stratigraphic sections containing similar lithic and fossil subdivisions. Fossil zones A and B are recognizable only in the limestones, and the lower limestone of section Y contains an unknown biota (U). H = hiatus (i.e., an unconformity). Part (a) represents "normal" correlations as might be initially proposed by many geologists. Part (b) represents what might be the actual rock (solid lines) and time (dashed line) correlations between the two sections. (After A. B. Shaw, 1964, pp. 214–215.)

not always an exact guide to contemporaneity, one could further suggest that strictly speaking faunal content only establishes biostratigraphic correlations. Such biostratigraphic correlations, however, closely approximate chronocorrelations and usually are interpreted as such. The important point here is that in this example A. B. Shaw (1964) is not so much distinguishing correct from incorrect correlations per se, but noting that one type of correlation (for example, a biostragiraphic correlation) does not necessarily imply another type of correlation (for example, a lithostratigraphic correlation), and that undue extrapolation of a particular correlation may not be justified. A. B. Shaw (1964, p. 216) summarized his conclusions on this topic succinctly, as follows: "It has been the burden of this book [A. B. Shaw, 1964] to demonstrate the independence of fossil and rock-unit correlations. Lateral relations among rock bodies must be established by study and tracing of the rocks themselves. The establishment of time correlations by means of fossils cannot demonstrate either the continuity or discontinuity of rock units."

Perhaps the stratigraphic units alluded to by B. R. Shaw (1982) could be interpreted as natural, or scientifically meaningful, stratigraphic units. The nature of scientifically meaningful stratigraphic units may be dependent upon the questions to be answered in any particular stratigraphic study. "Implicit in the choice of stratigraphic unit[s] is the reason for conducting the evaluation. If lithology and structure are primary objectives in an analysis, then, lithostratigraphic units are appropriate. If the time relationships between sections are of interest, a biostratigraphic unit may be employed. However, such clarity of purpose has not always had complete agreement among stratigraphers. In fact, the lingering disenchantment with procedures may be the cause of many problems in attempting stratigraphic correlation" (B. R. Shaw, 1982, p. 8). Rightfully, Shaw notes that clarity of purpose is often problematic. In this quotation Shaw suggests that chronocorrelation (temporal correlation) be carried out utilizing biostratigraphic units. Likewise, A. B. Shaw (1964) took as a basic premise of his work that the use of fossils is the only practical way of arriving at time correlations of most nonvolcanic sedimentary rocks, and that lithostratigraphic (rock) units are universally diachronous. In many cases biostratigraphic units may approximate chronostratigraphic units, and biocorrelation may approximate chronocorrelation; but why not simply use chronostratigraphic units in chronocorrelation? A possible answer would seem to be that there are no chronostratigraphic units per se, only other types of stratigraphic units that may be interpreted chronostratigraphically. Likewise, the type of correlation one is dealing with—chronocorrelation, biocorrelation, or some other type—is primarily a function of the interpretation placed by the investigator on the proposed matches or correspondences among features of rocks.

Time stratigraphy (correlations), paleontologically based stratigraphy (correlations), and rock stratigraphy (correlations) all can be interpreted as equivalent (cf. the proverbial universal lithologic successions of A. Werner, Chapter 1; irreversible faunal successions through time). However, lithologic correspondences (correlations), paleontologic correspondences (correlations), or any other types of correspondences need not be interpreted as indicating time equivalence (correlation). Likewise, time correspondence does not necessarily predict lithologic correspondence or paleontological correspondence (A. B. Shaw, 1964; B. R. Shaw, 1982). Returning to the original or restricted definition of correlation, correlations are correspondences seen in rock features or characteristics of stratigraphic units that are interpreted as temporally significant. Correspondence of such features is not correlation until interpreted as such. Using the newer terminology of lithocorrelation, biocorrelation, and chronocorrelation (and other forms of correlation), the interpretive aspect of much of correlation seems to have lost some ground to the "objective" matching of similar or identical features. Yet even to propose a lithocorrelation involves some interpretation—that there are some significant attributes shared by the units or strata so lithocorrelated. But when A. B. Shaw (1964, p. 82) stated that "rock correlations [and so on] . . . are all as much forms of correlation as is any form of time correlation," he was wrong. At least as he seems to use the terms, time correlation (traditional correlation, chronocorrelation) is not another type of correlation, but potentially incorporates all other forms of correlation and imbues them with a particular, unifying interpretation. Or, using the terminology of the

Code and the *Guide,* we might say that chronocorrelation is the interpretation of other types of correlations (often specifically biocorrelation or magnetocorrelation) as being indicative of temporal relationships between strata.

Unlike B. R. Shaw (1982), some investigators (e.g., Rodgers, 1959; see above) would not insist that stratigraphic units per se are always necessary in proposing correlations. Perhaps in proposing lithocorrelations one normally utilizes lithostratigraphic units (whether they are formal or informal units), and in proposing biocorrelations one utilizes biostratigraphic units; but in proposing chronostratigraphic correlations one could suggest that a single point (or plane, surface, bed, layer, or unit) correlates with another single point (or plane, and so on)—one need not utilize stratigraphic units. In chronostratigraphy, rocks or strata are interpreted as representative of points or intervals of time, and one could consider that it is the interpretations that are actually correlated with one another. Such correlations when related back to the rocks may be independent of any stratigraphic units or boundaries established otherwise in the rocks, including what are considered chronostratigraphic units and boundaries.

B. R. Shaw (1982; adopted by Boggs, 1987) suggests that a distinction should be made between formal (= direct) and indirect (= informal) stratigraphic correlation (correlation *sensu lato*), analogous to formal and informal stratigraphic units. "A formal correlation, one that can be 'demonstrated,' must be represented physically, that is, physical tracing of a stratigraphic unit is the only method of showing correspondence unequivocally" (B. R. Shaw, 1982, p. 9). Yet note that this concept of formal correlation is best applicable to lithocorrelation (the walking out of beds of similar or identical lithology) and totally nonapplicable to chronocorrelation because chronocorrelation is not based on simple observations of physical attributes and their continuity, but on the interpretation of features. Indeed, all correlations are subject to interpretation, even lithocorrelations that can be traced physically and continuously in the field. In contrast to formal correlations, B. R. Shaw (1982) would distinguish informal (indirect) correlations based on well logs, seismic traces, fossil assemblages in separated sections, and so on—that is, any correlations that cannot be physically walked out. "An informal correlation is the indirect tracing of units through projections of attributes" (B. R. Shaw, 1982, p. 9). Surely, Shaw's concept of informal or indirect correlation includes the vast majority of correlations proposed in the science; formal and indirect correlations grade into one another; all correlations are more or less subject to interpretation; and there are various levels of confidence or realibility associated with different correlations. Yet Shaw's terminology may be of some use in referring to extremes of the spectrum of procedures used in arriving at correlations.

Within his concept of informal or indirect correlation, B. R. Shaw (1982) labels correlations proposed on the basis of visual comparisons (of well logs, for example) arbitrary, whereas computer-generated analyses of numerous attributes simultaneously are termed systematic. From Shaw's terminology one might receive the impression that systematic correlations are perhaps more reliable; but systematic correlations among a set of data can vary according to various factors, such as the computer programs used, the attributes emphasized, and so on. Interpretive judgment on the part of a human investigator may in some cases be of greater importance than the advantages of sophisticated technology. Within the category of systematic correlations, B. R. Shaw (1982) distinguishes between monothetic correlations and polythetic correlations. In monothetic correlations the criterion to establish a particular correlation is rigid and requires that both units or items to be correlated possess a unique set of attributes; it is considered both necessary and sufficient that both contain all the attributes in order to establish a correlation. In contrast, polythetic correlations are based on statistical measures of various characters shared among the units or items that may potentially be correlated with one another; no single attribute is either essential or sufficient in order to propose a correlation.

Evidence and Methods of Correlation

All correlations would ultimately appear to be based on similarity or identity of features. Similarity of fossil content is the primary evidence used in biostratigraphic correlation; similarity or identity in lithologic characteristics and attributes is used in lithologic correlation; similarity or identity in temporal sequences (that is, abstract time) is utilized in chronocorrelation. However, there are fundamental differences and implications among the various activities currently grouped under the broad heading of correlation. Lithostratigraphic correlations *sensu stricto* involve primarily the correlation of geographically separated parts of a single lithostratigraphic unit. Such correlations may be proposed, in practice, on the basis of close lithic similarities between rocks cropping out and exposed at separated localities, but lithocorrelation implies that there is, or once was, a physical continuity of rock from one exposure to the next. That is, any now separated exposures of a presumed lithostratigraphic unit that are correlated with one another should at one time have formed a continuous body of rock without breaks. The preserved rocks may retain the continuity in subsurface and be traceable in closely spaced wells or boreholes; they may retain the continuity on the surface, and the stratigrapher may be able simply to walk out the units (trace their lateral extent directly); portions of the once continuous body of rock may have been lost to postdepositional processes such as erosion; or adjacent parts of the lithostratigraphic unit may have been structurally separated, such as by faulting or rifting. Likewise, allocorrelations of allostratigraphic units and unconformity-bounded correlations of unconformity-bounded stratigraphic units are based on the observed or inferred direct continuity of bounding unconformities or discontinuities. Similarly, pedocorrelations are based on the observed or inferred continuity of the disparate parts of the single pedostratigraphic unit being correlated.

In contrast to lithocorrelation, and related forms of correlations mentioned in the last paragraph, biocorrelation does not necessarily imply that the units, strata, or bodies of rock were once component parts of a continuous whole. Biocorrelation usually is based only on some similarity or identity in biotic content among the rock bodies being correlated. It may well be inferred that the organisms whose remains form the basis for the particular correlation had a geographically continuous distribution between any two exposures of rock now being correlated. However, this is different from suggesting that there was once a continuous body of rock connecting the two exposures that contain the fossil remains of the particular organism. Likewise, magnetostratigraphic correlation or magnetostratigraphic units may be based only on observed or presumed similarities in magnetic signature, and need not imply that there was once a continuous body of rock connecting the exposures being correlated that bore the magnetic signature upon which the magnetocorrelation is based.

Chronocorrelation (chronostratigraphic correlation), temporal correlation, or more simply correlation as classically used (see discussion above) is based just on the inferred or presumed temporal relationships of the material bodies being correlated relative to one another. Chronocorrelation does not imply that the rock units, strata, or rock bodies ever were materially connected, nor does it imply that they necessarily bear any particular physical or material attributes in common, or even share any general similarities. Furthermore, it is not the case (at least in practice) that chrono-correlation is applied only to chronostratigraphic units. Many stratigraphers infer or propose chronostratigraphic (temporal) correlations for disparate biostratigraphic, lithostratigraphic, magnetostratigraphic, or other stratigraphic units. One might in theory suggest that chronocorrelations can, or should, be proposed for chronostrati-graphic units, and when chronocorrelations are proposed for other types of strati-graphic units, those units are in fact being treated as chronostratigraphic units. However, chronostratigraphic units, while based on material referents, have by def-inition synchronous boundaries (and, some would argue, are therefore often unrec-ognizable with precision in real rocks). It may be readily acknowledged by a particu-lar stratigrapher that a lithostratigraphic unit, for instance, does not have synchronous boundaries, yet chronostratigraphic correlations still may be proposed for the unit. In other words, the lithostratigraphic unit presumably may be datable. Chronocorrelations typically are based on any and all evidence that is presumed to be of time significance or time-indicative; biostratigraphic and paleontologic evi-dence, lithologic evidence, magnetostratigraphic evidence, the evidence of isotopic dates, stratigraphic sequences and positional relationships, geophysical characters, and so on, may all be used. In some cases, of course, various lines of evidence may lead to differing chronocorrelations being proposed; in such cases, some judgment must be posited as to which line of evidence to prefer. In general, all other things being equal, it usually is agreed that the more characters or attributes two strati-graphic units have in common, the more confidence we can place in a proposed correlation, be it a lithocorrelation, a biocorrelation, or a chronocorrelation. How-ever, in few circumstances are all other things equal; this is especially true in chrono-correlation. It is often critical to understand the nature of the characters used in proposing a correlation in order to assess their relative merits and value to a particu-lar type of correlation. For this reason, multivariate statistical analyses, and similar techniques, applied to problems of correlation in stratigraphy have not found uni-versal approval among those composing the stratigraphic community. As an exam-ple of this thesis, McLaren (1978, p. 5) stated that "the evidence from a few selected fossils may be far more significant than analyses of very large faunas." Ultimately there seems to be no body of purely chronostratigraphic evidence, although some might suggest that isotopic and other numerical dating techniques come closest to providing such a category.

The Ulrich and Grabau Controversy

Differing general philosophies of stratigraphic correlation, particularly different emphases on the importance or acknowledgment of contemporaneous facies, can be manifested as very real differences in final stratigraphic interpretations, correla-tions, syntheses, and compilations. Dunbar and Rodgers (1957) recount the story of two conflicting basic philosophies of correlation that were prevalent among many American stratigraphers in the early twentieth century. Typifying the two opposing schools of thought were the eminent stratigraphers Edward Oscar Ulrich (1857–

1944) and Amadeus William Grabau (1870–1946). It is worthwhile to briefly relate this case example here.

On the basis of much detailed work, particularly on the stratigraphy and paleontology of the lower Paleozoic of eastern and central North America, Ulrich (e.g., Ulrich 1911, 1916; see B. Conkin and J. Conkin, 1984) developed his own philosophy and principles of stratigraphic correlation that became extremely influential. Ulrich greatly downplayed the importance of facies, believing that at least in the areas of his work facies were on the whole unimportant. Rather, Ulrich believed that every distinct fauna (group of fossils that he could distinguish as a separate entity) corresponded to a distinct rock unit and presumably a distinct interval of time. If in a particular local section a particular fauna was absent, then the corresponding rock unit and time interval were also unrepresented. These absences of fauna and rock units were believed to be represented by unconformities due primarily to nondeposition; Ulrich believed that almost all rock units in any local section were separated by fairly major unconformities represented by rock elsewhere. Thus, at no single locality can a relatively complete section be found even for a small part of the composite stratigraphic scale. Any composite stratigraphic column for a region must be composed of numerous stratigraphic units from different localities dovetailed together.

Based on his stratigraphic philosophy, Ulrich (see Dunbar and Rodgers, 1957) reinterpreted many older correlations. In many cases laterally adjacent but often distinct stratigraphic (lithostratigraphic) units had been correlated with one another in the belief that they represented contemporaneous lateral facies of a broad nature deposited in a single large sedimentary basin. Ulrich systematically reinterpreted such units (facies) as representing series of separate deposits that are of limited extent and were deposited in separate small basins at different times. Ulrich saw each distinct lithology (unit) and fauna as representing a single advance of a shallow sea over the continental interior. Through time the continents were subject to a number of advances and retreats of the sea, each successive advance generally covering a different area from that of the previous one. These shifting patterns of advance and retreat Ulrich termed oscillations. Dunbar and Rodgers (1957, p. 285) provide an example of the type of reinterpretations propounded by Ulrich: "For the Appalachians, this led to the concept of troughs and barriers, according to which deposits of, for example, sandstone, shale, shaly limestone, and pure limestone, thought by earlier workers to be an orderly succession of facies, were held to have been deposited separately and successively in one or another of four or five parallel troughs, in such manner that no two deposits were contemporaneous."

Ulrich's stratigraphic philosophy led him to recognize many more formations (and other stratigraphic units) than previous workers had, but in recent times succeeding workers have relegated many of Ulrich's units to synonymy. Ulrich also compiled composite stratigraphic sections for various regions that are now generally considered much too thick (up to 100% too thick). Furthermore, on the basis of his stratigraphic principles and philosophy, Ulrich even named additional systems, such as the Ozarkian, which was believed to fall between the Cambrian and the Ordovician (= Canadian of Ulrich).

Contrasting with the stratigraphic philosophy of Ulrich was that of Grabau. Though an American (Grabau was born in Wisconsin, studied at the Massachusetts Institute of Technology and at Harvard University, and taught at Columbia University from 1901 until 1919 when he moved to the National University in Peking [Beijing], China), Grabau apparently was deeply influenced by the European philosophy of stratigraphy that emphasized facies relationships and strict chronocorrelations. Grabau's ideas are expressed in his massive *Principles of Stratigraphy* (1913, dedicated by Grabau to Johannes Walther) and in his *Textbook of Geology* (1920–21; see also Grabau, 1936, and discussion in B. Conkin and J. Conkin, 1984).

Grabau's vision of the geologic column was broadly synthetic; he sought correlations among many different stratigraphic units of approximately the same age, justifying them on the basis that they represented broad, contemporaneous facies. Except for major and obvious unconformities (perhaps such as those separating systems), Grabau tended to downplay the importance of unconformities. Grabau believed that many of the deposits normally encountered in the stratigraphic record were the result of relatively widespread and semipermanent seas that very slowly and gradually advanced and retreated over whole continents (see Dunbar and Rodgers, 1957). A single long rhythm or pulsation, an advance and/or a retreat, might correspond to a time interval on the order of magnitude of a period. Following Walther's law (see above), according to Dunbar and Rodgers (1957), the mere juxtaposition of two different lithologies in a single local section was sometimes enough to suggest to Grabau that the lithologies (and perhaps units they represented) might be lateral equivalents. In his textbook, Grabau (1920–21) suggested many facies-based correlations without good evidence, but in numerous cases Grabau's general predictions have been confirmed (see Dunbar and Rodgers, 1957), even if in some cases he erred in the direction of oversimplification.

In the early twentieth century Ulrich was extremely influential in American stratigraphy, and his school of thought, although in conflict with many European ideas, appears generally to have dominated the scene. At least by the middle and late 1930s, however, many of Ulrich's specific correlations and interpretations were found to be in error, the importance of facies relationships was discovered or acknowledged by American stratigraphers, and Ulrich's basic philosophy eventually was rejected by the majority of investigators. Slowly the general school represented by the work of Grabau came to dominate American stratigraphy. Yet a generation of stratigraphers had been greatly influenced by what were later viewed as basic misconceptions. Dunbar and Rodgers (1957) suggested that many of the advances made in American stratigraphy in the period of approximately 1927–57 proceeded in direct opposition to the views and teachings of Ulrich. Further, these authors wrote: "Even though his major conclusions are now no longer accepted, the debris of his system is widespread in our stratigraphic terminology and even in some of our thinking, hampering the objective discovery and description of stratigraphic facts" (Dunbar and Rodgers, 1957, p. 288). Concerning Ulrich and his philosophy in particular, this is probably less true today than it was in 1957, but it is perhaps still the case to a lesser extent concerning Ulrich's general beliefs. Furthermore, it is certainly true of other stratigraphic philosophies and assumptions expounded by stratigraphers since Ulrich.

Stratigraphy, like any science, cannot proceed in a philosophical or mental vacuum. One does not just go to nature and discover and describe the simple facts or truth. A fact cannot exist outside the context of some sort of philosophical, theoretical, or mental (one can use whatever name one prefers) framework. We are all strongly influenced by philosophical or theoretical considerations, whether we acknowledge their existence or not. There is no guarantee that the theoretical frameworks and philosophies preferred by many stratigraphers today will not in the future come under the same type of harsh criticism that Ulrich's philosophy has been subjected to. Certainly Ulrich was not attempting to deviously mislead a generation of stratigraphers; he was merely expounding his beliefs and putting them into practice, as we all do. Anyone who dares to offer an opinion or interpretation is subject to criticism. Perhaps one might justly criticize Ulrich in his tenacity in maintaining his opinions and conclusions in the face of growing evidence that they required modification. On the other hand, if Ulrich was not convinced by the evidence against his ideas, he was personally justified in upholding them.

Chapter 4

CODES AND CONVENTIONS OF STRATIGRAPHIC NOMENCLATURE AND CLASSIFICATION

There is no single international body or organization that regulates stratigraphic nomenclature analogous to the International Commission on Zoological Nomenclature and no single code of stratigraphic nomenclature comparable to the *International Code of Zoological Nomenclature* (Ride et al., 1985). Rather, there are numerous national and regional stratigraphic codes (listed in Hedberg, 1976), which do not all agree among themselves. The International Subcommission on Stratigraphic Classification (ISSC—originally designated the International Subcommission on Stratigraphic Terminology, or ISST) of the International Commission on Stratigraphy (ICS), a commission of the International Union of Geological Sciences (IUGS), developed the *International Stratigraphic Guide: A Guide to Stratigraphic Classification, Terminology, and Procedure* (Hedberg, editor, 1976), which was at least provisionally endorsed by many regional and national stratigraphic committees, but was never adopted by the ICS as a "statutory policy document" (Cowie et al., 1986, p. 1). Since its publication there have been additions to the *International Stratigraphic Guide* (ISSC, 1979, 1987a,b,c, 1988), and at this writing the *Guide* is being revised (Cowie et al., 1986; Salvador, pers. comm., 1987). In North America most stratigraphers follow the specifications of the current *North American Stratigraphic Code,* developed by the North American Commission on Stratigraphic Nomenclature (NACSN, 1983; reprinted in this book as Appendix 1).

The *International Stratigraphic Guide* is intended to serve specifically as a guide, and not a code. The guide avowedly has no formal or informal legal (or quasi-legal) authority in matters of stratigraphic classification, terminology, and nomenclature. "There is no intention that any individual, organization, or nation should feel constrained to follow it, or any part of it, unless convinced of its logic and value. As reiterated in prefaces to all of its reports, the Subcommission believes that matters of stratigraphic classification, terminology, and procedure should not be legislated. Real and lasting progress will be achieved only as geologists in general agree voluntarily on the validity and desirability of certain principles, procedures, and terms. The purpose of the *Guide* is to inform, to suggest, and to recommend; and it must continually evolve in keeping with the growth of geologic knowledge" (Hedberg, 1976, p. 4). However, the fact that the *Guide* exists and was assembled by an international subcommission has imbued it with an air of authority, and in practice an individual may well be constrained to follow it, at least in part, or risk not being published; referees, reviewers, and editors now have the authority of the *Guide* to support particular views and biases as to the classification and nomenclature of the strata of the earth. Here it cannot be overemphasized that the *International Stratigraphic Guide* is concerned with much more than the determination of the correct name to be applied to units or categories once such units are decided upon by a particular investigator. Rather, whether intentionally or unintentionally, the *Guide* is codifying and reifying the very principles used in pursuing the science of stratigraphy; in the process the *Guide* is putting forth a single view as to the nature of the science of stratigraphy. An analogous situation in zoology might be imagined.

At present the sole purpose of the *International Code of Zoological Nomenclature* is to determine the correct name (when using the Linnaean system of nomenclature) of a taxon. "The Code consists of provisions and recommendations designed to enable zoologists to arrive at names for taxa that are correct under particular taxonomic circumstances. . . . The Code refrains from infringing upon taxonomic judgment, which must not be made subject to regulation or restraint. . . . Nomenclature does not determine the rank to be accorded to any group of animals but, rather, provides the name that is to be used for the taxon at whatever rank it is placed" (Ride et al., 1985, p. xiii). The *International Code of Zoological Nomenclature* in no way regulates what characters, procedures, or methodologies are used in distinguishing taxonomic units (for example, species, genera, families, orders) in zoology. Likewise the *Code* does not regulate how many or few species may be included in a genus, or how many genera make up a family, and so on. Given a group of individual organisms, any investigator is free to classify them in any arrangement that he or she feels is appropriate. For a particular group of individual organisms, one investigator might distinguish separate species and genera for each individual, whereas another investigator might classify all of the individuals under consideration in one species and genus (this hypothetical example is not so absurd as it may appear to the nonsystematist; such can be the case when one is dealing with certain little-known and poorly understood organisms, such as fossil hominids). Once it is decided what individual organisms will be grouped into what units (taxa), the *Code* merely sets forth rules for the unambiguous application of names to the taxa. Given that a certain group of individuals are to be united in a certain taxon, using the rules of the *Code,* any competent zoological systematist can determine the correct name for the taxon. It must be stressed, however, that one person's *Homo erectus* (as an example) need not, and often will not, correspond to another person's *Homo erectus;* that is, the concept of the taxon *Homo erectus* may include different groups of individuals, depending on the investigator. The same holds true for all taxa of the zoological hierarchy.

If the *International Code of Zoological Nomenclature* were analogous to the present *International Stratigraphic Guide* and many of the various national and regional codes (including the *North American Stratigraphic Code*), then the *Code of Zoological Nomenclature* might legislate the type of evidence as well as the principles to be used in interpreting such evidence in order to classify animals. For example, the *Code* conceivably could recommend (or legislate) that eye color, wing length, and wing veination patterns be of primary concern in classifying insects, and that these characters be interpreted utilizing the principles of numerical taxonomy in order to arrive at classifications of insects (cf. Schoch, 1986b). Furthermore, the *Code of Zoological Nomenclature* might mandate that once a group of organisms were classified in a particular way by a particular investigator (the units of the classification were defined in a particular manner), the classification could not be changed, at least not without great pains. In effect, the real situation in stratigraphic classification (as reflected in the *International Stratigraphic Guide* and the *North American Stratigraphic Code*) is analogous to this hypothetical situation in zoology. As is discussed further below, the stratigraphic *Guide* (and *North American Stratigraphic Code*) have the very real effect of legislating and mandating the types of evidence to be used, at least in particular cases, in stratigraphic classification, as well as mandating the principles to be followed in pursuing the science of stratigraphy. Furthermore, priority of definition has played a large role in stratigraphic classification, whereas it plays essentially no role currently in zoological classification; rather, in zoological classification one is concerned only with the priority of names in order to develop a stable and internationally understood nomenclature, not with priority of definition (how the group that the name was originally applied to was originally defined). This comparison of the codes and guides of stratigraphic classification and nomenclature to the *International Code of Zoological Nomenclature* is not capricious, but is made because stratigraphers (many of whom are also traditionally paleontologists) have both explicitly and consciously, and to some extent unconsciously, modeled their codes and guides on those used in biological nomenclature (botanical nomenclature follows essentially the same principles as zoological nomenclature). Also, it should be noted that one need not contest or disagree with the specific principles and practices put forth in the *International Stratigraphic Guide* or the *North American Stratigraphic Code* to question whether it is appropriate or healthy for a science to codify such concerns per se. One has to wonder whether we risk turning a science into a dogma.

In the preface to the *International Stratigraphic Guide,* Hedberg (1976, p. v) writes: "Agreement on stratigraphic principles, terminology, and classificatory procedure is essential to attaining a common language of stratigraphy that will serve geologists worldwide. It will allow their efforts to be concentrated effectively on the many real scientific problems of stratigraphy, rather than being wastefully dissipated in futile argument and fruitless controversy arising because of discrepant basic principles, divergent usage of terms, and other unnecessary impediments to mutual understanding." But one can wonder what the real scientific problems of stratigraphy are if they are not the development, modification, and refinement of the basic underlying principles of the science of stratigraphy. If various investigators are utilizing discrepant basic principles, maybe there are valid disagreements as to the basic principles of stratigraphy—and if stratigraphy is to lay claim to being a science, such controversy cannot simply be eliminated by mandating that only certain principles will be considered valid. Yet the view put forth in the *Guide* could be interpreted as espousing certain principles to the exclusion of others; and a majority consensus is the means used to decide what principles are to be espoused (in effect, what principles are valid). "The purposes of the *Guide* are to promote international agreement on principles of stratigraphic classification and to develop a common internationally acceptable stratigraphic terminology and rules of stratigraphic procedure—all in the interests of improved international communication, coordination, and understanding, and thus of improved effectiveness in stratigraphic work throughout the world. The recommendations of this first edition of the *International Stratigraphic Guide* are based on the current consensus of members of the Subcommission" (Hedberg, 1976, p. 1).

It also could be suggested that the use of a diversity of terminology, although perhaps confusing and inconvenient, is a sign that a science is healthy and alive; perhaps some agreement and standardization should be sought, but a stifling rigidity is not necessarily the best answer. Terminology, nomenclature, classification, and taxonomy are at the heart of any science pursued by human beings. Terminology is a measure and reflection of the very conceptual categories we use to think with, and thus to carry on scientific activities. Arguments and controversy over terminology are not necessarily fruitless or futile; they may, in fact, bring fundamental issues and considerations under close scrutiny.

Reading the *International Stratigraphic Guide* may leave one with the sense that the authors of the *Guide* believe classification in stratigraphy does not and perhaps cannot reflect any real, natural units or order; rather, all classification in stratigraphy is relatively arbitrary. "Diverse as are the strata of the Earth and their properties, they are certainly no more diverse than are the natures and characters of the persons who study them. All of our classifications and terminologies of natural bodies are no more than an attempted ordering contrived by human beings for the purpose of aiding our own imperfect conception and understanding of the infinite complexities of nature; and as such they have all the weaknesses of the human minds in which they have originated. Classification and terminology of rock strata are no exception" (Hedberg, 1976, p. v). It is a valid point of view to suggest, for instance, that there are no natural stratigraphic units, and that all stratigraphic classification is somewhat arbitrary. However, it is another matter entirely to allow such an interpretation to pervade, even subtly, an *International Guide* or regional codes. Such guides and codes, if truly meant for use by all stratigraphers, ideally should not favor (however slightly) any perconceived notions or biases. In contrast to the quotation from the *International Stratigraphic Guide* immediately above, some investigators may espouse the theory that as classifications are refined and approach more closely categories actually existing in the real world (that is, as natural classifications are striven for), such classifications may rise above, or go beyond, the weaknesses and limited view (or former weaknesses and former limited view) of the human mind.

However, the ISSC does distinguish, at least nominally, between "real" and "unreal" concepts; and I agree with the ISSC that terms and concepts should be defined explicitly and precisely. "Clarity of definition is as essential to stratigraphy as to any other science. The use of unreal concepts, vague and cloudy terms, and imprecise definitions is often defended because such procedure seems easier, simpler, more attractive, or more traditional than a more rigorous approach. However, if a concept or a term cannot stand attempts at precise definition, it is usually of dubious value" (Hedberg, 1976, p.5).

The *International Stratigraphic Guide* (Hedberg, 1976) makes no claim to being complete or all-encompassing. Whereas, according to the *Guide,* rocks (strata) can be classified in many ways, the *Guide* discusses only a few major systems of rock units and classification: lithostratigraphic units (Hedberg, 1976; ISSC, 1987b), biostratigraphic units (Hedberg, 1976), chronostratigraphic units (Hedberg, 1976), magnetostratigraphic polarity units (ISSC, 1979), and unconformity-bounded stratigraphic units (ISSC, 1987a,c, 1988). The *Guide* also presents basic recommendations concerning the principles of stratigraphic classification, definitions of certain terminology, procedures for establishing, naming, and revising stratigraphic units, and a chapter on the subject of stratotypes (Hedberg, 1976). The various recommendations put forth in the *Guide* are discussed throughout this book.

The *North American Stratigraphic Code* (NACSN, 1983; reprinted as Appendix 1) is similar in overall spirit to the *International Stratigraphic Guide,* but carries more weight as a legal document in North America than the *Guide* does in any country—as is indicated by the fact that one is called a *Code* and the other a *Guide.* In North America the recommendations of the *Code* are generally followed by various geological organizations, surveys, and editorial departments of scientific journals. The history of the current *North American Stratigraphic Code* is briefly recounted in that document (see Appendix 1). In general, the recommendations of the *North American Code* are consistent with those of the *International Stratigraphic Guide,* but there are differences as discussed throughout this book. The *North American Code* also formally recognizes some types of stratigraphic units not discussed in the *International Guide* (for example, pedostratigraphic, allostratigraphic, and diachronic units), whereas the unconformity-bounded units of the *Guide* do not find an exact equivalent in the *North American Code* (allostratigraphic units are similar to, but not exactly the same as, unconformity-bounded units).

Any authority invested in the *Code,* the *Guide,* or similar documents is gained primarily from the influence of the supporters of the documents, the tacit consent of many more stratigraphers, and willingness of editors and publishers to adopt such recommendations and impose them upon prospective authors. One hopes that such documents also will be authoritative by virtue of the rigor of their logic, internal consistency, and high scientific standards. As already noted above, the *North American Stratigraphic Code* is a quasi-legal document among North American stratigraphers, but the *International Stratigraphic Guide* was never adopted as a statutory policy document by the International Commission on Stratigraphy, and the "Guidelines and Statutes" of ICS (Cowie et al., 1986) take preference over the current *International Stratigraphic Guide* wherever differences arise.

STRATOTYPES

A tradition that has developed in stratigraphy is the use of stratotypes (type sections) as the material basis and standard of definition of stratigraphic units; this is somewhat (but not entirely) analogous to the use of type specimens in zoological nomenclature (Ride et al., 1985; see further discussion in Chapter 7). Often the stratotype or type section, rightly or wrongly, is regarded as "typical" of the particular stratigraphic unit it defines. The *International Stratigraphic Guide* (Hedberg, 1976, p. 24) defines a stratotype, or type section, as "the original or subsequently designated type of a named stratigraphic unit or of a stratigraphic boundary, identified as a specific interval or a specific point in a specific sequence of rock strata, and constituting the standard for the definition and recognition of the stratigraphic unit or boundary" (see also the *North American Stratigraphic Code,* Article 8, reprinted here in Appendix 1).

Stratotypes are applicable to various categories of stratigraphic units, especially lithostratigraphic (*sensu lato*) and chronostratigraphic units; however, certain types of stratigraphic units, for instance, some biostratigraphic units, may not have stratotypes (see Chapter 6). The *Guide* distinguishes three basic forms of stratotypes for those units to which the concept of a stratotype is applicable: unit-stratotypes, boundary-stratotypes, and composite-stratotypes. A unit-stratotype is simply the type section of strata that defines the content, and normally also the upper and lower boundaries, of a particular stratigraphic unit (such as a formation). Commonly a unit-stratotype consists of a single very complete section of the stratigraphic unit under consideration where both the lower boundary of the unit, and subjacent strata, and the upper boundary of the unit, and superjacent strata, are exposed. Ideally all intervening strata (between the upper and lower boundaries) are well exposed, but in reality some of the intervening interval may be covered by soil development and so forth (Fig. 4.1).

A boundary-stratotype is the specific point in a particular sequence of rock strata that defines a particular stratigraphic boundary. Thus a unit-stratotype normally is bounded by two boundary-stratotypes marking the lower and upper boundaries of the stratigraphic unit. Boundary-stratotypes also may stand alone, independent of unit-stratotypes. A particular stratigraphic unit (such as a high-ranking chronostratigraphic unit, for example, a stage or system) may be defined as composed of the strata stratigraphically above a particular boundary-stratotype, marking the lower boundary of the unit under consideration, and the strata below a second boundary-stratotype, marking the upper boundary of the unit (or more commonly marking the lower boundary of the next higher, superjacent, unit). The two boundary-stratotypes marking the lower and upper limits of such a unit need not occur in the same stratigraphic section, or even in the same geographic region.

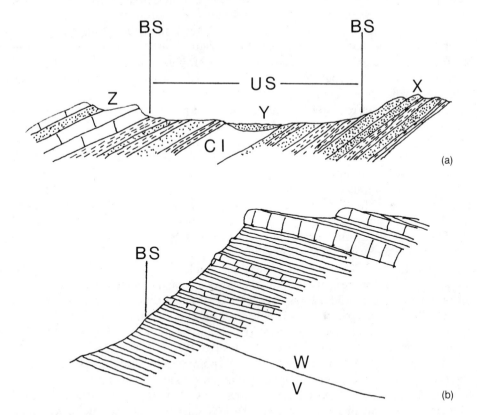

Figure 4.1. Diagram illustrating the concepts of unit-stratotypes (US) and boundary-stratotypes (BS) for (a) lithostratigraphic units X, Y, and Z, and (b) chronostratigraphic units V and W (CI = covered interval). The lower boundary of chronostratigraphic unit W is the upper boundary of chronostratigraphic unit V, and so on. (After Hedberg, 1976, p. 25.)

A composite-stratotype is "a unit-stratotype formed by the combination of several specified type intervals of strata known as *component-stratotypes*" (Hedberg, 1976, p. 24, italics in the original). The idea behind a composite-stratotype is that a particular stratigraphic unit may not be exposed completely in a single stratigraphic section; perhaps two (or more) sections will be needed to represent the total stratigraphic development of the unit. A particular stratigraphic section might be designated as the type of the lower part of the unit, and a second section might be designated as the type of the upper part of the unit. In such an instance, according to the *Guide,* one of the component-stratotypes would be considered a holostratotype, and the other would be considered a parastratotype (see below). A stratigraphic unit of higher rank may in some instances be composed of a combination of previously established constituent units of lower rank; in such a case, the stratotype of the higher-ranking unit may be considered a composite-stratotype composed of the stratotypes of the constituent units. According to the *Guide* (Hedberg, 1976, p. 25), given that "the components of a composite-stratotype are already established formal stratigraphic units, it is unnecessary to distinguish one as a holostratotype and others as parastratotypes."

The *Guide* (Hedberg, 1976, p. 26) recommended the use of such prefixes as holo-, para-, neo-, lecto-, and hypo- (analogous to terms used for different categories of type specimens in biological nomenclature, such as holotypes, paratypes, neotypes, and lectotypes) to be used with the term stratotype in order to add "precision to the designation and description of stratotypes." Accordingly, a holostratotype is an original stratotype that was designated by the original author(s) when first establishing a stratigraphic unit or a boundary of a stratigraphic unit. A parastratotype is a supplementary "stratotype" section designated by the original author in the original publication when establishing a new unit or boundary. A lectostratotype is a stratotype that is chosen after the fact (after the original publication establishing a stratigraphic unit or boundary) if an original stratotype was not adequately designated. A neostratotype is "a new stratotype selected to replace an older one which has been destroyed or nullified" (Hedberg, 1976, p. 26). A hypostratotype is essentially a reference section or auxiliary reference section that is designated after the original publication of the unit or boundary under consideration, and usually in a different geographic area or facies from that in which the primary stratotype was established. As is evident, only holostratotypes, lectostratotypes, and neostratotypes are true types of stratigraphic units or boundaries; parastratotypes and hypostratotypes are merely supplementary or reference sections.

Recently the ICS (Cowie et al., 1986) has condemned the use of these various categories of stratotypes, at least for the purposes of the current activities of the ICS in establishing boundary stratotypes for a global chronostratigraphic scale (see below). "It is considered preferable not to use parabiological analogies which imply unsound analogies and cause confusion (e.g. holostratotype and parastratotype) but to confine nomenclature, for ICS candidates [see section below], to two categories of stratotype: (a) global stratotype section and point (GSSP) and (b) auxiliary stratotype point (ASP)—the latter will be particularly useful in drawing upon stratigraphic correlation between markedly different facies, e.g. New Red Sandstone contrasted with marine Triassic or Devonian neritic facies contrasted with pelagic facies" (Cowie et al., 1986, p. 5). Note that the auxiliary stratotype point of the ICS mentioned above is essentially a point in a supplementary reference section, in a different geographic region or facies from the actual global stratotype section and point, and is not a primary stratotype point or boundary per se. It is essentially equivalent to a boundary-hypostratotype of the *Guide,* and gains any authority it may have only on the basis of the accuracy of a correlation with the global stratotype section and point.

Besides stratotypes, or type sections, the concepts of type localities and type areas or type regions have found common use in stratigraphy. A type locality is the original geographic locality where a particular stratigraphic unit, feature, or boundary was defined or named. If a unit or boundary initially was based on a stratotype, then the type locality is by definition the locality where the stratotype is found. If, as was often the case in the past, a unit or boundary was defined or named without the designation of a specific type section, and a type section subsequently is designated for the unit or boundary, it is possible that the subsequently designated stratotype may fall outside the original type locality. Type area and type region are general terms used to specify the general geographic territory surrounding a type locality. According to the *North American Stratigraphic Code* (NACSN, 1983, p. 853) a type area for a geologic/stratigraphic unit based upon a nonstratiform body (such as an igneous intrusion) may serve as the unit stratotype (not hyphenated in the *Code*) of the unit.

As noted above, the stratotype concept commonly is applied to both lithostratigraphic and chronostratigraphic units and boundaries. As pointed out by the *International Stratigraphic Guide,* in order for stratotypes to perform their function as material standards for the recognition and definition of units and boundaries, there must be some consensus among the appropriate stratigraphers concerning such stratotypes. The function of a stratotype is compromised if, after being established by a particular investigator, it is not accepted by the general stratigraphic community. In its function as a stratotype, a section is useful only insofar as it is accepted. "The usefulness of a stratotype is directly related to the extent to which it is generally accepted or acknowledged as *the type*. It is, therefore, always desirable, and in due course to be expected, that the designation of a stratotype be submitted for approval to the geological body having the highest standing in any particular case" (Hedberg, 1976, p. 29, italics in the original). In the case of most newly described lithostratigraphic units or boundaries, which are of purely local significance, the designation of stratotypes still may be the prerogative of the particular investigator(s). The designation of stratotypes for widely used (on either a regional or a global scale) chronostratigraphic units is another matter, however. "Stratotypes for chronostratigraphic units or boundaries of international or worldwide application should be approved by appropriate bodies of the highest international and worldwide geological standing" (Hedberg, 1976, p. 29). For example, the current International Commission on Stratigraphy has as a prime concern the establishment of boundary-stratotypes for the global stages, series, and systems of the chronostratigraphic scale (discussed further below). "On the other hand, stratotypes of units of only local extent and interest may require approval from no more than local or national surveys or stratigraphic commissions" (Hedberg, 1976, p. 29).

This brings us to the concept of the "golden spike." Initially many major chronostratigraphic units were based on unit-stratotypes that included upper and lower boundary-stratotypes. In many cases, theoretically adjacent units, such as stages, defined by means of unit-stratotypes in different localities, were not compatible with one another. Unless the mutual boundary between two adjacent units was defined in a single stratigraphic section, it would be very unlikely that the base of the upper stage would correspond precisely to the top of the subjacent stage. In some cases there might be a gap between two adjacent units, whereas in other cases there might be some degree of overlap between the stages (Ager, 1981; Hedberg, 1976). These types of situations led to many disputes, and, Ager (1981) has suggested, also promoted the search for natural divisions between adjacent units (what Ager [1981, p. 75] terms "the quest for the golden horizon"). Often major unconformities (perhaps expressed as faunal breaks) might be interpreted as such natural divisions, but further problems may ensue if a major unconformity is taken as the natural boundary between two stratigraphically adjacent units, and later strata corresponding to the unconformity are identified. Do the newly identified strata pertain to the upper unit, the lower unit, or a new unit between the two formerly recognized units? In this context the *Guide* (Hedberg, 1976) has suggested that if an unconformity is to be used loosely as the boundary of the unit, the unconformity per se must not be taken as the boundary-stratotype, but rather a stratigraphic horizon that abuts the unconformity is to be considered the boundary of the unit.

In order to deal effectively with the problem of gaps and overlaps between nominally adjacent (vertically or stratigraphically) chronostratigraphic units, the concept of defining the bases only (it could just as easily have been the tops only) of major chronostratigraphic units by means of single designated points in rock ("golden spikes" or "golden pegs") in boundary-stratotypes was introduced at least as early as the 1960s (Holland, 1986a). In this scheme the base of the unit so defined is taken as the lowest grain of sediment above the golden spike, the golden spike ideally being a single, infinitely small (dimensionless) geometric point (not a plane) in a particular stratigraphic sequence. The base of any chronostratigraphic unit so defined automatically defines the top of the subjacent unit, thus eliminating any possibility of gaps or overlaps between adjacent units. The term golden spike has been used to refer to the points of boundary-stratotypes used globally, and currently being established by the ICS and its committees, subcommissions, and working groups (described below); whereas the term silver spike has been used to refer to boundary-stratotypes of regional significance (Harland, 1978). Silver spikes subsequently might be promoted to the status of golden spikes. (I assume that the term golden spike used in stratigraphy may have been derived from the procedure, whether mythical or real, of early North American railway builders, who presumably drove in a "golden spike" to mark the completion of a major railroad track [Holbrook, 1967]. Holland [1986a] has expressed the same opinion.)

The concept of stratotypes, particularly unit-stratotypes, comes under occasional criticism from various stratigraphers. Often the suggestion is made that the very concept of a type section (stratotype) for a particular stratigraphic unit is fallacious because for many stratigraphic units there can be no "typical" or representative stratigraphic section. However, this criticism indicates a lack of understanding of the current concept of a stratotype. Certainly at one time in the history of stratigraphy it may have been thought that a type section (stratotype) was truly typical of the stratigraphic unit that it defined (hence the term "type section"), but this idea has long since been abandoned as naive. A stratotype currently serves only to define unequivocally what is and what is not considered to be a particular stratigraphic unit. It need not be typical of the unit, even though it is most convenient if it is typical of the unit; when describing a new unit, all other considerations being equal, the most typical exposed section probably should be designated as the stratotype. Likewise, it may be inconvenient (and less than ideal) if a unit-stratotype contains gaps or unconformities within its body, but this does not invalidate the concept of the stratotype. Assume that a chronostratigraphic unit, for instance, defined by a unit-stratotype and lower and upper boundary-stratotypes, is found to include within its limits a major unconformity. Strata subsequently discovered that correlate to the major unconformity still will be included within the said chronostratigraphic unit even if there is no material correlative of such strata in the unit-stratotype of the unit.

The concept of stratotypes also has come under fire from some investigators, not because there is anything inherently wrong with the concept but because it has perhaps been misused (especially in the past). For example, Alcide d'Orbigny (see Chapter 6) derived many of his Jurassic stage names from English place names even though he never visited England (Ager, 1981). Consequently, these English localities were taken as the type localities of the stages, and it was naturally thought that they should contain the appropriate type sections; yet the sections at the English localities did not necessarily correspond exactly to the rocks d'Orbigny and other continental workers referred to the given stages. It was partly in reaction to this situation that the concept of the "golden spike," that is, defining only stratigraphic boundaries, was initiated in Britain (Ager, 1981; Holland, 1986a).

Common sense dictates that, to fulfill its functions as a standard for the recognition and definition of a stratigraphic unit or boundary, any stratotype must be accessible to all interested stratigraphers. Such accessibility can be considered to be twofold. First, detailed information concerning the stratotype should be available in the literature, including all relevant information on the stratigraphy, paleontology, mineralogy, lithology, geochemistry, magnetics, and so on; and it should be conveyed by appropriate means, such as verbal descriptions, charts, diagrams, drawings, and photographs. Second, ideally any stratigrapher at any time should be allowed access to the actual physical stratotype, regardless of political, ideological, or other circumstances. This means that the stratotype must be preserved in its natural state with the rocks well-exposed. It cannot be bulldozed under and destroyed, or built upon. Also, documented samples from the stratotype may be housed in the permanent collections of appropriate museums, universities, governmental surveys, and similar institutions and thus made available to investigators.

PROCEDURES FOR THE DESCRIPTION OF NEW STRATIGRAPHIC UNITS

A set of procedures and requirements is outlined in both the *International Stratigraphic Guide* and the *North American Stratigraphic Code* for the recognition, establishment, description (definition), and naming of new formal stratigraphic units and for the revision, redefinition, or amendment of previously named formal stratigraphic units.

Establishment of a new formal stratigraphic unit (or modification of an existing formal stratigraphic unit) requires publication of such an intention in a "recognized scientific medium" with adequate description of the new unit. According to the *North American Code,* publication means that the work must be printed in ink on paper; be issued as a permanent, public, and scientific record; and be readily obtainable (either at no charge or for a fee). New stratigraphic units are published mainly in issues of scientific journals, periodicals, and irregular series (such as monograph series). Both the *Code* and the *International Stratigraphic Guide* dictate that a new stratigraphic unit cannot be established solely on the basis of an abstract, or as part of a legend on a map. Description and use of a new unit in an open-file report or release, a thesis, or a dissertation does not constitute publication. Also, it is generally agreed that publication in a commercial or trade journal, the popular press, or a legal document or publication does not render a new stratigraphic unit available. Unfortunately, in many particular instances it may be a matter of judgment whether a particular medium is adequate for establishing a new unit. Consequently, it is best to publish all new formal stratigraphic units (and modifications of old units) in journals or series whose status is unequivocal. Under the *North American Code* publication cannot be anonymous, but the *Code* does not elaborate upon the meaning of anonymous publication. I interpret it to mean that the individuals responsible for the publication must be indicated by name within the publication; publication of a new stratigraphic unit by the "Attleboro Geological Society," without listing the individual members, would constitute anonymous publication.

In spirit and intent, neither the *North American Code* nor the *International Stratigraphic Guide* is retroactive. Even if an older formal unit was originally defined inadequately, and in a publication that today would not be considered a valid and recognized scientific medium, if the unit under consideration has been adopted and used, it is not rejected retroactively as not properly established and invalid on that account (such a unit, however, may require amendment, revision, or redefinition).

The published statement establishing a new formal stratigraphic unit (or modifying a previous formal unit) generally should include some discussion of the topics listed below (Hedberg, 1976; NACSN, 1983). However, as is described elsewhere, not all of the topics that follow actually are utilized in the definition of formal stratigraphic units (for instance, inferences as to the genesis of a unit usually should be included but normally play no part in the definition and formal recognition per se of most formal stratigraphic units).

1. Statement of Intention. It should be stated explicitly within the publication under consideration that it is the intention of the author or authors to establish a new formal stratigraphic unit or modify an existing formal stratigraphic unit, or to modify an existing informal stratigraphic unit. After pursuing the publication, the reader should detect no ambivalence or ambiguity as to the intentions of the author.

2. Type and Rank of the Stratigraphic Unit under Consideration. It should be stated explicitly what type or category of formal stratigraphic unit is being erected or modified; for example, a lithostratigraphic unit, a chronostratigraphic unit, an unconformity-bounded stratigraphic unit, and so on. A particular rank also should be specified for the unit; for example, group, formation, member, and the like, for a lithostratigraphic unit.

3. Name. The formal name of the stratigraphic unit should be stated unambiguously, and the derivation of the name should be discussed. There is no simple rule or algorithm for the coining of new names for formal stratigraphic units. As is discussed in the appropriate sections of this book, different categories of formal stratigraphic units bear different types of names. Biostratigraphic units often bear names derived from the Linnaean names of fossil forms, whereas many lithostratigraphic units bear names derived from geographic localities.

4. Historical Background. Regardless of whether a new unit is being proposed or an existing unit is being modified, the history of classification of the unit (both informal and formal classification) and the rocks composing the unit should be thoroughly discussed and referenced. Any potential synonymies or priorities should be explored.

5. Stratotype or Equivalent. For units that require stratotypes, a stratotype should be specified precisely (see section above on stratotypes). For any unit, whether such a unit requires a formal stratotype or not, it is advisable to specify reference sections. Any stratotype must be described in detail both geographically and geologically; any competent investigator must be able to locate the stratotype in the field. Verbal descriptions normally are supplemented by such devices as maps (geologic and geographic), cross sections, columnar sections, photographs, drawings, and various graphs or charts. Invariably a detailed description of a measured section or sections should accompany any intention to erect or modify certain formal stratigraphic units—particularly those requiring stratotypes, in which case one of the described measured sections should be the stratotype itself.

It is imperative that the position of any boundaries of units in a stratotype be located exactly and unambiguously. The *International Stratigraphic Guide* emphasizes the recommendation that artificial permanent markers physically locating boundaries be installed on the actual stratotypes. Even if this is done, adequate information must be published so that any such markers can be restored if they should be removed or damaged.

Certain formal stratigraphic units discussed in the *International Stratigraphic Guide* do not lend themselves to typification via stratotypes and do not require stratotypes for their erection—for instance, certain biostratigraphic units. In such cases, the basis for definition, characterization, and recognition of the particular stratigraphic unit must be clearly published or cited. For many biostratigraphic units the diagnostic taxa should be thoroughly illustrated and described, or references in the literature to such descriptions and illustrations must be cited. In such cases the material standard for the unit may be specimens preserved in some adequate facility (such as a museum, university, or archive of a geological survey) rather than a stratotype or reference section.

6. Description of the Unit. The description of the unit and its stratotype and reference sections, if any, should be as clear and thorough as possible. Any and all relevant, or potentially relevant, information should be given. This includes not only the thickness, lithology, paleontology, and so on, of the unit, but such information as the structural attitude of the sections, nature of exposures and geomorphological expression, relationship to other units and rock bodies, geophysical properties, inferred age, and so forth. From the description alone a subsequent investigator should be able to recognize the unit.

7. Regional Aspects. The geographic extent and variations in thickness and composition of the unit, as far as is known, should be noted. Also included should be information on the relationship of the unit under consideration to adjacent (both horizontally and vertically) units, zones, and horizons, to other categories of stratigraphic units (some of which may pertain to the same rocks), and to other ranks of stratigraphic units, as well as the nature of the boundaries between the unit and other stratigraphic units in various geographic areas and suggested criteria to be used in extending the unit geographically away from its type locality.

8. Genesis. All observations and inferences (clearly identified as such) concerning the origins, conditions, and context of the formation of a rock body should be included in the publication. Wherever and whenever possible, the significance of a stratigraphic unit with respect to local and regional geologic history, paleogeography, and paleoclimatology should be discussed.

9. Correlations with Other Units. Any information and inferences as to the correlations (of whatever nature—temporal, spatial, or otherwise) of the unit under consideration with other units should be discussed, along with the criteria used in arriving at such correlations.

10. Geologic Age. Any information regarding the age of the stratigraphic unit, and the basis for the assignment of such an age, should be included in the publication.

11. References. All pertinent references and citations to the previous literature should be included.

Both the *International Stratigraphic Guide* and the *North American Stratigraphic Code* include specific provisions for the naming of formal stratigraphic units based on subsurface sections and exposures, such as in boreholes, wells, mines, or tunnels. In establishing and modifying subsurface-based units, the same general procedures apply as for surface-based units. For subsurface units, a borehole, mine, well, or tunnel serves as the stratotype. If a unit can be adequately based on either a surface stratotype or a subsurface stratotype, the surface stratotype usually is preferred, primarily because surface exposures are in general more easily accessible than subsurface sections to the majority of investigators. Many units, however, never are exposed at the surface under natural conditions and must be based entirely on subsurface sections.

The hole, mine, well, or tunnel must be located precisely, and all available information should be given, such as the name of the owner, details as to the company or agency operating the site (in the case of mines or wells, for instance), dates of drilling or cutting, and so on. The surface elevation at the site, the depths to which a site was drilled or tunneled, the depths of the boundaries of the unit, and the depths where the unit is exposed should be recorded. If a unit is based on offshore or subsea exposures, the depth of the sea floor and information on the project, platform, or vessel should be recorded. Any geophysical logs or records are also of value.

Reference materials relative to a subsurface unit should be curated in a permanent depository (such as the facilities of a museum, geological survey, university, or similar institution), to which all relevant investigators will have adequate access. The depository should be specified in the original publication. Materials to be placed in such a depository include core samples, rock samples, fossil samples, geological and geophysical logs and profiles, and any other relevant materials.

STATUS AND ADOPTION OF FORMAL STRATIGRAPHIC UNITS

A new formal stratigraphic unit should not be casually designated or established. Such a unit should be erected only after careful consideration of the consequences of erecting such a formal unit; if it is questionable whether a new formal stratigraphic unit should be erected, then it is often advisable to utilize an informal stratigraphic unit (that is, not formally named) that can be formalized at a later date.

Even after a new formal stratigraphic unit is duly erected (or an existing unit is modified) in a valid publication, the unit lacks clear status until it has been adopted for use by other investigators. On the other hand, once a new unit has been proposed (or an existing unit has been modified), the *North American Stratigraphic Code* mandates that it cannot simply be ignored by subsequent workers. "The decision not to use a newly proposed or a newly revised term requires a full discussion of its unsuitability" (NACSN, 1983, p. 852).

Once a formal stratigraphic unit has been duly proposed and defined in a certain way, it requires just as much justification and documentation to revise or abandon the unit as it does to establish a new formal unit (Hedberg, 1976; NACSN, 1983). It is interesting to note that in contrast to this situation in formal stratigraphic classification, in systematic zoology one is not required by the *Code of Zoological Nomenclature* (Ride et al., 1985) to justify one's decision to reject another investigator's classificatory scheme although any formal zoological names proposed by a previous investigator retain their priority. In this respect the formal stratigraphic units are reified in the eyes of the *Code* and *Guide,* by virtue of having been duly published, to a greater extent than are classificatory concepts in the eyes of the *International Code of Zoological Nomenclature.*

As generally used in stratigraphy, the terms redefinition, revision, and abandonment have specific definitions (see Hedberg, 1976; NACSN, 1983). The *North American Stratigraphic Code* (NACSN, 1983, p. 854) states: "Redefinition of a unit involves changing the view or emphasis on the content of the unit without changing the boundaries or rank, and differs only slightly from redescription." Thus a lithostratigraphic unit originally may have been defined as being predominantly composed of gray calcareous shales, but subsequently may be redefined as a type of limestone. In both cases exactly the same rocks in the field, the same material referent, would be referred to the unit; therefore, this is considered redefinition under the *Code.* In a sense a redefinition is a formal redescription of the composition of an existing unit, perhaps for the purpose of providing a basis or set of criteria for extending and correlating the particular unit under consideration. In most instances, redefinition does not require all of the elaborate and detailed justification and information (outlined above) needed in establishing, revising, or abandoning a formal stratigraphic unit.

Revision of a formal stratigraphic unit, as the term is used in the *Code,* "involves either minor changes in the definition of one or both boundaries of a unit, or [a change in] the unit's rank" (NACSN, 1983, p. 855). Note that if major changes are required in one or both of an existing unit's boundaries in order to satisfy the needs of a subsequent investigator, perhaps the unit should be formally abandoned and a new unit established in its place. A change in the rank of a unit requires careful consideration of the consequences of such an action, but normally does not involve revision of its boundaries, and the formal name of the unit does not change. The most common incidents usually involve raising the rank of a unit and then subdividing it into component formal stratigraphic units. Revision requires as much justification and documentation as does establishment of a new formal stratigraphic unit. As used in the *International Stratigraphic Guide,* redefinition variously includes redefinition *sensu* the *Code* and revision in the sense of involving minor changes to the boundaries of a stratigraphic unit.

Once established, a formal stratigraphic unit may be formally abandoned, but abandonment requires as much justification and thorough documentation as is required for the erection of a new stratigraphic unit. The *Code* lists several reasons for abandoning a formal stratigraphic unit, such as disuse, widespread confusion as to the application of the unit, demonstration of synonymy, demonstration of homonymy, assignment of the unit to an improper category, and violation of the stratigraphic code or procedures in use at the time of the original erection of the unit. Although formally abandoned, a unit later may be reinstated (following the same general procedures used in establishing a new formal stratigraphic unit).

Both the *International Stratigraphic Guide* and the *North American Stratigraphic Code* recognize a general concept of priority, but the rule of priority is not strictly enforced. "Priority in publication of a properly proposed, named, and defined unit should be respected" (Hedberg, 1976, p. 19). However, priority is not the only consideration one must take into account when dealing with stratigraphic classification and nomenclature; of more importance is the usefulness and general acceptance of a unit. Well-established and commonly used names are not normally to be displaced by poorly known and infrequently used names, all other considerations being equal, by the rule of priority.

In many instances an investigator can determine the general correlations of certain strata, but it is difficult or impossible to assign a particular group of rocks or strata to one of two formally named units. To handle such situations, the *International Stratigraphic Guide* (Hedberg, 1976, p. 21) has proposed the adoption of the punctuational conventions used in the following examples:

Silurian? = the rocks under consideration are doubtfully of Silurian age (doubtfully referable to the Silurain System/Period).

Rhode Island? Formation = the rocks under consideration are doubtfully referable to the Rhode Island Formation (Shaler, Woodworth, and Foerste, 1899).

Wamsutta-Rhode Island formation = the rocks are intermediate in position, either horizontally or vertically, between the two named formations; such rocks usually exhibit characteristics of both named formations such that they cannot be assigned decisively to either formation (eventually they may be used to establish a new formation).

Carboniferous-Permian = one part of the section or rocks under consideration is Carboniferous and another part is Permian.

Carboniferous or Permian = the rocks are questionably referable to either the Carboniferous or the Permian.

Carboniferous and Permian (undifferentiated) = the rocks are referable to both the Carboniferous and Permian, but it is not currently possible to definitively distinguish the particular rocks referable to one System/Period versus the other.

GLOBAL BOUNDARY STRATOTYPE SECTIONS AND POINTS

A major task of the current International Commission on Stratigraphy is the development of a chronostratigraphic framework for the global geologic time scale through the selection and definition of boundary-stratotypes. Adopting a strict chronostratigraphic point of view (see Chapter 7), the ICS is currently attacking the problem of precise definition of the boundaries of the systems, series, and stages of the global stratigraphic column (reviewed in part in Chapter 1, and see Appendix 3). The ICS has instituted a formal procedure for the consideration and rejection, or acceptance and ratification, of proposed boundary stratotypes for use by the international geological community (see Cowie, 1986; Cowie et al., 1986; see also Schoch, 1988). Global boundary stratotype sections and points (= GSSP) are not decided upon by individuals working in relative isolation. The ICS sets up committees, working groups, and subcommissions to study boundary problems of global importance. After much study not more than three candidates for a particular boundary stratotype section and point are selected and then considered in more detail, ultimately to be voted upon by voting members of the appropriate committee, working group, or subcommission. If 60% or better of the voting membership of the applicable body agrees upon a single stratotype section and point, the proposed stratotype section and point can be formally submitted to the ICS for consideration. If the submission is in adequate form (fully documented, and so on), it can be voted upon by the voting members of the ICS; a 50% plus one vote majority is required for approval of the stratotype section and point. Finally, the submission must be sent to the executive committee of the IUGS for final approval and ratification. As of this writing, as a result of the work of the ICS, boundary stratotypes have been established for a number of the major chonostratigraphic divisions of the global geologic time scale, including the boundaries of the Ordovician-Silurian, the Silurian-Devonian, the Pliocene-Pleistocene, the series and stages of the Silurian System, and the series of the Devonian System (Bassett, 1985; Holland, 1984).

ICS decisions in regard to global boundary stratotypes are not irrevocable (Cowie et al., 1986). If a majority of voting members of the ICS see fit, a new working group can be set up to evaluate the situation.

The ICS has published a number of specific guidelines and requirements to be followed when candidates are proposed for a boundary stratotype section and point (Cowie, 1986; Cowie et al., 1986, pp. 6–7). The proposed section should be characterized by continuity of sedimentation through the boundary interval; preferably the section should be composed of a marine succession without any facies change. The ICS also suggests that a series of "rapidly alternating and repeating facies changes" (Cowie et al., 1986, p. 6) could be suitable for a global boundary stratotype section, but a monofacial section appears preferable.

The proposed stratotype section should exhibit a significant amount of exposure, and a significant thickness of sediments, both above and below the GSSP and also lateral from the GSSP. The proposed stratotype section should be characterized by an abundance and diversity of well-preserved fossils. In this context, the guidelines of the ICS clearly state that "appearances and disappearances of single fossil species can be expected to be diachronous and therefore a bad guide for the location of a GSSP" (Cowie et al., 1986, p. 7). Rather, multiple species that are relatively independent of facies control should be present at the proposed boundary level. Furthermore, every attempt should be made to ensure that the GSSP is placed where the fossils, fossil horizons, and surrounding rocks are strictly contemporaneous; there should be no possibility that the fossils are reworked clasts in the rock matrix. To this end GSSPs should not be located in conglomerates, breccias, turbidites, or other deposits containing reworked clasts.

The proposed boundary stratotype section should be free from structural complications, metamorphism, or any other alteration. The ICS suggests that locating a GSSP in an exotic terrain might not eliminate the usefulness of the GSSP. The proposed stratotype section must be as free as possible from any types of unconformities (classic unconformities, disconformities, paraconformities) or any time breaks in sedimentation. The ICS warns that if a proposed boundary is obvious, it is suspect. Any obvious or marked change in lithology or biota suggests that a time break may be present. By these criteria most, if not all, of the classic type sections for the systems and series are eliminated from consideration as global boundary stratotypes because originally they were founded on major faunal, floral, or lithologic breaks. However, following ICS guidelines (Cowie et al., 1986), one is not restricted to the original "type" areas of the subjacent and superjacent units involved when defining a boundary stratotype section and point. Finally, any proposed GSSP should be suitable for magnetostratigraphic and geochronometric studies.

Among points insisted upon by the ICS is that a particular boundary stratotype point (a golden spike) within the boundary stratotype section at the stratotype locality be explicitly specified. "Insistence on a Boundary Stratotype Point is in order to define without doubt an instant of geological time. A horizon will, at the GSSP [= Global Boundary Stratotype Section and Point] locality, contain the Point but the horizon may, traced laterally, be diachronous (cutting across time-planes) and may drift away from the instant of time defined by the point thus vitiating the unique concept. The correctly selected GSSP gives an actual point in rock and is not an abstract concept—all other methods can prove to be diachronic" (Cowie et al., 1986, p. 5). The boundary stratotype point should be marked by a permanent artificial marker (at least once it is ratified) and also should be "described in position in words and visually by drawings and photographs so that removal by vandals or others does not prevent accurate restoration" (Cowie et al., 1986, p. 6). The ICS also discusses the need for reasonable guarantees of preservation and accessibility of the stratotype boundary section and point.

According to the philosophy of the ICS, "Boundary Stratotype Definition is a normative question which can be settled by a vote" (Cowie et al., 1986, p. 6). It is not necessarily important where a boundary is finally placed (given that an adequate stratotype section and point are defined), as long as all workers can agree to abide by the placement once it is decided upon.

SOVIET STRATIGRAPHIC CLASSIFICATION AND TERMINOLOGY

Soviet stratigraphers (at least as represented by such organizations as the Interdepartmental Stratigraphic Committee, USSR; the Committee on Stratigraphic Classification, Terminology, and Nomenclature, USSR; the United Stratigraphic Committee of the USSR) have repeatedly opposed the basic theoretical underpinnings and principles of stratigraphic classification espoused by the majority of individuals and organizations that have composed the International Subcommission on Stratigraphic Classification (and its predecessor, the International Subcommission on Stratigraphic Terminology). Thus the relevant Soviet organizations did not vote in favor of publication of the *International Stratigraphic Guide* (Hedberg, 1976) or the ISST's *Definition of Geologic Systems*. The stratigraphic philosophy and principles espoused by the majority of Soviet stratigraphers (at least as represented by the above-mentioned official organizations) is briefly summarized and discussed in Hedberg (1961, pp. 31–33) and ISST (1964, pp. 24–26) and more fully developed in various Soviet publications on stratigraphic classification and terminology, such as those of the Interdepartmental Stratigraphic Committee (= ISC-USSR, 1956, 1965).

In offering views that seemingly diverge from the views of the majority of stratigraphers, the Soviet point of view (whether one agrees with the Soviet ideas or not) has the potential for offering a new perspective on stratigraphy that would be valuable in honing the principles and theoretical underpinnings adopted by any investigator. Furthermore, the Soviet views are perhaps not so divergent from other views as may initially appear to be the case—in some senses the Soviets have retained and formalized many classic concepts from nineteenth-century stratigraphy that, in fact, are periodically resurrected by non-Soviet stratigraphers. Additionally, as holds true for the work of any group of investigators, one must possess a basic understanding of the philosophy and principles upon which such work is founded in order adequately to understand the classification and terminology developed in the course of such work. Soviet stratigraphy cannot simply be ignored because it might be founded upon a different set of theoretical underpinnings. For all these reasons, we will briefly review here the concept and philosophy of stratigraphy as espoused by Soviet investigators. The discussion that follows is based primarily on the early report of ISC-USSR (1956; see also ISC-USSR, 1965, and Donovan, 1966). This review is not meant as a guide to current Soviet stratigraphic nomenclature, but as a discussion of traditional Soviet stratigraphic philosophy.

O'Rourke (1976) suggests that the Soviet philosophy of stratigraphy is derived from dialectical materialism:

> The theory of dialectic materialism postulates matter as the ultimate reality, not to be questioned. All matter is in motion, which is its mode of existence. Motion involves not only a change of place but also a change in quality, an ascending development. Evolution is more than a useful biologic concept; it is a natural law controlling the history of all phenomena and can be deduced from such fundamental relationships as the indestructibility of matter, the constant sum of motion, and the conservation of energy. . . . The dialectic "Law of the Transformation of Quantitative into Qualitative Changes" says that small changes accumulate slowly until they reach a critical point, then pass quickly through a transformation of state, for example, heated water becomes steam. It is manifest in geology when slow repetitive processes culminate in world-wide orogenies with mass extinctions. These breaks constitute natural divisions of the geologic record. . . . (O'Rourke, 1976, p. 51)

The primary goal of Soviet stratigraphy is to identify natural stratigraphic units or subdivisions that "correspond to real historical steps [elsewhere referred to as natural stages] in the geologic development of the Earth as a whole or of its separate regions" (ISC-USSR, 1956, p. 23). Soviet stratigraphy acknowledges only one set of stratigraphic units, roughly equivalent to the Western concept of chronostratigraphic units. "As stratigraphic units of different ranks, in each specific case one distinguishes real geologic bodies, definite assemblages of sedimentary, igneous, and metamorphic rocks composing the Earth's crust, with their material constitution and with all their inherent attributes, relations, and special characters" (ISC-USSR,

1956, p. 24). Recently Sorokin (1984, p. 219) has summarized the major concern and goal of much of Soviet stratigraphy as follows. "One of the most urgent and difficult tasks of stratigraphy is the establishment and tracing of natural chrono-stratigraphic units based on the total combination of all characteristics that correspond to [a] commensurable stage in geological history; determination of the scope, rank and exact (in [a] geological sense) isochronous boundaries; determination of the hierarchy of chronostratigraphic units of different ranks and their relation with the units of rhythmo-, cyclo-, tectono-, eustatostratigraphic scales based on a specific characteristic [here Sorokin is referring to stratigraphic units based on rhythms and cycles of sedimentation, transgression, regression, tectonic changes, eustatic sea level change, and so on], on the one hand, and bio-, litho-, facio- and ecologostratigraphic [approximately ecostratigraphic or paleoecostratigraphic] scales, on the other." The Soviet stratigraphers regard all of their units, by virtue of the fact that they are material units, to be representative of temporal categories; these stratigraphers recognize a geochronologic scale equivalent and corresponding to their stratigraphic scale.

The Soviets are strongly opposed to different types of "stratigraphic scales" and differing ways of classifying rock units (for example, lithostratigraphy, biostratigraphy, chronostratigraphy, and so on), as elaborated upon in the following quotations (presented here at length in an attempt to do justice to the Soviet ideas, and because these ideas are of interest in their own right):

> It is impossible to agree with the initial premisses [sic] in the article by H. D. Hedberg, presented by him at the XIXth International Geological Congress [Hedberg, 1954, p. 206] and offered as a basis for discussion to those interested in the problem. As one of these premises appears the assertion that "The classification of rocks into various kinds of stratigraphic units, the hierarchical ranking of such units, and the naming of the units are essentially procedures of convenience" and that "In some respects they may be looked upon as barren and unproductive exercises necessitated only by the limitations of the human mind."
>
> The acceptance of such tendencies would promote the growth of a purely subjective and arbitrary attitude in the treatment of stratigraphic sections and would lead to a loss of historical perspective and to mistakes in the methodology of stratigraphic research.
>
> The distinguishing of stratigraphic subdivisions ought as far as possible to be stripped of elements of subjectivity and arbitrariness. It ought not to proceed from the extremely unsteady and conditional principle of utility and convenience but to pursue the aim of discovering as objectively as possible the actual course of geologic history. (ISC-USSR, 1956, p. 23)

(Here the present author must sympathize with the point that scientific classifications are not necessarily convenient or utilitarian, but should reflect meaningful concepts and entities [that is, scientific classifications should in some sense be natural, reflecting aspects of nature]; but the issue of whether the Soviet stratigraphers have discovered the basis for naturally subdividing and classifying the stratigraphic column is open to question, and not a matter for dogmatic assertions.)

Some geologists in the USSR [apparently there is no universal acceptance among stratigraphers in the USSR of the principles of the ISC-USSR] and in other countries consider that it is necessary to have not one general (unified) stratigraphic scale but at least two (general or international and local or regional) or three (general, provincial, and local). In the United States for instance three basic scales are accepted: (1) rock-stratigraphic [that is, lithostratigraphic], (2) biostratigraphic, and (3) time (chronostratigraphic), and moreover a whole series of factual stratigraphic scales, equal among themselves, which are distinguished on the basis of separate, arbitrarily selected characters: (1) on authigenic or fragmental minerals; (2) on chemical composition; (3) on the color of the rocks; (4) on cycles of sedimentation, etc.

It is impossible to agree with this. A unified stratigraphic scale ought to be accepted, based on the complex of historical-geological principles, on the distinction of definite steps in the history of the geological development of the Earth, and not on separate, arbitrarily selected characters of rocks.

Stratigraphic subdivisions are objective categories reflecting real steps in the geological development of the Earth as a whole or of its separate regions, and not artificial conventional conceptions, as sometimes held. Each stratigraphic unit represents an assemblage of sedimentary, volcanic, or metamorphic formations, or combinations of these, corresponding to a definite step in the development of the Earth or of a particular region. (ISC-USSR, 1956, p. 27)

Addressing the Soviet concept that all stratigraphic units represent age categories, the ISC-USSR promotes their position using the following logic:

[T]he scale of local subdivisions officially accepted in the USA, which is even called the rock-stratigraphic scale, and all its subdivision[s] ("group," "formation," etc.) in the judgment of the American Stratigraphic Commission (Hedberg) do not in general appear to be geologic age conceptions, for they are based on objective physical data and their age and the interval of time corresponding to them can change from place to place, as much as an entire geologic period.

The designated premises [*sic*] appear to be mistaken and cannot be accepted. In particular, it must be observed that precisely because the "rock-stratigraphic" subdivisions appear to be material units, because they represent real objectively existing geologic bodies, they cannot avoid being also at the same time age categories.

From a correct premiss [*sic*], that the limits of a formation and other lithologic subdivisions can intersect age boundaries, has here been drawn a false conclusion, that these subdivisions in general are not temporal. (ISC-USSR, 1956, p. 32)

Thus the ISC-USSR appears to acknowledge that the boundaries of some stratigraphic units are conceivably diachronous, at least on a very fine scale, but this consideration does not nullify their concept that stratigraphic units in general represent (are equivalent to) temporal units. Perhaps for the Soviet stratigraphers temporal units can, to a certain extent, be diachronous (cf. the concept of diachronic units, Chapter 5 and Appendix 1). Furthermore, the concepts of synchroneity of stratigraphic boundaries and the isochroneity of stratigraphic units do not appear to be so strict as they are for some Western stratigraphers; in the Soviet literature one finds references to isochroneity "in a geological sense" or "geologically" speaking (see, for example, Sorokin, 1984).

The ISC-USSR (1956) adopted the following units for the hierarchical subdivision of "the unified stratigraphic scale" (listed in order of decreasing inclusiveness): group (gruppa), system (sistema), division (otdel), stage (yarus), and zone (zona). The geochronological units corresponding to these stratigraphic units are: era (era), period (period), epoch (epokha), age (vek), and time (vremya), respectively. The ISC-USSR (1956) also recognizes a set of "auxiliary local (regional) stratigraphic subdivisions" that can be utilized in cases where the rocks of a region or district cannot, for whatever reason, be correlated confidently at a low hierarchical level to the unified stratigraphic scale. The local units used are, from most inclusive to least inclusive, series (seriya), suite (svita), subsuite (podsvita), and packet (pachka). A local unit of variable rank is the horizon.

In order to subdivide the stratigraphic column, and to distinguish stratigraphic units that reflect natural stages in the development of the earth, the ISC-USSR recommends the use of the general characteristics associated with (or caused by) phenomena such as tectonic movements extending over wide geographic areas, paleogeographic changes, changes in the processes of sedimentation and denudation, regional igneous activity, regional metamorphism, and changes in the organic world (evolutionary changes). Particular stress is placed on paleontological data, as reflective of other developments on the surface of the earth, for the actual recognition and correlation of stratigraphic units.

The Soviets advocate upholding the principle of priority in stratigraphic nomenclature, and they mandate that each stratigraphic unit should have a designated type section. In line with their philosophy about the nature of stratigraphic units, however, they do not appear to believe that the boundaries of a stratigraphic unit are fixed arbitrarily, either by the first person to describe or define a stratigraphic unit or subsequently by a reviser of the stratigraphic unit. Rather, because the stratigraphic units recognized ideally should correspond to natural units but are recognized only empirically and perhaps nonperfectly, the character, content, and boundaries may change with increasing knowledge to more perfectly reflect nature.

Groups are the largest subdivisions of the unified stratigraphic scale of Soviet stratigraphers. As groups the ISC-USSR (1956) recognized the Archean, Proterozoic, Paleozoic, Mesozoic, and Cenozoic. Groups are of worldwide extent, a single group uniting several adjacent geologic systems that are unified by the characteristics of tectonic activity, igneous activity, sedimentation, and organic content. Boundaries between adjacent groups often are marked by unconformities (particularly angular unconformities), reflecting strong and abrupt tectonic movements (such tectonism was manifested as orogenies, continental uplift and regression, and heightened igneous activity).

A single group is divided into several systems, and a system in turn is divided into two or three divisions. Systems are of worldwide extent and are the physical record or expression of the "major pulsatory movements of the Earth's crust" (ISC-USSR, 1956, p. 29) that leave their mark as major transgressions and regressions. At the boundaries of adjacent systems stratigraphic breaks, angular unconformities, abrupt faunal and floral changes, and igneous activity often are observed over wide geographic regions. Any particular system is easily characterized by its distinctive fauna and flora. In general, the systems of the Soviet stratigraphers are equivalent to the systems of non-Soviet stratigraphers (for example, Cambrian, Ordovician, Silurian, and so on).

Divisions are also usually regarded to be of worldwide extent, as the expression of tectonic movements of somewhat lesser extent (in magnitude or geographic area affected) than those represented by systems as a whole. The two or three divisions of a particular system are given names corresponding to their relative positions within the system: thus the Lower Jurassic, Middle Jurassic, and Upper Jurassic, and the Lower Cretaceous and Upper Cretaceous. Any particular division is characterized by a distinctive faunal and floral content said to reflect tectonic and paleogeographic conditions of the epoch during which the rocks of the division were formed. The boundaries between adjacent divisions, while perhaps being of somewhat more restricted geographic extent than the boundaries between adjacent systems, are also often marked by stratigraphic breaks, angular unconformities, and abrupt changes in the fauna and flora. In general, the divisions of the Soviet stratigraphers are more or less equivalent to the series of non-Soviet stratigraphers. The sediments of the divisions of the Quaternary system are the result of glaciations, interglacials, and related deposits formed in extraglacial regions.

A division of the Soviet unified stratigraphic scale is composed of a number of stages. A stage ought to be of extremely wide or universal geographic extent and is based primarily on a unique typical assemblage of organisms and assemblages of organisms that were synchronous with the "type" assemblage. In some cases it appears to be permissible, according to the ISC-USSR (1956), to erect a "stage" or "stages" for a local geographic region when the deposits of the geographic region under consideration cannot be correlated precisely with the universal scheme of stages of the unified stratigraphic scale.

In the deposits that, because they belong to particular biogeographic provinces, are not accessible to precise comparison with stages widely distributed on the Earth's surface, one may distinguish as a stage a combination of deposits corresponding in turn to a definite step in the geologic development of the given province, in the first instance of its fauna and flora. In the majority of cases, such stages will correspond more or less in their compass to stage-subdivisions [presumably this refers to subdivisions of the stratigraphic record at the level of stages, not subdivisions of stages] of deposits widely distributed on the surface of the Earth.

It is quite inadmissible to distinguish as new stages temporary preliminary units of the local stratigraphic scale, liable in the future to be replaced by some other stages of already established schemes. (ISC-USSR, 1956, p. 30)

A particular stage is subdivided into a number of zones, the smallest unit of the unified stratigraphic scale. Distinguished by a unique faunal or floral assemblage, a zone usually extends throughout a biogeographic region or province (and can range in extent from a significant portion of a province to encompassing several regions or provinces). Zones may be extended geographically by identifying synchronously formed assemblages and deposits. However, in sharply differing biogeographic provinces, separate zonal divisions may be applied to rocks of the same age.

The ISC-USSR (1956) recognized that not all of the divisions of the general global unified stratigraphic scale can be recognized in all regions. In particular, the subdivisions of lower rank often are not applicable to districts or regions geographically far removed from the region (primarily western Europe) where the unified stratigraphic scale was initially developed. However, in any district one can always recognize the local stratigraphic units: "the real geologic bodies here developed, definite assemblages of sedimentary, igneous, and metamorphic rocks, clearly marked off from adjacent assemblages, lithologically easily identified in the field [note, however, that these local units are not lithostratigraphic units; they are of the same nature as the units of the unified stratigraphic scale], well worked out and having a sufficiently wide area of extent" (ISC-USSR, 1956, p. 32). Such units, even if they cannot be correlated precisely to the unified stratigraphic scale, represent real steps in the development of the crust in the local region in which they occur—they are locally natural stratigraphic units. These local units can be formally distinguished as auxiliary (local) subdivisions of the stratigraphic column and are designated as series, suites, subsuites, packets, and horizons. In any particular region the auxiliary subdivisions are combined with the subdivisions of the unified stratigraphic scale. An investigator should apply the subdivisions of the unified scale to a particular region, as far as is objectively possible, and then subordinate the auxiliary subdivisions to the finest (most detailed) applicable subdivisions of the unified stratigraphic scale.

A series is the largest commonly used auxiliary stratigraphic unit of the Soviet scheme, although if necessary several stratigraphically adjacent series may be united into a complex. A series corresponds approximately to a division of the general unified stratigraphic scale, and can be subordinate to a system or division.

A suite, the next lower unit of the auxiliary stratigraphic scale, usually is restricted horizontally to the limits of one region, district, or major sedimentary basin. In the absence of fossils, a suite may be distinguished primarily (but usually not solely) on the basis of lithologic characters (and so appears to approach the concept of a lithostratigraphic unit used by non-Soviet stratigraphers). A suite may correspond to a part of a stage, several stages, or a division of the unified stratigraphic scale.

A suite may be divided into lower, middle, and upper subsuites. There should be no substantial stratigraphic breaks between the subsuites of a particular suite, but each subsuite should be distinguishable by its lithologic or paleontologic characters.

Minor lithologic-facies stratigraphic subdivisions of a suite or subsuite may be distinguished as packets, which are not formally named but may be given arbitrary designations of numbers or letters combined with a lithologic description: for example, packet 7 (gray limestone).

Finally the ISC-USSR (1956, p. 35) recognized the somewhat nebulous unit of a horizon. A horizon unites over a horizontal area several contemporaneous suites, parts of suites, or deposits of different facies compositions. A horizon can correspond (in terms of rank or hierarchy) to anything from a suite to a zone of the unified stratigraphic scale.

The ISC-USSR (1956) notes that difficulties often are encountered in attempts to classify formations or units composed of igneous rocks. Whenever possible, such units should be classified within the schema of the unified stratigraphic scale. However, in many cases one may have to resort to auxiliary or local subdivisions. Extrusive volcanic rocks, in particular, should be divisible into either standard stratigraphic units or auxiliary stratigraphic units, according to the same principles that are followed in dividing and classifying sedimentary sequences. Intrusive rocks, however, may not be readily classifiable using any of the stratigraphic concepts outlined above; in such instances, these rocks may be classified as "intrusive assemblages." The rocks commonly united in a single intrusive assemblage are believed to have originated from a single parent magma and accordingly should exhibit various geologic, mineralogic, petrographic, geochemical, and other characteristics, indicating that this is the case. Intrusive assemblages also are said to be connected with definite stages of tectonic activity (ISC-USSR, 1956). Intrusive assemblages are given geographic names.

The ISC-USSR (1956, p. 36) also recognized what it terms volcanic cycles and tectonic-magmatic cycles (= orogenic cycles). A volcanic cycle is composed of a group of genetically and temporally related intrusive and extrusive igneous rock bodies. A tectonic-magmatic cycle is "a grouping of magmatic formations connected with the development of a single mobile belt (geosyncline)." Tectonic-magmatic cycles are of particular importance in the recognition of the two main subdivisions of the Precambrian recognized by the ISC-USSR: the Archean Group and the overlying Proterozoic Group.

Below the formal level of zone, various Soviet stratigraphers have recognized different stratigraphic units. Sorokin (1984), for example, working on the Upper Devonian of the northwestern part of the Russian platform, has recognized a hierarchy of units below that of the zone (referred to by Sorokin as chronozone): superstratohorizons, stratohorizons (= regional stages), beds (labeled with geographic names), and "rhythms of small order" (labeled using numbers and letters). The various units have their natural basis primarily in the "close association between the evolutionary periods of sedimentation and eustatic fluctuations of the world ocean's level and epirogenic movements of the Earth's crust" (Sorokin, 1984, p. 221), which produces rhythms and cycles of various magnitudes in the stratigraphic record. In contrast, Menner (1984) has suggested that the following hierarchy of stratigraphic units below the level of the zone be adopted, particularly for use in Pleistocene stratigraphy: subzone division, link, superstep, step, "etap," "stadial," and "phasial" (= oscillation). According to Menner (1984), whereas the higher-ranking stratigraphic units of group (= eratem) to zone (= chronozone, = Oppelzone) of Soviet stratigraphy are generally based on the evolution of organic life (or at least are recognized on this basis), the lower-ranking units listed above have their general basis in eustatic and climatic fluctuations (cf. the now obsolete geologic-climate units of the American workers; Chapter 7 and Appendix 2).

Menner (1984) has also advocated the adoption of a hierarchy of formal units of rank above the group (eratem), as follows, from most inclusive to least inclusive: megatem (= megatema, = megathema; with geochronological equivalent of megachron), acrotem (= acrotema, and so on), eonotem (= eonotema, and so on), and phytem (= phytema, and so on). The general basis for these stratigraphic units, according to Menner (1984), are "tectono-magmatic cycles." In terms of duration in years, the geochronologic equivalents of these stratigraphic units would be approximately 4000 million years for a megachron, 2000 million years for an acrochron, 1000 million years for an eonochron, and 250 to 350 million years for a phychron (Menner, 1984, his Table 1). As actual examples of these units, Menner (1984) considers the entire Precambrian to be a megatem, the Archean and Proterozoic to be acrotems, and the Rhiphean (consisting of a good portion of the upper half of the Proterozoic) to be an eonotem. Menner (1984) regards the Paleozoic (usually considered an eratem) as a phytem that perhaps should be subdivided into eratems. The Mesozoic is unequivocally an eratem (= group of the ISC-USSR, 1956), according to Menner (1984).

Chapter 5

LITHOSTRATIGRAPHY AND RELATED SUBDISCIPLINES

LITHOSTRATIGRAPHIC UNITS

Lithostratigraphic classification is the subdivision, classification, and organization of rock strata on the basis of their lithologic characters (Hedberg, 1976; NACSN, 1983). That is, lithostratigraphic units are based on the kind of rock (such as sandstone, limestone, shale, conglomerate, and so on). The term physical stratigraphy sometimes is used as a synonym of lithostratigraphy (e.g., Wheeler, 1959); however, at present the concept of physical stratigraphy generally is considered more inclusive than lithostratigraphy. Such units as unconformity-bounded stratigraphic units, allostratigraphic units, pedostratigraphic units, and the associated branches of stratigraphy also are included under the auspices of physical stratigraphy but are distinguished from lithostratigraphy *sensu stricto.*

Under the *North American Stratigraphic Code* (NACSN, 1983) a lithostratigraphic unit can be composed only of a body of sedimentary, extrusive igneous, metasedimentary, or metavolcanic strata. In general, such lithostratigraphic units are stratified, are tabular in form, and conform to the principle of superposition. According to the *Guide,* "A lithostratigraphic unit may consist of sedimentary, or igneous, or metamorphic rocks, or of an association of two or more of these," that is, all rock types can be classified lithostratigraphically (Hedberg, 1976, p. 31; see also ISSC, 1987b), and lithostratigraphic units need not conform to the principle of superposition. Under the *North American Code,* intrusive igneous rocks and highly deformed or metamorphosed rocks are distinguished as lithodemic units (see below). Lithostratigraphic units may be composed of either consolidated or unconsolidated rock.

Lithostratigraphic units and lithostratigraphic classification commonly are regarded as the fundamental or basic stratigraphic units of general geology. In investigating a new geographic area the first approach usually is to devise, or revise, a lithostratigraphic classification of the rocks encountered. Such units are extremely useful in such pursuits as interpreting the structural geology of an area, delineating mineral resources, and attempting a preliminary reconstruction of the geologic history of a region. However, lithostratigraphic units often are found to be time-transgressive, and a single lithostratigraphic unit may cut across genetic depositional units, such as the sequences of Vail and his colleagues (see below). As studies in a particular region are continued and refined, lithostratigraphic units may be replaced, at least for some purposes, by other stratigraphic units, such as units be-

lieved to refer to genetic suites or units believed to have some type of chronostratigraphic significance.

In America there is a strong tradition of relying heavily on lithostratigraphic
units, but in Europe, for instance, biostratigraphic units and various types of chronostratigraphic units may be used almost exclusively. Thus some Europeans in particular consider lithostratigraphic studies not even to be part of stratigraphy per se,
but to represent what one does of necessity prior to undertaking true stratigraphic
(that is, chronostratigraphic) studies; thus lithostratigraphy, along with some related
studies, has been relegated to what is termed prostratigraphy (see McLaren, 1977;
Schindewolf, 1970a,b; some Soviet stratigraphers do not recognize lithostratigraphic units in any form—see Chapter 4). To give one example of this view, in
summarizing the content of a discussion that took place at a meeting of the International Subcommission on Stratigraphic Terminology (Copenhagen, 18 August
1960), John Rodgers wrote (reported in Hedberg, 1961):

> One group wished to see stratigraphy defined as the study of the age relations of
> rocks and its aim as the elucidation of the succession of geologic landscapes that
> constitute the past history of the Earth. They would restrict the term "strati-
> graphic unit" to the "chronostratigraphic units" of the Subcommission
> Circulars. . . . Certain persons advocated that pure stratigraphy should reject the
> "bizarre" new methods devised by the petroleum geologists [presumably, for ex-
> ample, various well-logging techniques, many of which are essentially lithostrati-
> graphic; seismic studies; and certain micropaleontologic studies] as having led to
> an unacceptable philosophy of terminology and classification, and urged return
> to the "noble" stratigraphy of the past, concerned with the correlation of stages
> and zones by paleontologic methods. But most members of this group acknowl-
> edged the great value of lithologic and biologic or ecologic studies not directly
> concerned with age questions, especially in the study of new countries where the
> stratigraphy is only just being worked out. They contrasted these working meth-
> ods, however, with the aim of stratigraphy to work out geologic history, and
> they were inclined to deny that such studies were properly a part of stratigraphy,
> preferring to consider them as "prostratigraphy" or as temporary auxiliary means
> to the main purpose. (Hedberg, 1961, p. 9)

Lithostratigraphic units *sensu stricto* are defined and distinguished solely by lithologic features (primarily rock composition) supplemented by stratigraphic position,
and such features should normally be of such a nature that they can be readily
recognized in the field by a knowledgeable geologist. Normally a lithostratigraphic
unit will exhibit some degree of overall lithologic homogeneity throughout the unit,
even if that unity includes several different alternating or intermingled lithologic
types. Inferences concerning geologic history, genesis, age, environmental factors,
and so on, do not enter into the definition of a lithologic unit, although the *North
American Stratigraphic Code* (NACSN, 1983, p. 856) suggests that "nevertheless,
considerations of well-documented geologic history properly may influence the
choice of vertical and lateral boundaries of a new unit," and that whenever feasible
it is a good idea to define lithostratigraphic units so that they correspond to genetic
units. In fact, under the present *Code,* classifying a rock body as a lithostratigraphic
unit automatically implies knowledge about the genesis and/or geologic history of
the rock body; only sedimentary, extrusive igneous, metasedimentary, and metavol-

canic strata are classified as lithostratigraphic units under the *Code*. In this respect the *Code* is somewhat contradictory. Likewise, a lithostratigraphic unit may not be defined solely on the basis of the organic species it contains, though fossils in the role of clasts may serve as lithic characters distinguishing a lithostratigraphic unit. A lithostratigraphic unit extends geographically only so far as it can be recognized by physical continuity (or presumed original physical continuity; for example, erosional remnants may now be separated from one another) and retention of its diagnostic characteristics.

The *International Stratigraphic Guide* (Hedberg, 1976) uses the term lithostratigraphic zone (or lithozone) to refer to an informal lithostratigraphic unit; for instance, the shaly zone of the Rhode Island Formation or the conglomeratic zone exposed north of Attleboro, Massachusetts. The *Guide* uses the term lithostratigraphic horizon (or lithohorizon) to refer to a surface that is distinguished by its distinctive lithostratigraphic (lithologic) character, or to refer to a surface of lithostratigraphic change (perhaps, but not necessarily, the boundary of a lithostratigraphic unit). A lithostratigraphic horizon may also refer to a very thin marker bed rather than a true surface or horizon per se.

Boundaries of lithostratigraphic units, or boundaries between adjacent (either laterally or vertically) lithostratigraphic units, are placed at points of lithologic change. Most commonly such boundaries will be placed at naturally occurring sharp or distinct contacts between differing lithologies, but they also may be fixed arbitrarily within zones of lithologic gradation or intertonguing (particularly where no sharp lithologic contacts are evident). In some cases if the zone of intergradation or intertonguing between two different lithologic types (each representing a distinct lithostratigraphic unit) is well developed (sufficiently thick and extensive over a reasonable geographic area), the zone of intergradation or intertonguing may be designated as a third lithostratigraphic unit independent of the two units represented by the two lithologic types on either side of the zone of gradation. In instances of lateral gradation between two formally named lithostratigraphic units, the *Guide* (Hedberg, 1976, p. 38) suggests that "the rocks of intermediate or mixed lithology may . . . be assigned to an informal provisional unit representing the lateral transition between two formations and bearing the names of both, separated by a hyphen—Alpha-Beta formation."

Key beds, marker beds, and unconformities may form the boundaries of lithostratigraphic units, but only as long as the body of rock making up the actual lithostratigraphic unit is characterized by a distinctive lithologic unity and can be distinguished from adjacent lithostratigraphic units. Thus rocks on either side of an unconformity should not normally be recognized as distinct lithostratigraphic units if they cannot be readily distinguished from one another on the basis of lithology. ("However, the union of adjacent strata separated by regional unconformities or major hiatuses into a single lithostratigraphic unit should preferably be avoided even if no more than minor lithologic differences can be found to justify the separation" [Hedberg, 1976, p. 39].) Likewise, if key beds, marker beds, or unconformities can be traced laterally into areas of rock types that differ markedly from the lithology of a lithostratigraphic unit as defined and originally described, then a new lithostratigraphic unit should be established for the differing rock type even though the new lithostratigraphic unit corresponds or correlates with the previously described lithostratigraphic unit. If the marker beds are not diachronous, all of the rocks found between a pair of marker beds may be considered to belong to a single stratigraphic unit; however, that stratigraphic unit may not be a lithostratigraphic unit (especially if differing lithologies are found in different places between the marker beds) but may be some form of chronostratigraphic unit (see Chapter 7; see also the concept of parastratigraphic units advocated by Krumbein and Sloss, 1963 [discussed below]).

When a lithostratigraphic unit is first established, investigators may have a range of choices for boundary placement. After all other considerations are taken into account, the final selection of boundaries may have to satisfy criteria that are not of a direct lithostratigraphic nature, such as topographic, physiographic, or geomorphic expression (especially erosion morphology) of the rocks in the field, fossil content, well log characters, and so on. Defining the boundaries of certain lithostratigraphic units to correspond to such characteristics may greatly aid the recognition and lateral extension of the unit in future studies. However, such boundaries cannot be designated if they contradict the basic requirements for establishing a lithostratigraphic unit (that is, that the lithostratigraphic unit and its boundaries be distinguished and delimited on the basis of lithology).

Both the *International Stratigraphic Guide* (Hedberg, 1976) and the *North American Stratigraphic Code* recognize the following basic hierarchy of ranks of formal lithostratigraphic units (from highest rank to lowest): supergroup, group, formation, member, and bed. Of this hierarchy, the fundamental unit of lithostratigraphic classification is the formation. A formation is composed of a body of rock that is defined by lithologic character and stratigraphic position; it is usually tabular, and it is mappable on the surface of the earth and/or traceable in subsurface boreholes, wells, or sections. The primary criterion used in determining the validity of a formation is its mappability. A formation is a lithostratigraphic unit that is readily mappable at a scale appropriate for the region in which it is proposed or is to be used. An appropriate scale for a particular region is determined to a great extent by the scale of geologic mapping previously practiced in the particular region. There is no standard for the stratigraphic thickness of formations; as the *Guide* notes, an acceptable formation may be less than a meter thick or several thousand meters thick. Very thick and inclusive formations often are utilized for initial geologic mapping in a new region. With later, more detailed studies, these formations may be raised in rank (perhaps to the status of groups or supergroups) and themselves be subdivided into more finely discriminated formations. According to the recommendations of the *International Stratigraphic Guide,* the stratigraphic column everywhere should be divided completely into formations, which are the only mandatory formal units in lithostratigraphy.

A formation may be divided, either completely or incompletely, into members (although such division is not required). A member is a part or portion of a formation that is distinguished by lithologic characters from adjacent (horizontally or vertically) parts of the formation; in many instances, the members of a single formation are separated from each other, or adjacent to each other in a vertical succession, but laterally equivalent parts of a formation may be designated as distinct members if they are lithologically distinct from one another. A member always must be a part of at least one formation; however, one formally named member may extend laterally from one formation to another formation. There are no standards with respect to thickness or extent of a member, but like all formal stratigraphic units a member should not be formally established without good reason. Members need not be mappable, but often they are mappable. A lithostratigraphic unit should not be raised in rank from member to formation merely because it is fully mappable.

Following the *North American Stratigraphic Code* and the *International Stratigraphic Guide,* a member of geographically restricted extent that is lens-shaped and enclosed completely within a formation is termed a lentil, lens, or lenticle. A tongue is a wedge-shaped member that projects out or extends out from the main body of a formation or unit; it may pinch-out within another formation. Under the *North American Stratigraphic Code,* organic reefs, carbonate mounds, and carbonate buildups may be formally recognized as members within a formation, or as formations in their own right.

The smallest formal lithostratigraphic unit is the bed (if a single layer or bed is named as a single lithostratigraphic unit; for example, the Alpha Coquina Bed) or beds (if a sequence of contiguous layers or beds is named as a single lithostratigraphic unit; for example, the Beta Limestone Beds). Beds may be recognized and named within formations or members. As used in the sense of recognizing formal beds, a bed is a layer (or layers) of rock that may range in thickness from about a centimeter to a few meters (Hedberg, 1976, p. 33). According to the *International Stratigraphic Guide,* layers of thickness less than about a centimeter are termed laminae. The terms bed and beds are often also used in an informal manner, coupled with a geographic name, to refer to informal lithostratigraphic units that, if or when formalized, might be considered to pertain to formal units of any rank (from group or above to bed). In some cases, units that traditionally have been designated as beds, even in a formal sense, are equivalent to units that now generally are considered to be of the rank of member or formation. The formal designation of beds usually is pursued sparingly. Many key beds or marker beds are named as informal units. In most instances where formal beds are designated, only particularly distinctive beds are named, and all intervening beds are left unnamed. A bed may extend beyond the lateral limits of a particular formation or member.

As used by the *North American Stratigraphic Code,* a flow is the extrusive igneous (volcanic) rock equivalent (equivalent in rank) of a bed or beds among sedimentary rocks. A flow is a lithostratigraphic unit, but according to the *Code* apparently need not be distinguishable solely on the basis of traditional lithologic criteria (such as composition or texture). For instance, a flow may be disitnguishable solely by order of superposition or paleomagnetism.

The lithostratigraphic unit most commonly used, which is next in rank higher than a formation, is the group. A group generally is composed of an aggregation or sequence of contiguous formations, or intervals of rock that may eventually be designated as formations, that share some lithologic features. A group may be entirely divided into named formations, or, particularly in the case of rocks where the stratigraphy has yet to be thoroughly elucidated, it may contain intervals of rocks that have not been referred to formalized formations. However, if a group in a certain area has not been even partially subdivided into component formations, then the said group is essentially a formation, and the *International Stratigraphic Guide* suggests that it be referred to simply as a formation. A group in one area, bearing a formal name of geographic origin, may be laterally equivalent to a formation in a different area that bears the same formal name of geographic origin.

The formations composing a group need not be the same from one geographic area to another. A formation may laterally wedge-out within a group, or it may extend laterally from one group to another. The type sections or reference sections of a group normally are considered to be the type or reference sections of the component formations of the group; thus a group may not have its own type section per se. In rare instances, however, a group is initially named prior to the naming of any of the component formations, with the expectation that the rocks of the group will be subsequently subdivided into formations; in other cases, a group may originally be designated as a formation. In such instances there may be a type section specifically for the group per se. When a single formation is subdivided into formal lithostratigraphic units of formational rank, the old formation is raised to the rank of a group, and new names are applied to the new component formations (the old name is not normally applied in a restricted sense to one of the new component formations).

Groups sometimes are subdivided into subgroups; a subgroup has all of the essential characteristics of a group, but is of semewhat lesser scope. Subgroups are not formally discussed in the *North American Stratigraphic Code* but are listed in the *International Stratigraphic Guide.* Two or more contiguous groups, or a group of groups and formations (some of the formations may not belong to a formally named group), may be associated into a supergroup. As the *Code* notes, the term series sometimes has been employed in a formal sense to designate an assemblage of formations and/or groups. In formal lithostratigraphic terminology, such series now usually are referred to as groups or supergroups. Informally, the word series still may be used occasionally in the older sense, but as a formal term series is strictly only a chronostratigraphic unit and rank.

The *International Stratigraphic Guide* (Hedberg, 1976, p. 35) uses the term complex as a formal lithostratigraphic unit of variable rank (usually equivalent to a group, formation, or member) that is "composed of diverse types of any class or classes of rocks (sedimentary, igneous, metamorphic) and characterized by highly complicated structure to the extent that the original sequence of the component rocks may be obscured." The term basement complex is also commonly used in an informal sense to refer to predominantly igneous and metamorphic rocks that have been highly deformed and underlie predominantly sedimentary sequences in many areas. The *North American Stratigraphic Code* (NACSN, 1983; see Appendix 1) uses the term complex in a similar sense to that of the *Guide,* but as a lithodemic term (see below).

In including under the auspices of lithostratigraphy all rock types or classes (intrusive and extrusive igneous rocks, metamorphosed rocks, and sedimentary rocks), some of which do not conform to the principle of superposition, the 1976 version of the *International Stratigraphic Guide* noted some particular difficulties in attempting to classify certain highly deformed metamorphic and intrusive igneous rocks lithostratigraphically. Extremely deformed and structurally complicated packets of such rocks may be relegated to the status of complexes. Another problem concerns rock bodies that are intrusive into preexisting (and therefore older) rock bodies; such intrusions may be of an igneous or sedimentary nature (for example, mobilized sandstone or salt deposits). In cases of intrusions or extrusions of sedimentary rocks, the *Guide* suggests that the rock body under consideration either may be referred to as a displaced part of the stratum from which it originated, or may be considered a new lithostratigraphic unit in its own right. The *Guide* (Hedberg, 1976, p. 36) dismisses similar cases involving igneous intrusives by simply stating: "These cross-cutting igneous bodies do not in themselves constitute stratigraphic units but they are, of course, an important part of the lithologic picture. They may conveniently be referred to as 'associated' with the strata of the lithostratigraphic units that they transect.''

Recently the ISSC (1987b) has published further discussion and guidelines concerning the stratigraphic classification and nomenclature of igneous and metamorphic rock bodies. The current guidelines and discussion of the ISSC (1987b) summarize consensus opinions reached by its members and by various geologists polled by the ISSC, including particularly geologists concerned or familiar with the study of igneous and metamorphic rock bodies. As yet the guidelines of the ISSC in these matters remain fairly general (they do not form a "code" in the sense of the *North American Stratigraphic Code*); the hope is expressed in the 1987 discussion (ISSC, 1987b) that more specific and detailed recommendations can be developed in the future.

The ISSC strongly asserts that all rock bodies, including those composed of intrusive igneous rocks and bodies of metamorphic rocks of undetermined origin (that is, generally nonstratiform rock bodies) are part of the domain of stratigraphy; even in a strict sense, metamorphic rock bodies have unequivocal stratigraphic significance, are therefore subject to the rules of stratigraphic classification and nomenclature, may be considered to form stratigraphic units, and may be legitimately classified lithostratigraphically. The ISSC (1987b) also points out that traditionally all sedimentary rock bodies, even if nonlayered (such as some reefs) have been included within stratigraphy *sensu stricto* and have been classified stratigraphically. In these opinions the ISSC (1987b) supports and elaborates upon the concepts developed in the first edition of the *International Stratigraphic Guide* (Hedberg, 1976; at this writing a revised version of the *Guide* is being prepared by the ISSC):

> That nonstratiform rock bodies fall within the scope of stratigraphy stands to reason if it is agreed that one of the most important objectives of stratigraphy is to determine the relationship in space and the succession in time of rocks of the Earth's crust and to fit them into a chronology which allows the reconstruction of geologic history. Nonlayered igneous and metamorphic rocks often provide particularly valuable information to attain this objective. Not only are they a source of geochronometric (numerical) ages determined by isotopic methods, but they often provide critical age information through the establishment of their cross-cutting and boundary relationships with layered and/or nonlayered rocks with which they are associated. An intrusive body, for example, is younger than the rocks it intrudes, and older than those that intrude it. An eroded intrusive or metamorphic rock body is older than the strata that were laid down unconformably upon it, and older, too, than the clastic sedimentary deposits containing the products of its erosion. (ISSC, 1987b, p. 441)

The ISSC (1987b) recommends unequivocally that all igneous and metamorphic rocks bodies (not just extrusive igneous, metasedimentary, and metavolcanic bodies) be classified lithostratigraphically (not lithodemically). Formal units should be named with "an approriate local geographic term combined with either (1) a unit-term indicating its rank—group, formation, member—or (2) a simple field lithologic term—granite, gneiss, schist—indicating its predominant rock type, or both" (ISSC, 1987b, p. 440; these recommendations conform with those of the *International Stratigraphic Guide,* see below). Such units traditionally have been named and mapped as lithostratigraphic units whether they were explicitly considered litho-stratigraphic or not. The ISSC (1987b, p. 442) acknowledges that "most geologists, for instance, may agree that the terms 'group,' 'formation,' and 'member' imply stratification and position within a stratified sequence showing original layering, and may not want, therefore, to apply them to nonstratified intrusive igneous rocks." If so, then such rocks still can be classified lithostratigraphically, even if essentially unranked, by simply designating them by a simple field lithologic term (for instance, "granite"). The term complex (a term of variable rank in the litho-stratigraphic hierarchy) may be used for igneous, metamorphic, and sedimentary rock bodies, or any combination of rock bodies of differing types, as needed (ISSC, 1987b; this term is particularly useful for a rock body composed of various litholo-gies that have been strongly deformed and/or metamorphosed).

The opinions of the ISSC on these matters are somewhat in contrast to those expressed in the *North American Stratigraphic Code* (NACSN, 1983) where predom-inantly intrusive and highly deformed and/or highly metamorphosed rock bodies are relegated to a separate class of "stratigraphic" units termed lithodemic units (see below). Furthermore, even though included within a stratigraphic code, it is not absolutely clear that the lithodemic units of the *Code* are considered to be strati-graphic units *sensu stricto,* although they certainly are considered stratigraphic units *sensu lato* by the *Code* and as a class of stratigraphic units (even if not an acceptable class) by the ISSC (1987b). Here it should be pointed out that the *Code* includes other types of units, such as geochronologic units and geochronometric units, that are explicitly excluded from being considered stratigraphic units by the *Code.*

LITHODEMIC UNITS

In 1983 the NACSN (1983) proposed a new formal procedure for classifying and naming rock bodies that do not conform to the principle of superposition. Unlike the 1976 version of the *International Stratigraphic Guide* (see discussion and the quotation concerning cross-cutting igneous bodies, above), the NACSN (1983) treats all such rock bodies as capable of being classified as valid stratigraphic units. How-ever, instead of using lithostratigraphic units for such rock bodies as proposed by the ISSC (1987b), the *North American Stratigraphic Code* (NACSN, 1983) intro-duced a new category, lithodemic units. Lithodemic units are composed predomi-nantly or wholly of igneous intrusive, highly metamorphosed, and/or highly de-formed rocks that often do not conform either in part or in full to the law of superposition (such rocks generally lack primary stratification—either because it was never present, or because it has been obliterated or destroyed). Like lithostrati-graphic units, lithodemic units are defined, distinguished, and delimited on the basis of lithic (lithologic, rock) characteristics (such as rock type, composition, texture, and so on). Inferences as to genesis, or geologic history, are not to be incorporated into the definition of a lithodemic unit although in some cases age may enter into the definition. According to the *Code* (NACSN, 1983, p. 859), rock masses that are "lithically similar but display objective structural relations that preclude the possibility of their being even broadly of the same age . . . should be assigned to different lithodemic uints." As is the case with boundaries between, and delimiting, lithostratigraphic units, the boundaries of lithodemic units are based on lithologic characteristics, and may be abrupt or placed arbitrarily within zones of gradation.

The fundamental unit of lithodemic classification is the lithodeme; in the litho-demic hierarchy it is comparable to the formation of the lithostratigraphic hier-archy. Like formations, lithodemes must be mappable on the surface or traceable in the subsurface. According to the *North American Stratigraphic Code,* lithodemes are not formally divisible into units of lower rank. Informally, however, they may be divisible into zones (for example, the pegmatized zone of a particular formal lithodeme) or other informal divisions.

A suite is the unit or rank next higher than a lithodeme in the lithodemic hier-archy. Comparable to a group in lithostratigraphy, a suite is composed of two or more associated lithodemes (either named or unamed, but usually at least some are named) that are of the same general class of rocks (for instance, metamorphic). Analogously to groups, the component lithodemes composing a suite may change from place to place, and if locally a suite loses its subdivisions (perhaps they no longer are readily distinguishable), it may be referred to as a lithodeme while retain-ing the same formal name. When a unit originally regarded as a lithodeme is subdi-vided into component lithodemes, the original lithodeme normally is raised to the rank of a suite.

The ISSC (1987b, p. 442) has objected specifically to the use of the term suite proposed by the *North American Stratigraphic Code.* "[T]he use of the term 'suite' for nonlayered igneous and metamorphic rock bodies seems inadvisable. The term has been commonly used for associations of apparently comagmatic intrusive ig-neous rock bodies of similar or related lithology and close association in time, space, and origin. It has also been used widely in the Soviet Union for local stratigraphic units corresponding approximately to a formation" (see Chapter 4).

A supersuite is a lithodemic unit composed of two or more suites or complexes (as used by the *North American Stratigraphic Code,* see below). The component units of a supersuite, the suites or complexes, should have "a degree of natural relationship to one another, either in the vertical or the lateral sense" (NACSN, 1983, p. 860). The meaning of this last statement is not further clarified by the *Code;* presumably the intention is that the suites or complexes united in a supersuite should be adjacent to one another (either vertically or horizontally) and share some physical characteristics and/or a common geologic history.

The term complex is used in the *North American Stratigraphic Code* (NACSN, 1983) in a manner similar to its use in the *International Stratigraphic Guide* (Hed-berg, 1976; see above), except that the *Code* considers the complex to be a litho-demic unit. A complex, composed of an assemblage of igneous, sedimentary, or metamorphic rocks, is unranked but usually equivalent or comparable to a suite or supersuite. A complex may be composed of, in part, lithodemes or lithostratigraphic units (such as formations, and so on). In such instances, it may be that a complex originally was named and mapped during early stages of the geologic investigation of a particular area; later, more detailed investigations may have identified compo-nent lithodemes or formations while the original concept of the complex was re-tained. Two special types of complexes that the *North American Stratigraphic Code* recognizes are volcanic complexes and structural complexes.

Just as the use of the term series to designate a lithostratigraphic unit now is considered improper (see above, and NACSN, 1983), the term series should not be applied to lithodemic units. Lithodemic units that have been designated series in the past are usually of the rank of suite, supersuite, or complex.

FORMAL NOMENCLATURE OF
LITHOSTRATIGRAPHIC AND LITHODEMIC UNITS

Under both the *Guide* and the *Code* (see Appendix 1 for the legalities of nomenclature established by the *North American Stratigraphic Code*), formal lithostratigraphic and lithodemic units usually bear compound names consisting of a geographic name and either a rank term (for example, formation, group, complex, lithodeme, suite) or a descriptive lithic (or related) term naming the dominant rock type of the unit (for example, limestone, shale, coal, schist, gabbro, dike, sill, pluton, intrusion). In a few instances the name of a formal unit may include a geographic name, a lithic term, and a rank term.

The geographic component of the name of a formal lithostratigraphic or lithodemic unit is derived from some named geographic feature (natural or artificial; for instance, a river, mountain, town) in or near the type locality of the unit. The feature after which a stratigraphic unit is named should, if possible, be the name of a relatively permanent feature; the name should be found on professional, official, or otherwise comparable maps of the area (for example, state, county, forest service, topographic maps). If necessary, the geographic name from which the name of the stratigraphic unit is derived may be described and identified (preferably on a map) in the same publication as that describing the stratigraphic unit. It is often customary for higher-ranking stratigraphic units (which themselves generally cover wide areal extends) to bear names derived from more extensive geographic features than those of lower-ranking stratigraphic units. No two stratigraphic units should bear the same geographic name. This principle applies also to a unit that is a subdivision of a larger unit; for instance, the Rhode Island Formation should not contain a Rhode Island Member. In some cases identical geographic names are applied to minor units that occur within one vertical sequence (such as the lower Attleboro coal, the Attleboro sandstone, the upper Attleboro coal), but such nomenclature is informal. Once a geographic name is applied to a stratigraphic unit, the use of that geographic name for another stratigraphic unit generally is precluded (see Article 7 of the *North American Stratigraphic Code* [Appendix 1]; various indexes and lexicons of geologic names are published frequently, such as Keroher, 1970; Luttrell et al., 1981, 1986; Wilmarth, 1957; and Wilson, Keroher, and Hansen, 1959, for United States geologic names). The only exception to this general rule is that, according to the *Guide* (Hedberg, 1976, p. 21), in some instances a name may be duplicated if "geographic separation precludes confusion."

Once the spelling of the geographic component of a stratigraphic name is established (that is, generally accepted and used) in the stratigraphic literature, the spelling is not changed because the spelling of the name of the original geographic feature has been changed, or because the geographic name has been repeatedly misspelled in the stratigraphic literature (that is, the commonly used spelling, or misspelling, may be retained). However, if the name of a stratigraphic unit repeatedly has been spelled in various ways in different stratigraphic publications, its spelling may be standardized to conform to that of authoritative geographic studies and publications. Likewise, the name of a stratigraphic unit is not changed because the name of the geographic feature after which it is named has been changed. The name of a stratigraphic unit is retained and considered valid even if the geographic feature after which it was named no longer exists. When a stratigraphic unit crosses political or linguistic boundaries, the geographic portion of the name of the stratigraphic unit remains the same (it is not translated, and the spelling is the same), although the lithologic or rank component of a stratigraphic name may be translated from one language to another.

It generally is considered inappropriate to use a particular geographic name for a stratigraphic unit if the name suggests, or is homonymous with, another better-known locality, region, or area (political, artificial, or natural). Thus, to give an example cited by both the *Guide* and the *Code,* the name "Chicago Formation" should not be applied to a stratigraphic unit in California, even though there is a Chicago in California (USA), because most persons associate the name Chicago with a large city in Illinois (USA). Of course, deciding which is the better-known locality of two or more homonymous localities can be a matter of judgment; in such instances, it may simply be a question of which locality takes priority in having a stratigraphic unit named after it (given that more than one locality could lend its name to a stratigraphic unit).

The *International Stratigraphic Guide* (Hedberg, 1976, p. 42) notes that lithostratigraphic units established on the basis of offshore subsurface units (for example, units known only from offshore wells), or in the context of underwater mapping, may pose nomenclatural problems. In such cases, it may be difficult to find adequate geographic locality names after which the units may be designated. The *Guide* suggests that such units may be named after an offshore well if the well has in turn been given a name derived from "coastal, oceanic, or other features"; otherwise, it may "be necessary to use purely arbitrary nomenclatural designations" (Hedberg, 1976, p. 42).

Among lithostratigraphic units (NACSN, 1983), a formational name may be composed of a geographic name component and the term formation or a lithic descriptive term (such as sandstone, limestone, or tuff). Groups are always named by combining a geographic name with the term group. Members are named by combining a geographic term with the word member; optionally, an intervening lithologic designation may be inserted between the geographic name and the term member. Beds and flows combine three terms—a geographic term, a lithologic designation, and the appropriate word from the following: bed, beds, flow, flows.

The name of a lithodeme (NACSN, 1983) is composed of a geographic term and a lithologic designation (such as granite) or descriptive term (such as sill, dike, intrusion). The formal name of a suite is composed of three terms; a geographic name, an adjective relating to the character of the suite (for example, metamorphic, plutonic, intrusive), and the word suite. The name of a supersuite simply combines a geographic name with the word supersuite.

The first letter of each word of a formal lithostratigraphic or lithodemic unit is standardly capitalized.

The lithologic or descriptive component (where used) of a lithostratigraphic or lithodemic name should be relatively simple and well known; compound terms and specialized lithologic nomenclature not generally used by nonspecialists usually are unacceptable. The *Guide* (Hedberg, 1976, p. 43; see also ISSC, 1987b), in discussing lithostratigraphic nomenclature, suggests that only purely lithologic (lithic, or lithographic) terms expressing the dominant rock type of a particular unit should be used. Lithogenetic terms (for instance, turbidite or flysch) should be avoided. The *Guide* recommends that terms such as "'dike,' 'stock,' 'pluton,' 'batholith,' and other similar names or more general terms such as 'intrusion,' are neither lithologic nor stratigraphic terms; accordingly, the names of such intrusive igneous bodies as the Idaho batholith or the Ordubad pluton should not be considered stratigraphic terms.'' These and similar terms, however, are acceptable as formal lithodemic terms under the *North American Stratigraphic Code.*

In formally establishing or modifying a lithostratigraphic or lithodemic unit, the same general procedures must be followed as those used in establishing or modifying any formal stratigraphic unit (see Chapter 4 and Appendix 1). For lithostratigraphic units a type section (stratotype) and type locality are essential, often supplemented by reference sections. A lithodemic unit must be based on type and reference localities (where applicable).

PARASTRATIGRAPHIC UNITS

The *North American Stratigraphic Code* and the *International Stratigraphic Guide* distinguish between formal and informal lithostratigraphic units, and understandably both documents place more emphasis on the formally defined and named units. In practice, however, many extremely useful stratigraphic units may never be formalized, or may not even fit into any of the units recognized by the conventions of the *Code* or the *Guide.* Such a unit, if it fits the concept of some conventionally recognized stratigraphic unit, may be considered an informal stratigraphic unit that could be formalized. An example of such an informal unit would be a body of rock that fulfills all the requirements necessary to be dubbed a formation (it is distinguishable on the basis of lithologic criteria from adjacent formations, is characterized by lithologic homogeneity, is mappable, and so on), but has never been formally named. However, some "stratigraphic units" do not readily meet the requirements of any conventionally recognized stratigraphic units—for instance, a unit composed of all the rock (no matter how heterogeneous) between two marker beds. In the context of a discussion of lithostratigraphic classification, Krumbein and Sloss (1963) have referred to unconventional units as parastratigraphic units. Using the terminology of the *Guide,* such units may be considered various types of informal zones.

Parastratigraphic units are defined by Krumbein and Sloss (1963, p. 333) as including "a variety of groupings of strata, identified by objective lithologic criteria but lacking in either mappability or, most commonly, in lithologic homogeneity and constancy." As an alternative term, Krumbrein and Sloss (1963, p. 338) referred to parastratigraphic units as operational units. Here I would suggest that the term parastratigraphic unit, if used, be restricted to the latter half of Krumbein and Sloss's (1963) definition. Under the current *Code,* a lithostratigraphic unit that is not mappable (at least, not at the scale of formations) may be designated a member. Thus a parastratigraphic unit is a body of rock distinguished by some physical (that is, lithologic) criterion, but lacking in the lithologic homogeneity or uniformity normally associated with formal lithostratigraphic units.

Within their category of parastratigraphic units, Krumbein and Sloss (1963) distinguished two basic kinds of parastratigraphic units: attribute-defined units and marker-defined units. An attribute-defined parastratigraphic unit is distinguished or based upon some special attribute(s) of the component strata forming the parastratigraphic unit, but the special attribute(s) is not of the nature of those attributes generally considered in establishing conventional or formal lithostratigraphic units. For example, an attribute-defined parastratigraphic unit may be established on the basis of trace-element content, heavy-mineral composition, insoluble residues, or geophysical characteristics of strata. As Krumbein and Sloss (1963) note, in some cases (particularly over limited geographic areas) the boundaries of units defined by such characteristics may well coincide with the boundaries of conventionally or formally recognized lithostratigraphic units. In other cases, however, such parastratigraphic units may be markedly discordant with conventional stratigraphic units, especially if a parastratigraphic unit can be extended laterally beyond the boundary of conventional or formal lithostratigraphic units. Krumbein and Sloss (1963) cite the seismicity-based units utilized in seismic surveying as sometimes forming extreme examples of parastratigraphic units. However, in modern terminology, some such units may form conventional unconformity-bounded stratigraphic units (see below).

A marker-defined parastratigraphic unit is a unit or body of rock composed of all the rock between a pair of upper and lower key beds or marker horizons, but not all such units are necessarily parastratigraphic units. If the strata between two key beds are characterized by homogeneity, constancy, or uniformity of lithic characteristics, then the unit thus defined may be treated as a formal lithostratigraphic unit. If each bounding key bed is considered to be a synchronous surface, then the unit thus defined may be considered a chronostratigraphic unit. However, if the strata between the bounding key beds are heterogeneous (lack lithologic constancy), and the bounding key beds are considered to be time-transgressive, then the unit fits the concept of a marker-defined parastratigraphic unit. The key beds or horizons may be anything from thin lithologic units (such as thin limestones, sandstones, or bentonite beds) to reflecting horizons on seismic sections. In many instances, especially for mapping, stratigraphic, and structural analysis on a regional scale, marker-defined parastratigraphic units may be extremely important. Often such units have been considered, perhaps incorrectly to correspond to or closely approximate chronostratigraphic units. Various designations that have been applied more or less synonymously to the concept of marker-defined parastratigraphic units include "lithozone," "format," and "assise."

As should be apparent from the preceding discussion, the concept of parastratigraphic units in some ways forms a conceptual bridge or link between the conventionally or formally recognized lithostratigraphic units and other types of stratigraphic units reified by the *Code* and the *Guide*. Certain marker-defined parastratigraphic units may form a distinct class of stratigraphic or quasi-stratigraphic units in their own right, whereas some attribute-defined and marker-defined parastratigraphic units imperceptibly grade into various conventionally recognized stratigraphic units.

LITHOCORRELATION

Lithocorrelation involves the demonstration of equivalency or correspondence of rock bodies, strata, or stratigraphic units on the basis of lithologic characteristics and stratigraphic position. The *North American Stratigraphic Code* (NACSN, 1983, p. 851, italics in the original) defines the term as follows: "*Lithocorrelation* links units of similar lithology and stratigraphic position (or sequential or geometric relation, for lithodemic units)." The concept of lithocorrelation involves the lateral continuity of strata and stratigraphic units. Thus, as Krumbein and Sloss (1963) among others have noted, the easiest and most accurate way to establish lithocorrelations is by tracing lithostratigraphic units from one area to another along continuous outcrops on the surface (a method often referred to as "walking out" a unit or correlation). Similarly, but with a lesser degree of accuracy, a lithostratigraphic unit may be traced laterally through data from wells and boreholes and other subsurface data. When walking out a unit, one may follow a contact between a subjacent and a superjacent unit, or one may concentrate on tracing a key bed of particularly distinctive lithology. In many cases it may be possible to trace a certain unit even through soil-covered areas, as characteristic soil and vegetational patterns may be developed above the unit under consideration; in this respect aerial photographs may be particularly helpful. Certain lithostratigraphic units also may be recognized by a characteristic geomorphological expression in the area under study.

When a lithostratigraphic unit is traced laterally, it may be found to intertongue or grade into a unit representing a different facies. Even if such units are of very different lithologies, they still may be of similar stratigraphic positions (perhaps they both overlie a single formation and are themselves overlain by another formation) and thus be considered to be lithocorrelatable. However, if two lithostratigraphic units were of virtually identical lithologies and shared innumerable physical properties, they would not be considered to be lithocorrelatable if they were of widely differing stratigraphic positions. For example, if two lithostratigraphic units occur in the same section and are separated stratigraphically by a unit of dissimilar lithology, then the two never would be considered to be lithocorrelatable because they are not of equivalent stratigraphic position (of course, they both may be included within a single lithostratigraphic unit of higher rank). In other words, lithostratigraphic correlation always involves a consideration of stratigraphic position; commonly this is stratigraphic position relative to other lithostratigraphic units, but it also may involve stratigraphic (sequential) position relative to any geologic features, such as unconformities, distinctive stratigraphic horizons, and in some cases even such phenomena as igneous intrusions, faults, folds, metamorphic effects, and so on.

In a general way, considerations of stratigraphic position may translate into a consideration of the temporal relations of the rock bodies involved. In practice, in certain regions at isolated exposures it may be very difficult to determine what formation is present on the basis of lithology alone. However, in a given region formations of extremely similar lithology may differ widely in age, and a few fossils used as an indicator of the general age of the particular rock under consideration may serve immediately to narrow the possibilities of what formation may be represented and to allow identification of the formation exposed in the particular outcrop under consideration. Yet it is important to note that two rock bodies or units may be lithocorrelatable even if they do not share a single time plane in common—that is, the entire unit in one area may have formed before the beginning of deposition of the lithocorrelatable unit in another area—if they both bear the same relative stratigraphic positions and presumably can be continuously traced, one to the other, laterally (or at least at one time they formed a continuum.).

It should also be pointed out that in some stratigraphic sections in particular areas cyclical successions of repeating lithologies may be encountered, for example, some of the cyclothems of the Upper Carboniferous (Pennsylvanian). Because the fine details on one cyclothem may be extremely similar to the details of another cyclothem, in such cases it may be extremely difficult for an investigator to lithocorrelate individual beds or lithologies (perhaps named as members or formations) through discontinuous exposures even when attempting to take relative stratigraphic position into account. In attempting lithocorrelations in such areas, one usually must treat the cyclothems themselves as units and correlate the cyclothems relative to one another before attempting to lithocorrelate individual lithologies or units within the cyclothems.

Lithocorrelations made on the basis of mechanical or geophysical logs of wells and boreholes (see Chapter 2) rely heavily on the concept of using relative stratigraphic position (position in a sequence) in establishing lithocorrelations. Points on such logs typically are proposed as correlatable, or as being corresponding points, not because they demonstrate the same absolute value of the parameter being measured, but because they represent similar or identical patterns of relative highs or lows (for example) within the context of a continuous recording corresponding to stratigraphic position. The term interval correlation has been applied to the correlation of marker horizons, such as those inferred or traced on the basis of distinctive points or kicks on continuous logs (although the same term can be applied to the correlation of any marker horizons or key beds, whether biostratigraphic, lithostratigraphic, or otherwise recognized), and the strata between the horizons correlated. Often the strata between the correlated horizons may be of very dissimilar lithology when traced laterally, and constitute a marker-defined parastratigraphic unit (Krumbein and Sloss, 1963; see above).

Apparently using the concept of lithocorrelation in an even broader sense than that above, or perhaps not realizing that hydrostratigraphic and similar units are not necessarily synonymous or coextensive with conventional lithostratigraphic units, Krumbein and Sloss (1963) have suggested that the behavior of fluids (such as water, oil, or gas) may be used in establishing lithostratigraphic correlations, or minimally in demonstrating some sort of lateral or quasi-lateral continuity between strata. As an example, Krumbein and Sloss (1963, p. 340) propose the following: "If withdrawal of oil from a producing well affects the bottom-hole pressures and gas–oil ratios in other wells, an interconnection is indicated and the correlation of the productive unit throughout the area may be considered established." Likewise these authors suggest that the correlation of aquifers can be established on the basis of similar drawdown effects, or perhaps even by introducing dyes or chemical or isotopic tracers into waters and then recording in what aquifers they are subsequently found to occur.

In some cases correlations between two or more named and defined lithostratigraphic units may be problematic or obscured simply because even though the actual sequence of strata (lithologic types) is very similar in two adjacent regions, and lithocorrelations can be established precisely, the boundaries of the formally (or, for that matter, informally) recognized units do not correspond. Differing definitions of units within essentially the same broad sequence of rocks may be simply a function of differing emphases that the original definers of the respective units placed on particular lithologic characters used in determining the boundaries of units. As discussed elsewhere, the boundaries of many lithostratigraphic units (the same holds true for many other types of stratigraphic units) may be defined relatively arbitrarily, but with equal validity (at least in the eyes of the *Code* and the *Guide*), at a number of different points.

PEDOSTRATIGRAPHIC UNITS

The current *North American Stratigraphic Code* (NACSN, 1983; see Appendix 1) recognizes a type of stratigraphic units, distinct from all others, termed pedostratigraphic units. A pedostratigraphic unit is, in a loose sense, a buried soil horizon or buried paleosol. As the *Code* points out, however, a pedostratigraphic unit is not synonymous with any paleosol. As generally used, a paleosol refers to any soil that formed on an ancient landscape. Such a soil may have been buried and remained buried, it may have been buried and then exhumed, or it may remain on the surface as a relict soil. A pedostratigraphic unit can be thought of as a paleosol, or part of a paleosol, that has been buried and preserved enclosed in rock. A pedostratigraphic unit is defined as composed of rock that consists of, or represents, one or more pedologic horizons (that is, essentially soil horizons); so, in this respect, recognizing and establishing a pedostratigraphic unit explicitly requires some degree of inference and interpretation. This is in contrast, for example, to the recognition and establishment of lithostratigraphic units, which ideally does not require any interpretive inference. The pedologic horizons that pedostratigraphic units may include are the A horizon and/or B horizon of an ancient soil profile (depending upon what is preserved); O horizons (organic debris on top of, or forming the uppermost layer of, a soil) are by definition (NACSN, 1983) excluded from pedostratigraphic units, and C horizons (formed from weathered rock) may be included in whole, in part, or not at all in these units. Unlike the boundaries of some stratigraphic units, which may be placed arbitrarily within gradational zones (for instance, lithostratigraphic units), the boundaries of pedostratigraphic units must be placed at definite physical boundaries. The lower boundary of a pedostratigraphic unit, in particular, must be placed at the lowest definite physical boundary of a pedological horizon; thus if there is a clearly defined physical boundary between the ancient B and C horizons, but the base of the C horizon is not clearly defined, then the base of the pedostratigraphic unit must be placed at the base of the B horizon, and the C horizon must be excluded from the pedostratigraphic unit. A C horizon, or saprolite, alone may not be considered a pedostratigraphic unit. Of course, a saprolite may form the immediate parent material upon which a soil is developed, and eventually the soil may become buried in the rock record and be preserved, serving as the basis of a pedostratigraphic unit.

A formal pedostratigraphic unit is established only within the context of a lithostratigraphic, lithodemic, or allostratigraphic classification of the rocks under consideration. By definition (see Appendix 1) a pedostratigraphic unit must be developed in a lithostratigraphic, lithodemic, or allostratigraphic unit and must be overlain by a formal lithostratigraphic or allostratigraphic unit. A pedostratigraphic unit and/or its boundary with another stratigraphic unit may be traced laterally from one lithostratigraphic or allostratigraphic unit to another.

In the field, pedostratigraphic units are recognized by such characteristics as color, texture, concretions, stainings, structure, accumulations of organic matter, and so forth. In some instances laboratory work, such as analysis of the mineralogy and micromorphology of the material, may be required for correct identification of a particular pedostratigraphic unit. A formal pedostratigraphic unit should possess some degree of lateral traceability.

The only formal unit of pedostratigraphic classification is the geosol, referring to a single buried soil or a portion of a single buried soil now preserved in the rock record. The formal name of a geosol consists of a geographic name followed by the word geosol (both are capitalized when used in a formal sense); the same general procedures used in establishing or modifying any formal stratigraphic unit are followed in establishing or modifying a formal geosol (see Chapter 4 and Appendix 1). The *North American Stratigraphic Code* suggests that a pedostratigraphic unit be based on a composite stratotype (in order to characterize the potential range of physical and chemical properties that may distinguish such a unit) within a type area. In many cases, a pedostratigraphic unit may be recognized but named informally by reference to the rocks that lie below and above the unit. If several buried soils are merged or "welded" (NACSN, 1983, p. 865) together such that the individual geosols cannot be distinguished, then the *Code* mandates that formal pedostratigraphic nomenclature be abandoned and informal names be used.

The immediate predecessor of the *North American Stratigraphic Code,* the *Code of Stratigraphic Nomenclature* (ACSN, 1970), recognized a class of soil-stratigraphic units that have since been abandoned. The old soil-stratigraphic units were similar to, but not identical with, pedostratigraphic units. "A soil-stratigraphic unit is a soil with physical features and stratigraphic relations that permit its consistent recognition and mapping as a stratigraphic unit. Soil-stratigraphic units are distinct from both rock-stratigraphic and pedologic units" (ACSN, 1970, p. 10). The single formal unit of soil-stratigraphic classification was the "soil." The boundaries of a soil could be placed at sharp contacts or in zones of gradation. According to the *Code of Stratigraphic Nomenclature* (ACSN, 1970, p. 10), the difference between a pedologic unit (a soil unit that is not a stratigraphic unit, perhaps as used by some soil scientists and engineers) and a soil-stratigraphic unit was that "stratigraphic relations are an essential element in defining a soil-stratigraphic unit but are irrelevant in defining a pedologic unit. A soil-stratigraphic unit may comprise one or more pedologic units or parts of units." There was no discussion in the 1970 code in regard to the relationship of a soil-stratigraphic unit to other types of stratigraphic units; presumably a soil-stratigraphic unit could be based on any kind of paleosol (see above) as long as it was physically recognizable or traceable (mappable) and of stratigraphic significance (had consistent stratigraphic relations). Soil-stratigraphic units were formally named in a manner similar to that of pedostratigraphic units; they took geographic names (derived from a feature of their type area) combined with the word soil. Some old soil-stratigraphic units may be redefinable as pedostratigraphic units.

UNCONFORMITY-BOUNDED AND ALLOSTRATIGRAPHIC UNITS

As has been pointed out (see Chapter 1), many of the units of the current chronostratigraphic scale originally were recognized and defined as suites of rocks bounded by distinct, abrupt, and major lithologic/faunal changes, breaks, unconformities, or discontinuities. The boundaries of such units have since been redefined (or are being redefined) to fit current notions of chronostratigraphy. Likewise, units that are recognized primarily as being bounded by unconformities have been used at all scales and ranks (from the level of supergroup and higher to that of member). Yet until recently such unconformity-bounded stratigraphic units have on the whole remained relatively informal, a situation that can be attributed to various factors. What were perhaps originally unconformity-bounded units (for example, many of the systems as originally defined) were redefined and interpreted chronostratigraphically. This occurred as the concept of synchronous worldwide episodes of diastrophism and tectonicism, thought to be responsible for major natural breaks in the stratigraphic record, fell into disfavor. There was no formal recognition of unconformity-bounded units in the *International Stratigraphic Guide* (Hedberg, 1976; the *Guide* only mentions [p. 92] the concept of such units in passing) or in the various American codes of stratigraphic nomenclature prior to the 1983 code. Various workers of the mid-twentieth century, using the concept of major units ("sequences") distinguished by bounding unconformities, considered such units to be lithostratigraphic (rock-stratigraphic) units and not distinct and separate types of stratigraphic units (e.g., Krumbein and Sloss, 1963; Sloss, Krumbein, and Dapples, 1949). Some workers recognized what are essentially unconformity-bounded units ("sequences") but extended such sequences beyond the bounding unconformities, where such units were bounded by "conformities" believed to be correlative with the boundary unconformities (Mitchum, Vail, and Thompson, 1977; see ISSC, 1987a,c, 1988).

Recently there has been renewed interest in the concept of groups of strata that are defined and identified solely on the basis of bounding discontinuities or unconformities (e.g., Chang, 1975; ISSC, 1987a; Vail et al., 1977) as well as the "natural" units of time and strata that might be delineated by such boundary unconformities. This has resulted in proposals for the formalization of unconformity-bounded stratigraphic units by both the ISSC (1987a) and the NACSN (1983). Relatedly, the NACSN (1983) has formalized the concept of diachronic units, as discussed below.

As the ISSC (1987a) uses the term:

an unconformity-bounded unit is defined as a body of rock bounded above and below by specifically designated, significant, and demonstrable discontinuities in the stratigraphic succession (angular unconformities, disconformities, etc.), preferably of regional or interregional extent. Being bounded by stratigraphic discontinuities is the single diagnostic criterion used to establish and recognize these stratigraphic units. An unconformity-bounded unit should be extended only as far as both of its bounding unconformities are identifiable. . . . For the purpose of establishing and recognizing unconformity-bounded units, an *unconformity* is defined as a surface of erosion and/or nondeposition between rock bodies, representing a significant hiatus or gap in the stratigraphic succession caused by the interruption of deposition for a considerable span of time. (ISSC, 1987a, p. 233, italics in the original)

As the ISSC notes, what exactly constitutes a significant hiatus or discontinuity is open to subjective interpretation. However, such a term is generally understood by most stratigraphers. J. G. Johnson (1987, p. 443, italics in the original) objected to the use of the term surface in the above quotation, stating: "A *surface* is the outer boundary of an object or body. A *contact* is the junction of two surfaces. Without two surfaces, there can be no lack of conformity. Therefore, an unconformity is not a *surface;* it is a *contact.*" In reply the ISSC (1987c, p. 444) noted that both in geological terminology and in standard English (according to the *Oxford Universal Dictionary*) an unconformity (or a similar entity) can be referred to as either a contact or a surface; an unconformity is represented by both.

The basic unconformity-bounded unit of the ISSC is the synthem; hierarchically, a synthem can be divided into two or more subsynthems, and two or more synthems

can be united as a supersynthem. A miosynthem is "a relatively small, minor synthem within a larger synthem, but not a component of a hierarchy of unconformity-
bounded units" (ISSC, 1987a, p. 236). Other terms that have been proposed as
unconformity-bounded units or units that are sometimes bounded by unconformities, of varying hierarchical rank, include mesothem, interthem, cyclothem, and sequence (see discussion by the ISSC, 1987a). In particular, the term sequence sometimes has been used in a sense approaching the same meaning as the term synthem,
but it has also been used as a chronostratigraphic term (equivalent to erathem:
Moore, 1958; Weller, 1960), as a lithostratigraphic term of extremely high rank
(larger than the supergroup: Krumbein and Sloss, 1963; Sloss, 1963), and as a hybrid
term for units that are in part unconformity-bounded and in part chronostratigraphic where such units are said to be bounded by isochronous horizons or conformities correlative with the unconformities (e.g., Mitchum, Vail, and Thompson,
1977; Ramsbottom 1977, 1978, 1979: see ISSC, 1987a). The term sequence has
found common application in seismic stratigraphy (see Chapters 2 and 7).

The ISSC (1987a) also points out that there are many processes that can be
recorded as unconformities in the stratigraphic record, but the recognition of
unconformity-bounded units is not based on the interpretation of the genesis, or
causal processes, behind unconformities or the deposits bounded by unconformities:

> Orogenic episodes, epeirogenic cycles, and eustatic sea-level changes are com-
> monly recorded by unconformities in the stratigraphic succession. Unconformity-
> bounded units, for this reason, have sometimes been considered to be equivalent
> to "sedimentary cycles" or tectonically controlled stratigraphic units: stratotec-
> tonic, tectostratigraphic, tectono-stratigraphic, or tectogenic units, tectonic
> cycles, tectosomes, structural or tectonic stages, and so on. All of these types
> of units, however, have a definite *genetic and causal meaning* and require for
> their recognition an *interpretation* of the observed stratigraphic relationships.
> Unconformity-bounded units, in contrast, entail no such genetic or causal inter-
> pretation. They are objective, noninterpretive units. Calling a unit a "tectonic
> stage," for instance, implies that the unconformities bounding the unit are the
> result of tectonic events; unconformity-bounded units, on the other hand, are
> established and recognized without any regard to the cause of their bounding
> unconformities, whether they are the result of orogenic events, epeirogenic epi-
> sodes, eustatic sea-level changes, or any combination of them. (ISSC, 1987a,
> p. 234, italics in the original)

Similarly, lithostratigraphic units are recognized to be independent of interpretations as to genesis or geologic history (NACSN, 1983).

The ISSC considers unconformity-bounded units to form a separate, distinct,
kind or category of stratigraphic units; they are separate from such categories as
lithostratigraphic, biostratigraphic, chronostratigraphic, magnetostratigraphic, and
other units. The rocks of an unconformity-bounded unit can simultaneously be classified using any other stratigraphic units; that is, an unconformity-bounded unit can
include any other kinds of stratigraphic units. The rocks composing an unconformity-bounded unit may be of any class, type, or combination of classes (sedimentary, igneous, metamorphic).

The ISSC (1987a, p. 234) suggested that "unconformity-bounded units have a
certain amount of chronostratigraphic significance because all of the rocks below

an unconformity or disconformity are older than all of those above, and time lines do not cross unconformity surfaces'' (cf. the chronostratigraphic significance of sequences, Chapter 7). Taking a narrow view of "chronostratigraphic significance," J. G. Johnson (1987, p. 443, italics in the original) criticized this statement of the ISSC stating: "There is, in fact, no chronostratigraphic significance to an unconformity-bounded unit, because its lower and upper surfaces are necessarily diachronous. There is no significance to an unconformity for the same reason. It is no more possible to have a *certain amount* of chronostratigraphic significance than it is to be a certain amount pregnant." J. G. Johnson (1987) went on to suggest that there can be instances where an unconformity can migrate over an area with time such that not even all of the beds above the unconformity are necessarily younger than all of the beds below it (e.g., Cohen, 1982). In response to Johnson's comments, the ISSC (1987c, p. 444) has written: "It has been claimed that in the process called 'ravinement' a very low angle unconformity can develop progressively landward with time so that some landward beds below the unconformity may be younger than some basinward beds above it. It is possible, however, that if studied in detail, ravinement may be found to represent a combination of multiple unconformities, not a single unconformity. In the great majority of cases, however, all of the rocks below an unconformity are indeed older than all those above it. For this reason, unconformity-bounded units provide evidence to establish superposition of stratigraphic succession and sequence in time—both of chronostratigraphic significance." From the last quotation, it seems to be clear that the ISSC intends that a single unconformity above and a single unconformity below bound an unconformity-bounded stratigraphic unit; however, this leaves open the question of how one can always separate and distinguish unequivocally single unconformities. Perhaps in instances where single unconformities cannot be established with a fair degree of certainty, formal unconformity-bounded units should not be erected. Furthermore, an unconformity-bounded unit need not have chronostratigraphic significance to be considered a valid unconformity-bounded unit.

Like most formal stratigraphic units, an unconformity-bounded unit is given a formal name based on the name of a geographic feature in an area where the unit is well developed. The procedures for establishing, or otherwise describing or amending, an unconformity-bounded unit follow the general procedures to be used in establishing any formal stratigraphic unit (see Chapter 4; Hedberg, 1976). In particular, stratotypes for the bounding discontinuities that define the boundaries of the unconformity-bounded unit must be specified; different sections may serve as the stratotypes for the lower and upper boundaries.

The NACSN also has proposed a formalization of stratigraphic units bounded by unconformities, labeling such units allostratigraphic units. "An allostratigraphic unit is a mappable stratiform body of sedimentary rock that is defined and identified on the basis of its bounding discontinuities" (NACSN, 1983, p. 865). Whereas synthems of the ISSC are of regional or interregional extent and may include rocks of any class, allostratigraphic units of the NACSN are in general of much more limited extent and encompass only sedimentary rocks. Allostratigraphic units are comparable in scale to lithostratigraphic units. The basic allostratigraphic unit is the alloformation. Allogroups and allomembers are of higher and lower ranks, respectively, than alloformations. Allostratigraphic units are named, defined, redefined, or revised according to the same principles and procedures adopted by the NACSN for lithostratigraphic units.

Allostratigraphic units are closely analogous to lithostratigraphic units. The major difference between the two types of units is that lithostratigraphic units are primarily distinguished by lithic characteristics (normally a formation has some degree of internal lithic homogeneity), whereas allostratigraphic units, distinguished only on the basis of bounding discontinuities, may be lithologically (also paleontologically, chemically, and so on) diverse (in both the lateral and vertical dimensions) or homogeneous. On the one hand, a single allostratigraphic unit could be recognized for a diverse assemblage or lithologies bounded above and below by unconformities that would otherwise demand the recognition of numerous lithostratigraphic units. On the other hand, a series of distinct allostratigraphic units could be recognized for a lithologically homogeneous (thus precluding subdivision into lithostratigraphic units) sequence of rocks that was interrupted by a series of unconformities.

Somewhat akin to recognizing stratigraphic units on the basis of bounding unconformities (where the unconformities are generally extremely diachronic), the NACSN recently has formalized diachronic time units that may correspond to the time of deposition or formation of any material stratigraphic (except lithodemic units) referent that does not have isochronous boundaries. "A diachronic unit comprises the unequal spand of time represented either by a specific lithostratigraphic, allostratigraphic, biostratigraphic, or pedostratigraphic unit, or by an assemblage of such units" (NACSN, 1983, p. 870). The boundaries of a diachronic unit at any geographic locality are marked by the time of first instance of deposition of the material referent preserved at that locality and the last instance of deposition of the material referent preserved at that locality. Within this time interval there may be one or more smaller time intervals during which the material referent was not being deposited (see Figs. 10 and 11 of NACSN, 1983 [reprinted as Appendix 1]). Diachronic time units appear to fulfill the need for natural geologic time units corresponding to natural geologic units and events that formed those units (Watson, 1983; Watson and Wright, 1980). Diachronic units in stratigraphy are analogous to many historical time periods that vary in their boundaries and durations from one geographic locale to another. To give a generalized example, eighteenth-century furniture of the Queen Anne style was produced first in England and then spread to America, where it continued to be produced after the style was superseded in England (Williams, 1963). Therefore, the Queen Anne period is a diachronic (but natural) unit, beginning and ending later in America than in England.

The basic and nonhierarchical diachronic unit is a diachron. When a hierarchy of diachronic units is desired, the following units (from most inclusive to least inclusive) can be utilized: episode, phase, span, and cline. As with most stratigraphic and related units (diachronic units are nonmaterial temporal units), the name of a formal diachronic unit should be derived from a geographic locality. The full formal name is composed of the geographic name followed by the term diachron or a rank term, the first letters of both being capitalized. In cases where the diachronic unit is the temporal equivalent of a single formal stratigraphic unit that bears a geographically derived name, the diachronic unit may also bear the same geographic name. Otherwise, the name of the diachronic unit must not duplicate that of any other formal stratigraphic unit (see NACSN, 1983). In some cases well-established and often used names and units that are treated as if they were geochronologic units but are actually time-transgressive may be redefined as diachronic units while retaining their names. Thus if the base of the Rancholabrean "Age" (Quaternary) in North America is time-transgressive, as suggested by Harington (1984; see also Schoch, 1984, and

Woodburne, 1987, on North American land mammal ages), perhaps it should be redefined as the Rancholabrean Diachron or the Rancholabrean Episode.

Diachronic units, being temporal units, do not have stratotypes per se. However, the material units on which they are based generally will be based in turn on stratotypes and/or reference sections. Furthermore, the code specifies that when proposing a formal diachronic unit, one must review the stratigraphic units that serve as the material referent(s) for the diachronic unit, and must designate and describe multiple reference sections relative to the diachronic unit.

HYDROSTRATIGRAPHY

Groundwater (hydrogeologic) studies are, of necessity, often intimately concerned with the lithostratigraphy of a particular region and its relationship to hydrogeologic units. Tóth (1978, p. 807; quoted in Galloway, Henry, and Smith, 1982, p. 27) defines a hydrogeologic unit as a "single stratum or combination of strata that function in bulk as either a water-bearing or a water-retarding rock complex relative to adjacent strata." Such units are essentially stratigraphic units defined and distinguished by their water-bearing properties (or lack thereof); they may be termed hydrostratigraphic units, and the study of their relationships and properties hydrostratigraphy (cf. Galloway, Henry, and Smith, 1982).

In the groundwater literature two basic types of hydrostratigraphic units commonly are distinguished: aquifers or aquifer systems and confining beds or confining systems. An aquifer can be defined as simply "a lithologic unit or combination of units capable of yielding water to pumped wells or springs" (Domenico, 1972, p. 17). Classically, however, it often has been used in a more restricted, or more practical or "economic," sense to refer to a rock formation or stratum that actually serves (is not just potentially capable of serving) as a source of water to wells and springs. In this sense the definition of an aquifer is somewhat analogous to that of an ore. A rock or an aggregate of minerals is an ore only as long as it is profitable to the miner to extract some product from it; the same rock can be an ore one day and not one the next, depending on the market. Treating an aquifer in a similar manner, Meinzer (1923) wrote:

> A rock formation or stratum that will yield water in sufficient quantity to be of consequence as a source of supply [of water] is called an "aquifer," or simply a "water-bearing formation," "water-bearing stratum," or "water bearer." It should be noted that the term "water-bearing formation," as here defined and as generally used, means a water-yielding formation—one that supplies water to wells and springs, one that contains gravity ground water. The term has no reference to the quantity of water that the formation may contain but will not yield to wells and springs. It is water-bearing not in the sense of holding water but in the sense of carrying or conveying water. Few if any formations are entirely devoid of gravity ground water, but those that do not contain enough to be practical sources of water supply are not considered to be aquifers; they are not called water-bearing formations. Hence it may happen that in a region underlain by strong aquifers a formation yielding only meager amounts of water will not be classed as water bearing; whereas in a region nearly destitute of available water a similar formation may be a recognized aquifer tapped by many wells. (Meinzer, 1923, pp. 52–53)

A single large aquifer or a series of interrelated aquifers (there is a continuity of water flow from one aquifer to another) may be termed an aquifer system.

Units of rock that are of relatively low permeability or transmissivity often are referred to as confining beds. Confining beds bound an aquifer and, according to Domenico (1972), may also be found interspersed or interstratified within an aquifer. A set or region of confining beds that separates major aquifer systems or forms a boundary of a single major aquifer system often is termed a confining system. Obviously confining beds and confining systems cannot be defined or delimited independently of aquifers and aquifer systems, except perhaps in the situation where beds of relatively low permeability within an aquifer are considered to be confining beds. By defining and delimiting the boundaries of an aquifer or an aquifer system, one also delimits confining beds or a confining system (the rocks that are not part of an aquifer or an aquifer system but form the boundaries of the aquifer).

As Domenico (1972, p. 17) states, "an aquifer [or an aquifer system] may be coextensive with a geologic formation, a group of formations, or a part of a formation; or it may cut across formations so diversely as to be essentially independent of any geologic unit." Likewise, confining beds or confining systems may correspond in part, as a whole, or not at all to standard lithostratigraphic units. Given the same lithostratigraphic and hydrostratigraphic units, their mutual relationships may vary laterally. Galloway, Henry, and Smith (1982) provide a specific example where this is the case. A single aquifer system, the Jasper aquifer system of the Texas Coastal Plain, is in some places coincident with a particular lithostratigraphic formation, the Oakville Sandstone of Miocene age, and the Fleming Formation above the Oakville forms the Burkeville confining system. However, in other areas the Jasper aquifer system occupies only the lower half of the Oakville Formation, while the Burkeville confining system corresponds to the lower part of the Fleming Formation and the upper part of the Oakville Formation. In still other areas the Jasper aquifer system occupies the lower part of the Oakville Formation and the upper part of the underlying Catahoula Formation.

Aquifers, aquifer systems, confining beds, and confining systems commonly are named by hydrostratigraphers and ground water specialists in a manner analogous to that of lithostratigraphic units (as the example cited above from Galloway, Henry, and Smith, 1982, indicates). In contrast to most stratigraphic units, aquifers and related units can vary dramatically over relatively short periods (over time spans of the order of magnitude of the durations of human cultures). For instance, a unit may or may not be designated an aquifer, depending on the demands of the local human population for water. Once depleted of its water, a former aquifer may no longer be considered an aquifer. However, extremely short-term fluctuations (for instance, over days or years) in water supply from a particular rock body may commonly be ignored in defining and delimiting an aquifer or aquifer system.

The topic of hydrostratigraphy, primarily a specialized field within the discipline of hydrogeology, is of little concern and rarely dealt with by most stratigraphers. As is the case with many other units of primarily industrial or economic value, hydrostratigraphic units per se receive no formal status under either the *International Stratigraphic Guide* (Hedberg, 1976) or the *North American Stratigraphic Code* (NACSN, 1983). Indeed, the *Guide* (Hedberg, 1976, p. 35) refers to aquifers as lithostratigraphic units, although as demonstrated above they are not necessarily lithostratigraphic, or at least they are no more lithostratigraphic than are magnetostratigraphic, allostratigraphic, or unconformity-bounded units (to give just a few examples). "Lithostratigraphic bodies that are recognized more for utilitarian purposes than for their lithologic unity, such as aquifers, oil sands, quarry layers, and orebearing 'reefs,' are considered informal units even if named" (Hedberg, 1976, p. 35). Ironically, the tenor of the *Guide's* treatment of lithostratigraphic units implies that in many cases formally recognized lithostratigraphic units also may be primarily of utilitarian purpose. Lithostratigraphic units are said to be "defined by observable physical features and not by inferred geologic history or mode of genesis" (Hedberg, 1976, p. 31); that is, they are relatively "theory free." Furthermore, among the many reasons for formally recognizing lithostratigraphic units is that they are useful "in investigating and developing mineral resources" (Hedberg, 1976, p. 30). Likewise, hydrostratigraphic units are defined by observable physical features and are characterized and delimited by a unity that, while it may not be strictly lithologic, is in proper context of vast significance. If a higher percentage of stratigraphers in general were intimately concerned with hydrostratigraphy, then hydrostratigraphic units probably would receive recognition under the *Guide* and the *Code*.

ARCHEOLOGICAL STRATIGRAPHY

Stratigraphic considerations are, or should be, of prime importance to the archeologist. The interrelationships between archeology and stratigraphy (and geology more generally) are discussed in numerous scattered works; Godwin (1981), Harris (1979), Heizer and Graham (1967), and Rapp and Gifford (1985) provide introductions to the subject. Of course, any distinctions made below concerning geological stratigraphy and archeological stratigraphy blur when we consider the archeology and geological context of our Pleistocene forebears.

Stratigraphy in archeology differs from stratigraphy in geology in that the archeologist is often primarily concerned with stratification produced by human intervention. It is clear that such man-made or artificial stratification is produced knowingly when holes are dug and filled or mounds are erected artificially, but humans unwittingly produce a stratigraphy in any place inhabited for a significant length of time. Addressing this topic, the classical British archeologist Sir Charles Leonard Woolley (1931) wrote:

> How do houses and cities sink below the earth's surface? They do not: the earth rises above them, and though people do not recognize the fact, it is happening all around them every day. Go no further than London. How many steps does one have to go down to enter the Temple Church? Yet it stood originally at ground level. The mosaic pavements of Roman Londinium lie twenty-five to thirty feet below the streets of the modern City. Wherever a place has been continuously occupied the same thing has happened. In old times municipal scavenging did not amount to much, the street was the natural receptacle for refuse and the street level gradually rose with accumulated filth; if it was re-paved the new cobbles were laid over the old dirt, at a higher level, and you stepped down into the houses on either side. When a house was pulled down and rebuilt the site would be partly filled in, and the new ground-floor set at or above street level; the foundations of the older building would remain undisturbed below the ground (Woolley, 1931, pp. 15–16)

Archeological stratigraphy generally differs in scale from that of geological stratigraphy. Within a small archeological site, perhaps measured in only tens of meters on a side, hundreds or thousands of small stratigraphic units may be distinguished. Archeological stratigraphy may include much "microstratigraphy," for example, the microstratigraphy of paint layers successively deposited on the floors, walls, and so on, of buildings. Thus historical archeologists (perhaps grading into architectural historians) may sepak of the "paint history" of features of buildings or other structures (e.g., Kirk, 1984). Harris (1979) has suggested that the principles of geological stratigraphy provide a necessary foundation, but are not adequate for the study of archeological stratigraphy.

Harris (1979) distinguishes geological stratification (the object of study for geological stratigraphy) from archeological stratification. Geological stratification is the result of natural, nonhuman agencies, whereas the strata studied by the archeologist can be the result of both natural and human agencies. According to Harris (1979), with the advent of humankind three major changes in the processes of stratification took place, at least in the immediate vicinity of areas occupied by humans. First, humans developed a series of inanimate objects (the equivalent of fossils for the archeologist) that do not conform to the processes of organic evolution. In particular, man-made objects can go "extinct" and then potentially reappear in identical form, whereas once an organic species genuinely goes extinct, it will never reappear exactly in the same form. Here it should be noted that many human objects/ artifacts have, for practical purposes, undergone a unidirectional evolution and thus can be used as fossils are in standard stratigraphic paleontology. A case in point is the evolution of beer bottles and beer cans that have been used to date old mining camps in the American West (cited in Ager, 1981; Hunt, 1959). Second, humans preferentially use certain areas and erect boundaries around areas—in effect establishing artificial basins of deposition that might arise as topographic highs (such as hills, mounds, walls, structures) above the mean level of the surface of the earth in the local area. Stratification thus induced can work against, or in spite of, the laws of gravity, which have a great influence on the formation of strata under natural conditions. Third, humans have often dug into the earth by cultural preference (for example, to provide for burial of the dead, foundations for buildings, erections of walls, and construction of wells), and have produced stratigraphic features with no equivalent in the natural world.

According to Harris (1979, p. 36), three classes of stratification are important to the archeologist. First are layers of material that are deposited in more or less horizontal planes, one above the other. This class Harris subdivides into natural strata and man-made layers. Natural strata, no matter what the origins of the materials of which they are composed (natural strata can be formed of man-made or man-transported materials), are formed under natural conditions of deposition, such as deposition from or arrangement by running water. In contrast, the materials composing man-made layers are transported and deposited by human action. In forming such layers, people might transport materials great distances and against the natural forces of gravity, water, or wind currents; examples of such man-made layers include pavements, floors, and fillings of holes, pits, valleys, or other depressions in the earth. Harris (1979) failed to note, however, that in practice it may not always be possible to distinguish unequivocally between natural strata and man-made layers. A second important class of stratification is features that are cut into other horizontal layers (such as pits or graves), labeled a feature interface by Harris

(1979). Third are features or constructions that present topographic relief aroud which stratigraphic layers can accumulate; labeled upstanding strata (Harris, 1979), they can be considered man-made strata that are vertical rather than horizontal.

Many archeological excavations of the past have paid relatively little heed to stratigraphic units or layers (Harris, 1979). Rather, a site was excavated by arbitrary levels or spits that might cut across stratigraphic units, and there was little attempt to delimit actual stratigraphic units. The major goal of many early archeological excavations (particularly in the eighteenth, nineteenth, and early twentieth centuries) was to recover antiquities (artifacts) and reconstruct major structures, such as buildings and fortifications. The original positions of important objects sometimes were recorded using a three-dimensional grid system relative to fixed geographic and elevation datums, but objects might not necessarily be associated with particular stratigraphic layers. It even might be assumed that invariably those objects coming from a higher absolute elevation must be relatively younger than objects coming from lower elevations at the same site (Harris, 1979, p. 21). Of course, this assumption ignored possibilities such as that the inhabitants of a site at later times may have dug into the layers already deposited, and buried or dropped younger objects into lower elevations. Fortunately modern archeological excavations tend to be more sophisticated than this stratigraphically.

Of prime importance to archeological stratigraphy in particular is the concept of the interface. An interface is either the surface of a stratum (such as the surface of a road) or a surface formed by the removal of a stratum or strata (such as a hole or ditch dug down into older strata). The former concept often is referred to as a layer interface (Harris, 1979) and is somewhat analogous to a bedding plane in geology. The latter concept is referred to as a feature interface and is analogous to an unconformity. Unlike typical bedding planes in geology, however, layer interfaces in archeology can be either horizontal or vertical (upstanding) surfaces (such as the sides of a wall that was buried in rubble). Finally, a period interface in archeology is essentially the sum total of horizontal and vertical surfaces in use at one particular time; it can be composed of both feature interfaces and layer interfaces. A particular feature or layer interface may form a part of more than one period interface, as might occur if the same floor or wall were used during two or more distinct archeological periods. Of course, the entire period interface of any particular time may not be preserved in the archeological record. Just as the geologic stratigraphic record is incomplete, so too is the archeological stratigraphic record.

Archeological stratigraphers face many of the same problems encountered by their geological counterparts, plus some problems limited primarily to their own specialty. Archeological stratigraphic units tend to be relatively small and usually limited to one specific site. It is generally very difficult or impossible to correlate such units from one archeological site to another. In some instances geologic stratigraphic units may be present at an archeological site, such as debris from a mudflow or a volcanic ash fall, that are more readily correlatable to other sites and a larger context; but this is relatively rare. Rather, for intersite correlations the archeologist is typically dependent upon either methods of absolute dating (such as ^{14}C, obsidian hydration, and so on; see Chapter 7) or more commonly the use of preserved cultural objects. Just as the paleontological stratigrapher must worry about the possible diachroneity of fossil species in different areas, the archeological stratigrapher faces similar concerns in correlating by artifacts (even barring concerns of infiltration or reworking, discussed below). Ethnographic studies have documented that objects can continue to be manufactured and used in one area long after they have ceased such use in other areas. Even in the twentieth century, certain cultures continue to use stone implements thousands of years after such use was discontinued in other areas. Given the relative shortness of the archeological record (a few tens of thousands to a few million years if one considers archeology to extend back to the first appearance of our genus on the earth), diachroneity of a couple of centuries or a millenium—which may be insignificant, to the point of undetectable, to the average geological stratigrapher—is extremely significant to the archeological stratigrapher. Furthermore, cultural objects may continue in use long after the production of such objects has stopped. Even if the date of an object's manufacture is known precisely, and it is found in situ at an archeological site, that does not mean that it can be used to date accurately the layer in which it occurs. To give one example of this phenomenon, Imperial Roman coins (ca. first to fifth centuries A.D.) continued to circulate in parts of Europe well into the nineteenth century (Clain-Stefanelli, 1965), and even today one can occasionally find coins that are up to 2000 years old mixed with modern coins in junk shops, flea markets, and the like (author's personal observations made in Massachusetts, Virginia, and Italy). This concept of man-made objects continuing to circulate long after their manufacture is quite different from the norm in paleontological stratigraphy. In the latter case, once a species goes extinct, it is no longer found in strata (unless it forms a reworked clast, which often can be distinguished as such).

A major problem for acheological stratigraphers concerned with long-range correlations between various sites and different cultures is that the archeological (cultural) record in one area may have developed in total isolation from that of any other area. For instance, for tens of thousands of years the human cultures of the Western Hemisphere appear to have developed in nearly total isolation from all other human cultures in the rest of the world (that is, for nearly the whole span of the archeological record in the Western Hemisphere). Therefore, correlations during this time period between the Western Hemisphere and the rest of the world are impossible if based on cultural objects. Rather, such correlations must be based on geological criteria (for instance, climatic changes) and "absolute" dating.

Archeologists use the designation "indigenous remains" to refer to objects that are approximately contemporaneous with the stratigraphic layer in which they are found. The archeological equivalent of the geologist's "reworked" is "residual." Residual remains are significantly older objects found in younger strata. Such residual remains literally could have been reworked up into younger strata, in a geological sense, or they could be items that remained in circulation for a significant time after their manufacture (such as the Roman coins mentioned above). Infiltrated remains (or contamination) are objects that were produced at a later period in time and have found their way into earlier layers, perhaps by means that no longer can be detected readily (Harris, 1979).

Analogous to the concept of overturned strata in geology, some archeologists have identified what they call "reversed stratigraphy" at some archeological sites. The basic concept of reversed stratigraphy is that when some form of hole is dug by human activity, often the spoils are dumped in one place as the hole is dug. In digging the hole the uppermost layers of the earth are dug through first, of course, and are the first to be deposited on the spoil heap. Subsequently, lower layers of the earth are broached and placed on the heap. Once the hole and the spoil heap are completed, the oldest layers of earth dug from the hole will be on the top of the spoil heap and the youngest layers on the bottom; thus it is said that the spoil heap exhibits a reversed stratigraphy. Harris (1979, p. 95) has argued, however, that the concept of reversed strata is erroneous. In a purely geological context, when a large block of strata is overturned, the strata lose none of their original characteristics (except for being turned upside down), and no new strata are formed. In the archeological case this is not true. In digging a hole a block of strata is not typically removed en masse and turned upside down; rather, the soil is eroded away a shovelful at a time and redeposited somewhere else. In the archeological context, new strata are formed from the materials of older strata. The new strata so formed may share some characteristics, in reversed order, with the older strata from which they were formed, but they are not simply reversed or overturned strata. Objects that were perhaps indigenous to the original strata will become residual remains within the new strata so formed, and in general these residual remains may appear in the newly formed strata in reversed spatial order. This is an example of how one must be careful in applying valid geological concepts superficially, and erroneously, to archeological stratigraphy.

In much of geological stratigraphy the principle of superposition holds true as an indicator of the relative age of strata. This principle applies primarily to horizontal-lying strata. In archeology, as noted above, it is also common for "strata" to be deposited vertically (as against the face of a wall) as well as horizontally, or at any angle between horizontal and vertical. This must be taken into account when one is interpreting the relative ages of archeological strata. Also it is probably much more common in archeology than in geology for underlying "strata" to have absolutely younger ages than overlying strata, even when the strata are not overturned or reversed. This is so because in many archeological situations the lower strata may have been replaced in the course of human activity, whereas the upper strata remained intact. An example would be replacement of the foundation of an old building without disturbance to the upper portions of the structure.

LUNAR AND EXTRATERRESTRIAL STRATIGRAPHY

An exciting new field has just begun to be developed appreciably in the last quarter century: the application of the stratigraphic tradition and approach to the Moon and to other extraterrestrial Earthlike bodies. Progress made relative to applying stratigraphic principles and methods to the geology and history of the Moon in particular is reviewed in Mutch (1972) and Wilhelms (1987). Very little work has been carried out to elucidate the stratigraphy of other extraterrestrial objects, mainly because there is little information upon which to base such studies. As space exploration accelerates in the future, however, we might predict that the study of such planets as Mercury, Venus, and Mars will be particularly amenable to the stratigraphic approach. As Mutch (1972) pointed out, we have had over two hundred years to develop and perfect the science of stratigraphy on one planet, the Earth. With twentieth-century (and future) explorations of the Moon, we essentially are presented with a second "planet," up until a few decades ago virtually unknown stratigraphically, upon which to test the science. In the future we may hope to gain access to more planets for detailed stratigraphic studies.

With regard to lunar studies specifically, initially there was some resistance to the application of standard stratigraphic methods and principles to the Moon. However, by at least the early 1960s it was realized by some investigators that "the fundamental stratigraphic approach (Albritton, 1963) was, in fact, readily transferable to the partly familiar, partly exotic deposits visible on the lunar surface (Mutch, 1970; Wilhelms, 1970)" (Wilhelms, 1987, p. vii).

Current stratigraphic reconstructions of the Moon's surficial geology are based primarily on remote sensing techniques, particularly telescopic and photogeologic studies, supplemented by rock samples gathered, and surface studies undertaken, by the half dozen American Apollo manned landings on the Moon made during the years 1969–72 and by various unmanned American and Soviet missions to the Moon in the 1960s and 1970s. The entire surface of the Moon has been photographed, via orbiting spacecrafts, and at least preliminary geologic maps attempted of all areas. Photogeologic techniques are relatively easily applied to the Moon because it lacks any cloud cover, and the surface geology is not obscured by such features as vegetation cover, ice caps, or bodies of water. Yet no matter how clear such photographs might be, stratigraphic interpretations formed solely on the basis of such photographs will remain uncertain. As Wilhelms (1987) has written: "Our present perception of the Moon has emerged from the interplay between sampling studies and stratigraphically based photogeology. These two approaches are complementary: Photogeology contributes a historical context by viewing the whole Moon from a distant vantage point, whereas the samples contain information on rock types and absolute ages unobtainable by remote methods. Neither approach by itself, even the most elaborate program of direct surface exploration, could have yielded the current advanced state of knowledge within the relatively short time of two decades (Greeley and Carr, 1976)" (Wilhelms, 1987, p. vii).

A modern understanding of lunar stratigraphy was reached only when the rocks on the surface of the Moon were interpreted in terms of three-dimensional stratigraphic units (analogous to formations and other lithostratigraphic units). Previously, many geomorphic features on the surface of the Moon had been interpreted primarily in structural terms; for example, as floods, faults, and fissures that were the product of endogenic deformation (see Wilhelms, 1987). Such authors as Mutch (1972) and Wilhelms (1987) now commonly treat lunar surface features as indicative of stratigraphic units whose relative histories of emplacement can be elucidated by the stratigraphic principle of superposition; "simply put, younger units overlie and thus modify older units" (Wilhelms, 1987, p. 17).

Lunar stratigraphy is not identical in its general nature to terrestrial (Earth-based) stratigraphy, however. The Moon is characterized by many features, with distinct geomorphic expressions, that have few if any terrestrial analogues; examples of such features are basins (ringed, crater-like depressions), maria (singular—mare; dark, smooth plains filling the basins), terrae (singular—terra; rugged uplands or highlands), and ubiquitous craters of all scales and sizes. General models of the stratigraphy and genesis of these categories of features were required before a comprehensive lunar stratigraphy could be envisioned. Current ideas about the stratigraphy of such features are summarized in Mutch (1972) and Wilhelms (1987).

A major controversy in the early literature on stratigraphic studies of the Moon concerned the nature of the origin of craters; are they of exogenic or endogenic origin—that is, are they due to meteorite (and the like) impacts or volcanic events? It now generally is agreed that most lunar craters are impact structures, and the patterns of lunar crater deposits can be interpreted in terms of impacts and the resulting ejecta and impact breccia. Such craters can be compared to experimentally modeled impact and explosion craters, and also to known or presumed craters of meteoritic origin on Earth.

Another major problem in elucidating lunar stratigraphy has concerned the origins and relationship of lunar basins and the maria that commonly fill such basins. Initially it was thought that a basin and its mare were formed by the same process, whether that process was exogenic or endogenic. It is now hypothesized by many investigators that most basins are of impact origin, like craters, and are perhaps the result of asteroids or similar bodies colliding with the Moon. The maria filling basins, however, now usually are viewed as of volcanic origin (perhaps consisting primarily of basaltic flows) and as postdating, in some cases by considerable amounts of time, the formation of their basins. Thus the stratigraphy of volcanic flows on Earth is of considerable importance as an analogue for interpreting the maria on the Moon. Fundamental questions concerning the formation of maria filling basins remain to be answered, however. For example, are basins commonly filled with mare materials from one or a few centrally located volcanic vents, or are such vents located around the margins of the basins? These two different models could produce different internal stratigraphies of the maria (see Mutch, 1972). Interpretations of the terrae by various investigators have included hypothese of volcanic origin and impact origin; I would suggest that perhaps some combination of both origins formed the terrae.

Standard stratigraphic nomenclature, involving lithostratigraphic, chronostratigraphic, and geochronologic units, has been applied, more or less, to lunar stratigraphy. Various formations, systems, and series have been named on the basis of lunar materials, but it is questionable whether these lunar stratigraphic units really are comparable to their terrestrial equivalents. The first explicit discussion of lunar stratigraphy (Shoemaker and Hackman, 1962) dividend the surficial rocks of the lunar surface, representing some 4.5 billion years or more of lunar history, into five temporally arranged groups or sets, the four youngest of which were considered "systems." Most recently the stratigraphy of the Moon has been reviewed by Wilhelms (1987), who recognizes five lunar systems: from oldest to youngest, the pre-Nectarian (an informal system), Nectarian, Imbrian, Eratosthenian, and Copernican (Appendix 4). Among the systems Wilhelms (1987, p. 123) recognizes four informal "major sequences" (shown in Appendix 4) and also formally subdivides the Imbrian into Lower and Upper Series. Formal lithostratigraphic formations are recognized among the rocks filling some of the lunar basins.

The details of the *North American Stratigraphic Code* (NACSN, 1983), the *International Stratigraphic Guide* (Hedberg, 1976), and so on, cannot for the most part be fulfilled when describing formal lunar stratigraphic units. Currently potential stratotypes for such units are not directly accessible to Earth-bound stratigraphers, but type areas for lunar stratigraphic units are specified, and the units are described and interpreted as thoroughly as possible.

Chapter 6

BIOSTRATIGRAPHY AND MAGNETOSTRATIGRAPHY

THE DEVELOPMENT OF BIOSTRATIGRAPHY

Traditionally biostratigraphy has provided the most precise and accurate means of correlating (primarily in the sense of chronocorrelation; see Chapters 3 and 7) Phanerozoic rocks. It has been primarily through biostratigraphy that the Standard Global Chronostratigraphic Scale (Hedberg, 1976) has been developed. In many instances biostratigraphy remains the preferred method of dating (relative to other rocks, not numerically) and correlating rock sequences. The magnetostratigraphic scale has been worked out in detail only back to mid-Jurassic times, and the uncertainties associated with isotopic and other forms of numerical dating methods are often much greater than the durations of biostratigraphic zones, especially during the Paleozoic (cf. D. G. Smith, 1981b). Moreover, not all rocks are suitable to magnetostratigraphic or radiometric analysis, whereas any rocks that contain fossils are potentially amenable to biostratigraphic interpretation. Fossils also have the practical advantage that in many cases no sophisticated technological techniques are required to utilize them in determining correlations. Fossils only need be studied by a competent paleontologist; in some cases they are readily identified in the field. This section touches on a few major themes and issues in biostratigraphy; for more detailed coverage of this important branch of stratigraphy the reader should consult various references cited below and Kauffman (in prep.).

The conventional beginnings of modern biostratigraphy date to the discoveries of William Smith in the 1790s (see Chapter 1), and the discipline has flourished ever since. In a sentence, Smith discovered that one can identify and correlate rocks on the basis of their fossil content, the essence of biostratigraphy. However, for Smith fossils remained intimately tied to a lithostratigraphic framework; formations of greatly differing lithologies were not routinely correlated on the basis of similar fossils, and lithologically homogeneous formations were not subdivided solely on the basis of fossil content. It was for others to elaborate and expand upon this basic concept.

The next major development in the principles of biostratigraphy was the concept of facies (in the sense of laterally equivalent environments represented by differences in lithology and fossil content), and the use of certain fossils to correlate differing lithofacies. By the first quarter of the nineteenth century it was understood by at least some investigators that rocks of differing lithology can be of the same age, that certain types of fossils may be found only in particular lithologies, and that differing faunas need not necessarily imply differing ages (B. Conkin and J. Conkin, 1984). As Hancock (1977) has pointed out, in the 1820s and 1830s (and perhaps even earlier, see B. Conkin and J. Conkin, 1984) various workers implicitly recognized the concept of lithofacies (e.g., De La Beche, 1839; Young and Bird, 1822), and some even used fossils in attempting correlations among differing lithofacies (e.g., Brongniart, 1823, 1829; Fitton, 1827; Phillips, 1829). It was only with the work of the Swiss geologist Amanz (or Amand: Hancock, 1977) Gressly (1814–65: Debus, 1968), however, that the concept of facies was named and discussed in detail. Gressly (1838, p. 11, italics in the original; translated in B. Conkin and J. Conkin, 1984, pp. 137–138 [for another translation see Hancock, 1977]) defined facies in the following manner: "At once, two principle facts everywhere characterize the assemblages of modifications that I call *facies* or *aspects of terrain:* The one consists in that a *given petrographic aspect of any terrain necessarily implies, wherever it occurs, the same paleontologic assemblage; the other is that a given paleontologic assemblage rigorously excludes fossil genera and species frequent in other facies.*" Gressly was well aware that his various facies represented different environments of deposition and the organisms living and dying there.

Contemporaneous with the early descriptions and discussions of facies was the development of the concept of subdividing, correlating, and classifying rocks solely on the basis of their contained fossils (independent of lithologic considerations). As has been discussed (see Chapter 1), Lyell (1833, using the data of Deshayes) proposed the epochs of the Tertiary solely on the basis of the contained fossils. Earlier Deshayes (1830) and Bronn (1831) had already undertaken similar studies (Hancock, 1977). Conceptually this was an extremely important breakthrough: "In theory, at least, stratigraphy was now free from a lithological control, and the possibility of a pure biostratigraphy had been attempted" (Hancock, 1977, p. 9: cf. Arkell, 1933; Blow, 1979).

The French paleontologist Alcide d'Orbigny (1802–57) is credited with being the first person to recognize a systematic series of typical fossil assemblages, independent of lithological or facies considerations, covering the whole of the known fossiliferous portion of the geologic column, and to use such assemblages to correlate strata globally. This was the origin of the concept of the stage.

D'Orbigny (1842–49) began by subdividing the known Jurassic rocks into ten successive subdivisions (which he referred to as "stages" or "zones") on the basis of their fossil content and then correlating these divisions globally (Arkell, 1933; B. Conkin and J. Conkin, 1984; Hancock, 1977). Of D'Orbigny's original ten Jurassic stages, many of the names are still in use: for example, Sinemurian, Toarcian, Bajocian, Bathonian, Callovian, Oxfordian, Kimmeridgian. In his later work he extended the system to cover the entire fossil record (Blow, 1979). The original theoretical basis behind D'Orbigny's stages appears to have been a belief in a series of catastrophic destructions of all life on earth followed by repeated new creations. Each stage represents the interval between a creation and a destruction; thus concerning his stages, D'Orbigny (1842–49; translation from Arkell, 1933, p. 10) wrote that "there is nothing arbitrary about them and that they are, on the contrary, the expression of the divisions which nature has delineated with bold strokes across the whole earth." Blow (1979, p. 237) has argued, however, that D'Orbigny had somewhat less catastrophic views: "The idea of overwhelming, total catastrophe, with complete extinction and subsequent total re-creation (of the school of Cuvier [note that Cuvier himself probably held views more similar to those attributed by Blow to D'Orbigny rather than the extreme catastrophic views held by some of his later followers and often attributed to Cuvier by his detractors: see Rudwick, 1976]) was replaced by the concept of sudden world-wide marine regressions, subsequent re-population of the areas involved (with the appearance of many new genera and species) occurring during violent transgressions."

In describing and defining his stages, he based them entirely on their fossil con-tent. For D'Orbigny a stage was a body of rock strata, which might be found in any part of the world, containing at least some of a set of diagnostic fossils. For this reason Hancock (1977, p. 19) has stated that there is no question that stages, at least as originally used, were biostratigraphic. However, as discussed elsewhere (see especially Chapter 7), the stage has since been adopted for other purposes, especially as a chronostratigraphic unit. Some workers essentially have insisted that the stage has been a chronostratigraphic rather than a biostratigraphic unit all along. Thus in discussing their concept of the stage, also in the context of a brief review of the historical background of stratigraphy and a discussion of D'Orbigny's work, Storey and Patterson (1959, p. 709) wrote the following: "A stage is commonly and incor-rectly held to be a paleontological unit because it is recognized universally by fossil criteria. This view is not complete, however, as it omits to indicate that the paleonto-logical unit varies in stratigraphic (vertical) range in different places, and that the total range must be established by adding up all of the different local ranges. Fur-thermore, owing to the natural discontinuous nature of facies criteria, certain non-fossiliferous strata which are contemporaneous with the faunal unit must also be included in order to recognize the presence of strata representing the stage in ques-tion."

In his publications (see B. Conkin and J. Conkin, 1984, and Hancock, 1977, for examples) D'Orbigny would list typical genera found in a certain stage, and the first and last appearances that occur during the duration of the stage (in the rocks attrib-uted to the stage). D'Orbigny's stages often were named after a particular town or other geographic feature where characteristic rocks of the stage were exposed (hence the concept of a type locality), but he would also list numerous other "types" (from various countries and regions) that were correlative and included in the stage. D'Or-bigny also listed synonymies for his stages: rock name and stratal terms that he considered to be equivalent to, correlative with, or included within his stage.

The introduction of D'Orbigny's stages met with some opposition, as recounted by Arkell (1933) and Hancock (1977). In particular, the influential Quenstedt (1856) criticized D'Orbigny's synthesis as premature. "Of what avail is it if a man has seen the whole world, and he does not understand aright the things which lie in front of his own doors? . . . To compare faithfully two beds, each a hand in height, one on top of the other in their true order, can effect a more fruitful development of science than the use of stratigraphic catalogues from the furthest regions of the earth. Right from the outset one has to admit that such records are not reliable" (Quenstedt, 1856, pp. 23–24; translated in Arkell, 1933, p. 14, and corrected in Hancock, 1977, p. 14). Or as Arkell (1933, pp. 13–14) wrote: " . . . d'Orbigny's scheme of classifica-tion was premature. The ideal would have been to wait until a tolerably complete record of the succession of Jurassic faunas had been elucidated, by piecing together the results obtained by workers in different parts of Europe, and then, bearing in mind the original scope of the old terms [terms used by D'Orbigny to name stages], to divide up the column so that each division should carry the name of the locality or region where its particular fauna or group of faunas was best developed." Writing further, Arkell (1933, p. 10) considered D'Orbigny to have been "unfettered by too much knowledge." But in D'Orbigny's defense, his stages have paved the way for the development of a global stratigraphic classification below the level of the system. It is perhaps all too easy to criticize D'Orbigny in hindsight and thus belittle his real contribution—the concept that systematic stratigraphic correlations are possible at the level of the stage independent of local lithologies and facies.

D'Orbigny used the terms stage and zone somewhat interchangeably (see discussion in Blow, 1979, and Hancock, 1977). Sometimes he used the term zone as a synonym for stage; sometimes he used it in a nontechnical or geographic sense to refer to a band, belt, or area (such as the deep ocean zone); and Blow (1979) has interpreted that D'Orbigny sometimes used the term zone for a subdivision of a stage (although according to Blow, most of D'Orbigny's stages were comprised of only one zone). As interpreted by Arkell (1933), D'Orbigny used the term zone to refer to the paleontological content of a stage. A few years after D'Orbigny's work, the German Albert Oppel (1831–65) gave an explicit and definite meaning to the word zone. Besides studying the rocks of his native land, Oppel toured the Jurassic regions of Switzerland, France, and England and attempted to correlate the rocks in great detail. Oppel (1856–58) divided Jurassic rocks into 33 zones based on their fossil content, explaining his methodology as follows:

> Comparison has often been made between whole groups of beds, but it has not been shown that each horizon, identifiable in any place by a number of peculiar and constant species, is to be recognized with the same degree of certainty in distant regions. This task is admittedly a hard one, but it is only by carrying it out that an accurate correlaion of a whole system can be assured. It necessarily involves exploring the vertical range of each separate species in the most diverse localities, while ignoring the lithological development of the beds; by this means will be brought into prominence those zones which, through the constant and exclusive occurrence of certain species, mark themselves off from their neighbours as distinct horizons. In this way is obtained an ideal profile, of which the component parts of the same age in the various districts are characterized by the same species. (Oppel, 1856–58, p. 3; translated in Arkell, 1933, p. 16)

For Oppel a zone was a bed, group of beds, or set of strata identified by an assemblage of fossils (Arkell, 1933); the term zone was not used solely for an assemblage of fossils, but was a stratigraphic term. Oppel usually cited 10 to 30 species that were characteristic of a zone, and the zone was named after one (or a few) typical "index" species. However, there was nothing sacred about the index fossil, and Oppel suggested that zones could as easily have been named after geographic places, as D'Orbigny had named his stages. Oppel fitted his zones into D'Orbigny's stages, but at times a particular zone might straddle a stage/stage boundary. As Hancock (1977, p. 13) has pointed out, the genesis of Oppel's classification was not to subdivide D'Orbigny's stages into zones, but to develop his system of zones and then build larger groupings (which might correspond to D'Orbigny's stages) from the zones.

A concept that is perhaps difficult to realize in hindsight is that D'Orbigny, Oppel, and other investigators of the nineteenth century were not explicitly thinking in terms of time or building up a time scale per se when developing their stratigraphic systems. Terms such as stage and zone, as discussed above, referred primarily to spatial rather than chronological relations. Here a comment by Arkell (1933, p. 19, italics in the original) is pertinent: "It will be noticed that in all these definitions and uses of the word zone there is no mention of *time*. It is, of course, implicit in Oppel's and all other authors' tables of zones, stages or strata, that they occupied a certain time in forming, and that the subdivisions probably occupied a lesser time than the major divisions. Before 1893, however, we find no attempt to formulate any strictly chronological ideas, or to construct a time-scale independent of strata, whereby might be compared the relative time taken to deposit strata in different localities. No vocabulary for any such conceptions existed."

In 1893 Sydney S. Buckman explicitly introduced a concept of time into stratigraphic geology; hence the reference to 1893 in the quotation above. Buckman's work was extremely influential and important, but in fairness an explicit time term had been introduced into stratigraphy some years earlier. As Arkell (1933, p. 21) himself notes, at the International Geological Congress at Bologna in 1880 the term moment was proposed as the time-equivalent of a zone. Likewise it was decided at the same congress that stratigraphic divisions would form the following hierarchy (beginning with the most inclusive): group, system, series, stage, substage, assise, stratum. The geologic time equivalents of the first four terms were era, period, epoch, and age, respectively (Hancock, 1977). Yet these developments seem to have received very little attention, and Buckman's 1893 proposal was ground-breaking:

> *The term "Hemera."*—It is for a palaeontological purpose . . . that I propose the term "hemera" . . . Its meaning is "day," or "time" and I wish to use it as the chronological indicator of the faunal sequence. Successive "hemerae" should mark the smallest consecutive divisions which the sequence of different species enables us to separate in the maximum developments of strata. In attenuated strata the deposits belonging to successive hemerae may not be absolutely distinguishable, yet the presence of successive hemerae may be recognized by their index-species, or some known contemporary; and reference to the maximum developments of strata will explain that the hemerae were not contemporaneous but consecutive.
>
> The term "hemera" is intended to mark the acme of development of one or more species. It is designed as a chronological division, and will not therefore replace the term "zone" or be a subdivision of it [in the same paper Buckman used the term bed for a subdivision of a zone], for that term [zone, also bed] is strictly a stratigraphical one. Our present "zones" give the false impression that all the species of a zone are necessarily contemporaneous; but the work of Munier-Chalmas in Normandy, and my own labours in other fields, show that this is an incorrect assumption. The term "hemera" will therefore enable us to record our facts correctly; and its chief use will be in what I may call "palaeobiology." (Buckman, 1893, pp. 481–482)

Buckman's new term, hemera, was misunderstood at the time; many workers merely regarded it as a subdivision of a zone. "The conception of a time-scale entirely independent of deposit, so that it could be said that each hemera must have passed everywhere, whether any deposit was formed or not, seems to have been too much for many of the geologists of the time" (Arkell, 1933, p. 20). In a sense, as Arkell (1933) notes, Buckman's (1893) concept of a hemera is a subdivision of the time taken to deposit a zone. In 1902, however, Buckman (1902, p. 556) redefined a hemera: "the hemera is the time during which a certain piece of work, namely, the deposition of what is called 'the zone,' was done." Arkell (1933) found this second definition of hemera irreconcilable with the first, and perhaps it is, unless one considers that in 1902 a zone for Buckman represented the observable acme of a species in the stratigraphic record (zones for Buckman in 1902 were more restricted than they were for him in 1893), and the hemera was the time corresponding to a zone so distinguished (Blow, 1979). Arkell (1933) continued to use the term hemera as a subdivision of the time interval during which a zone was deposited.

A proliferation of new terms, and redefinitions of old ones, took place from the late nineteenth century into the twentieth century (for further discussions see Arkell, 1933; Blow, 1979; Hancock, 1977; Hedberg, 1976; NACSN, 1983; Weller, 1960; and references cited therein). Once Buckman redefined his term hemera to be the time-equivalent of a zone, it was then a synonym of the earlier term moment. Some workers adopted the term moment with essentially this meaning, or used variations on the term such as time-moment or zone-moment (Arkell, 1933). Furthermore, a 1901 report arising from an International Geological Congress in Paris suggested that the term phase be used as the time-equivalent of a zone, and Jukes-Browne (1903) coined the new term secule to refer to the time taken to deposit a zone (Jukes-Browne considered a hemera to be a synonym of a subzone). Trueman (1923) introduced the term epibole as a stratigraphical term referring to the deposits accumulated during a hemera. Using Buckman's 1893 concept of a hemera, the term epibole might be useful (Arkell, 1933), but by Buckman's 1902 concept of a hemera the term epibole is merely a synonym of the term zone (Blow, 1979).

In his 1902 paper Buckman introduced two other new terms, which have since gained wide currency (but with various, sometimes contradictory, meanings) in biostratigraphy: biozone and faunizone. In the process Buckman also modified the use of the term zone itself:

> To sum up, a *zone* indicates the horizontal extension of species and so is a geographical term; a *biozone* is the range of any organism or group of organisms in geological deposits, so it may be said to indicate vertical extension; *faunizones* are, to paraphrase Mr. Marr, "belts of strata, each of which is characterized by an assemblage of organic remains," with this provision, that faunizones are independent of lithic structure—the strata of a faunizone may vary horizontally or vertically, or the strata may not vary and yet may show several successive faunas. So faunizones are the successive faunal facies exhibited in strata. Lastly, *hemera* is a time term—the subdivision of an "age"; it indicates the period of time from the rise of one dominant species to the rise of the next. It is the time term which corresponds to "faunizone" as the stratigraphical term, just as "age" corresponds to "stage." (Buckman, 1902, p. 557, italics in the original)

In the above passage, Buckman's (1902) rationale for redefining the zone, as he does, is that this is how it is used in zoology. As may not be clear from the quotation, Buckman (1902, p. 556) considered biozone to be a time term, not a stratigraphic term, "to signify the range of organisms in time as indicated by their entombment in the strata. . . . thus the biozone of the Trilobita would be, say, from Cambrian to Carboniferous [Permian by present knowledge]." The faunizone of Buckman (1902) is essentially the zone as used previously in stratigraphy. Commenting on these terms and their distinctions in the early twentieth century, Arkell (1933, p. 22) writes: "The distinction between the biozone of a given species and its hemera may be expressed as the difference between its absolute duration (as indicated by its total range) and its acme; but since in practice neither can be very accurately defined, owing to the imperfections of our collecting ('collection-failure'), the two units become virtually synonymous."

Despite what appears to be Buckman's (1902) intention to introduce biozone as a time term, it has generally been adopted as a spatial or stratigraphic term (Arkell, 1933; Blow, 1979) signifying the total stratigraphic range of a particular taxon. However, the total stratigraphic range of a particular taxon at one locality may not correspond to its stratigraphic range at another locality. The total theoretical range of a taxon, its biozone, is not something that is simply observed, but it is deduced from numerous observations in different sections and is always subject to further revision. "For this reason, Pompeckj (1914) introduced the term 'part-zone' (as 'Teilzone') [Hancock, 1977, states that Frebold introduced this term in 1924] to denote an interval recognized by the total observed range of a particular taxon in a particular locality or area; thus, in any one area, the 'Teilzone' or 'part-zone' could be used for the recognizable part of the total, deduced Biozone" (Blow, 1979, p. 238). The sum of the teilzones (assuming that all are completely known) of a taxon is equal to its biozone, whether the term biozone is used in a temporal or a stratigraphic sense (although the taxon may have existed during some time interval when no fossils of it were formed). As Buckman (1902, see above) originally defined biozone, it is the time term corresponding to the total stratigraphic range (not the total time interval during which the taxon existed on earth); if used as a stratigraphic term it is merely the total stratigraphic range of a taxon. The only reason we currently do not know absolutely the biozone of any taxon (although some are extremely well approximated) is that not all stratigraphic sections have been thoroughly studied.

If biozone is used as a stratigraphic term, which is the more common usage, there is still another term that was originally coined for the same concept as biozone when used as a time term. In 1901 Wiliams coined the terms geochron and biochron explicitly as time terms (even if his definition includes some stratal terms) to distinguish the "time-value" of a lithologic body and the duration of a taxon, respectively. ' It is essential to distinguish the geochron (expressed in terms of feet thickness of stratified sediments of uniform lithologic constitution) from the biochron (expressed in terms of presence in the sediments of fossils of the same species, genus, or family). Thus the time-value of the Hamilton formation would be spoken of as the Hamilton geochron; while the time-value of the species *Tropidoleptus carinatus* would be the *Tropidoleptus* [*carinatus*] biochron" (Williams, 1901, pp. 579–580; quoted in Arkell, 1933, p. 22).

It should also be mentioned that, as recounted by Arkell (1933), Buckman grouped his hemerae into ages and in 1896 gave these ages the same names as the corresponding stages of D'Orbigny. In 1898 the Council of the Geological Society of London objected to this usage, believing that the use of the same names for stratigraphic stages and chronological (temporal) ages would lead to confusion. In compliance, Buckman (1898) named ages (and also epochs), corresponding to stages (and series), after ammonites. In later publications Buckman increased the number of his ages and sometimes fit several ages into the interval of a stage. This practice was followed by some into the first third of the twentieth century, but not universally (Hancock, 1977).

CURRENT BIOSTRATIGRAPHIC UNITS

As is well recognized, and should be apparent from the above discussions, zone and many other terms used in biostratigraphy (and stratigraphy more generally), have had a long and confusing history with multiple usages and redefinitions. In recent years both national (NACSN, 1983) and international (Hedberg, 1976) groups have attempted to clarify the meanings of such terms (or at least mandate definitions for commonly used terms), as will now be discussed.

The International Subcommission on Stratigraphic Classification adopted a very nebulous definition of the term zone. "**Zone.** The term is commonly used for a minor stratigraphic interval in any category of stratigraphic classification. Thus there are many kinds of zones depending on the stratigraphic characters under consideration—lithozones, biozones, chronozones, mineral zones, metamorphic zones, zones of reversed magnetic polarity, and so on. When used formally, the term zone is given an initial capital letter (Zone) to distinguish it from its informal use" (Hedberg, 1976, p. 14).

The ISSC considered the term biozone to be a short alternative term for the longer phrase biostratigraphic zone. The NACSN (1983) considers the biozone to be the basic unit of biostratigraphic classification. A biostratigraphic zone or biozone can signify any kind of biostratigraphic unit ("a body of rock defined or characterized by its fossil content" NACSN, 1983, p. 862), and can be simply called a zone if it is clear from the context that it is a biozone or that a biostratigraphic unit is being referred to. Thus a biozone could range from a thin local bed to a unit several kilometers thick and extending worldwide, as long as such units were unified by their paleontological character and differentiated from other units on the basis of fossils (Hedberg, 1976). Biozones can be grouped into larger units that the ISSC terms superbiozones (superzones) or subdivided into subbiozones (subzones). Subbiozones can in turn be further subdivided into zonules.

A barren interzone is an informal term for an interval lacking fossils found between two successive biozones. A barren interval within a biozone is a barren intrazone (Hedberg, 1976). One may also distinguish intervals between biozones that need not be barren. Any rock unit within which one or more biozones are distinguished need not be completely subdivided into biozones. Biohorizons are surfaces of "biostratigraphic change or of distinctive biostratigraphic character; preeminently valuable for correlation (not necessarily time-correlation); commonly used as a biozone boundary, though often recognized as horizons *within* biozones. . . . Biohorizons have been called surfaces, horizons, levels, limits, boundaries, bands, markers, indexes, datums, datum planes, datum levels, key horizons, key beds, marker beds, and so on" (Hedberg, 1976, p. 49, italics in the original).

Within the general category of biozones, the ISSC recognizes four types in particular: assemblage-zone, range-zone, acme-zone, and interval-zone (= interbiohorizon zone). (In comparison, the NASC [1983] recognizes interval, assemblage, and abundance biozones; see discussion of the NACSN's biostratigraphic units below.) The various types of zones are not mutually exclusive; a single section may be subdivided in differing manners using different types of zones. Furthermore, the same set of strata and fossils as recognized in the field may alternately be defined as more than one type of zone; for instance, in a particular case, an assemblage-zone, a lineage-zone, or a range-zone might all incorporate precisely the same strata (see following discussion).

An assemblage-zone is "a group of strata characterized by a distinctive *natural assemblage* of all forms [fossils] present or of the forms present of a certain kind or kinds" (Hedberg, 1976, p. 50). Assemblage-zones often are referred to as ceno-zones or faunizones. The fossils defining or characterizing an assemblage-zone must at least be found in association together in the strata; they often are interpreted as having lived together, or at least died together, or as having been accumulated together shortly after death (Hedberg, 1976). The basis of an assemblage-zone can be the entirety of the fossils found in the particular strata, or only the fossils of certain groups (either taxonomic groups or groups defined by other means); thus, for example, one could recognize different assemblage zones based on crinoids, microfossils, benthic forms, or floral elements. According to the ISSC, assemblage-zones are particularly important as environmental indicators, and the same assemblages may recur many times in a particular stratigraphic sequence (where the recurring assemblages do not vary significantly in age and thus show little, if any, evolutionary change).

For any particular assemblage-zone, defined as containing a series of elements (fossil taxa), not all of the defining elements need be present in order for strata to be assigned to the assemblage-zone. Likewise, the range of any constituent taxon may extend beyond the limits of the assemblage-zone.

From the wording of the *International Stratigraphic Guide* (Hedberg, 1976), it appears that a stratotype for an assemblage-zone is not mandated, but is highly recommended. The name of an assemblage-zone is derived from the names of two or more of the taxa that are considered diagnostic or characteristic of the assemblage-zone (although their presence is not necessarily required to assign taxa to the assemblage zone); for example, *Pyxilla oligocaenica-Coscinodiscus descrescens* Assemblage-zone (see Fenner, 1985). The ISSC suggests that recurrent assemblage-zones, or differences within an assemblage-zone, can be distinguished by letters, numbers, or terms such as lower, middle, and upper.

A range-zone (also sometimes termed an acrozone, although the ISSC does not recommend using the latter term) is "a group of strata representing the *stratigraphic range of some selected element* of the total assemblage of fossil forms present" (Hedberg, 1976, p. 50 italics in the original). Here the term range refers to both the horizontal and the vertical extent (range) of the selected element. The element may be anything paleontological; for example, a taxon or grouping of taxa, or a lineage or segment of a lineage. In regard to range-zones (the same may hold true for any paleontologically based stratigraphic unit), the ISSC notes (Hedberg, 1976, p. 53): "While the accuracy of identification and biologic description of the taxons [it is ironic that the ISSC, a body presumably attempting to promote international standardization, rejects the plural form of taxon, taxa, endorsed by the International Zoological and Botanical Codes] on which the zone is based is critical to its value, there is always a degree of subjectivity and impermanence involved in taxonomic identification. Also a considerable variation in the range of a taxon may depend on whether its limits are defined morphotypically or by statistical population studies." Within the category of range-zones the ISSC distinguishes four principal kinds: taxon-range-zone, concurrent-range-zone, Oppel-zone, and lineage-zone (phylo-zone).

"A *taxon-range-zone* is the body of strata representing the total range of [established] occurrence (horizontal and vertical) of specimens of a particular taxon (species, genus, family, etc.)'' (Hedberg, 1976, p. 53). What have sometimes been termed genus-zones or species-zones are types of taxon-range-zone. The ISSC (Hedberg, 1976, p. 55) suggests that such terms as teil-zone, local-range-zone, and topozone, which are used to express the range of a taxon in a particular section or area, not be used because a "range in a local area is not meaningful unless the name of that area is given"; instead the ISSC prefers that one use something similar to a statement such as "the range-zone of *Equus* in North America" or "the range-zone of *Conoryctella pattersoni* in section C of Kutz Canyon, San Juan Basin, New Mexico." It seems to me that teil-zone and similar terms are of value; they are certainly economical of words and should be unambiguous if used in proper context.

A taxon-range-zone is simply named after the taxon upon which it is based. The concept of a taxon-range-zone is dependent solely on the concept of the taxon (the name of the taxon is tied to a type specimen that must have come from some section if it is a fossil); therefore, taxon-range-zones do not have stratotypes per se. It is suggested that reference sections are useful, however (Hedberg, 1976).

"A *concurrent-range-zone* is defined as the concurrent or coincident parts of the range-zones of two or more specified taxons selected from among the total forms contained in a sequence of strata" (Hedberg, 1976, p. 55). This form of zone has been referred to as an overlap zone or a range-overlap zone. By definition, all of the taxa named in the definition of a concurrent-range-zone should be recognized before strata can be assigned to the given concurrent-range-zone. In practice, as the ISSC points out, such zones often are recognized only on the basis of the joint occurrence of some substantial number of the index taxa.

A concurrent-range-zone is named for two or more of the taxa that are diagnostic of the zone; for example, *Globigerinoides altiaperturus-Catapsydrax dissimilis* Concurrent-range-zone (see Iaccarino, 1985). A concurrent-range-zone does not have a stratotype, but reference sections should be cited.

The primary use of concurrent-range-zones is for chronocorrelation; the basic concept is that the ranges of two taxa, if used simultaneously (that is, where they are coincident), will have more time significance (be a more dependable tool in time correlation) than the range of any one taxon in isolation. However, the statement that "the use of two or more taxons whose range-zones overlap reinforces the time significance of an individual taxon-range" (Hedberg, 1976, p. 56) is not a fact or an observation in every case, but represents a hypothesis to be tested. One can imagine scenarios in which the overlapping of range-zones does not reinforce time significance; perhaps two taxa consistently overlap in a certain area because of ecological factors.

"The Oppel-zone may be defined as a zone characterized by an association or aggregation of *selected* taxons of restricted and largely concurrent range, chosen as indicative of approximate contemporaneity" (Hedberg, 1976, p. 59, italics in the original). The Oppel-zone, named after Albert Oppel, is thought to approximate his usage of the term zone. An Oppel-zone is almost a concurrent-range-zone, but is more loosely defined, more subjectively identified, and therefore more easily applied. Boundaries of an Oppel-zone in particular rely to a great extent on the subjectivity of the investigator's judgment. By definition, not all of the taxa considered diagnostic of the Oppel-zone need to be present in order to identify the zone, whereas strictly speaking they should all be present to assign rocks to a concurrent-range-zone.

The name of an Oppel-zone is based on the name of a single prominent taxon, although that named taxon need not be present throughout the entire zone. Oppel-zones are not based on stratotypes, but reference sections should be designated.

"A lineage-zone [or phylozone] is a type of range-zone consisting of the body of strata containing specimens representing a segment of an evolutionary or developmental line or trend, defined above and below by changes in features of the line or trend" (Hedberg, 1976, pp. 58-59). Other names that have been applied to lineage-zones include phylozones, phylogenetic zones, morphogenetic zones, and evolutionary zones. Suffice it to say that the theoretical basis of lineage-zones is open to dispute; it may well be the case that ancestor–descendant relationships (the basis of most lineage-zones) cannot be recognized in the fossil record (Schoch, 1982, 1986a,b; see also Kohlberger and Schoch, 1985, 1986, and the discussion of the polyhemeral system below). A lineage-zone is essentially nothing more than a taxon-range-zone clothed in the context of a subjective evolutionary scenario.

Lineage-zones are named after some key taxon in the lineage. Lineage-zones are not based on stratotypes, and the ISSC makes no mention of reference sections in regard specifically to lineage-zones; however, reference sections are always useful for any type of stratigraphic unit.

Another type of biozone, the acme-zone, is a "group of strata based on the *abundance* or development of certain forms, regardless of either association or range" (Hedberg, 1976, p. 50, italics in the original). An acme-zone represents the "acme or maximum development—usually maximum abundance or frequency of occurrence—of some species, genus, or other taxon, but not its total range" (Hedberg, 1976, pp. 59-60). The way the term "maximum development" is interpreted may vary significantly from one investigator to another; it can refer to commonness of specimens of a certain taxon, or even to the number of species (or other taxonomic category) within a higher taxon. Acme-zones have also been called epiboles (where a hemera is the time interval equivalent to the acme of a taxon), peak-zones, and flood-zones. The name of an acme-zone is derived from the name of the taxon under consideration. Acme-zones are not based on stratotypes, and the ISSC makes no specific statement concerning reference sections in regard to acme-zones.

An interval-zone (= biostratigraphic interval-zone, biointerval-zone, or interbiohorizon zone) is "the stratigraphic *interval* between two biohorizons" (Hedberg, 1976, p. 50, italics in the original). The two biohorizons that form the bottom and top boundaries of the interval-zone may be any biohorizons that are identifiable, such as the first or last occurrences of taxa or the bases or tops of any other biozones. The name of an interval-zone can be derived from the names of the boundary horizons, in which case the name for the lower boundary horizon precedes that for the upper boundary horizon. Alternatively, an interval-zone can be named after some taxon typical of the zone. The ISSC does not mention stratotypes or reference sections specifically with regard to interval-zones.

The biostratigraphic units recognized by the *North American Stratigraphic Code* (NACSN, 1983) are similar to, but differ in some details from, those recognized by the *International Stratigraphic Guide* (Hedberg, 1976). The NACSN (1983, p. 862) defined an interval zone in a somewhat more restricted manner than the ISSC's interval-zone. According to the *North American Code*, the boundaries of an interval zone are marked solely by the documented lowest and/or highest occurrences of single taxa; the boundaries of an interval zone cannot be any two biohorizons. Used in this sense, the interval zone of the NACSN includes the taxon-range-zone of the ISSC, the lineage-zone of the ISSC, and some forms of interval-zones and concurrent-range-zones of the ISSC (see the *North American Stratigraphic Code,* reprinted in this volume as Appendix 1). The assemblage zone of the NACSN includes the assemblage-zone of the ISSC, the Oppel-zone of the ISSC, and some forms of the concurrent-range-zone of the ISSC. The abundance zone of the NACSN is the acme-zone of the ISSC.

In contrast to the guidelines of the ISSC, which do not require stratotypes for many biozones (in fact the ISSC argues that many biozones cannot be typified by stratotypes), the NACSN explicitly requires the designation of a stratotype and reference sections for all new biostratigraphic units (NACSN, 1983, p. 863).

SPATIAL BIOSTRATIGRAPHIC UNITS

Some authors (e.g., Ludvigsen et al., 1986; Valentine, 1963, 1977) have consistently confused biostratigraphic units as described in the last section with temporal units. "Naturally, most geologists have assumed that biostratigraphy is concerned only with the temporal significance of fossils in rocks; that is, with the *age* of rock units, with the drawing of *time lines* in strata, and with the establishment of relative *time scales*. The spatial significance of fossils in rocks has seldom been satisfactorily integrated into biostratigraphic analyses, but this aspect is potentially as significant as the traditional temporal aspect (Kleinpell, 1938, p. 32)" (Ludvigsen et al., 1986, p. 139, italics in the original). The basic unit of temporal biostratigraphy would be the zone, whereas the basic unit of spatial biostratigraphy would be the biofacies. To present these authors' ideas in their own words (Ludvigsen et al., 1986):

> Biostratigraphy may be defined as that branch of stratigraphy that is concerned with recognition and mapping of fossil units in rock, and with their temporal and spatial significance. From this perspective, a major task of biostratigraphy is to separate, as clearly as possible, the temporal and spatial controls on the distribution of fossils in rock. This is best done with separate units—zones and biofacies, respectively.
>
> Zones are best defined by species range data, simply because species have the short temporal durations necessary for the establishment of fine divisions. For biostratigraphy, biofacies should be defined at generic or higher levels in order to produce units with significant stratigraphic ranges which may, in turn, be used to gauge the degree of environmental association of biotas. Thus, segregation of the spatial and temporal components is based on analyses of fossil collections at different taxonomic levels—genera for biofacies and species for zones. (Ludvigsen et al., 1986, p. 139)

The hierarchy of temporal biostratigraphic units, according to these authors, from least inclusive to most inclusive, would be zones, stages, series, and systems, whereas the hierarchy of spatial biostratigraphic units would be biofacies, provinces, and realms.

In presenting their system of "dual biostratigraphy," Ludvigsen et al. (1986) ignore several important points. It is now generally agreed that whereas some biostratigraphic units may approach chronostratigraphic or temporal units, biostratigraphic units are not strictly temporal units (see especially Chapter 7 of this book). Stages, series, and systems originally may have been defined, at least for practical purposes, as biostratigraphic units, but they have now been adopted universally as chronostratigraphic units. This, however, does not mean that on a theoretical level chrono-stratigraphic units are merely biostratigraphic units. It is too late now to attempt to redefine stages, series, and systems as biostratigraphic units. Furthermore, certain types of biostratigraphic zones as used by the general stratigraphic community are simple three-dimensional units, not simply vertical, or, on a hypothetical basis, temporal units as implied by Ludvigsen et al. (1986). These authors apparently fail to understand the nature of biostratigraphic units (as discussed in the last section) and the theoretical distinction between biostratigraphy and chronostratigraphy. They revert to the concept that chronostratigraphic units either do not exist or are identical to biostratigraphic units (Ludvigsen et al., 1986, p. 143; see also Jeletzky, 1956; Johnson, 1979; and Chapter 7 of this book), a position that was perhaps appropriate in the past but becomes increasingly untenable with the advancement of sophisticated techniques of chronocorrelation.

BIOSTRATIGRAPHIC CORRELATION

As Boggs (1987) notes, the biostratigraphic units described above (and any similar units) generally are considered to be observable, objective units that can be traced or matched from one locality or stratigraphic section to another, just as can lithostratigraphic units to a certain extent. Furthermore, it often is assumed that many biostratigraphic units have a stratigraphic significance in the sense that they always occur stratigraphically above certain units and below certain other units. Thus one can correlate biostratigraphic units and propose biostratigraphic correlations or biocorrelations. Virtually all of the various types of biostratigraphic units described in the last section were proposed so that they might be biocorrelated with like units. Biocorrelation often takes the form of various types of matchings of biotic content of the biostratigraphic units being utilized.

In the past it often has been assumed that biocorrelations are equivalent to, or at least approximate to some degree, temporal equivalencies or chronocorrelations. Although this is still often held as a general rule or first approximation, in recent years there has been a disparaging of the view that fossils are a totally reliable guide to time planes and synchroneity, especially at a very fine level of analysis (see, e.g., Hedberg, 1948, 1976; Wheeler, 1959; Chapter 7). Certainly on a theoretical basis any biostratigraphic unit can transgress time planes to a greater or lesser degree when traced laterally. Yet in practice identical fossils still are most often viewed as occurring more or less contemporaneously, and thus the rocks that enclose them are hypothesized to be contemporaneous. With this thought in mind, A. B. Shaw (1964, p. 90, italics in the original) made the following observation: "There could be built up a perfectly consistent body of stratigraphic reasoning based upon the opposite assumption that species found from place to place do *not* indicate anything regarding contemporaneity of their enclosing matrices. Such a system would not, however, be the one we are accustomed to. Ever since the first elucidation of the succession of faunas by William Smith it has been a basic tenet of geology that faunal identity indicates, in some degree or other, contemporaneity."

Buckman's Polyhemeral System
of Biostratigraphic Correlation

In the late nineteenth and early twentieth centuries Buckman in particular used what Arkell (1933) referred to as the "polyhemeral system" for biostratigraphically corre-

lating rocks in different sections and regions. Buckman's work was widely influential and deserves consideration here. In the discussion that follows (based primarily
on Arkell, 1933) the terms epibole and hemera are used in the sense of Arkell (1933);
both are based on the oberved local acme of a species (in the case of Buckman and
Jurassic strata, usually an ammonite species) where the epibole is the stratal term
referring to the deposits recording the acme, and the hemera is the time during
which the strata consituting the epibole were formed.

Numerous English workers (as recounted in Arkell, 1933) built up hemeral tables
for the Jurassic based on the epiboles observed in many sections throughout Great
Britain. "Thus Dr. Lang has been able to divide up the Lower Lias of Dorset into
38 epiboles, each representing the hemera of a certain ammonite. These epiboles
average between 9 and 10 ft. [2.7 to 3 meters] in thickness, but some are much
thinner. In the Belemnite Marls, for instance, there are 13 epiboles, with an average
thickness of 5–6 ft. [1.5 to 1.8 meters]. Some of them may be mere layers no thicker
than the ammonites. The boundaries have to be fixed arbitrarily, for the fossils
often occur only in thin seams, separated by barren clays, which cannot be assigned
with certainty to any particular epibole" (Arkell, 1933, p. 26).

When sequences of epiboles, the hemeral successions, were compared from different areas, however, they often were found to differ in details although they were of
the same general age and contained some epiboles in common. Thus one section
might contain a whole series of epiboles between two particular epiboles, but that
series might be lacking in a section from a far removed locality. Likewise, the same
situation might hold true in other instances for the second section relative to the
first. If one assumed that there is theoretically a single ideal, complete sequence of
epiboles (and corresponding hemerae, where the various epiboles occur in only one
unique order), then this ideal sequence is very poorly and incompletely represented
in any particular section. Buckman and his supporters believed this to be the case.
At any one locality only a small fraction of the hemerae were represented by epiboles; the stratigraphic and paleontologic record at any one locality and section is
extremely incomplete.

Buckman assumed that ammonite species (or faunas) were essentially ubiquitous
during the times of their acmes, and that an ammonite species would attain its acme
simultaneously (at least "simultaneously" relative to the scale of geologic time) over
its entire geographic range. The validity of such assumptions was challenged, particularly on the basis of the ecology of extant organisms. It was argued, for instance,
that dissimilarities in fossil faunas from different localities that are of approximately
the same age can be explained by the fact that living marine organisms today often
have restricted distributions. One may today find different organisms living in different parts of the seas simultaneously.

As Arkell (1933) has demonstrated, Buckman was not unaware of such considerations; he simply did not consider them valid. Arkell (1933, p. 27, 29) relates an
exchange on this very subject that he had with Buckman [in the following quotation
it is also interesting to note Buckman's opinion of Lyell—an opinion that may not
have been uncommon in his day, and is still worthy of consideration]: "In the course
of an argument over his insertion of an excessive number of hemerae into the time-
table of the Corallian [upper Jurassic] rocks, I attempted to justify my view (which
I still hold) that many of his hemeral indices lived side by side on the same sea-bed,
by reminding him of Lyell's principle that the present is a [note Arkell's use of "a"
instead of "the"] key to the past and instancing the sporadic distribution of many

modern sea-shells around our coasts. He caused considerable provocation by remarking with a smile 'Ah! So you have been reading Lyell: a most misleading book.' At the time there seemed no more to be said, so outrageous was the heresy. But the remark often recurred to me and I realized that in this particular connexion there was at least a germ of truth in it.''

The major difference between stratigraphers such as Buckman and his school and the ecology-minded stratigraphers was their differing conceptions of the vastness of geologic time. Those of an ecological bent analogized hemerae to very short periods of time, as observed in the present day—where even several centuries or a thousand years is short. On the other hand, Buckman and his school, while considering a hemera to be very short relative to geologic time, easily envisioned it as comprising up to a million years. Given such time, a free-swimming organism such as an ammonite, or even a sessil organism such as a brachiopod, could deposit its remains over an extremely wide area. In other words, the basic argument of Buckman and his school was that ecological observations made in the present cannot necessarily be extrapolated and applied wholesale to geologic phenomena—a surprisingly modern-sounding argument.

Buckman's polyhemeral system faced more serious objections than those theoretical objections based on modern ecology. It was discovered that not only were epiboles missing from sections in some cases, but one could find two sections with a number of epiboles in common where the epiboles were preserved in differing orders in different sections (see Arkell, 1933). Such occurrences could not be explained by hypothesizing that any particular section is just very incomplete. Here it should be specifically pointed out that it was the acmes (the epiboles, the hemerae) that appeared in differing orders in various sections; the total documented ranges of species were usually much greater than their acmes.

The way this new information was interpreted by Arkell (1933) and other opponents of the polyhemeral system was to suggest that the acme of a species as observed in any particular locality and section need not correspond to the complete acme of the species over the entire area in which it was distributed or migrated. Observed acmes record local events and therefore are not of value in correlating strata, except over extremely short distances.

In place of epiboles and hemerae, based on the acmes of species, Arkell (1933) suggested that theoretically one should attempt to use biozones and biochrons (based on the total durations of species). According to Arkell, all we usually know are local species ranges in particular sections (teilzones, and their temporal equivalents the teilchrons), so that is what we must use, always respecting the limitations of the data (cf. A. B. Shaw, 1964). In some cases, however, Arkell (1933) believed that we can know the absolute range (biozone) of a species through the strata. "When we can detect lineages running up through a stratified series and so can get the antecedents and descendants of a species, then only can we define its range and be satisfied that the visible range is also the absolute range. For obviously the biozone of a species cannot embrace either antecedent or descendant species of the same lineage, and so bounds are set to the biozone (Arkell, 1933, p. 33).'' If this were true, then biochrons (where known) perhaps would compose the ideal tool for correlation. However, what Arkell considers obvious, that ancestral forms cannot coexist with descendant forms, is an a priori assumption that may well be incorrect (Kohlberger and Schoch, 1985, 1986; Schoch, 1986b)—at least as far as we can determine on the basis of our evidence (fossil morphologies in stratigraphic successions). Fossil taxa such as species can be recognized only on the basis of their intrinsic morphology (at least without introducing circularity into one's reasoning), and there is no reason why a taxon of relatively primitive morphology may not coexist in the stratigraphic record with what appears to be its morphological descendant. Indeed the concept of persistent primitive forms (e.g., "living fossils": Eldredge and Stanley, 1984) is well known. Furthermore, there is no a priori reason to believe that a lineage will be traceable up any single section; this very notion undercuts some of the arguments used by Arkell against the polyhemeral system (that is, that acmes preserved in particular sections are local events, and species migrated widely).

Current Methods of Biocorrelation

Biostratigraphic correlation, or biocorrelation, is in the simplest sense merely an expression of "similarity of fossil content and biostratigraphic position" (NACSN, 1983, p. 851). As discussed above, it usually is assumed that biocorrelations will approximate temporal correlations to at least some extent, but this need not always be the case. Some biocorrelations simply may reflect particular environments of deposition and their associated facies. Assume that in a certain region a terrestrial sandstone is overlain by a freshwater shale, which in turn is overlain by a marine limestone, and the three lithologies bear terrestrial, freshwater, and marine fossils, respectively. Furthermore, assume that all three lithologic units are strongly time-transgressive. Given several geographically separated sections that record all three lithologies and biotas, a stratigrapher might biocorrelate the terrestrial faunas with one another, the freshwater faunas with one another, and the marine faunas with one another on the basis of both biotic similarity and stratigraphic position, even if no species in common are shared between the terrestrial faunas exposed in different sections, and so forth. In practice, however, most biostratigraphers attempt biocorrelations that are based on identical species (or other low-ranking taxonomic units) with the idea in mind that biocorrelations will approximate chronocorrelations.

Any of the biostratigraphic units described above may form the basis of biocorrelations. Various types of fossil assemblages (such as assemblage-zones) may be biocorrelated with one another on the basis of identical or closely similar species shared among the assemblages being correlated. In practice it has long been common to correlate and date strata on the basis of guide fossils or index fossils (Weller, 1960). An index fossil usually is considered to be a species (or perhaps genus, or some other taxon) that is believed to be especially typical and useful in identifying strata of a certain interval (biostratigraphic interval or time interval). Such index fossils have been recognized since the beginning of the discipline of biostratigraphy, and in many respects can be considered to constitute the simplest (or some might suggest most naive) form of biocorrelation. If different strata or stratigraphic units share an index fossil in common, then they are considered to be biocorrelatable at some level. However, the label of index fossil actually incorporates several different concepts that must be distinguished from one another.

Initially, the concept of an index or guide fossil appears to have originated in the naming of assemblages of fossils (such as assemblage-zones and Oppel-zones) after a striking or significant species included within the assemblage or zone (A. B. Shaw, 1964). Understandably, in some investigators' minds the organism (fossil) after which a zone or assemblage was named came to be regarded as typical of the zone or assemblage, or even as necessary and/or sufficient for recognition of the zone or assemblage (however, as is discussed above, the named fossil need not necessarily be present in order for investigators to recognize an assemblage or zone). Likewise, index fossils would be used to recognize epiboles. Thus, in this sense, one might use the identification of an index fossil in the field to extrapolate or to assume that the particular zone or assemblage was represented; or the concept of index fossils might be viewed as simply a shorthand notation for zones and assemblages. Essentially the use of index fossils, in this sense, is synonymous or compatible with the use of the traditional biostratigraphic units from which the index or guide fossils are derived and to which they refer; the index fossil is simply the indicator of a zone.

The term index fossil also has been used in a fundamentally different way by North American stratigraphers, particularly in the early and mid–twentieth century (A. B. Shaw, 1964; Grabau and Shimer, 1909–10). Rather than being a zonal indicator, the index fossil is considered to indicate a lithologic (lithostratigraphic) formation. At one time, as part of their formal education, many would-be geologists had to take a course that consisted primarily of memorizing the names of hundreds of index fossils and the formations for which they are the indices (A. B. Shaw, 1964). Whether it was explicitly acknowledged or not, this procedure implied that a formation could be recognized not only on the basis of its lithologic characters and continuity from local section to local section (see Chapter 5), but also on the basis of its fossil content. Rock units containing the same index fossils were in some cases given the same formation name despite the fact that they might occur in separate geographic areas, that they might be composed of differing lithologies, and that there was no demonstrated lateral continuity between the rock bodies under consideration. As A. B. Shaw (1964) points out, this American usage or concept of index fossils was based on assumptions such as that the index fossil is restricted to only a single formation, but occurs throughout that formation. Shaw attributes these assumptions in part to the primarily nineteenth- and early twentieth-century American view that formations are generally time-parallel. Many American paleontological and biostratigraphic studies of the past have been woefully inadequate in documentation of the distribution of fossil taxa within formations. In many cases, species have been cited as merely present in or absent from a particular formation, the apparent underlying and unwritten assumption being that if a particular species is present anywhere in a formation (or smaller lithostratigraphic unit), then it is present throughout the formation (or smaller lithostratigraphic unit).

Any taxon represented by fossils can be used as the basis of a taxon-range-zone (see above). Traditionally, particular taxon-range-zones have been considered of high value in biocorrelation, and the taxa upon which they are based have been dubbed index fossils or guide fossils. A basic principle underlies the use of index fossils as indicators of taxon-range-zones—namely, that any stratum or rock body containing a particular index fossil, given that the fossil is found in situ, must have been deposited during the biochron of that particular species. The temporal durations of most species (at least those forming good index fossils) are relatively short; so two rocks or stratigraphic units deposited during the biochron of such a single species must be approximately contemporaneous. Series or successions of index fossils have been compiled for various regions so that, by use of such an index fossil, a single stratum or unit can be correlated to a composite stratigraphic section for the area, and in turn this local or regional composite section usually can be related to global composite sections.

Weller (1960) and many other authors have made several observations relative to the successful use of index fossils in this sense (as indicative of taxon-range-zones). In order to procure maximum benefit from the use of index fossils, the identification of such fossils should be accurate and consistent. Unfortunately, as is too well known, identifications of fossils may vary from investigator to investigator. The limits of many species and other taxa—that is, clear definitions of what specimens are regarded as belonging to the taxon under consideration—can vary from worker to worker. Paleontological taxa are periodically revised, redefined, lumped, or split. In many cases, whether working with index fossils or larger assemblages in attempting biocorrelations, one cannot simply consult and compare previously published lists of the species found in the rocks at various levels in different sections, but one must have some background as to how the identifier was delimiting and recognizing the species named. As Weller (1960, p. 549) astutely notes, the use of different specific names does not guarantee that the specimens are actually distinguishable, and likewise the use of the same name does not demonstrate that the actual specimens are identical. Weller (1960) has also noted that when utilizing index fossils, one should consider their presence to be of prime importance; the mere absence of an index fossil does not prove that the strata under consideration were formed during an interval other than that corresponding to the biochron of a particular species. Of course, a second species may be found in the strata under consideration whose biochron is known not to have overlapped with that of the first species; then one may fairly safely conclude that the first species is absent because the strata under consideration were formed beyond the range of its biochron. However, such a conclusion is based not on the mere absence of the first species per se, but also on the presence of the second species.

It is often stated that the ideal index fossil should have the following attributes: In order to promote widespread use of the index fossil by many stratigraphers and general geologists, it should be easily recognized and distinguished from all similar fossils, and it is helpful if it is commonly represented by abundant specimens that are easily located in the field. In order to yield fairly precise and accurate biocorrelations, the ideal index fossil should be restricted vertically (and correspondingly temporally) to a fairly narrow stratigraphic interval. The ideal index fossil should have an extensive lateral (geographic) distribution. Finally, it should be an organism that was adapted to, could survive in, or had its remains deposited and preserved in a wide range of environments such that it can be found by the stratigrapher in many different types of rocks. In practice there are few, if any, ideal index fossils. It is often suggested that theoretically this should be the case; any species that is geographically widespread and abundant was also probably extremely successful in terms of its evolutionary adaptations and would therefore be expected to be fairly long-ranging in a temporal or stratigraphic (vertical) dimension. Locally or regionally a species of long temporal and stratigraphic duration may have been (or may be believed to have been) restricted to only a subinterval of its total chron or zone; thus its precision as an index fossil locally or regionally may be increased.

The range of index fossils (or any fossils—all fossils are potentially index fossils at some level), both horizontally and vertically, necessarily is controlled by numerous factors, such as environmental conditions prevalent during the species's existence, the time and place of origination of a particular species, barriers to the migration or dispersal of the species, time of extinction of the species, preservation factors, and so on. Within the vertical range of a species, ecological control is often extremely important in determining the detailed distribution of preserved specimens of the species in rocks.

The use of any particular index fossil in correlating strata generally can be no more precise than the limits of its biozone or biochron. Given two stratigraphic units that are biocorrelated relative to one another on the basis of a single index fossil, there may be no temporal overlap between the interval of time when one formed and the interval of time when the second formed. One unit may have formed during the time interval equivalent to the very beginning of the biochron of the index species, whereas the second may have formed during the time interval equivalent to the very end of the biochron of the species. For these reasons, detailed biostratigraphic correlations (the usual purpose of which is to approximate chronocorrelations) generally employ the use of numerous species simultaneously; this leads to the use of various biostratigraphic units (discussed above) in biocorrelation, which are defined on the basis of the occurrence of multiple species. Thus various types of assemblage zones, range zones, acme zones, and interval zones have been used with varying degrees of success in attempts to establish accurate biocorrelations (which in most cases ultimately are interpreted in terms of chronocorrelations). The basic idea is that the overlap in range of two or more species (or the acme of a single species, or the overlapping interval of the acmes of two or more species, and so on) often will define a smaller interval of time than the range of a single species and thus will lead to more refined (more precise or accurate) correlations.

It has become a common practice for stratigraphers to develop assemblages and assemblage zones, which in turn form the basis of biocorrelations, from the raw data of fossil taxa distributions arranged in range charts (see Figs. 4 and 5 of Appendix 1). In a basic range chart the local range of each pertinent taxon (usually species) is plotted against a generalized stratigraphic section drawn to scale or against an axis representing stratigraphic position relative to some datum point. On some range charts the relative abundances or acmes of the various taxa plotted are also indicated graphically, by widening the line used to plot the local range of the species under consideration (the wider the line, the more abundant the occurrence). When the stratigraphic occurrence of a particular species is uncertain, this may be indicated by some convention such as dashed lines, dotted lines, or question marks. Based on such compiled range charts, the stratigraphic column can be divided into a number of successive and nonoverlapping assemblage zones. In many cases, in local areas there may be breaks or shifts in environments and biotopes represented by the fossils such that locally "natural" assemblages are readily distinguished by the eye. In other cases, the column may not be readily separated into neat packets of assemblages, and the boundaries between assemblages may be delineated more or less arbitrarily, or on the basis of only the local ranges of a small subset of the fossils potentially usable to subdivide the column. Multivariate statistical analysis also has been applied to the problem of recognizing and delineating assemblage zones (e.g., Hazel, 1977).

Haq and Worsley (1982) suggest that currently biostratigraphy, biocorrelation, and biochronology (*sensu* Berggren and Van Couvering, 1978; see Chapter 3) are based on three basic types or categories of information derived from the biostratigraphic record: (1) the stratigraphic appearances and disappearances of morphotypes—that is, recognizable and named types of fossils—such as what are usually labeled species or subspecies; (2) changes in the abundance of any particular morphotype; (3) changes in the morphology of a morphotype. Obviously, these three categories grade into one another. The initial general appearance of an organism in the stratigraphic record certainly relates to its being abundant enough to be found by the stratigrapher, and likewise for all practical pusposes the last appearance of an organism is that point at which its abundance has diminished sufficiently that it will not be located by the stratigrapher. Attempting to determine abundances otherwise may be fraught with difficulties, as has been discussed above (see also further discussion below). Changes in the morphology of a morphotype seen up or down a stratigraphic column also may be expressed in terms of appearances and disappearances of more finely distinguished morphotypes. Even if there is a true gradation between two end-point morphotypes, the series still can be divided into a number of "submorphotypes" (even if they are distinguished from each other somewhat arbitrarily).

For considerations such as those outlined in the last paragraph, and because it is usually easiest to communicate that a certain organism (species, subspecies, morphotype, and so on) is either present or absent at a particular point in a section, biostratigraphers have come to record faunal and floral sequences in terms of strict presence/absence of species (or any other taxonomic unit; from here on I will simply use the term species to refer to the lowest level of morphotype analysis used in any particular biostratigraphic study). In most stratigraphic sections, at normal levels of analysis and resolution for the purposes of biostratigraphy, a particular species first appears at a certain stratigraphic level, and then appears more or less contiuously in higher strata until it finally no longer appears in any higher strata. That is, there is a single interval of strata in which the species is present; furthermore, it usually is assumed that if enough rock was searched (either in the local section or in chronocorrelative strata), there will be no gaps within the interval where the species was not present. In terms of evolutionary theory, a species does not appear (evolve), disappear (go extinct), reappear (re-evolve), and disappear (go extinct) again. Any particular species only appears once, plays out its duration, and then disappears forever. (If a "species" genuinely seems to appear, evolve, disappear, reappear, re-evolve, and so on, then one is probably not dealing with a species per se, but with ecophenotypes that are of limited value in biocorrelation—see below.) Given this situation, the biostratigrapher may record and communicate the distribution of a species in a stratigraphic section by two points: the first appearance datum (FAD) and the last appearance datum (LAD). FADs and LADs are the primary datum events currently used in biostratigraphy and biochronology on regional and global scales (Boggs, 1987; Berggren and Van Couvering, 1978; Haq and Worsley, 1982).

The FAD of any particular species in a certain geographic area usually is considered due to either the in situ evolution of the species from an ancestral form or the immigration of the form from another area. The tacit assumption, barring any evidence to the contrary, is generally that an FAD represents the immigration of a species shortly after its origination (evolution) in some unspecified area. An LAD is equated with the local extinction of a taxon. For many purposes, FADs and LADs are treated as if they were globally instantaneous events, that is, globally synchronous. However, it is acknowledged that neither FADs nor LADs are necessarily exactly synchronous—a plane or surface defined by all of the FADs or LADs of a specified species in all local sections where the species is found would be diachronous or time-transgressive. This phenomenon is due both to the nature of biological evolution, immigration, and extinction (assuming a perfect biostratigraphic record where all individuals are preserved as fossils in context) and to inherent imperfections in the stratigraphic record (displacement and mixing of fossils from one stratigraphic level to another, accidents of preservation of fossils, collection biases, and so on).

Once a species has evolved, it may take a certain amount of time to spread throughout what will eventually be its total geographic range, and it may also take a certain amount of time to become sufficiently numerous in colonized areas that its remains will be consistently preserved in the stratigraphic record. Based on modern analogies, however, it has been asserted that the immigration or spreading out (also known as prochoresis) of most taxa is extremely rapid on the scale of geologic time. Even at rates of prochoresis/immigration on the order of 1 to 10 km a year, an organism could spread over a very large region (perhaps a good percentage of the globe) in a couple of millenia. Likewise, planktonic microorganisms (extremely important in global biostratigraphy) such as "radiolaria, diatoms, coccoliths, tintinnids, and forams could have been distributed across oceanic areas to the limits of their adaptive range by the normal mixing and meandering of current gyres in a few tens or hundreds of years, whatever their mode of origin" (Berggren and Van Couvering, 1978, p. 42). In a similar manner, sessil organisms that bear planktonic larvae could be distributed at relatively rapid rates. These suggestions, however, probably are readily applicable only to a species spreading out within a certain climatic and ecological niche corresponding to a distinct geographic range. Beyond this normal tolerance, range or zone dispersal might be much slower, and possibly diachronous even on the scale of normal geologic time. Consequently the use of fossils for correlation may be less certain at the limits of their geographic ranges.

Some writers suggest that most extinction events, seen as LADs in the biostratigraphic record, occur less rapidly than origination and prochoresis events (e.g., Boggs, 1987). A species may undergo local extinction within part of its geographic range (perhaps due to adverse environmental deterioration) and yet continue to flourish elsewhere.

The duration, diachroneity, or nonsynchroneity of an FAD or LAD also may be due to processes other than those inherent in the distribution of biological organisms. A "true" FAD or LAD in a local section may differ from the apparent or recorded FAD or LAD by an interval of several centimeters or more. Referring specifically to microfossils gathered ("picked") from cores obtained from the Deep Sea Drilling Project (DSDP; see Boggs, 1987, pp. 7–8), Berggren and Van Couvering (1978, p. 42, italics in the original) wrote: "Even on the stratigraphic level, bioturbation, accidents of preservation, collection methods, and analytical bias combine to make a given 'pick' uncertain by several centimeters *at least* in any given stratigraphy. This represents uncertainty on the order of thousands of years at oceanic depositional rates under the best of conditions." Similarly, Haq and Worsley (1982, p. 25, italics in the original) assert that "the effects of slightly displaced fossils (i.e. *via* bioturbation, penecontemporaneous slumping, selected winnowing, etc.) seriously perturb the fine-scale fidelity of the palaeoclimatic and palaeoenvironmental signal contained in the stratigraphic record. For example, deep-sea sediments accumulate at an average rate of about 1 cm/10^3 yr and deep-sea benthonic organisms burrow to an average depth of 30 cm. These factors have the effect of defocusing the stratigraphic record in average deep-sea sections to a resolution of about 30,000 years." For some types of fossils, such as terrestrial mammal remains in Cenozoic continental sequences where the fossil specimens are primarily collected in gullies after having weathered out of the rock, the finest resolution possible may be on the order of several meters of stratigraphic section. One way to attempt to locate FADs and LADs more accurately is to use sections that are characterized by extremely high rates of continuous sedimentation, or to use sections that have been minimally disturbed (for instance, sequences deposited under anoxic conditions such that bioturbation is minimized). Yet such sections may exhibit other problems, such as a scarcity of fossils. Furthermore, such "ideal" sections are rare, and a "good" section in California may be of little use when one has a group of rocks in Massachusetts that one wishes to correlate with similar strata in Rhode Island or Connecticut. Despite the conditions that compromise the absolute synchroneity of FADs and LADs generally, Berggren and Van Couvering (1978) and Boggs (1987) suggest that FADs and LADs for some well-documented planktonic microfossils may vary on the order of only 10,000 years.

As Haq and Worsley (1982) have remarked, there has been much concern among the paleontologic and biostratigraphic community over the problem of the effects on correlation and analysis of displaced fossils, observed ranges of fossils that do not correspond to the true ranges, and so on; that is, there is a general concern over the adequacy of the fossil record. There are numerous means by which fossils can come to be found out of context in stratigraphic sections; for example, older fossils may be reworked and thus deposited in younger sediments, whereas younger fossils may be worked into older sediments by bioturbation. Biostratigraphy utilizing FADs and LADs is based on detecting species at the ends of their ranges (thus it is sometimes called "end-point" stratigraphy), but this is exactly where it is often the most difficult to find the species (Haq and Worsley, 1982; A. B. Shaw, 1964). Numerous workers have addressed the problem of the general completeness and adequacy of the fossil record, the reliability of stratigraphic ranges of species, and the establishment of the correct sequences of biostratigraphic datum events (namely, FADs and LADs). The classic work in this field is Alan B. Shaw's *Time in Stratigraphy* (1964); other important works include the contributions by Ager (1981), Edwards (1978, 1982a,b, 1984), Harper (1980, 1981), Hay (1972), Miller (1977), Paul (1982), Sadler (1981), and Southam, Hay and Worsley (1975). This field of endeavor has recently been reviewed in Schoch (1986b, pp. 199–231, including comments on "ecostratigraphy"; see also Brenner and McHargue, 1988, p. 257, and Boucot, 1983, 1984a,b), and such a review need not be repeated here. A basic conclusion of much of this work is that whereas some fossil specimens and datum events (FADs and LADs) may occasionally be displaced in local sections, in many cases the displacement is great enough that it can be recognized. In other instances where displaced fossils and datum events cannot be detected, this is so because the displacement is less than the current level of resolution of biocorrelation and biochronology; in such cases slight displacements will not materially affect biocorrelations.

Returning to the concept of using changes in abundance of a species (or morphotype), such changes in the abundance of a certain species through time can be quantified and plotted as an abundance profile, or "wiggly line" (Haq and Worsley, 1982), against time or stratigraphic position (analogous to a plot of a continuous geophysical well log, see Chapter 2). In some cases abundance profiles have been plotted for differing stratigraphic sections and correlations then attempted on the basis of matching peaks and troughs of the curves. What such correlations represent, however, is problematic. The underlying assumption behind such correlations appears to be that any particular species will reach its peaks of abundance simultaneously throughout its range, but on biological and ecological grounds this may not be the case (cf. the controversy between Buckman and Arkell recounted above). Changes in abundance of a species in a local section may be due primarily to local environmental parameters. Furthermore, abundance curves for particular species may be calculated as relative abundances; therefore, the abundance of any one species will depend in part on changes in the abundance of other species in the fossil biota (Haq and Worsley, 1982). An accurate abundance curve for any particular stratigraphic section also may be very tedious to derive. The sampling frequency in the section must be smaller than, or on the same order as, the frequency of peaks and troughs on the "true" (and initially unknown) abundance curve; otherwise a fallacious curve will ensue.

It has been suggested repeatedly (cf. Haq and Worsley, 1982; Hedberg, 1976) that the use of continuous morphologic change within a species (morphotype, taxon) through a stratigraphic section (and therefore through time), interpreted as an evolutionary lineage of ancestors and descendants, should be an extremely useful tool in biostratigraphic correlation. However, the actual recognition of ancestors, descendants, and evolutionary lineages is fraught with difficulties, and much presumed evolutionary change seen in stratigraphic sections may be indistinguishable from ecophenotypic variation (noninherited responses to local environmental conditions: see discussion in Schoch, 1986b, on these topics). Biocorrelations on the basis of similar ecophenotypes within a particular species will tend to reflect similarities in environmental conditions, but may be temporally diachronous (at least on a fine level of resolution).

Even if accurate lineages of species can be established (that is, species A gave rise to species B, which in turn gave rise to species C), one does not need to know the evolutionary scenario to biocorrelate like species. Strata containing species B in geographic area X can be correlated with strata containing species B in geographic area Y, whether or not A gave rise to B or B gave rise to C. It has sometimes been suggested (e.g., Arkell, 1933) that if we can establish an ancestor–descendant relationship, then we can know with relative certainty the LAD of the ancestor and the FAD of the descendant. However, there is now good evidence that a morphologic ancestor may temporally coexist with its morphologic descendant (Schoch, 1986b).

Undetected unconformities may seriously compromise methods of biocorrelation. If in a local section FADs or LADs should occur in strata that are temporally equivalent to the strata and time missing or unrepresented at an unconformity, there may be a pileup of FADs and LADs at the unconformity (Haq and Worsley, 1982). That is, a number of species may all exhibit FADs and LADs in the local section exactly at the surface of the unconformity, whereas in a more complete section these same FADs and LADs would be distributed throughout a significant interval of strata. Indeed such pileups (which classically have been used as the boundaries of various biostratigraphic units) may provide an important way to detect otherwise unrecognized unconformities.

In attempting to use a lineage approach to biocorrelation, Haq and Worsley (1982) suggest that unconformities will manifest themselves in the biostratigraphy as what appear to be "evolutionary bursts" (cf. the punctuations in evolutionary patterns of the fossil record: Eldredge and Gould, 1972; Schopf, 1981). An abundance profile for a species plotted against stratigraphic position may completely fail to detect an unconformity.

Biocorrelation, like many other forms of correlation, usually becomes increasingly tenuous and uncertain as correlations are attempted across larger geographic distances (taking into account shifting lithospheric plates and changing paleogeographies). The standard procedure investigators use in attempting large-scale, lateral biocorrelations is based on mutually interfingering, interlocking biozones (Hedberg, 1976; see Chapter 7). As mentioned above, it is always important to remember that the reliability of biocorrelations based on any particular species is probably highest within its normal geographic range and may diminish at the extreme limits of the widest area it inhabited. As Haq (1973) and Haq and Worsley (1982) have noted, with changing climatic regimes the geographic range of various organisms will expand and contract, thus changing the biocorrelations that are feasible. For example, during periods of general global warming, low-latitude assemblages of organisms may expand into higher latitudes and thus render low- to high-latitude biocorrelations more precise.

In attempting global biocorrelations across latitudes, for example, between low and high latitudes (and outside of what was probably the "normal" geographic range of any one species), it has been suggested that latitudinal shifts of fossil assemblages relative to paleobiogeography can be used in establishing correlations (Haq, 1980; Haq and Lowman, 1976). "The fundamental assumption in this concept is that the times of maximum environmental change, recorded by the major floral and faunal shifts through latitudes, are essentially contemporaneous. This scheme is analogous to the method of local time correlation by position within bathymetric cycles recorded in transgressive-regressive stratigraphic sequences" (Haq and Worsley, 1982, pp. 27–28; see Chapter 7 of this book).

Resolution in biocorrelation within any wide area during any particular interval is in large part a function of the number of datum events one has available upon which to base biocorrelations. Thus greater resolution may be possible among sections that exhibit high species diversity and rapid species turnover. Likewise, given a large suite of various organisms in a particular geographic region through a certain broad stratigraphic interval, certain species will tend to be more useful than others in developing refined biostratigraphic systems of classification and correlations. In particular, it may be extremely useful to be able to predict which species will be characterized by rapid evolutionary turnover and/or widespread dispersal within certain environmental conditions. Such species then may be singled out and used in developing initial high-resolution biostratigraphic schemes and correlations for the area. This mode of analysis is discussed in particular by Kauffman (1977; see also Kauffman, 1970, and Kauffman's book currently in preparation).

MAGNETOSTRATIGRAPHY

"*Magnetostratigraphy*—the element of stratigraphy that deals with the magnetic characteristics of rock units" (ISSC, 1979, p. 579, italics in the original): by this definition magnetostratigraphy encompasses any and all magnetic characteristics of rocks. In general, the main concern of magnetostratigraphy is the recognition, interpretation, and correlation of the remanent magnetism of rocks.

The global magnetic field of the earth (geomagnetic field) is composed of two dominant components: a dipole field and a non-dipole field. Apparently both of these have their origination below the mantle, but exactly how they originate and what accounts for their properties (for example, the variations through time discussed here) remains enigmatic (see Barendregt, 1984, for a brief review of geomagnetic dynamo theory). In general, it is believed that fluid motions in the core that are driven by convection produce a dynamo action that generates the magnetic field (Harland et al., 1982). The dipole field of the earth presently is centered in the earth and inclined to the spin axis by 11 degrees; it appears to have remained virtually stationary over the last two centuries (Watkins, 1972) although it is believed that over time the field wobbles several tens of degrees about its basic dipole direction (Barendregt, 1984). The non-dipole field has an average value of about 5% of the main field of the earth, but can constitute from zero to one-third of the total field, depending on geographic position. It is presently drifting westward at a rate of 0.2 to 0.3 degree of longitude per year (Barendregt, 1984; Watkins, 1972).

Many rocks are characterized by the property of remanent magnetism; magnetic iron oxide minerals, in particular, will become aligned with a magnetic field (imprinted) during either deposition of a sediment or the cooling and solidification of an igneous rock from a molten state (Kennett, 1980). Early in this century it was discovered that some rocks are magnetized in a reversed direction with respect to the present main geomagnetic field (e.g., Brunhes, 1906), wich led to the realization that the magnetic field of the earth has not been stable through geologic time but has undergone a number of polarity reversals (see Kennett, 1980, and papers reprinted and cited therein). Although there were earlier attempts or suggestions made that paleomagnetic data could be used in stratigraphic correlation (Khramov, 1957; Matuyama, 1929; Mercanton, 1926; Rutten and Wensink, 1960), it was in the early 1960s that magnetostratigraphy was developed into the basic form that it takes today (reviewed in detail by Watkins, 1972).

It was early realized that the potential importance of geomagnetic (dipole) reversals is that they are (at least in theory) truly synchronous events of global extent that are thus in many ways ideal for chronocorrelation. Of course, a drawback to magnetostratigraphy, especially initially, is that any particular magnetic signature preserved in a rock is not unique; there have been many times of normal polarity and likewise many times of reversed polarity in the history of the earth. The only way to attempt to uniquely identify and correlate a sequence solely on the basis of polarity signature is to use the ratios of the lengths of a number of polarity zones and then search for a similar match in other sections or against a global composite scale. In order to create a feasible stratigraphic tool, a detailed global polarity time scale first must be developed, and then the paleomagnetic sequence of any particular section has to be worked out in sufficient detail, and over a sufficient length of section, to find a unique match with the global polarity time scale. To date, a fairly detailed history of geomagnetic reversals has been established for the end of the Jurassic to the Recent (Harland et al., 1982; Palmer, 1983; see Appendix 3). This history has been developed by numerous investigators on the basis of the terrestrial and marine sedimentary records, as dated by isotopic methods, and the marine magnetic anomalies produced on the ocean floor as new crust is formed along mid-ocean ridges and thence carried away from the ridge by sea-floor spreading (see papers and references in Kennett, 1980; also Harland et al., 1982; Heirtzler et al., 1968). Unfortunately, the marine ocean floor anomaly record only extends back to the late Jurassic (Oxfordian); the record exposed on land prior to that time has not yet been worked out in detail.

In order to determine the relative length of a normal or a reversed sequence in a local section that one is attempting to relate to the global scale, it might be necessary to posit assumptions, for example, concerning average rates of sediment deposition in the section. Also, unconformities, whether detected or undetected, may seriously hamper magnetostratigraphic correlation. Commonly other criteria, such as paleontological, superpositional, or structural data, are used in conjunction with paleomagnetic data to bracket the date of the section under consideration. In some cases, however, this could leave one open to criticisms of circularity. To give a hypothetical example, a particular section is of Late Miocene age according to the fauna; this is corroborated by the paleomagnetics, which in turn was correlated with the Late Miocene initially on the basis of the fauna. In fact, on the basis of the paleomagnetics alone the section might show just as good a match to the Middle Miocene of the standard global polarity sequence.

By convention the polarity of the earth's magnetic field is defined as normal when, as at present, the field is directed toward the north (a standard magnet or compass needle seeks or points approximately toward the North Pole), and the inclination (essentially the plunge of the field: imagine magnetic lines of force emanating from the South Pole, traveling north and entering the North Pole) is directed down (toward the surface of the globe) in the Northern Hemisphere and up in the Southern Hemisphere. When the polarity is reversed, the field points toward the south, and the inclination at any point is changed by 180 degrees. The direction of magnetization of a rock or other object is defined as its "north-seeking magnetization." "If the north-seeking magnetization points toward the Earth's present magnetic north pole, the rock is said to have 'normal magnetization' or 'normal polarity.' On the other hand, if the north-seeking magnetization points toward the present-day south magnetic pole, the rock is said to have 'reversed magnetization' or 'reversed polarity'" (ISSC, 1979, p. 578).

It is not the case, however, that all rocks recording magnetization can be unequivocally established as having either normal or reversed magnetism. Rocks formed in a transition zone (formed during the time when the geomagnetic field was undergoing a reversal from one polarity to the other) may show intermediate polarities. Another problem arises because the paleomagnetic poles (dipole field poles) have not remained stable relative to crustal materials (for example, the cratons) on the surface of the earth; this problem is especially acute for Proterozoic and early Paleozoic time (Harland et al., 1982). Addressing this topic, the ISSC states (1979, pp. 578–579, italics in the original): "A problem arises because the north paleomagnetic pole [presumably the ISSC is here referring to the ancestor of our present North Pole, regardless of whether is was characterized by normal or reversed polarity; approximated by the spin axis of the earth] is believed to have crossed the equator in Paleozoic time, so that for some lower Paleozoic and older rocks it is unclear which is the direction of the 'north pole' and which of the 'south pole'. Polarity must therefore be defined with respect to the 'apparent polar wander' (APW) path for each crustal plate. If the direction of magnetization of a rock unit indicates a paleomagnetic pole [that is, points toward a north pole] that falls on the APW path that terminates at the present north pole, the rock unit has *normal* polarity. If the magnetization is directed 180 degrees from this, it has *reversed* polarity." Still, in some cases it may as yet be impossible to accurately reconstruct the path of the apparent polar wander so as to trace it back to the present. In a sequence of Proterozoic rocks, for instance, one may be able to determine that some of the rocks are characterized by magnetization of a particular direction that is reversed in other rocks of the sequence, but one might have no way of knowing which magnetization corresponds to the definition of normal polarity (Harland et al., 1982).

The geomagnetic field exhibits a number of variations at different scales (summarized in Stupavsky and Gravenor, 1984), some of which are (or potentially are) useful in stratigraphic correlation. The earth's magnetic field displays small variations (well less than 1% of the main dipole field) on a time scale of milliseconds to decades. These are of external (for instance, solar storms, meteorological activity) origin, geographically are of continental to global extent, and thus far have not been useful in stratigraphic correlations. The phenomenon termed secular variation has its origin in the non-dipole field, is of continental extent, and consists of small-amplitude variations (changes of approximately 30 degrees in declination and 10 or 15 degrees in inclination of the magnetic field) that last for a few hundred to a few thousand years. Secular variation has some application in regional stratigraphic correlation. Major variations in the main dipole field that are of global geographic extent traditionally (but see below) have been labeled polarity excursions, polarity events (now termed subchrons), and polarity epochs (now termed chrons). A time interval that lasts on the order of 10^5 to 10^7 years and is characterized by one primary polarity (it shows this primary polarity for most of its duration) is a polarity epoch (chron). During a polarity epoch (chron), the magnetic field may have undergone one or more short reversals (lasting on the order of 10^4 to 10^5 years); these are termed polarity events (subchrons). Changes between polarity epochs and/or events involve variations in the magnetic field direction of approximately 180 degrees declination. Polarity excursions are global (or possibly large-scale regional) amplitude deviations of greater than 30 degrees declination that last for less than 10^4 years (Stupavsky and Gravenor, 1984). Harland et al. (1982) suggest that polarity excursions may approach 180-degree changes and have durations of only 1000 years. Polarity excursions are of short duration, are not always global phenomena, and may (because of their short duration) be missing in many sections. For these reasons they are usually not included in composite global magnetostratigraphic time scales (see further discussion of excursions below).

The limits of magnetostratigraphic polarity units (for example, "epochs" or chrons) and their included subunits ("events" or subchrons and possible excursions) are, of course, arbitrary and determined by mutual agreement. Changes in the geomagnetic field are also not instantaneous although they are of relatively short duration. The time required for a full polarity transition has been estimated as on the order of 1000 to 10,000 years (Barendregt, 1984; Fuller, Williams, and Hoffman, 1979; Harland et al., 1982, and the ISSC, 1979, state that a polarity transition takes about 5000 years), although the geomagnetic pole can shift by 10 degrees or more in less than 200 years (Verosub, 1979).

In interpreting the magnetization of rocks, one must always be aware that the magnetization may be due to factors other than the geomagnetic field existing when the rock was originally formed (discussed further below). Excursions can be extremely important in the detailed correlation of stratigraphic sequences, particularly in the late Pleistocene and Holocene, but reported particular polarity excursions can be problematic (Stupavsky and Gravenor, 1984; Verosub and Banerjee, 1977). In order to be useful, an excursion must have internal and spatial (lateral) consistency. In other words, the purported excursion should be reproducible within the rocks from which it is first reported, and it should be present in rocks known to be of approximately the same age as those rocks. If these minimal requirements are not met, there is a good possibility that the purported excursion is spurious.

Remanent Magnetism in Rocks

Volcanic rocks often have a very strong remanent magnetization due to the orientation of magnetic minerals in the ambient magnetic field during cooling and solidifcation. This is commonly termed thermoremanent magnetism (TRM). Plutonic rocks also acquire remanent magnetism according to the ambient magnetic field; their magnetism is acquired after crystallization, but before the K-Ar clock is set in their biotite (Harland et al., 1982). In fine-grained sediments that include grains of ferromagnetic materials, during deposition or lithogenesis the grains may be oriented by the geomagnetic field; this is termed detrital remanent magnetization (DRM). Secondary magnetization(s) of a rock sample may be due to such factors as weathering, secondary mineralization, or even being struck by lightning. Secondary magnetization, termed viscous remanent magnetization (VRM), tends to weaken or destroy the DRM. Any particular rock sample usually contains several natural magnetizations, the sum of which is termed the natural remanent magnetization (NRM). Various techniques can be used to "clean" a sample and separate out various components of magnetism; the two most widely used methods at present are alternating field demagnetization and thermal demagnetization, both described by Collinson (1983):

> The underlying principle of the routine magnetic cleaning techniques is based on the generally lower stability of secondary magnetizations relative to those acquired by the primary processes of chemical remanence and thermoremanence. This allows the preferential removal of secondary NRM by the application of sufficient energy [either magnetic energy or thermal energy in alternating field demagnetization or thermal demagnetization, respectively] to overcome the magnetostatic energy of alignment within particles carrying secondary remanence, leaving them magnetically randomly oriented. In practice the intensity of the primary component is often decreased also, but since it is the direction of the primary NRM which is usually of interest for palaeomagnetic interpretation the intensity decrease is only important if it significantly affects the accuracy with which the surviving NRM can be measured. (Collinson, 1983, p. 308)

Using paleomagnetics the probability of successfully dating a sedimentary sequence may be relatively low: Stupavsky and Gravenor (1984) estimate the probability to be about 20% for deep-sea cores; the remaining 80% are magnetically overprinted and reflect the present geomagnetic field. Many factors could cause the detected remanent magnetization of a sediment or rock sample to reflect something other than the geomagnetic field in which it was formed (Barendregt, 1984; Stupavsky and Gravenor, 1984; Verosub 1975, 1977a,b; Verosub and Banerjee, 1977). The DRM commonly is acquired as magnetically susceptible sedimentary grains falling through a water column orient themselves with the ambient magnetic field. However, the DRM may not reflect the geomagnetic field exactly because of such factors as the size, shape, orientation, and preferred packing of the particular grains in a sedimentary regime (for instance, flat grains may inherently pack in a preferred orientation, even if free from the influence of external forces, whereas spherical grains might not). Currents may orient the grains, or the slope of a local bedding plane may have an influence on their orientation. There may be magnetic particles in the sediment that are too coarse to be aligned by the geomagnetic field. According to Stupavsky and Gravenor (1984), the accuracy of the DRM as a record of the geomagnetic field is relatively poor; typically the recorded inclination may deviate from that of the ambient field by as much as 20 degrees, and the declination may deviate by as much as 40 degrees.

Once a sediment is deposited, it may acquire a post-depositional remanent magnetization (PDRM) by various diagenetic effects and externally produced influences such as dewatering, bioturbation, sediment deformation, or earthquakes. The PDRM will, of course, reflect the geomagnetic field when the causative process took place. Stupavsky and Gravenor (1984) suggest that most sediments are characterized by a PDRM; in the case of deep-sea sediments, the PDRM is acquired on the order of 10^3 to 10^4 years after deposition of the sediment, whereas for lake sediments it is acquired approximately 10 to 100 years after deposition. Chemical remanent magnetization (CRM) may occur in a sample because of the production of chemicals in a sediment. "Possible self-reversal, as a result of post-secondary alterations in the ionic ordering of the crystal framework of the magnetic constituents of the sediment (Irving, 1964)" (Barendregt, 1984, p. 113) poses another problem.

In some cases problems that will compromise paleomagnetic results are evident simply from careful visual inspection (for instance, sediment deformation or bioturbation). In other cases, however, it may be necessary to sample two (or more) parallel, closely situated cores or time-equivalent profiles. If the two series exhibit large deviations from one another, then it can be suspected that one or both do not accurately reflect the history of the geomagnetic field in that area. However, even if they agree, that does not guarantee that they do accurately record the paleomagnetics; they both may record spurious results due to other common factors.

Errors may be introduced during sampling and specimen preparation in the laboratory. A sample may be misoriented in the field, either because it was incorrectly labeled, or because, in coring, the rock may have been broken, twisted, or otherwise scrambled. Furthermore, when a sediment (for instance, in a deep-sea core) is subjected to mechanical shock, its viscosity may decrease momentarily, and the remanent magnetism of the sample may be partially or wholly reset (Francis, 1971). This may occur naturally, for instance, as the result of an earthquake, or it may occur during the sampling process. Sediments that are susceptible to such shock resetting can be readily identified by a laboratory shock test (Symons et al., 1980).

Magnetostratigraphic Nomenclature

In the 1960s and 1970s it became common practice in the magnetostratigraphic and related literature (for example, in archeology or Quaternary geology) to refer to time intervals based on the magnetic polarity of stratigraphic sequences as "epochs" and "events." These epochs and events were designated by names, and variously treated informally or as formal chronostratigraphic/geochronologic units. Thus the last 5.2 million years was divided into four successive epochs on the basis of characteristic dominent magnetic polarities: Gilbert Reversed Epoch, Gauss Normal Epoch, Matuyama Reversed Epoch, and Brunhes Normal Epoch (from oldest to youngest; the names are derived from distinguished contributors to the science of paleomagnetics). Any particular epoch might contain shorter periods when the earth's field was opposite that of the primary polarity during the epoch; these shorter intervals were termed events and also designated by names (often of geographic origin).

The ISSC has vigorously objected to the above-described terminology and practices. The consensus of the ISSC is that these practices represent misuse of the terms epoch and event in stratigraphy. The epoch is the formal geochronological equivalent of a series in the Standard Global Chronostratigraphic (Geochronologic) Scale (ISSC, 1979: see Chapters 1 and 7 of this book), whereas an *"event is a happening* and not an interval [either] of time or of rock strata" (ISSC, 1979, p. 581). In the vast magnetostratigraphic literature (a representative sample of which is reprinted in Kennett, 1980) there appears to have been little, if any, real confusion of polarity epochs with formal epochs or events with happenings.

The ISSC (1979) introduced a recommended terminology for magnetostratigraphic polarity units and geochronologic and chronostratigraphic equivalents. Some major definitions proposed by the ISSC (1979) are as follows:

> *Magnetostratigraphic polarity-reversal horizons* are surfaces or very thin transition intervals in the succession of rock strata, marked by changes in magnetic polarity. (In practice, the term *magnetostratigraphic polarity-reversal horizon* may be allowed to stand for a transition interval of the order of thickness of 1 m. [note: it is unclear on what basis the ISSC arrived at this thickness, and whether this is a formal part of the definition or not].) Where the polarity change takes place through a more substantial interval of strata, the term *magnetostratigraphic polarity transition-zone* should be used. Magnetostratigraphic polarity-reversal horizons and magnetostratigraphic polarity transition-zones may be referred to simply as *polarity-reversal horizons* and *polarity transition-zones* if in the context it is clear that the reference is to changes in magnetic polarity. Polarity-reversal horizons or polarity transition-zones provide the *boundaries* for polarity stratigraphic units, although they may also be contained *within a unit,* where they mark an internal change subsidiary in rank to those at its boundaries.
>
> *Magnetostratigraphic polarity units* are bodies of rock strata, in original sequence, unified by their magnetic polarity . . . which allows them to be differentiated from adjacent strata. (ISSC, 1979, p. 579, italics in the original)

The formal term for the basic magnetostratigraphic polarity unit is the magnetostratigraphic polarity zone. Polarity zones are bounded by polarity-reversal horizons or polarity transition-zones. Magnetostratigraphic polarity zones also may be designated by differing hierarchical ranks: polarity megazone, polarity superzone, polarity zone, polarity subzone, and polarity microzone.

The ISSC (1979) notes that magnetostratigraphic polarity units are objective units based on a property of the rocks (namely, their magnetic polarity). In this respect they are allied with lithostratigraphic and biostratigraphic units (indeed, one could perhaps make a case for a magnetostratigraphic polarity unit in a sense being a form of lithostratigraphic unit: Harland et al. [1982, p. 65] even refer to such units as "magnetic lithostratigraphic intervals"); however, polarity units differ from lithostratigraphic and biostratigraphic units in being theoretically and potentially of worldwide extent. In this sense, magnetostratigraphic polarity units are similar to chronostratigraphic units, and this has led to some confusion over the distinctions between the two. However, even though polarity units *"may closely approximate* chronostratigraphic units, . . . they are *not* chronostratigraphic units because they are defined primarily *not by time* but by a specific physical character—*polarity of remanent magnetism*" (ISSC, 1979, p. 581, italics in the original). Indeed, in some cases the identifiable boundaries of a polarity unit may not be strictly isochronous, perhaps because of detected or undetected imperfections in the rock record (missing strata, rock deformation, and so on).

The geochronological equivalent of a magnetostratigraphic polarity unit is a chron; that is, a chron refers to the time interval during which a polarity unit formed. Hierarchical ranks can be associated with the term chron: megachron, superchron, chron, subchron, microchron (highest to lowest). A chron, not capitalized, is an informal subdivision of geologic time (if a chron is formalized, then it must be capitalized). The chronostratigraphic equivalent of a magnetostratigraphic polarity unit is a chronozone. A chronozone encompasses the entire body of rock strata (regardless of any remanent magnetism such rocks may possess) formed during the corresponding chron. The geochronological equivalent of a magnetostratigraphic polarity-reversal horizon is a moment.

With regard to the "epochs" mentioned above (for example, the Matuyama Epoch), the ISSC (1979) suggests that the entrenched and useful names be preserved, but that they be treated primarily as polarity zones (hence the Matuyama Polarity Zone—capitalized if used in a formal sense) and not as formal chronostratigraphic or geochronologic units. The chronostratigraphic equivalent of the old "epochs" should be informal chronozones (for example, Matuyama chronozone), and the geochronologic equivalents should be informal chrons (for instance, Matuyama chron). The traditional magnetostratigraphic "events" can be treated as polarity subzones with corresponding subchronozones and subchrons.

The ISSC (1979) recommends that the same general procedures be followed in establishing, revising, or redefining magnetostratigraphic polarity units as are used in establishing any stratigraphic units. Ideally a polarity unit should be based on a stratotype section with designated boundaries (preferably identified by permanent artificial markers in the stratotype section: ISSC, 1979, p. 580). As Harland et al. (1982) point out, however, the stratotype concept is of only limited use in magneto-stratigraphy. Because of variations in sedimentation rates, the ratios of lengths of polarity zones vary from section to section; and, because of hiatuses (perhaps undetected), some subzones may be absent from any particular section. "Adopting any one section as a stratotype would generally result in an international standard lacking a polarity fine structure that is potentially very important for correlation. For these reasons the most generally accepted time scale has evolved as a composite based on the global consistency of many different magnetozones, most of which are found in oceanic crust" (Harland et al., 1982, p. 71).

In naming formal magnetostratigraphic polarity zones, the ISSC (1979) suggests that geographically derived names are preferable, and that numbers and/or letters should be used only in an informal or quasi-stratigraphic sense (as for linear magnetic anomalies of the ocean floor).

The procedures and recommendations of the North American Commission on Stratigraphic Nomenclature (1983, see Appendix 1) with regard to magnetostratigraphic units and their chronostratigraphic and chronological equivalents follow closely the recommendations of the ISSC (1979). The NACSN, however, formally recognizes polarity-chronostratigraphic units and polarity-chronologic units, whereas the ISSC (1979) regards such units as primarily informal. For polarity-chronologic units the NACSN has adopted the following units: polarity superchron, polarity chron, and polarity subchron (in contrast to the units superchron, chron, and subchron of the ISSC, 1979).

Since the 1960s an informal numbering system has been used to refer to the major characteristics of the composite global magnetic polarity sequence: major anomalies (periods of normal polarity) are numbered serially, beginning with the present. This system originated when marine geophysicists numbered, from 1 to 32, the prominent anomaly peaks appearing on magnetic profiles over ocean basins. Temporally, this numbering scheme extended back to the late Cretaceous (Maastrichtian) and was adopted by many magnetostratigraphers. However, the original scheme did not number all anomalies; and since the original scheme was devised, various workers have modified it by interpolating letters, decimals, and so on, in order to label all the details of the developing composite scale accurately. In their recent magneto-stratigraphic time scale Harland et al. (1982), for instance, label all of the chrons and subchrons that they distinguish informally with a composite numbering system that has its basis in the original scheme of the 1960s described above, and extends back to the Campanian with chrons 33 and 33r (= 33 reversed). A similar number-ing system, using the designations "M0" through "M29" generally has been adopted for the middle Jurassic (late Callovian) to early Cretaceous (early Aptian). Again, marine anomalies were designated serially in order of increasing age, but in this case the anomalies are zones or intervals of reversed polarity. Harland et al. (1982) have adopted and modified this system also. The interval between these two numbering systems, encompassing Santonian through Albian, and much of Aptian, time (that is, a good portion of the Cretaceous; about 31 million years) is character-ized by normal polarity with at most a few short and uncertain reversed intervals (Lowrie et al., 1980).

It has been documented that throughout geologic time the geomagnetic field has undergone long intervals (lasting on the order of 30 to 100 million years) during which there has been a "polarity bias" (Harland et al., 1982, p. 76). During normal polarity bias intervals the geomagnetic field is normal the vast majority of the time, during reversed polarity bias intervals it is reversed the vast majority of the time, and during mixed polarity intervals it alternates more or less evenly between normal and reversed polarity. A number of these intervals have been designated by name (see review in Harland et al., 1982), and most recently Harland et al. (1982) have designated the six best-characterized polarity bias intervals as polarity superchrons named after the periods during which they predominantly occur (for example, the Permo-Carboniferous reversed polarity superchron, the Cretaceous normal polarity superchron, and the Cretaceous-Tertiary-Quaternary mixed polarity superchron).

Chapter 7

CHRONOSTRATIGRAPHY AND GEOCHRONOLOGY

CHRONOSTRATIGRAPHIC UNITS

Chronostratigraphy is defined by the *International Stratigraphic Guide* (Hedberg, 1976, p. 66, italics in the original) as "the element of stratigraphy that deals with the *age* of strata and their *time* relations," and thus chronostratigraphic (= chronostratic) classification is "the organization of rock strata into units on the basis of their age or time of origin" (ibid.). However, with newer developments in stratigraphy these definitions are no longer strictly true. For instance, the diachronic units of the *North American Stratigraphic Code* (NACSN, 1984; see Chapter 5 and Appendix 1) also deal with time relations and ages of strata, yet are not considered chronostratigraphic units or a part of chronostratigraphy proper. One could argue that diachronic units, like geochronologic units (see the *Code,* Appendix 1) are not strictly stratigraphic units because they are temporal rather than material units; but certainly the material referent of a diachronic unit is a stratigraphic unit, and the material referent of a diachronic unit (what could be termed a diachronostratigraphic unit) deals with time and age but is not a chronostratigraphic unit in the traditional sense. The conceptual basis of the practice of chronostratigraphy is the division, classification, and correlation of rocks and geologic time (events in geologic history) on the basis of time intervals that are isochronous and time planes that are synchronous. In the past the terms isochronous and synchronous have been used as virtual synonyms. Here, as in the *Code,* the term synchronous is used in the sense of simultaneously in time. Thus a synchronous surface or boundary was formed everywhere simultaneously. The term isochronous means equal time duration, but as commonly used in stratigraphy it refers to bodies of rock that were formed during the same interval of time, the interval of time under consideration being bounded by synchronous time planes (surfaces). Returning to the concept of diachronic time units, such units are not generally considered chronostratigraphic because they are not bounded by synchronous time planes or surfaces, and the resulting units are not isochronous but rather cut across time lines (lines of equal time or synchroneity).

The terms of synchroneity and isochroneity are not always used in a literal sense by stratigraphers, but often with the idea of being "synchronous" or "isochronous" relative to the magnitude of geologic time. For example, Harland (1978, p. 14) stated: "Logically as well as practically we can do without the concept of anything observable being precisely synchronous." Rather in many instances sequences of geologic events and synchroneity/isochroneity are conceptually expressed as follows (cf. Harland, 1978): Given the universe of events *A, B,* and *C,* one possibility is that *A* occurred first, temporally followed by *B,* and then followed by *C.* Thus we can say that event *B* (such as the formation of a certain rock body) is post-*A* and pre-*C.* Given a fourth event, *D,* if we find that event *D* is also post-*A* and pre-*C,* but we cannot resolve any temporal sequence among *B* and *D,* it may be concluded that *B* and *D* are synchronous events. A fifth event, call it *E,* might be both pre-*B* and post-*B* (that is, it began before *B,* spanned *B,* and ended after *B*); *B* might be considered by some stratigraphers as synchronous with *E,* or at least synchronous with some part of *E* (whether or not this part of *E* is specified or identifiable; if it were identifiable, perhaps event *E* could be divided into events of shorter temporal duration that did not entirely span *B* and overlapped the end points of *B*). Recently, however, stratigraphers generally have become increasingly concerned with using the concepts of synchroneity and isochroneity in as precise and literal a sense as possible.

As should be evident, the subdiscipline of chronostratigraphy is highly conceptual and theoretical, depending as it does on a very specific notion of time and synchronous time planes. It also should be noted that chronostratigraphy often has been referred to as "time-stratigraphy," and that the 1970 *Code of Stratigraphic Nomenclature* (ACSN, 1970, p. 13) defined a chronostratigraphic unit (= chronostratic unit of some authors) in the following manner: "A time-stratigraphic unit is a subdivision of rocks considered solely as the record of a specific interval of geologic time." Here the assumption is that time intervals (at least the time intervals referred to by the 1970 *Code*) will be isochronous and bounded by synchronous surfaces. Such assumptions are no longer warranted; stratigraphers now also recognize other, nonisochronous time intervals. Chronostratigraphy and chronostratigraphic units should not be referred to as time-stratigraphy and time-stratigraphic units. If the terms time-stratigraphy and time-stratigraphic units are used, they conceivably can refer to any stratigraphic or related categories expressing or relating to geologic time or age. The current *North American Stratigraphic Code* (NACSN, 1983) recognizes the following "time-stratigraphic" categories: the materially based (and thus genuinely stratigraphic) chronostratigraphic and polarity-chronostratigraphic units, and the non-materially based (and thus purely conceptual and not stratigraphic *sensu stricto*) geochronologic, polarity-chronologic, diachronic, and geochronometric units. Certainly other time-stratigraphic units are conceptually possible, such as the unit composing the material referent (what I have termed above a diachronostratigraphic unit, but not formally distinguished as a time-related unit by the *Code*) of a diachronic (strictly nonmaterial or temporal) unit.

The *Code* and the *Guide* provide two similar, but slightly different, definitions of a chronostratigraphic unit; however, within the remarks of the *Code* a chronostratigraphic unit is redefined to conform more closely with the definition of the *Guide.* Thus the *Guide* (Hedberg, 1976, p. 67) defines a chronostratigraphic unit as "a body of rock strata that is unified by being the rocks [that is, any and all rocks] formed during a specific interval [that is, isochronous interval] of geologic time." In contrast, the *Code* (NACSN, 1983, p. 868; see Appendix 1) states that a "chronostratigraphic unit is a body of rock established to serve as the material reference for all rocks formed during the same span of time. Each of its boundaries is synchronous." Following this definition of the *Code* literally, a chronostratigraphic unit is limited to the type (or equivalent) section(s) of the unit, other rocks being perhaps correlatable to the chronostratigraphic unit but not part of it *sensu stricto*. The *Code* goes on to state, however, that "a chronostratigraphic unit is a material unit and consists of a body of strata formed during a specific time span. Such a unit represents [note that the term "represents" is used instead of "consists of" or a similar phrase] all rocks, and only those rocks, formed during that time span" (ibid.). Most practicing stratigraphers appear to use the term chronostratigraphic unit without sharply defining it and without worrying about the differences between the definitions of the *Code* and *Guide*. In this book the term chronostratigraphic unit will be used in the sense of the *Guide* unless otherwise stated.

At least part of the confusion over precisely what constitutes a chronostratigraphic unit derives from the question, which is primary, the rock interval or the time interval? Or is neither primary? Harland (1978) has labeled the two opposing views the "rock–time model" and the "time–rock model." In the rock–time model the boundaries of the chronostratigraphic unit (a body of rock, for example, a system) first are defined by points (or some would argue synchronous planes) in a type section or sections, and subsequently the geochronologic unit (for example, a period) is derived from the chronostratigraphic unit as the interval of time during which the rock composing the corresponding chronostratigraphic unit was formed. This is the consensus view of most Western, particularly American, stratigraphers. Alternately, in the time–rock model the geochronologic unit (for example, a period) first is defined as the interval of time from one particular instant to another instant in time. The chronostratigraphic unit (for example, a system) then is derived from the geochronologic unit as the rock formed during the specific interval of time under consideration. How are the instants in time marking the boundaries of the geochronologic unit identified? In practice they are specified by points (one point for each instant) in stratigraphic sections that are interpreted as instants in time. Thus the same reference points in rock sections can serve either interpretation—that the chronostratigraphic unit is primary, or that the geochronologic unit is primary. Perhaps neither is primary, or rather what is primary is the points in rock used to translate or convert time into a material entity. The points in rock are significant, in this context, only to the extent that they can be interpreted as points or events in time.

Both the *Code* (NACSN, 1983) and the *Guide* (Hedberg, 1976) recognize the following formal hierarchy of chronostratigraphic units (from most inclusive to least inclusive): eonothem, erathem, system, series, stage (the equivalent geochronologic terms, that is, names for the spans of time corresponding to the chronostratigraphic units [see below], are: eon, era, period, epoch, age). The prefixes super- and sub- commonly are applied to the terms system, series, and stage (or their geochronologic counterparts period, epoch, age; the term age also is sometimes used as the temporal equivalent of a substage) if additional ranks are found to be necessary. The term chronozone is used by both the *Code* and the *Guide* as a nonhierarchical (and, as used by the *Code,* commonly small) informal chronostratigraphic unit whose boundaries may be independent of the boundaries of any formally ranked chronostratigraphic unit. The *Guide* also uses the term chronozone (geochronologic equivalent is a chron) in a formal sense as a chronostratigraphic unit of rank lower than a stage.

Not a chronostratigraphic unit (either formal or informal), but intimately related to chronostratigraphy is the chronostratigraphic horizon (chronohorizon) of the *International Stratigraphic Guide* (Hedberg, 1976, p. 67), defined as "a stratigraphic surface or interface that is isochronous [that is, synchronous as the term is used in this book]—everywhere of the same age.'' Although theoretically or ideally a chronostratigraphic horizon is two-dimensional (without thickness), in practice horizons often are recognized as very thin distinctive intervals that are thought to be essentially isochronous and are thus very useful in chronocorrelation. Chronohorizons, or potential chronozones (the statement that a horizon or thin interval is a chronohorizon is a hypothesis to be tested), include such features as some bentonite beds (thought to result from volcanic ash falls), some biohorizons (see Chapter 6), certain electric log markers, some horizons of seismic reflectors (see Chapter 2), horizons of magnetic reversal, and so on. The corresponding geochronologic term for a horizon in the sense of a two-dimensional surface is an instant; if a so-called horizon actually is represented by a very thin bed, then the geochronologic equivalent is a moment. Since the publication of the 1976 edition of the *International Stratigraphic Code,* its chief architect, H. D. Hedberg, has emphasized the importance of chronohorizons in ordering geologic events in time. ''These chronohorizons may represent important geologic events or they may simply be based on points of particularly reliable and significant isotopic age determinations. They may mark important fossil extinctions or important magnetic polarity reversals. They are no less important features of the international geochronologic scale [actually, chronohorizons are part of a chronostratigraphic scale, their instants or moments being part of a geochronologic scale] and no less important to earth history just because they may not happen to coincide with classic chronostratigraphic unit boundaries'' (Hedberg, 1978, p. 38).

Just as chronostratigraphic intervals bounded by points in rock and time may be conveniently labeled (for example, the Devonian System/Period), it has been suggested that chronostratigraphic systems and scales could be established simply by directly naming points in rock corresponding to points in time—the ''alternate point system'' (Harland, 1978; Hughes et al., 1967). ''If a point in rock is sufficiently useful for correlation or if it represents a convenient event in local history it may usefully be labeled. This is done typically in measured sections where, for example, particular samples were collected or where a position is identified for relocation. Such a point in rock (if diagenetic and other changes are excluded) refers very precisely to the point in time of initial formation which is thereby defined. This is the chronostratic way of defining an instant in time'' (Harland, 1978, p. 14). Elaborating on how points could be named and integrated into the global chronostratigraphic sale, Harland (1978, p. 21) wrote: ''The alternate point system will almost certainly acquire a new set of names from the locality where the boundary stratotype is selected ([for example,] Klonk for the Siluro-Devonian boundary; McLaren, 1973). If care is taken on later boundaries, the exercise is likely to impress key names on memory and they will be available for use when preferred. At this point I would urge committees setting up such boundary reference points to adopt the most conveniently memorable and usable name.''

As used in a formal sense by the *Guide,* a chronozone is the lowest-ranking division of the formal chronostratigraphic hierarchy. A chronozone (whether used formally or informally) usually is based on some previously designated stratigraphic unit, for instance, a formation or a biozone. A chronozone is composed of all of the rocks formed during the total maximum time span that corresponds to the stratigraphic unit upon which the chronozone is based. Thus if a chronozone is based on a biozone, and the bottom of the particular biozone was marked by the first appearance of the species *Psittacotherium multifragum* (see Schoch, 1986a) and the top of the biozone by the last appearance of *P. multifragum,* then the chronozone would include all strata formed during the interval of time marked by the first and last appearances of *P. multifragum.* According to the *Guide,* if a chronozone is based on a stratigraphic unit that has a stratotype (for example, a lithostratigraphic unit such as a formation), then the chronozone can be defined in either one of two manners. The rocks included in the chronozone may correspond to the time interval represented precisely by the stratotype (the time interval whose beginning is marked by the earliest preserved deposit [theoretically the oldest preserved grain] of the unit at the stratotype, and whose end is marked by the latest preserved deposit [the youngest preserved grain] of the unit at the stratotype); or the chronozone may correspond to the total time span of the unit, taking into consideration all areas where the unit is developed (the chronozone corresponds to the interval from the time when the first grain [given that the grain is preserved] of the unit under consideration was deposited to the time of deposition of the last preserved grain of the unit). It is extremely important, when establishing a chronozone, to indicate precisely to what the chronozone is meant to correspond. If a chronozone corresponds to the time interval represented by a fixed stratotype, then the chronozone itself is permanently fixed. However, if the chronozone is considered to correspond to the total time interval represented by the total development of the unit, as further research (in either a positive or normative sense, see Chapter 3) extends or restricts the total body of rock that is believed to be referable to the unit, the chronozone will change. A chronozone based on the total development of a unit often will correspond to a larger interval of time than a chronozone based only on the stratotype; this is simply a function of the fact that many stratigraphic units are time-transgressive.

Theoretically a chronozone, like all chronostratigraphic units, is composed of all rocks formed during a certain designated interval of time; so all chronozones are potentially of global extent. In practice, however, the applicability of any particular chronozone may be limited to a certain area where diagnostic criteria (for example, the taxon or lithostratigraphic unit it is based on) are present such that the particular time interval can be identified or approximated in the rocks. Formal chronozones are named after the stratigraphic unit upon which they are based. Thus the chronozone based on the *Psittacotherium multifragum* Range-Zone would be the *Psittacotherium multifragum* Chronozone, or the chronozone based on the Rhode Island Formation would be the Rhode Island Chronozone. Where formal chronozones are used, they need not form a sequence of mutually adjacent, nonoverlapping units. Gaps or overlaps may be present among the formal chronozones used in a given area. The boundaries of chronozones also may straddle the boundaries of any higher-ranked formal chronostratigraphic units. In many areas formal chronozones are not recognized; there is no requirement that stages or other more inclusive chronostratigraphic units be divided into chronozones.

The *North American Stratigraphic Code* (NACSN, 1983) uses the chronozone in a formal sense in a manner similar to, but not identical to, that of the *Guide.* In the *Code* the chronozone is strictly nonhierarchical, but usually considered to be small, at least when used formally. According to the *Code,* a chronozone may be based on a biostratigraphic, lithostratigraphic, or magnetopolarity unit, and the *Code* suggests that the modifying prefixes litho- or bio-, or the term polarity, may at the discretion of the author be usefully combined with the term chronozone to indicate the basis of the chronozone (see Appendix 1).

As used in an informal sense, a chronozone is defined by the *Guide* as "a zonal unit embracing all rocks formed anywhere during the time range of some geologic feature or some special interval of rock strata" (Hedberg, 1976, p. 67). The basis of such an informal chronozone can be any other stratigraphic unit, series, or sequence of adjacent stratigraphic units, or any feature(s) of rocks or strata that has a time range; or the basis may be a purely arbitrary specification of a certain interval of strata. An informal chronozone may correspond to a time span of any length. To give a few examples, one can informally speak of the chronozone of the dinosaurs, the chronozone of life, the chronozone of the mammals, the chronozone of the Mattapan volcanic complex, the chronozone of the tills of Maine, or the chronozone of the Rhode Island Formation.

The stage is often considered the basic working group of intraregional and intra- or intercontinental chronostratigraphy, yet at least some stages are applicable, or potentially applicable, worldwide. In the hierarchy of formal chronostratigraphic units a stage is of relatively minor rank; the stage is the lowest level or unit generally recognized in proposed global chronostratigraphic scales (see below). In many cases higher-ranked chronostratigraphic units (eonothems, erathems, systems, and series) are defined, either directly or indirectly, in terms of the stages they contain. In such a context the boundary of a particular stage may be coterminous with the boundaries of any other coterminous divisions. To give an example, as used by Harland et al. (1982, p. 92), the lower boundary of the Hettangian Stage is also the lower boundary of the Lias or Early Jurassic Epoch and the lower boundary of the Jurassic Period. However, the entire Phanerozoic (above the base of the Cambrian) global chronostratigraphic scale has yet to be entirely subdivided into even tentatively agreed-upon stages (for instance, compare Harland et al., 1982, with Palmer, 1983).

Stages may be divided, partially or completely, into substages. Two or more adjacent stages may be grouped into a superstage, although in practice this is rarely done. The geochronologic equivalent of a stage is an age; the geochronologic equivalent of a superstage is a superage, while that of a substage is termed a subage or, somewhat ambiguously, simply an age.

The names of stages, superstages, and substages, if newly created, should be derived from the name of a geographic locality or feature in the vicinity of a type section or a type area. Stages usually are defined by two separate boundary stratotypes that may well be in widely separated geographic regions, namely, the lower boundary of the stage under consideration and the lower boundary of the next overlying stage, which automatically defines the upper boundary of the underlying stage. Many traditional stage names have been based on geographic localities, but others have been based on the names of lithostratigraphic units, and still others have been derived from various other sources. In the English language, stage names commonly, but not always, end with "-ian" (for example, Bajocian [Jurassic], Cenomanian [Cretaceous], Lutetian [Eocene]).

A series is a major subdivision of a system; it is expected that most series can be recognized worldwide (ideally all series should be recognizable globally—that is, wherever rocks of the appropriate age are preserved). Series usually, but not necessarily, are subdivided into stages. If a series is subdivided into stages, it usually is agreed that the lower boundary of the series is coincident with the lower boundary of its lowest stage, and likewise the upper boundary of the series is coincident with the upper boundary of its highest stage (which equals the lower boundary of the superjacent stage and series). If not subdivided into stages, a series has its own boundaries marked by boundary stratotypes. Series may be, but rarely are, united into superseries (two or more adjacent series) or subdivided into subseries. The geochronologic equivalent of a series is an epoch.

If newly established, the name of a series should be derived from that of a geographic feature or locality. Long-recognized and currently used names of series are in many cases derived from other sources, however. Some series names are composed of two terms, the first being a positional adjective and the second being the name of the system of which the series is a part; for example, the three generally recognized series of the Devonian are the Lower Devonian Series, Middle Devonian Series, and Upper Devonian Series (in such a case the corresponding geochronologic terms take adjectives appropriate to temporal intervals, that is, Early Devonian Epoch, Middle Devonian Epoch, and Late Devonian Epoch). Some series, such as those of the Cenozoic in particular, have classical derivations (such as Eocene, Miocene, and so on; see Chapter 1).

Systems are units of major rank in the chronostratigraphic hierarchy and are recognizable worldwide. Of all the units of the global chronostratigraphic hierarchy, systems "are probably the most widely recognized and the most widely used" (Hedberg, 1976, p. 73), at least for the Phanerozoic. In special instances supersystems and subsystems have been recognized, especially regionally. The geochronologic equivalent of a system is a period.

Ideally chronostratigraphic systems are defined by boundary stratotypes. If a system is subdivided into series and/or stages, as all Phanerozoic systems are, the lower boundary of the system is the lower boundary (at a boundary stratotype, via means of a "golden spike" or "peg"; see Chapter 4) of its stratigraphically lowest formal component unit (usually a stage); and likewise the upper boundary is the upper boundary of its uppermost component unit (equal to the lower boundary of the overlying unit). All of the commonly recognized Phanerozoic systems were established in the eighteenth and nineteenth centuries (see Chapter 1). When originally established, many of the systems were very imprecisely defined, and in some instances it was subsequently discovered that there were either gaps or overlaps at the presumed boundaries of adjacent systems. Furthermore, there has not always been universal agreement as to what component stratigraphic units (for example, series and stages) are to be included in a particular system; in some cases such controversy continues. To give one example, the Rhaetian Stage variously has been considered the uppermost stage of the Triassic or the lowermost stage of the Jurassic, and some authors do not even distinguish a distinct Rhaetian Stage (see, for instance, Arkell, 1933; Harland et al., 1982; Palmer, 1983). A major task of the current IUGS Commission on Stratigraphy is to refine and clarify the definitions of the commonly used systems (see Chapter 4).

The names of currently recognized systems have a diversity of origins (see Chapter 1) and do not bear standardized endings. The International Geological Congress held in Paris in 1900 attempted to introduce some standardization in the endings of most systems by proposing that the following names be adopted for the Phanerozoic systems, from oldest to youngest: Cambric, Siluric, Devonic, Carbonic, Triassic, Jurassic, Cretacic, Tertiary, and Modern. Note that the name Ordovician System, even though proposed by Lapworth in 1879, was not officially accepted by an International Geological Congress until 1960 (see Hedberg, 1976). In 1900 the Ordovician was considered part of the lower Silurian, and the Permian was considered the upper part of the Carboniferous. The term Modern used in 1900 was meant as a replacement for the term Quaternary. Needless to say, the conventions for system names proposed in 1900 were never generally adopted by the geological community and have now been regulated to obscurity.

An erathem is composed of several adjacent systems. Currently three erathems are recognized for the Phanerozoic: Paleozoic, Mesozoic, and Cenozoic (oldest to youngest). These terms refer to the development of life on earth and can be roughly translated as "old life," "intermediate life," and "recent life," respectively (Hedberg, 1976). The erathems of the Phanerozoic are defined primarily on the basis of what systems they include. Potentially eras earlier than those of the Phanerozoic could be defined chronometrically rather than chronostratigraphically; for example the Archean Era (if treated as an era instead of as an eon, see the next paragraph) might be defined as simply the time interval from exactly 4×10^9 years ago to 2.5 $\times 10^9$ years ago. At present there is no widely adopted set of erathem or era terms for rocks below the base of the Cambrian or time before the earliest Cambrian (see Harland et al., 1982).

The eonothem and its geochronologic equivalent, the eon, are the units of next higher rank above the erathem and era. The Phanerozoic (meaning evident life) Eonothem/Eon encompasses the rocks/time of the Paleozoic, Mesozoic, and Cenozoic Erathems/Eras. The lower boundary of the Phanerozoic Eonothem is the lower boundary of the Cambrian System, whereas the upper boundary of the Phanerozoic Eonothem is the type of rocks being formed at present (the Phanerozoic has not yet ended). For time and rocks before and below those of the Cambrian, various systems of eons and eonothems have been proposed. In most cases eons prior to the Phanerozoic have been defined chronometrically. To give an example, Harland et al. (1982) provisionally recognize three pre-Phanerozoic Eons, the Priscoan, the Archean, and the Proterozoic. In the scheme of these authors, the Proterozoic extends from exactly 2500 Ma (million years ago) to the beginning of the Phanerozoic, the Archean Eon is defined as spanning the time interval from exactly 4000 Ma to 2500 Ma, and the Priscoan is simply defined as pre-Archean time. The terms Precambrian, Archeozoic, and Cryptozoic have all been widely used by various authors to refer to the rocks lying below the oldest Cambrian rocks and/or to refer to the time before the beginning of the Cambrian. These terms have been regarded by such authors as erathems, eonothems, eras, or eons.

GEOCHRONOLOGIC AND
GEOCHRONOMETRIC UNITS

The concept of geochronologic units—temporal units (intervals of geologic time) corresponding to the time span of chronostratigraphic units—already has been introduced (Chapter 1). The concept of geochronologic units cannot be divorced from the concept of chronostratigraphic units, yet both the *Guide* and the *North American Stratigraphic Code* regard geochronologic units as distinct from stratigraphic units. Stratigraphic units must be material units (composed of rock bodies), whereas geochronologic units are purely conceptual, although the concept of a geochronologic unit is based on a material unit. However it could be argued that any stratigraphic unit is, as a unit per se, conceptual even if the concept is being applied to a material object(s). This is particularly true with regard to chronostratigraphic units whose boundaries, except perhaps in a type section, ultimately are distinguished on the basis of time and only can be approximated in the actual material rock record.

"The 'geologic system' exists in theory by extension of a conceptual time surface from the boundary points. It exists in practice too but as an alleged system which as soon as it is delineated is almost certainly not the true system" (Harland, 1978, p. 23). If chronostratigraphic units are considered stratigraphic units, then perhaps geochronologic units also should be considered valid stratigraphic units (at least as long as they are referring to rocks, in either their presence or their absence). Harland et al. (1982, p. 2) have considered chronostratigraphy and geochronology to be merely "different aspects of the single discipline of time-correlation."

Here it is appropriate to point out that there is no universal agreement as to what precisely constitutes geochronology. Sometimes geochronology is considered a part of stratigraphy per se, and sometimes a distinct but closely related discipline (see Chapter 1). The *Guide* (Hedberg, 1976, p. 15) defines geochronology as "the science of dating and determining the time sequence of events in the history of the Earth." By this definition there is broad overlap between geochronology and much or most of stratigraphy; indeed geochronology occasionally is used as a synonym of chronostratigraphy. The *Guide* (p. 15) defines geochronometry as "that branch of geochronology that deals with the quantitative measurement of geologic time (usually in years)." Alternatively, the terms geochronology and geochronometry have been used as simple synonyms (see Harland, 1978). Berggren and Van Couvering (1978, p. 39; see also Chapter 3 of the present book) have tied the concept of geochronology to "the ordinal progression which links a series of events in a system of irreversibly varying properties." These authors distinguish a number of types of geochronologies, each of which must have an underlying theoretical basis of ordinal progressions. Thus, the geochronologic system based on organic (biologic) evolution, and associated with the biostratigraphic record, is biochronology, whereas the geochronology based on unstable isotopes can be termed radiochronology. According to Berggren and Van Couvering (1978, p. 40) "there is, as yet, no accepted theory which specifies the existence and duration of each individual geomagnetic polarity interval, so that a true magnetochronology does not exist."

Both the *Guide* and the *Code* agree in recognizing the following basic hierarchy of geochronologic units (in order of decreasing rank): eon, era, period, epoch, and age (the equivalents of the chronostratigraphic eonothem, erathem, system, series, and stage). Under the *Guide* a chron is the equivalent of a chronozone; under the *Code* a chron is a nonhierarchical geochronologic unit.

By the conventions of the *Guide* and the *Code,* names of geochronologic units are in most cases identical to those of the chronostratigraphic units upon which they are based. However, in formal geochronologic nomenclature the terms Early, Middle, and Late (or similar temporal terms) replace the terms Lower, Middle, and Upper (or similar terms) used in formal chronostratigraphic nomenclature. The *Code* (NACSN, 1983, p. 869) notes that in some cases the names of eras and eons have been formed independently of the names of any corresponding chronostratigraphic units (indeed the same may hold true for geochronologic units at other ranks also); in such instances the geochronologic unit is equivalent to two or more adjacent chronostratigraphic units. These adjacent chronostratigraphic units subsequently may be united as a single higher-ranking formal chronostratigraphic unit that takes a name identical to that of the geochronologic unit. In other words, historically there have been occasions where the formal geochronologic nomenclature preceded the formal chronostratigraphic nomenclature—which serves as further evidence of the close association between chronostratigraphic and geochronologic units.

The *North American Stratigraphic Code* (NACSN, 1983, p. 872) recognizes a distinct category of units of geologic time termed geochronometric units. Like geochronologic units, geochronometric units are temporal abstractions and thus not considered stratigraphic units by the *Code*. Unlike geochronologic units, however, geochronometric units are not based upon any material body of rock (such as a chronostratigraphic unit), but rather are "established through the direct division of geologic time." For example, a geochronometric unit can be erected by defining the unit as the interval of time beginning at exactly 438 Ma and ending at exactly 408 Ma. The *Code* uses the same rank terms for geochronometric units as for geochronologic units; this practice could lead to confusion, but it also allows for "hybrid" units, those in which one boundary is defined chronostratigraphically while the other is defined geochronometrically (the Proterozoic is such a unit as effectively used by Harland et al., 1982). According to the *Code,* once a geochronometric unit is defined, one may recognize a corresponding chronostratigraphic unit. This implies that for our hypothetical geochronometric unit, comprising the interval 438 to 408 Ma (let us formalize it by calling it the Pseudosilurian Period), there is a formal chronostratigraphic unit, composed of all rocks formed during the interval 438 to 408 Ma, named the Pseudosilurian System. The Pseudosilurian System, based on a chronometric unit, lacks a stratotype or boundary stratotypes; in fact the exact boundaries of this system can only be approximated in the rocks. Of course, when we are dealing with a standard chronostratigraphic system, one defined on the basis of a stratotype or boundary stratotypes, the top of the system is definable only at the exact point of the upper boundary stratotype and the bottom of the system only at the exact point of the lower boundary stratotype. Everywhere else the boundaries of the system can only be approximated. However, the concept of geochronometric units does suggest the possibility, at least to some workers, that perhaps all or much of a standard global stratigraphic scale could simply be defined chronometrically.

THE STANDARD GLOBAL
CHRONOSTRATIGRAPHIC SCALE

An ideal advocated by the *International Stratigraphic Guide* (Hedberg, 1976, p. 76) is the erection of a "Standard Global Chronostratigraphic (Geochronologic) Scale" (essentially equal to the "Standard Stratigraphic Scale" of various authors, such as Harland et al., 1982; the "Geochronostratic Scale" of Harland, 1978; and geologic tables and scales of numerous other authors, e.g., Cohee, Glaessner, and Hedberg, 1978; Haq and Van Eysinga, 1987; Odin, 1982a; Palmer, 1983) that can "serve as a standard scale of reference for the dating of all rocks everywhere and for relating all rocks everywhere to world geologic history." Such a scale would consist of a universally agreed-upon hierarchy of systems, and if possible, series and stages, along with their geochronologic equivalents. The standard stratigraphic scale or geologic time scale, which is an ultimate goal of many stratigraphers, may, at least for the Phanerozoic (see, e.g., Harland et al., 1982; Obradovich, 1984), be composed of two intimately related aspects, the chronostratigraphic framework and the geochronologic scale that can then, at least in theory, be calibrated numerically to form a geochronologic/geochronometric scale (see below). As the *Guide* notes, at present there is general (but not universal) agreement as to the number and the names of the major systems that compose the Phanerozoic. However, the exact boundaries of many of the systems are currently in dispute. Disagreement can concern such basic issues as the problem of how systems should be defined in theory. Even among individuals who agree on the same basic principles, such as that systems should be defined by boundary stratotypes (which may be coterminous with the boundaries of stratigraphic units at other ranks), the best locations for type areas and the exact positions of the boundaries are disputed. In some cases there is not even universal agreement as to which stages should be included in a particular system.

Relative to the nomenclature of various types of geologic time scales, Harland (1978) suggested that any scale based on chronostratigraphic units may be termed a chronostratic (= chronostratigraphic) scale. Likewise, any scale of time that is devised or calibrated in terms of standard units of duration (such as years) would be termed a chronometric scale. A chronometric scale may or not have named divisions (named divisions being unnecessary, but perhaps convenient). A chronologic scale is composed of a sequence of phenomena (perhaps the boundaries of chronostratigraphic units) and may be calibrated chronometrically. Thus a chronologic scale may be composed of a chronostratic scale with chronometric estimates of the dates of the boundaries of the divisions. Examples of all of these types of scales may be of a local, regional, or global nature. Harland (1978) suggests adding the prefix geo- to any of these terms and capitalizing the first letters of each word to denote scales that are intended as global standards: Geochronostratic Scale, Geochronometric Scale, and Geochronologic Scale. Harland (1978) also suggests that the phrase "stratigraphic scale" can be used as a general term with appropriate modifiers as necessary.

According to the *International Stratigraphic Guide,* the first step toward the erection of a truly standard global chronostratigraphic scale will be the establishment of boundary stratotypes between all globally applicable units. At present such work is being undertaken by the IUGS Commission on Stratigraphy (see Chapter 4). Once established, it is imperative that such boundary stratotypes be respected and heeded by the international stratigraphic community, or the scale will not form a true standard. This may present serious problems for those stratigraphers who honestly cannot agree with the philosophy espoused by the *International Stratigraphic Guide* (Hedberg, 1976) and the current Commission on Stratigraphy concerning the concepts of chronostratigraphy in general and the erection of a standard global chronostratigraphic scale in particular (see below).

The *Guide* acknowledges that in many instances, particularly at lower ranks (such as at the level of the stage and below), a standard global chronostratigraphic scale may not be unambiguously applicable to rocks in areas far removed from the localities where the standard scale is defined. In such cases, stratigraphers may have to continue to utilize regional chronostratigraphic scales. "It is better to refer strata with accuracy to local or regional units rather than to strain beyond the current limits of time-correlation in assigning these strata [strata removed from the type areas of the global scale] to units of a global scale (Hedberg, 1976, p. 81)."

Here it should be noted that whatever standard or local scales are adopted, the choice of a scale or scales in no way affects the precision of stratigraphic correlations per se, although it can to a certain extent affect the statement of the age of a rock body or stratigraphic unit. To give an example, for a stratigraphic unit that currently falls near or at the Silurian–Devonian boundary, it may be stated that the unit is of (questionably) Silurian or Devonian age; whereas if the Silurian–Devonian boundary had been placed (arbitrarily) a bit lower, then one might have been able to state unequivocally that the unit was of Devonian age. In the latter case, however, the actual age and correlations of the unit are known no more accurately than before, yet superficially it may appear that the accuracy of the dating of the unit has been improved. Analogously, naming an object or concept does not increase our knowledge of it per se, but can help us in our communication and thinking concerning it. Likewise, if two or more separate scales are used simultaneously, such as a geochronostratigraphic scale and a geochronometric scale, or a magnetopolarity and a biostratigraphically based scale, the exact relations (correlations) between the scales may not be certain. One scale may be more useful for some rocks and units, whereas another scale may be better applied to other units. However, the hypothesized temporal relations of units dated relative to different scales may be only as accurate as the correlations between the framework scales utilized.

THE CONTROVERSY OVER CHRONOSTRATIGRAPHY: CHRONOSTRATIGRAPHY AND THE NATURE OF STRATIGRAPHIC UNITS

Chronostratigraphy, and all that it implies, has not been unanimously accepted by the stratigraphic community. On the contrary, an ardent minority has been opposed to the entire chronostratigraphic program.

It has been suggested that the modern concepts of, and controversies over, chronostratigraphy and chronostratigraphic units originated with the work of Hollis D. Hedberg in the early 1950s (see Hancock, 1977, p. 21, who further contends that "the current controversy over 'chronostratigraphy' is largely an atavism: over the years 1850–1862 the same problems were discussed"; it should also be noted that in 1925 Wilmarth, for example, commented on the difference, and sometimes confusion, between time and rock units, and in 1941 Schenck and Muller proposed the adoption of distinct, but corresponding, time and time-rock [time-stratigraphic] units although Hedberg, 1965, criticized these authors for not distinguishing between time-stratigraphic and biostratigraphic units). Essentially Hedberg (e.g., 1941, 1965, 1973a, b, and numerous references cited in ISSC, 1976) and his associates suggested that there be established a global chronostratigraphic scale composed of units whose boundaries are, by definition, synchronous. As the fundamental units of the chronostratigraphic scale, Hedberg adopted the units that had been used in biostratigraphy: system, series, and stage. As Wiedmann (1970, p. 41) has pointed out, the VIII Geological Congress in Paris at the turn of the century "unequivocally erected the sequence of

    system
        series
            stage
                zone

and ascribed this to biostratigraphy." On the whole, nineteenth-century workers may have believed that Silurian faunas in North America, for instance, were more or less contemporaneous with Silurian faunas in Europe; but as Huxley pointed out, this need not have been the case. "For anything that geology or palaeontology are able to show to the contrary, a Devonian fauna and flora in the British Islands may have been contemporaneous with Silurian life in North America, and with a

Carboniferous fauna and flora in Africa. Geographical provinces and zones may have been as distinctly marked in the Paleozoic epoch as at present, and those seemingly sudden appearances of new genera and species, which we ascribe to new creation, may be simple results of migration" (Huxley, 1862, p. xlvi; quoted in Hancock, 1977, pp. 17–18). Thus Huxley introduced his concept of homotaxy (similarity of the arrangement of ancient faunas and floras without implying that they are temporally equivalent) versus the concept of the synchrony or contemporaneity of similar ancient biotas (chronotaxy). But it is not clear that such issues were even of particular concern to most stratigraphers in the nineteenth century; the primary concern seems to have been to subdivide local stratigraphic columns into natural units and correlate (leaving this term undefined for the moment; see discussion in Chapter 3) these units from one place to another.

There arose the belief, however, that (at least at the appropriate scale) the units of biostratigraphy (if carefully defined, distinguished, and correlated) did mark periods of time that were essentially synchronous throughout the world. With the founding of modern chronostratigraphy it was logical to adopt (actually take over or steal) the major biostratigraphic units that indeed came closest to meeting the ideal of chronostratigraphic units in the real world (cf. Berggren and Van Couvering, 1978). Only the lowest-level unit, the one that most often might be significantly diachronous given how fine a scale it represents, the zone, was left to biostratigraphy. Certain biostratigraphers have also claimed that certain zones (classic examples are some ammonite and graptolite zones: see Hancock, 1977) are also of chronostratigraphic value.

With continuing work it has been either found, suggested, or suspected that many of the traditional boundaries of systems, series, and so on, are not synchronous. Many of these major units originally were defined and recognized as being bounded by major unconformities, which are not perfectly synchronous on a very fine level over wide areas. Thus, in order to ensure good chronostratigraphic units, the concept of the "golden spike" was introduced to mark permanently at one point only (in a type section) the instance in time that marked the boundary between two units. By definition, the chronostratigraphic unit boundary extended from this point by synchronous surfaces (whether such surfaces could, in fact, be recognized or not). In the process, what had once been natural stratigraphic units with nonarbitrary (or at least non-man-made) boundaries (that is, the boundaries corresponded to distinct changes in lithology or biota) and some kind of internal coherency were transformed into units whose boundaries were fixed completely arbitrarily. In fact, ideally the boundary between two such chronostratigraphic units should be set in a homogeneous section with a continuity of monofacial sedimentation. Boundary stratotype definition became "a normative question which can be settled by a vote" (Cowie et al., 1986, p. 6).

Objections to this form of chronostratigraphy have fallen into two broad, but related, categories. It has been argued that chronostratigraphy in practice is essentially synonymous with biostratigraphy (particularly as much of chronostratigraphy is based upon biostratigraphy) and therefore unnecessary as a distinct system (e.g., T.G. Miller, 1965; Wiedmann, 1970). After presenting a hypothetical case demonstrating that boundaries between both lithological and biostratigraphic units can be diachronous, especially when comparing incomplete stratigraphic sections (that is, sections containing diastems or unconformities, as most sections do), T.G. Miller (1965, p. 130, italics in the original) stated: "This model brings out the primacy of the biological evidence in geochronological analysis, and shows that sequential division must be in terms of *intervals,* which represent more or less temporal duration, rather than synchronous surfaces, which, although imaginable, and ideally present, cannot at present be detected in sufficient number to be practically useful.'' This leads into the second major objection to chronostratigraphy, that in geology most events are time-transgressive. The exact synchronous boundaries of the ideal formal chronostratigraphic units are not generally recognizable with any precision at a distance from their type areas (Watson and Wright, 1980). To quote two recent opponents of chronostratigraphy: ''The fact is that chronostratigraphic units are an imaginary entity of no value, and to claim an ammonite zone to be chronostratigraphic is to debase a practical stratigraphic unit, as well as to deny the biological characteristics of the origin, dispersal, and extinction of species populations—none of which is likely to be isochronous'' (Hancock, 1977, p. 19; similar arguments apply to stages, series, systems, and so on). ''Chronostratigraphy is thus more than useless. Rather than being 'an ultimate goal of stratigraphy' (Hedberg [ISSC], 1976, p. 96), chronostratigraphic classification is based on a hypothetical model that is shown by the test of stratigraphic research to override actual lithostratigraphic and biostratigraphic relations, to present a false picture of Earth history, and to give rise to pseudo-problems of correlation and inclusion'' (Watson, 1983, p. 177).

If chronostratigraphy is rejected, a question arises as to how we may place geologic events in an absolute temporal framework. A rejection of chronostratigraphy does not mean a rejection of absolute geologic time. Such real time can be approached through a number of means, such as isotopic dating and the paleomagnetic polarity reversal record or even traditional biostratigraphy (as long as one does not become entrapped by circular reasoning). Named geochronologic units (period, epoch, age, and so on) of the *North American Stratigraphic Code* currently are based on corresponding chronostratigraphic units, but this need not be the case. The *North American Stratigraphic Code* states that ''geochronological rank terms (eon, era, period, epoch, age, and chron) may be used for geochronometric units when such terms are formalized'' (Art. 97), and ''geochronometric units are units established through the direct division of geologic time, expressed in years'' (Art. 96). Thus if there is felt to be a need to establish and label a Devonian Period (a geochronologic/geochronometric unit) that has arbitrary synchronous boundaries, it could perhaps be agreed upon internationally that the Devonian Period is the unit of abstract time lasting, for example, from exactly 410 million years before A.D. 1950 to 360 million years before A.D. 1950 (cf. Lucas et al., 1981, for a definition of the Eocene in this manner). The Devonian System, a natural unit with an inherent consistency and cohesion in many places, can be left alone. In many localities the rocks representing the Devonian System may have been deposited during the Devonian Period, and in this way the Devonian System will temporally approximate the Devonian Period, but this need not be strictly true. In some places the very highest (stratigraphically) rocks of the Devonian System actually may have been deposited during what is arbitrarily defined as the Carboniferous Period. Alternately, the ''Devonian Period'' could be defined explicitly as a diachronic unit, the ''Devonian Diachron'' or ''Devonian Episode,'' rather than as a chronostratigraphic unit (see Chapter 5). (In analogy, cultural periods are not necessarily synchronous; the Bronze Age or the Renaissance did not begin and end everywhere at the same time). There is not necessarily any need to resort to a type section to define a geochronologic unit, unless (given as an example again) it is merely to say that the Devonian

Period lasted from the time represented by point A in section X to the time represented by point B in section Y. However, the system and period could temporally diverge from one another away from the type section and point. The question then arises: why not dispense with name periods (and other geochronologic units) altogether and merely use absolute ("years and millennia," Watson, 1985, p. 767) and relative (relative to certain fixed points) dates?

Stratotypes

Traditionally stratigraphic units (such as systems, series, and so on) have been based on the rocks of a typical locality that has served as the standard "type" or reference section of the unit (see Chapter 4). Originally most such units were distinguished in the belief that they represented important aspects or episodes in earth history; that is, they were viewed as natural units. In stratigraphy one school of thought has advocated strict adherence to the original type sections (when such exist) and original descriptions in defining and redefining major stratigraphic units. Stratigraphic boundaries are seen as primarily a matter of priority and convention, somewhat analogous to the situation in zoological and botanical nomenclature. From this school evolved the philosophy that stratigraphic boundaries (that is, of the global stratigraphic column) are arbitrary, mere conventions, that can be agreed upon by a majority vote. Such a philosophy inherently denies that such stratigraphic units have any objective meaning in nature—that such units are in any way natural. Potentially it also would force upon all workers the use of the same units, no matter how cumbersome or inappropriate they might be in particular situations; so no one could attempt to develop natural, meaningful stratigraphic classifications without having to set up opposing classifications. Any individual or small group of investigators attempting to arrive at some sort of natural stratigraphic classification will inherently be at a disadvantage if forced not only to pursue science but also to compete with an arbitrary—but supposedly internationally agreed-upon—classification. Progress toward such an arbitrarily defined global chronostratigraphic column is advancing rapidly (see Chapter 4; Cowie et al., 1986). The boundaries of what were perhaps once intended to be natural stratigraphic units are being arbitrarily defined for all of posterity at global boundary stratotype sections and points.

As Wiedmann (1970) has astutely pointed out, in zoological and botanical nomenclature only the names of units (in this case, taxa) are fixed by convention and priority, not the scope or classification of the units. Types in zoology and botany are single specimens that any researcher is free to classify in whatever manner he or she chooses. The only restriction is that if the researcher should group two or more types into the same unit (the same taxon), then the unit takes the oldest name attached to a type specimen. There is no international ruling (nor should there be) that states that lions, tigers, leopards, and jaguars must all be included within the genus *Leo,* and that there are exactly four species within this genus. Indeed, some workers may distinguish different numbers of genera and species within this group, or perhaps not even include all of these animals in a single grouping relative to other organisms; this is a matter of scientific research and judgment, not normative voting.

In stratigraphy at present we have a trend in a very opposite (and perhaps antiscientific) direction. Rocks between two arbitrary points, and all rocks that can be correlated with these rocks, are designated for all time as Devonian (as an example).

In the future there will no longer be any discussion as to the boundaries of the Devonian, unless an international commission votes to reinvestigate the matter. Furthermore the subdivisions of the Devonian, and their boundaries, also will be set. What will be left to the stratigrapher is the process of correlating to the standard sections, not discovering and refining a natural classification of strata and earth history. An analogous situation in zoology would occur if an international commission established a single classification of animals that all workers had to follow. Yet the way one classifies organisms can greatly affect the nature and validity of one's research, especially in dealing with historical aspects (phylogenetic and evolutionary) of the development of organisms (see Schoch, 1986b).

O. H. Schindewolf (cited in Erben, 1972, p. 94; see also Schindewolf, 1970a,b) has suggested that the holotype of a fossil index species is sufficient for fixing a stratigraphic unit. Erben (1972, p. 94) countered that "the best index fossil is worthless for stratigraphical purposes if the places of its occurrence and, therefore, its range within the rock sequence are not carefully and exactly identified." I would suggest that priority and convention could be preserved in stratigraphic classification and nomenclature if type points were designated for stratigraphic units (rather than the boundary points that the International Commission on Stratigraphy currently is designating). A single point in a type section in a type locality could be designated as, by definition, belonging to the particular named unit. Such points could be designated in the future by original authors when they named new units, and for previously established units such type (neotype?) points could be named after the fact by revisers or an international commission based on the material (sections) the original author actually studied along with his or her original descriptions and intentions. Such type points could be chosen as being truly typical of the named unit; perhaps they often would fall near the middle of such units. An original author would be free to define the boundaries of the unit as he or she saw fit, but later workers would be free to revise such boundaries as they thought necessary. However, when it came to naming units, any particular unit would take the oldest name among any type points contained within the recognized unit. If no type point were contained within the recognized unit, then it would be necessary to propose a new name (and type point) for the unit. Just as in zoological nomenclature particular type specimens and the names they bear can lapse into obscurity if they are lost, destroyed, found to be nondefinitive, or not used for a certain length of time (Ride et al., 1985), similar rules and guidelines could be established for stratigraphic nomenclature.

The details of such a system certainly could be worked out; such a system might be applied usefully not only to chronostratigraphic units but also to diachronic units, and to other types of stratigraphic and related units. For purposes of priority, nomenclature for different parts of the stratigraphic column, and also perhaps at different hierarchical ranks, could begin with certain internationally agreed-upon publications. Analogously, in zoology most nomenclature (there are some exceptions) begins with the tenth edition of C. Linnaeus's *Systema Naturae,* which is arbitrarily assigned the publication date of 1 January 1758. In paleobotany the nomenclature of many fossil plant groups dates back to G. K. Sternberg's *Flora der Vorwelt,* assigned the date of 21 December 1820. Because it is generally agreed that systems, series, stages, and so on, should not bear identical names (the Devonian System does not contain a Devonian Stage), provisions could be made to compensate for such problems. There might be different type points designated for different hierarchical ranks (for example, system type points and stage type points). Or it might be agreed that when type points were used, if more than one type point fell within any particular unit, the oldest named point (back to a certain agreed-upon date) would be used for the highest applicable hierarchical rank, and later named type points would be utilized for progressively lower ranks in the hierarchy.

The system of stratigraphic nomenclature based on type points outlined above is somewhat similar to the concept of naming boundary points between units mentioned earlier (see above; Harland, 1978; Hughes et al., 1967). However, the naming of boundary points, if done on the basis of international agreement, potentially still could impose an arbitrary classification upon all stratigraphers, rather than merely regulating nomenclature. The advantage of a system similar to the one I proposed above is that only nomenclature would be regulated (a truly normative question), not aspects of classification per se (which can include positive aspects of science— scientific principles). Investigators would be free to pursue either a natural classification of rock and time (however the investigator defines "natural") or purely arbitrary classifications, depending on their goals and underlying philosophical biases. Even if such a system were adopted, the current work of the International Commission on Stratigraphy of the IUGS in setting up an arbitrary standard global chronostratigraphic scale would not be wasted. Those investigators who believe that stratigraphic (or at least chronostratigraphic) classification is purely arbitrary, normative, and a matter of convention, and do not wish to set up their own stratigraphic classification, could adopt the standard global chronostratigraphic scale of the ICS. Also, as a matter of expediency, even those persons not convinced that stratigraphic classification is merely a matter of convention and convenience might still adopt the standard global chronostratigraphic scale pending the development of their scales.

Schindewolf's Views on Stratigraphy and Chronostratigraphy

The German paleontologist and stratigrapher Otto Heinrich Schindewolf has played a central role in a school of stratigraphic philosophy that does not accept the idea of defining the boundaries of stratigraphic units at stratotypes (see Chapter 4; Hedberg, 1961; McLaren, 1977; Schindewolf, 1970a, b). For Schindewolf, stratigraphy is restricted to the study of temporal relations among rocks, strata, and the events they record; that is, the stratigraphy of Schindewolf is essentially synonymous with the chronostratigraphy of other investigators. Likewise, the only correlations in stratigraphy recognized by Schindewolf are temporally based. In Schindewolf's view, lithostratigraphy is limited to relatively local rock sequences and is not necessarily reflective of true stratigraphic relations (that is, temporal relations). Although lithostratigraphy provides a necessary basis for actual stratigraphic work, it is not stratigraphy per se but rather what is termed "prostratigraphy."

The actual aims of stratigraphy, temporal correlations ideally on a worldwide scale, are pursued primarily by the methods of biostratigraphy, at least for Phanerozoic rocks. Thus for Schindewolf stratigraphy is, for many practical purposes, virtually identical with biostratigraphy. In the biostratigraphy practiced by Schindewolf and his school the primary stratigraphic unit is the zone, approximately equivalent to the taxon-range zone or taxon-range chronozone of the *International Stratigraphic Guide* (Hedberg, 1976; see Chapter 6). Any higher-ranking stratigraphic unit is merely the sum of its included zones, and thus the lower boundary of the lowest zone and the upper boundary of the highest zone define the lower and upper boundaries, respectively, of any higher-ranking stratigraphic unit.

According to Schindewolf and his school, any particular zone is documented or defined by the occurrence or presence of a particular species (or by contemporaneity with strata in which the particular species occurs). The earliest appearance (stratigraphically lowest occurrence) of the species defines the base of the zone, and the latest appearance (stratigraphically highest occurrence) of the species defines the top of the zone. Any species is itself ultimately tied to, documented by, or, one might say, defined by the holotype specimen. Thus the type of a zone is in essence the holotype of the fossil species upon which it is based, and the concept of a separate stratotype is unnecessary.

Schindewolf's views are subject to various criticisms and dissensions. As this school acknowledges, stratigraphy in practice is made virtually synonymous with biostratigraphy. Therefore, stratigraphic units of the Schindewolfian school, as described above, are strictly biostratigraphic units. Many investigators wish to recognize, as part of stratigraphy per se, units that are not biostratigraphic in nature. Furthermore, biostratigraphic units may, in practice, closely approach chronostratigraphic units (that is, the ideal stratigraphic units that Schindewolf appears to be striving for), and biostratigraphic correlations may closely approximate temporal (chronostratigraphic) correlations, but there is no guarantee that biostratigraphic units and correlations are chronostratigraphic in the strict sense. Indeed, the boundaries (or biohorizons) of biostratigraphic units may transgress ideal time surfaces. McLaren (1978, p. 4) has suggested that "the relatively crude systems of biostratigraphic correlation [applied to the Phanerozoic] developed in the past have resulted in faunal zones that may be accurate to within one or two million years," whereas more recent refinements (e.g., Kauffman, 1970, 1977) may lead to an accuracy "perhaps as much as 10 times greater" (McLaren, ibid.). According to the same author, Precambrian acritarch and stromatolite zones are two or three orders of magnitude less precise than common Phanerozoic biocorrelations interpreted as chronocorrelations. Thus some of the most accurate chronocorrelations based on biostratigraphy, under ideal circumstances, may still be off by a couple of hundred thousand years.

The boundaries of a zone or other stratigraphic unit defined by a particular species never are known with certainty. Even when the morphological range or limits of a species are precisely agreed upon, further research and collecting may refine the boundaries of the zone as determined by the highest and lowest known occurrences of the defining species. Furthermore, with taxonomic revision the morphological boundaries of a species—that is, which actual specimens a particular taxonomic or systematic paleontologist includes or excludes from the definition of the species— are subject to change. With any such change the concept and boundaries of the corresponding zone will change. Drooger (1974) has discussed these ideas in particular, suggesting that whereas the central portion of a stratigraphic unit defined on the basis of the taxon-range of a particular species (that is, a chronostratigraphic unit distinguished biostratigraphically) may be identifiable with relative certainty, due both to the very nature of species that evolve and change with time and to the precision and accuracy of taxonomic identification, the boundaries between vertically adjacent units are characterized by some degree of uncertainty. Thus, according to this mode of analysis, boundaries of such natural stratigraphic units are subject only to statistical approximation.

Some workers have argued that species should not be used to define stratigraphic units, at least not internationally recognized chronostratigraphic units, because species are ultimately subjective. Thus McLaren (1977, p. 28) has reasoned that "theoretically, if evolution were [is] true, then there could [can] be no way of distinguishing between species in a chronological succession that was [is] complete except by subjective and arbitrary decision." Or as mentioned above, even if theoretically species are considered to be real, distinct, natural entities (se Schoch, 1986b, for a discussion of various concepts of species in paleontology and neontology), the boundaries of a species (what specimens are included within a species) are subject to change with taxonomic revision. Perhaps more fundamental is the point, apparently not recognized by all investigators, that a holotype of an organismal species and the stratotype of a stratigraphic unit (as used by the chronostratigraphic school outlined in Hedberg, 1976, for instance) are not equivalent entities. The holotype of a species does not, even in theory, define a species. The holotype (or type specimen) in actuality need not even be typical of the average morphology of the individuals composing the species; the holotype even may be an extremely aberrant or deviant individual. The holotype is merely the name bearer of a species, nothing else. Holotypes are used only in nomenclatural issues, in establishing the correct name to be applied to a species once the scope of the species (what individuals are to be included within the compass of the species) is delineated. If a species, as defined by a particular investigator, includes within its boundaries two or more individuals that are holotypes, the oldest name applied to one of the holotypes normally has priority and gives its name to the entire species (see Ride et al., 1985; Schoch, 1986b). In stratigraphy, stratotypes are not just name bearers; unlike a holotype in zoology or botany, a stratotype serves to define precisely and unquestionably the exact limits or scope of a particular stratigraphic unit. Thus Schindewolf's concept of stratigraphic units based on holotypes of species and the concept of stratigraphic units defined on the basis of stratotypes not only might differ in their boundaries and the rocks and strata they encompass, but they are fundamentally different ideas.

As a last note, Schindewolf's concept of stratigraphy is based on the premise that biostratigraphy is essentially the only method of practical use in establishing stratigraphic (or at least chronostratigraphic) units and correlations. This may have been true early in the twentieth century, but in recent years many new techniques have been developed (and they continue to be developed) and have been widely applied. Cases in point are the quickly developing fields of magnetostratigraphy and the numerical (including isotopic) dating of strata directly.

CHRONOCORRELATION

Chronocorrelation, chronostratigraphic correlation, or time correlation is the process or activity of attempting to establish the time equivalence and temporal relationships of stratigraphic units, rocks and features (such as bedding planes, faults, intrusions, evidence of metamorphism, and so on). The ideal synchronous boundaries of formal chronostratigraphic units may be approximated, with greater or lesser degrees of accuracy, in areas away from their type sections. Likewise, one may attempt to delineate or trace synchronous time planes, surfaces, or boundaries (whether or not they are formally designated boundaries or horizons) through strata. In practice, chronocorrelation involves not only the establishment of the strict time equivalence of stratigraphic units but also the temporal relationships of units and features. Given two stratigraphic units, call them A and B, chronocorrelation can involve suggesting such possibilities as the following (Harland, 1978): A predates B; A postdates B; A spans B (that is, the formation of A began before the beginning of the formation of B, and the formation of A ended after the end of the formation of B); B spans A; A and B partially overlap; A and B are strictly isochronous (they have mutually synchronous upper [later] and lower [earlier] boundaries); A and B are coeval (they are approximately, but not exactly, isochronous); and so on.

There is currently no body of evidence, characteristics, or attributes of rocks or stratigraphic units that is exclusive to chronostratigraphy and widely applicable to all rock types, although numerical dating methods (discussed below), as they increase in precision and accuracy, certainly are approaching the status of "pure" chronostratigraphic evidence. Likewise, chronocorrelations are routinely proposed not only for chronostratigraphic units, but for many diverse types of stratigraphic units and features (see section on correlation in Chapter 3). Consequently, all possible lines of evidence that might help in establishing chronocorrelations are commonly being utilized. In this section I briefly touch on a representative sample of the methods and principles that have been applied, with varying degrees of success, to the problem of chronocorrelations; for further discussions the reader is referred to such sources as Boggs (1987); Cohee, Glaessner, and Hedberg (1978); Harland et al., (1982); Hedberg (1976); Odin (1982a); and Weller (1960).

For very short-range chronocorrelations the use of direct physical evidence continues to be among the most accurate and reliable methods (Weller, 1960). When strata are in direct physical contact with one another, or former contacts that are now lost (for instance, because of erosion or structural disturbance) can be restored, the mutual physical relationships of strata invariably should shed some light on their temporal relationships. According to the law of superposition, for instance, strata at the bottom of a section are necessarily younger than overlying strata. A simple bedding plane traced over a limited distance may be the best indicator or approximation of an ideal surface of synchroneity (Hedberg, 1976). A distinctive key bed, such as a bentonite layer presumed to have formed from an airborne volcanic ash, or an iridium layer presumed due to an extraterrestrial asteroid or meteorite impact, may be of potential chronostratigraphic use over considerable distances (see following section on event stratigraphy). More generally, stratal continuity and/or lithologic similarity (the basis of lithocorrelation, see Chapter 5) often is applied as a first approximation for chronocorrelations. This remains the case despite the fact that most stratigraphers now would agree with Weller (1960, p. 541; see also Hedberg, 1976, p. 87) that "lithologic similarity is a much more accurate index of similar genesis than it is an indication of contemporaneity. Within a single basin of deposition similar conditions may have prevailed simultaneously throughout large areas and they may have shifted gradually from one part to another, in either case being recorded by laterally continuous strata of similar characters and appearance. Also similar conditions may have occurred simultaneously in several isolated areas or may have recurred in different areas at different times, in either of these cases producing discontinuous bodies of similar strata." Yet initially lithostratigraphy may be all that one working in a region has to deal with, and it is tempting to interpret lithocorrelations chronostratigraphically.

Ager (1981, pp. 58–61) identified two basic ways of temporally interpreting lithocorrelations. One mode of interpretation, he suggests, is based on the concept of the "gentle rain from heaven"; that is, all kinds of different contemporaneous facies and environments were preserved more or less simultaneously in the stratigraphic record by a constant rain of sediment occurring everywhere. Ager suggests that this is a subconscious attitude toward sedimentation that has been, and is, held by many stratigraphers, but this type of sedimentation may apply only questionably to deep ocean basins (where the remains of pelagic organisms continually rain down) or to such cases as volcanic dust literally raining down over a large area. This concept can be termed vertical sedimentation.

In contrast, according to Ager (1981), we can think of sedimentation in terms of lateral sedimentation, or the "moving finger writes" model. Ager (1981, pp. 58–59) contends that the major contribution of studies of Recent (that is, Holocene) sedimentation "has been the demonstration of lateral rather than vertical sedimentation. Modern deposits are not, it seems, laid down layer upon layer over a wide area. They start from a particular point and then build out sideways as in the traditional picture of a delta. In other words, all bedding is likely to be cross-bedding, though often on so gentle a scale as not to be recognisable in the field." In many areas, over any significant amount of time, sediments may not even build up to any extent vertically. They may only be continuously shifted about laterally. Perhaps only with concurrent subsidence will the product of sedimentation be preserved. This fits with Ager's notion that at any one local stratigraphic section the "gaps" (unconformities) generally represent more time than the preserved rocks.

In any particular case (at a given scale of resolution), either of the extreme models of vertical or lateral sedimentation may be an oversimplification. As Ager (1981; see also A. B. Shaw, 1964) again has noted, in the early days of stratigraphy either one did not concern oneself with time correlations, or similarities in lithology implied equivalence in time. Then, with the recognition of facies, some stratigraphers considered that if rocks in different localities share the same lithology, that constitutes evidence that they may be of different ages (that is, units of single lithology tend to be diachronous, at least on a fine scale). Likewise, rocks that are composed of very different lithologies may still be of the same age (perhaps they represent facies formed in contemporaneous, but differing, environments).

Chronostratigraphic interpretations of lithocorrelations generally are accepted tentatively, pending more information. This procedure is reinforced by the *International Stratigraphic Guide,* which, after acknowledging that "the boundaries of all lithostratigraphic units eventually cut across isochronous [= synchronous] surfaces and vice versa, and that lithologic features are repeated time and again in the stratigraphic sequence," states that "a lithostratigraphic unit such as a formation always has some chronostratigraphic connotation and is useful as an approximate guide to a chronostratigraphic position. Individual limestone beds, phosphate beds, bentonites, volcanic ash beds, or tonsteins, for example, may be excellent guides to approximate time-correlation over very extensive areas. Distinctive and widespread general lithologic developments also may be significant of chronostratigraphic position" (Hedberg, 1976, p. 87). Here the idea appears to be that any single laterally and vertically continuous lithostratigraphic unit must have general temporal bounds on the time of its formation. The first grain of the unit (given that it is composed of sedimentary rock) was deposited at a certain instant in time, and the last grain of the unit at a later instant in time, and all other grains were deposited during the interval of time bounded by the two above-mentioned instants. Therefore, the lithostratigraphic unit and its contained strata carry some temporal connotation even if the boundaries of the unit, when traced laterally, transgress time planes, and there may be no temporal overlap between the actual material of the unit developed in one geographic area as compared to the unit developed in a second area.

As originally established, many of the systems (see Chapter 1) and their major subdivisions were meant primarily as broad lithostratigraphic units. The rocks of a system were first and foremost characterized by a distinctive unity of lithology; a system might be recognizable elsewhere (away from the general type locality) by the manifestation of the same or similar lithology. On the basic concept of a system as a large lithostratigraphic unit was imposed the concept that a distinctive lithology might characterize all, or most, of the rocks formed during a particular interval of geologic time representing a particular episode in earth history. The lithostratigraphic units and lithocorrelations might correspond directly to temporal correlations. This idea is now generally rejected, at least in an extreme form, yet chronostratigraphic interpretations of lithocorrelations persist in part because on a gross scale they sometimes do work. Some long-range chronocorrelations between Europe and North America, on the basis of gross lithology, are reasonably accurate at the level of systems/periods and series/epochs (Weller, 1960). This point is elaborated upon, with numerous examples, by Ager (1981) to the point that Ager sets out the following proposition, which he dubs the "Phenomenon of the Persistence of Facies": "at certain times in earth history, particular types of sedimentary environment were prevalent over vast areas of the earth's surface" (Ager, 1981, p. 14). Ager (1981) is fully cognizant, however, that the upper and lower boundaries of such units (facies), on a detailed level, are not synchronous. "The point about the Urgonian limestones [late Early Cretaceous in age; one of Ager's examples of a widespread and persistent facies], say, is that we know that they are of *about* the same age throughout Europe *in spite of* the fact that fossil evidence shows them to have started and ended at different times in different places" (Ager, 1981, p. 71, italics in the original). In some cases Ager suggests that the rocks of one of his single persistent facies began to form in many different centers at approximately the same time, spread out, and in some cases eventually coalesced.

On a very broad scale chronocorrelations may be based on the concept that the earth is undergoing a unidirectional historical development. The earth has been undergoing geochemical differentiation since its formation, and thus at least some characteristics of certain types of rocks have changed systematically with time; a case in point may be $^{87}Sr/^{86}Sr$ ratios. Classically, tectonism, diastrophism, and periodic worldwide orogenies (mountain builiding events) have been postulated and used on a coarse level to establish chronocorrelations. Similarly, paleogeographic reconstructions, especially as related to local, regional and global sea level changes, and paleoclimatic changes (reflected in, for instance, oxygen isotope ratios) have been utilized in chronocorrelation. These types of evidence are discussed further below.

Classically, many geologic systems are defined as bounded by major unconformities, and the concept that unconformities are of chronostratigraphic value persists (see the sections on unconformity-bounded units in Chapter 5 and below). Certainly unconformities, like lithostratigraphic units, may have time signifiance. Unconformities, such as surfaces of erosion, may themselves be more or less datable; so unconformities bounding strata may help to bracket the age of the enclosed strata. However, a surface of unconformity usually will vary in age from place to place; geographically widespread unconformities may develop over long periods of time

(Hedberg, 1976). Strata subsequently deposited immediately upon the surface of unconformity also may vary in age from place to place.

Structural deformation or development (that is, structural disturbances manifested in rocks, such as the folding or faulting of strata) and degree of metamorphism occasionally have been utilized in proposing chronocorrelations. In general such chronocorrelations are extremely unreliable, except perhaps over very short geographic distances. However, before the advent of radioactive dating this was a major method of correlating Precambrian rocks; high-ranking metamorphic rocks were correlated to the Archeozoic, whereas lower-grade metamorphic rocks were correlated to the Proterozoic. Unmetamorphosed Precambrian sediments were relegated to the uppermost Proterozoic. As Weller wrote in 1960 (p. 546), such correlations "are no longer taken very seriously."

For Phanerozoic strata the classic and most widely used means of chronocorrelation employs fossils. Biostratigraphic correlation already has been discussed (Chapter 6). The basis for interpreting biocorrelation as chronocorrelation is that empirically it has been found that sequences of fossils in rocks are nonrepeating, and any particular species is characterized by a finite stratigraphic range. The orderly sequences of fossils in rocks are attributed to progressive (in the sense of unidirectional and irreversible, not in the sense of heading toward some fixed goal) biological evolution through time. Using the notion of evolution and positing certain trends such as that life has become increasingly complex over time, one can even attempt to chronocorrelate rocks on the basis of the "stage of evolution" of the contained organisms. Two stratigraphic units may share no species in common, yet chronocorrelations still may be attempted between the units on the basis of the degree of complexity or evolutionary advancement of the enclosed fossils. Similarly, evolutionary lineages may be postulated for certain fossil organisms, and then the rocks containing the organisms may be temporally arranged or correlated on the basis of the hypothetical lineages. Such scenarios, however, may be very tenuous at best (see Schoch, 1986b); consequently, the chronocorrelations of strata so arrived at may be dubious.

While it is now widely recognized that biostratigraphic correlation is not necessarily equivalent to chronocorrelaton, it continues to be the case that the best biocorrelations often are believed to approximate true chronocorrelations. Any particular biohorizon (such as the First Appearance Datum of a particular species) may not be exactly synchronous for many reasons, such as migration rates, facies restrictions on the organism, nonpreservatioin of the remains of the organism, and so on. Furthermore, any particular species, or even a particular biota, normally had only a limited geographic extent. However, many investigators believe that extremely long-range, and accurate, chronocorrelations can be arrived at by utilizing numerous interlocking and mutually reinforcing species ranges and biozones. Laterally "interlocking biozones can be particularly helpful in providing a tie across major lateral changes in depositional environment. An example is the use of the land-to-ocean progression of terrestrial animals and plants, pollen, benthic marine organisms, and planktic and nektic marine organisms in the correlation between continental and marine deposits" (Hedberg, 1976, p. 88).

Fossils, however, are of no use where they do not occur (or at least are not found). Precambrian rocks are largely lacking in usable fossils, and there are also many Phanerozoic rocks that are not fossiliferous.

Of increasing importance to chronostratigraphy and chronocorrelation is the geomagnetic polarity time scale, discussed in Chapter 6. As noted, geomagnetic reversals may potentially be truly synchronous global events that can be recorded in diverse rock types. However, as previously pointed out, magnetostratigraphic polarity units are not chronostratigraphic units. Currently this is a field of active research within chronostratigraphy and geochronology.

Directly applicable to chronocorrelations are various means that have been developed to numerically date (that is, in years before the present) minerals, rocks, and strata, such as isotopic (radiometric) dating. A variety of these methods are discussed in a following section. One must always bear in mind that ages arrived at by numerical means are always apparent and may be subject to more or less deviation from the true age that one is attempting to obtain.

EVENT STRATIGRAPHY

Ager (1973, 1981) applied the term event stratigraphy to "the use of inferred geological events, rather than the intrinsic characters of rocks, for correlation" (Hallam, 1984b, p. 197). According to Ager (1981, p. 69), in event stratigraphy "we correlate not the rocks themselves, on their intrinsic petrological characters, nor the fossils, but the events such as the Triassic transgressions just discussed" (that is, just discussed by Ager in his 1981 book). By these definitions, many means of establishing conventional chronocorrelations fall under the aegis of event stratigraphy. For instance, the highest occurrence of a particular fossil species might be used as an intrinsic characteristic (biocharacteristic) of certain rocks and used to establish biocorrelations; but if it is interpreted or inferred to represent the extinction of the species, and furthermore if it is assumed that the extinction of the species was a more or less simultaneous event throughout a certain region, then the highest occurrence of the species in any one section might be inferred to signify the geological event of the extinction of the species. Correlations established on the basis of such an inferred event, the extinction of a particular species, might be considered to be event-chronostratigraphic. Likewise, lithocorrelations might be established on the basis of a bentonite layer without suggesting an interpretation for the genesis of the bentonite. However, if it is inferred that the bentonite layer represents a short-lived volcanic event (for all practical purposes instantaneous relative to geologic time) then the correlations based upon the bentonite might be considered chronocorrelations derived via the means of event stratigraphy.

Many workers have used the term event stratigraphy in the study of "rare" events within a stratigraphic context, such as volcanic eruptions, earthquakes, floods, storms, turbidity currents, climatic fluctuations, sea level changes, asteroid impacts, and so on (Seilacher, 1984). Seilacher (1984, p. 49) has suggested that we "call the level of research centered on the analysis of individual beds 'event stratinomy' in order to make the broader term 'event stratigraphy' available for the study of rare events at any scale." Storm deposits (tempesites and related deposits) in particular have been the focus of much analysis among event stratigraphers (e.g., Einsele and Seilacher, 1982a, b).

As Hallam (1984b) has pointed out, among the smallest-scale events that have been recognized in certain portions of the stratigraphic record are annual varves, for example, in glacial lakes or certain organic-rich shales and carbonates. Annual varves and storm beds (perhaps formed with a periodicity of one or two decades) may allow very fine chronocorrelation within single sedimentary basins. One must always be careful when interpreting deposits as short-lived storm beds, however; it has been suggested that some shell beds that mimic storm deposits may have been formed over longer intervals of time (see papers in Einsele and Seilacher, 1982a; Hallam 1984b; Seilacher, 1984). On a larger scale, some researchers have identified what they believe to represent cycles on the order of several millenia that can be utilized in event stratigraphy. "Many Mesozoic marine sequences contain regular cyclic alterations of laminated organic-rich shales, more or less barren of benthos, and non-laminated, bioturbated mudstones with benthic fauna, signifying regular alterations in degree of oxygenation of the bottom waters; they may be traceable over thousands of sq. km. The thickness of these cycles is usually in the decimetre to metre range, and on the assumption that the laminae are annual, cyclic alterations in bottom conditions of the order of 10,000 to 100,000 years can be inferred, suggesting possibly the operation of Milankovich-type climatic cycles" (Hallam, 1984b, p. 179).

Potential global "catastrophes," such as an asteroid impact at the end of the Cretaceous (Silver and Schultz, 1982), may be the ultimate material of event stratigraphy, but such events often may be based more on supposition than direct evidence and can be open to criticism. Similarly, presumed or inferred global changes in sea level, major unconformities, and orogenic cycles (perhaps placed within a modern plate tectonic framework; see discussions later in this chapter), have been, and continue to be, utilized in what may be termed event stratigraphy. These correlations may be very gross (that is, imprecise), and they also may be criticized for circular reasoning if they are not confirmed by independent evidence, such as the fossil record (cf. Hallam, 1984b).

CHRONOSTRATIGRAPHIC SIGNIFICANCE OF DEPOSITIONAL SEQUENCES

Sloss (1963, p. 93) recognized what he termed sequences as high-rank lithostratigraphic units bounded by unconformities: "Stratigraphic sequences are rock-stratigraphic units of higher rank than group, megagroup, or supergroup, traceable over major areas of a continent and bounded by unconformities of interregional scope." Mitchum, Vial, and Thompson (1977, p. 53, italics in the original; see Fig. 7.1) adopted Sloss's term and used it in their concept of the depositional sequence: "A *depositional sequence* is a stratigraphic unit composed of a relatively conformable succession of genetically related strata and bounded at its top and base by unconformities or their correlative conformities." The sequences of Mitchum, Vail, and Thompson (1977) differ from those of Sloss (1963) in that the former tend to be one or more orders of magnitude smaller (according to these authors their depositional sequences may range in vertical thickness from a few millimeters to thousands of meters, although those generally recognized in practice are tens to hundreds of meters thick), and depositional sequences may be bounded by unconformities or by equivalent conformities. Sloss's (1963) stratigraphic sequences, being defined and recognized solely by the bounding unconformities, essentially fit the definition of unconformity-bounded stratigraphic units (synthems) of the ISSC (1987a; see Chapter 5), whereas the depositional sequences of Mitchum, Vail, and Thompson (1977) do not. In this section the term sequence is used to refer specifically to sequences *sensu* Mitchum, Vail, and Thompson; however, the term sequence is used elsewhere

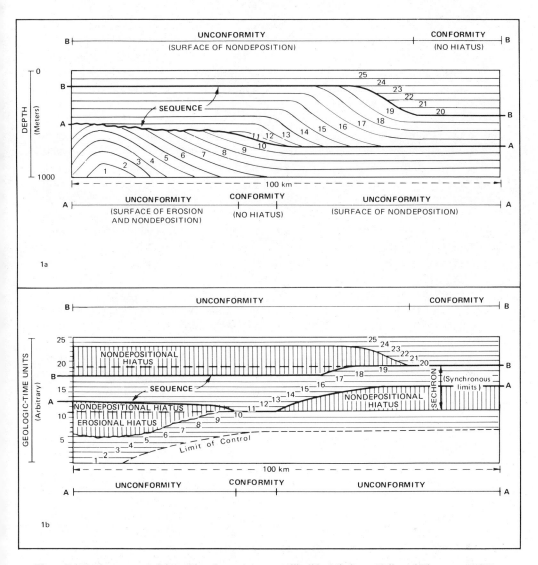

Figure 7.1. Basic concepts of depositional sequences, as utilized by Mitchum, Vail, and Thompson (1977, p. 54, reprinted by permission of the American Association of Petroleum Geologists):

"Basic concepts of depositional sequence. A depositional sequence is a stratigraphic unit composed of relatively conformable successions of genetically related strata and bounded at its top and base by unconformities or their correlative conformities.

"**1a.** Generalized stratigraphic section of a sequence. Boundaries defined by surfaces A and B which pass laterally from unconformities to correlative conformities. Individual units of strata 1 through 25 are traced by following stratification surfaces, and assumed conformable where successive strata are present. Where units of strata are missing, hiatuses are evident.

"**1b.** Generalized chronostratigraphic section of a sequence. Stratigraphic relations shown in **1a** are replotted here in chronostratigraphic section (geologic time is the ordinate). Geologic-time ranges of all individual units of strata [are] given as equal. Geologic-time range of sequence between surfaces A and B varies from place to place, but variation is confined within synchronous limits. These limits determined by those parts of sequence boundaries which are conformities. Here, limits occur at beginning of unit 11 and end of unit 19. A sechron is defined as maximum geologic-time range of a sequence."

in this book in a more informal sense to refer to any stratigraphic succession (or part of a succession) of rocks.

Mitchum, Vail, and Thompson (1977) assert that their sequences have chronostratigraphic significance in that the rocks of the sequence were deposited only during a certain interval of geologic time. Presumably the upper and lower limits of the age of the sequence are given where the upper or lower boundaries of the sequence are conformities rather than unconformities. Mitchum, Vail, and Thompson (1977, p. 55) assume that, at least for their purposes and at the scales with which they are working, any stratal or bedding surfaces are essentially isochronous boundaries; "The conformable part of a sequence boundary is practically synchronous because the hiatus is not measurable; the time span generally is less than a million years. The physical surfaces that separate groups of strata of individual beds and laminae within a sequence are essentially synchronous." A particular sequence must be just slightly younger than the youngest rocks that some part of the sequence conformably rests upon, and just slightly older than the oldest rocks that overlie conformably some part of the sequence. To represent this concept, Mitchum, Vail, and Thompson (1977) coined the term sechron—the total interval of time during which all parts of the sequence were deposited. All the strata within the sequence will be datable to some part of the sechron, although the age range of the strata of the sequence will vary within this time interval (the sechron), depending on locality, where the sequence is bounded by unconformities.

As Mitchem, Vail, and Thompson (1977) note, the hiatus that may be represented by an unconformity bounding a sequence is variable, often ranging from a million years to hundreds of millions of years. Yet these authors also contend that the unconformities bounding sequences are themselves chronostratigraphically significant. Their reasoning is that in general even if a surface of unconformity is not isochronous, still (in general, but one can imagine exceptions such as a migrating surface of erosion) all of the rocks above an unconformity are younger than all the rocks below an unconformity. Certainly this holds true in instances where all the rocks below the unconformity were deposited and then more or less simultaneously underwent a period of nondeposition and/or erosion before deposition resumed. In such cases the relationship of strata relative to a particular unconformity provides information, at least on a gross scale, as to the age of the strata. In instances where the hiatus represented by a surface of unconformity is relatively short, an unconformity may even serve as a guide to the location of chronostratigraphic boundaries.

As should be evident from the above discussion, Mitchum, Vail, and Thompson's (1977) idea as to the potential chronostratigraphic significance of depositional sequences and their bounding unconformities does not qualify such sequences as chronostratigraphic units as defined by either the *International Stratigraphic Guide* (Hedberg, 1976) or the *North American Stratigraphic Code* (NACSN, 1983). Depositional sequences are not bounded by isochronous surfaces, although in some areas a sequence may be bounded by conformable surfaces that approximate synchroneity. Indeed, carrying the logic of Mitchum, Vail, and Thompson (1977) to an extreme, any lithostratigraphic unit, even if it bears blatantly diachronous boundaries, has some chronostratigraphic significance in that the first grain of the unit was deposited at some point in time, and the last grain of the unit was deposited at some later point in time, so that all portions of the unit must fall within the particular interval of time defined by the points when the first grain and the last grain were deposited. Of course, the age range of the strata in our diachronous lithostratigraphic unit will vary from place to place, but the same is true for the strata composing many sequences that are said to be chronostratigraphically significant. Here the point is that a "true" chronostratigraphic unit is an ideal, and real material bodies of rock in three dimensions may have varying degrees of chronostratigraphic significance. Perhaps well-distinguished depositional sequences often, for all practical purposes (within our limits of useful resolution), approach chronostratigraphic units (although, of course, they seldom correspond to named formal chronostratigraphic units).

Mitchum, Vail, and Thompson (1977) also suggest that depositional sequences are natural, genetic units (often deposited during a relatively short, episodic event) that are of real significance in the geologic history of a region. In contrast, arbitrarily defined units (perhaps based on synchronous surfaces or arbitrary cutoffs between differing lithologies) may have little or no significance relative to the actual events that occurred during the geologic history of an area (cf. similar thoughts on natural versus artificial stratigraphic units discussed elsewhere in this book).

DIASTROPHISM

In the nineteenth and early twentieth centuries there was a common belief that most or all of the geologic systems were separated by relatively brief episodes of diastrophism (often manifested as full-scale orogenies—mountain building episodes). The major geologic systems originally were defined as bounded by major unconformities, and subsequently were correlated on the basis of unconformities. These unconformities thus were interpreted as recording major phases of worldwide diastrophism. By the late nineteenth century it was perhaps almost universally held that the overall history of the earth is a succession of periods of relative quiescence (recorded in the rocks as the geologic systems) separated by periodic episodes of brief but violent, extreme, and rapid global orogenic/diastrophic disturbances (recorded in the rocks as major unconformities). These brief events came to be known as "revolutions" (if of greater magnitude) or "disturbances" (if of relatively small magnitude). Each revolution or orogenic episode separating two adjacent systems was given a name; for example, the Appalachian Revolution (or Hercynian Revolution in Europe) was said to separate the Permian from the Triassic (Chamberlin and Salisbury, 1909; Pirsson and Schuchert, 1915: see also B. Conkin and J. Conkin, 1984; Simpson, 1970; Weller, 1960).

Around the 1920s the concept of periodic diastrophism began to be questioned (Weller, 1960). It was suggested that many of the so-called worldwide revolutions were not worldwide, or at least not synchronous worldwide. Some of the presumed revolutions were composed of several nonsimultaneous episodes of mountain building encompassing a much larger interval of time than previously imagined. Many of the orogenies on which the system was based were not definitively known outside of a localized area, and in other cases orogenies of equal magnitude were not included in the system because they did not correspond to traditional system boundaries. Finally, the dating of the orogenies was not so accurate as had been claimed; not all of the revolutions necessarily occurred exactly at the system boundaries as had been presumed. The final blow to the general belief in periodic diastrophism apparently came when Gilluly (1949; see also Gilluly, 1967) definitively demonstrated that the supposed worldwide synchronous orogenies are largely nonexistent; rather, various tectonic and orogenic events in various regions occur more or less randomly relative to one another.

Variations on the theme of using periodic diastrophism, and the resultant unconformities, as a basis of stratigraphic correlation (Chamberlin, 1909) are still espoused by some. In localized areas well-known unconformities may be extremely useful in establishing overall or coarse stratigraphic correlations, and one may still hear references to localized "revolutions," although this is becoming increasingly more unusual. To this day some stratigraphers, following Chamberlin (1909) and Grabau (1936), continue to maintain that "diastrophism is of prime importance in chronostratigraphy" (J. Conkin and B. Conkin, 1985, p. 490; see also B. Conkin and J. Conkin, 1984), and that small-scale unconformities may be used to determine chronostratigraphic boundaries (see Chapter 3). In this case, however, Conkin and Conkin are not correlating by means of large-scale diastrophic events such as orogenies, but by shallow-water transgressions and regression on a craton. Such transgressions and regressions are claimed to be very rapid in terms of geologic time and are controlled by changes in sea level, which ultimately are controlled by tectonic processes. Similarly, with the development of seismic stratigraphy there has been a resurgence of interest in developing sea level curves for use in stratigraphic correlation (see Chapter 2 and below, this chapter).

STABLE ISOTOPE STRATIGRAPHY

Since the middle of the twentieth century a new tool, stable isotope stratigraphy, has come into use as a means of postulating correlations, particularly chronostratigraphic correlations over large geographic areas (see general reviews in Boggs, 1987; Prothero, in press). The basic principle behind stable isotope stratigraphy is that apparently the ratios of stable isotopes for many elements have not been constant, but rather have fluctuated on a global scale over geologic time. These fluctuations may be recorded in the sedimentary record, and thus, analogous to the methodology of magnetostratigraphy, a sequence of isotopic fluctuations, excursions, or events recorded in a single stratigraphic section for an applicable element theoretically should be correlatable to a composite global curve of isotopic fluctuations for the particular element through time. Successively matching various stratigraphic sections to a composite global curve will in effect provide correlations among the actual stratigraphic sections.

General pitfalls of such methods, however, include the problems inherent in attempting to assemble composite global isotopic curves and then deriving true (or comparable) isotopic ratios for particular sections (not all lithologies, for instance, may be representative of regional or global trends in stable isotope ratios; some lithologies may give misleading results). Indeed, in some cases it could be argued that perhaps one must work with regional rather than global isotopic curves (see, for instance, the discussion of carbon isotopes below). Even if no regional or global curves are used, correlations between the rocks or stratigraphic units of two stratigraphic sections within a single area may be proposed on the basis of similarities in stable isotope fluctuations plotted against vertical thickness (representing time). In order to use this method to correlate (date) a particular part of a single stratigraphic section, in most cases the approximate correlation of the section must be established already on the basis of other data—single isotopic fluctuations are not necessarily unique. The most prominent isotope systems that have been applied to stable isotope stratigraphy thus far are those of oxygen, carbon, sulfur, and strontium.

Oxygen occurs in the form of three natural isotopes, ^{16}O, ^{17}O, and ^{18}O, where currently over 99.75% of the oxygen is in the form of ^{16}O, and a little over 0.2% is in the form of ^{18}O (based on the present isotopic composition of oxygen in seawater, given by Odin, Renard, and Grazzini, 1982, p.55, Table 2). In the middle of the twentieth century it was discovered that $^{18}O/^{16}O$ will fractionate with changes in temperature. In 1955 Cesare Emiliani demonstrated that the biogenic calcite formed in the marine realm during the Pleistocene (for instance, in the shells of particular foraminiferans) can exhibit differing ratios of $^{18}O/^{16}O$ through time and suggested that these ratios reflect changes in temperature (where lower $^{18}O/^{16}O$ ratios reflect higher temperatures).

Later work indicated that not only is temperature per se involved, but global ice volume also influences the $^{18}O/^{16}O$ ratio; in fact, this might be the primary determinant of the global $^{18}O/^{16}O$ ratio observed in sediments at any one time. Simplistically, the proposed mechanism (Matthews, 1984) is that as water evaporates from the oceans, it is relatively enriched in ^{16}O relative to ^{18}O (simply because ^{16}O is lighter, and so water containing it will enter into a gaseous state more readily), and thus the $^{18}O/^{16}O$ ratio of the ocean water increases if the evaporated water is not returned to the oceans. Likewise, as water travels in a gaseous state, any precipitation will tend initially to favor water that might contain ^{18}O; this effect will further ensure that atmospheric moisture carried far inland will be characterized by low $^{18}O/^{16}O$ ratios. When the glacial ice volume of the world is stable, as water evaporates from the oceans it either precipitates directly back to the oceans or precipitates over land and eventually finds its way back to the oceans; thus the overall $^{18}O/^{16}O$ ratio of the ocean remains constant. During times of active growth in continental and polar glaciers, the water bound up in the ice does not return to the oceans; and because it is ^{16}O-enriched, the $^{18}O/^{16}O$ ratio of the ocean increases. During times of deglaciation, when water that was bound up in glaciers is melting and returning to the oceans, the $^{18}O/^{16}O$ ratio of the oceans will decrease. Consequently, changes in the $^{18}O/^{16}O$ ratio of the world's oceans apparently reflect fluctuations in the volume of ice bound up in continental glaciers. Calcite formed in the shells of certain marine organisms appears to reflect the $^{18}O/^{16}O$ ratio of the ocean water when the organism was living (in actuality the $^{18}O/^{16}O$ ratio of any particular marine carbonate may be a function of both the temperature at which it was formed and the $^{18}O/^{16}O$ ratio of the seawater in which it precipitated; Odin, Renard, and Grazzini, 1982), and thus a record of these local $^{18}O/^{16}O$ fluctuations is recorded in many stratigraphic sections.

It has been suggested that the mixing time of the oceans may be on the order of a thousand years (Boggs, 1987; Faure, 1982; Odin, Renard, and Grazzini, 1982), so that any global changes in isotopic ratios of ocean water will be geologically instantaneous. Furthermore with the mechanism (discussed above) proposed to account for $^{18}O/^{16}O$ ratio fluctuations, these fluctuations should be taking place more or less simultaneously throughout various parts of the world. If this scenario is valid, that $^{18}O/^{16}O$ fluctuations reflect environmental changes of global magnitude, then correlations of stratigraphic sequences on the basis of such data should approach true chronocorrelations. Matthews (1984) also notes that if $^{18}O/^{16}O$ fluctuations correlate with changing ice volumes, then they should also correlate with changing sea levels and the positions of shorelines. Theoretically it should be possible to correlate and intercalibrate a diversity of phenomena in the stratigraphic record. However, in the real world not all organisms will necessarily be simulta-naeously exposed to seawater of identical oxygen isotope ratios at identical temperatures. Indeed, latitudinal or depth differences may affect the actual absolute values of oxygen isotope ratios; for example, benthonic foraminifera (and other deep-water organisms) may reflect changes in deep-water temperatures before regis-tering oxygen isotope changes of the seawater due to ice accumulation (Odin, Re-nard, and Grazzini, 1982) whereas planktonic forms naturally will be more strongly affected by local surface conditions. But the shapes of oxygen isotope curves for local stratigraphic sections may still parallel one another; corresponding trends and sequences of oxygen isotope curves for diverse sections may be interpretable in terms of global climatic changes affecting the global $^{18}O/^{16}O$ ratios of seawater.

In actual oxygen isotope studies, fluctuations in the $^{18}O/^{16}O$ ratio are recorded in terms of the deviation from an arbitrary standard; normally in dealing with calcites it is customary to employ the PDB standard used by the University of Chicago. The PDB standard is set as the $^{18}O/^{16}O$ ratio in oxygen of a fossil belemnite from the Peedee Formation (Cretaceous) in South Carolina. Another common standard, especially as a standard for natural waters, is SMOW (Standard Mean Ocean Wa-ter). The $\delta^{18}O$ value of any particular sample, given in per mil (parts per thousand, 0/00) relative to the standard can be calculated as follows:

$$\delta^{18}O = \frac{[(^{18}O/^{16}O) \text{ sample} - (^{18}O/^{16}O) \text{ standard}]}{(^{18}O/^{16}O) \text{ standard}} \times 1000.$$

A sample that gives a positive $\delta^{18}O$ value relative to the standard is enriched in ^{18}O, and a sample that is characterized by a negative $\delta^{18}O$ value is depleted in ^{18}O relative to the standard. The $^{18}O/^{16}O$ ratios used in the above equation can be determined by using a mass spectrometer. The basic methodological principles used in oxygen isotope stratigraphy also hold for stable isotope stratigraphy using other elements. Actual stable isotope values are calculated with respect to some arbitrary standard by using an equation analogous to that given above.

Oxygen isotope stratigraphy has found particular use in the Quaternary where detailed oxygen isotope curves have been derived and correlated, on the basis of cores from the Deep Sea Drilling Project, for the later part of the Matuyama mag-netically reversed period and the succeeding Brunhes normal period (that is, for about the last million years). For this period a number of oxygen isotope "stages" (by convention, numbered sequentially from youngest to oldest) have been identi-fied and correlated on the basis of cores taken in the Atlantic, Pacific, and Indian oceans and in the Mediterranean Sea. In turn, the oxygen isotope correlations/ curves have been compared to the paleomagnetic data obtained from the same or comparable cores. Interestingly, whereas there is a general correspondence between the paleomagnetic data and the oxygen isotope data from core to core, it is not perfect. For example, in two cores from the Pacific (from the work of Shackleton and Opdyke, 1976, and illustrated in Odin, Renard and Grazzini, 1982, p. 65, Fig. 17) the magnetic field reversal representing the boundary between the Matuyama and the Brunhes occurs at the base of isotopic stage 19 in one section but within isotopic stage 20 of another section. This could suggest that the boundaries between the isotopic stages are not exactly synchronous, or perhaps that the exact position of the magnetic field reversal has not been precisely identified in one or both of the sections (or, although it is difficult to imagine, one might suggest that the magnetic field reversal was not exactly synchronous).

Oxygen isotope stratigraphy is also finding increased use in pre-Pleistocene sections; for the Atlantic and Pacific in particular fairly complete records of oxygen isotope fluctuations for carbonates have been published (see references in Odin, Renard, and Grazzini, 1982).

There are two stable isotopes of carbon, ^{12}C with an abundance of about 98.9% and ^{13}C with an abundance of approximately 1.1% (the radioactive ^{14}C occurs in trace amounts). ^{13}C/^{12}C ratios (in terms of per mil deviation of δ^{13}C from a standard, such as PDB) have been calculated for carbonates found in marine stratigraphic sections and exhibit systematic variations with time that have been used in correlations. In general, fluctuations in ^{13}C/^{12}C ratios appear to reflect oceanographic circulation patterns and large-scale climatic change. The ^{13}C/^{12}C ratios of marine carbonates (such as the calcareous shells of organisms) apparently reflect the ^{13}C/^{12}C ratios of the carbon in CO_2 dissolved in the seawater in which the material was formed (see Boggs, 1987; Odin, Renard, and Grazzini, 1982; and Prothero, in press, for further discussions of carbon isotope ratios). The CO_2 dissolved in seawater at any one time and place may have some relationship to the general CO_2 on the earth and in the atmosphere at the time, but may also be greatly influenced by local conditions and the particular source of the carbon in the CO_2.

Various factors can influence the ^{13}C content of ocean (or other) water. Organic materials, for instance, tend to be relatively low in ^{13}C; and organic-rich waters may tend to be characterized by low ^{13}C/^{12}C ratios. Deep ocean water masses with a long residence time near the bottom may be relatively depleted in ^{13}C. Organic matter, characterized by low ^{13}C/^{12}C ratios, that sinks from the surface to such deep waters will be oxidized and the carbon depleted in ^{13}C trapped in these water masses. If, due to sudden (on the scale of geologic time) changes in oceanic circulation patterns, there is upwelling of such deep waters, then ^{13}C/^{12}C ratios may be depressed in surface waters. Or, changes in circulation patterns may cause bodies of water along shallow margins to be carried into deeper basins, with the result that ^{13}C/^{12}C ratios are changed. Furthermore, conceivably such factors as increased production of biomass on the continents, episodes of increased rates of erosion of organic-rich sediments, or increased rates of organic-rich sediment burial in the oceans could all influence ^{13}C/^{12}C ratios. Some of these effects might be of a global nature, whereas others would be restricted to a single ocean basin, or even part of a basin. However, as Odin, Renard, and Grazzini (1982) note, whatever the origin of carbon isotope shifts, they can be and have been used as stratigraphic markers. One need not necessarily know or fully understand the genesis of an attribute to utilize it in stratigraphic correlation (although few would deny the value of such information if available). At least partly because of the nature of the factors that determine carbon isotope ratios, this method of correlation has found its primary applicability in relatively local and regional stratigraphic studies, as opposed to global studies. However, a few variations of foraminiferal carbon isotope ratios have been identified that may be of global extent (reviewed in Odin, Renard, and Grazzini, 1982), such as an increase in benthonic foraminiferal δ^{13}C values during the Middle Miocene that has been believed due to the spreading of Arctic waters.

Sulfur has four stable isotopes (sulfur 32, 33, 34, and 36), with ^{32}S (approximately 95%) and ^{34}S (approximately 4.2%) being the most abundant (Odin, Renard, and Grazzini, 1982). ^{34}S/^{32}S values vary for different classes of materials (such as seawater sulfate, marine algae, rainwater sulfate, and so on) at present, and apparently have fluctuated through time within certain environments.

The sulfur ratios of the surface waters of the earth's oceans, as recorded in evaporite deposits, apparently have undergone major fluctuations since at least the late Precambrian (Odin, Renard, and Grazzini, 1982). In fact at least three major excursions, characterized by greatly increased $^{34}S/^{32}S$ ratios, can be identified: the Yudomski event (late Precambrian), the Souris event (Late Devonian), and the Röt event (Early-Middle Triassic; see Holser, 1977). In addition, two reverse events (characterized by lowered sulfur ratios) have been recorded: one during the Late Permian and another during the Late Paleogene (Oligocene: see Odin, Renard, and Grazzini, 1982). These events, and the sulfur curve in general, are potentially useful as a means of chronocorrelation of evaporite deposits in particular. The mechanism behind these changes in sulfur ratios is not well understood, but it has been suggested that the heavy sulfur peaks may have been caused by occasional mixing of ^{34}S-rich brines with surface waters (Holser, 1977). Odin, Renard, and Grazzini (1982) suggest that sulfur isotopic variations are greatly affected by bacterial sufate reduction. These bacteria preferentially metabolize light isotopes of sulfur so that the remaining sulfate becomes enriched in ^{34}S relative to ^{32}S; this can account for high $^{34}S/^{32}S$ ratios in some caprocks of saltdomes and relatively lower ratios of marine sulfates.

Other isotope stratigraphic methods also have been employed, such as that involving strontium (see, Faure, 1982, for a recent review of this method; also Prothero, in press). The $^{87}Sr/^{86}Sr$ ratio of the oceans (as recorded in the calcite of unreplaced fossil shells) has fluctuated through time, and has exhibited a steady increase since the middle Mesozoic. ^{87}Sr is a daughter product of the decay of ^{87}Rb, and the continued decay of the latter may be responsible in part for the increase in the former, although apparently there also have been times of decreasing $^{87}Sr/^{86}Sr$ ratios, such as during the late Carboniferous and Permian. If there is a true increase of the $^{87}Sr/^{86}Sr$ ratio with time through at least the majority of the Cretaceous and Cenozoic, this would mean that any particular $^{87}Sr/^{86}Sr$ value would correspond to a unique point in Cenozoic time. Thus, a single $^{87}Sr/^{86}Sr$ value from a single marine organism potentially could be measured and placed on the $^{87}Sr/^{86}Sr$ curve (given that such a global curve for strontium ratios of seawater through time had been accurately established with sufficient accuracy) in order to arrive at a date for the specimen. Of course, it must be known by other means that the approximate age of the specimen to be dated fits within the confines of a particular portion of the $^{87}Sr/^{86}Sr$ curve (that is, in the example given, that the specimen is referable to the Cenozoic); for instance, inspecting the curve of $^{87}Sr/^{86}Sr$ ratios in the world oceans provided by Faure (1982, Fig. 1), one notes that approximately the same strontium ratios occurred at times in the early Ordovician, the late Silurian–early Devonian, and the Miocene.

As with any potential dating or correlation method, the strontium geochronometer must be used with caution. Faure (1982) points out that, for this method to be used successfully, several conditions must be met or at least taken into account. The strontium ratio of the rock or mineral being measured must have been derived from seawater (and be representative of the strontium ratio of the seawater at the time of formation of the rock or mineral)—the strontium must not be derived from volcanic rocks or nonmarine rocks or minerals; the strontium in the oceans must have been isotopically homogenous at the time of the formation of the rock or mineral; the strontium ratio must not have been altered by the in situ decay of ^{87}Rb, or by diagenesis, dolomitization, recrystallization, metamorphism, weathering, and so on; and finally the strontium ratio of the authigenic marine minerals being analyzed must not be altered by the release of strontium from detrital minerals contained within the rock (such admixture or alteration of strontium could occur during laboratory preparation). Faure (1982) suggests that smectite and illite clays in particular may release strontium into solution during the dissolution of carbonate minerals with hydrochloric acid in the laboratory.

Faure (summarized in his 1982 paper, and see references to his earlier work cited therein) has suggested that the strontium composition of the oceans is controlled primarily by strontium input from three sources: Phanerozoic marine carbonate and evaporite rocks; volcanic rocks, of both the continents and the ocean basins; and Precambrian sialic rocks (that is, silica and aluminum–based rocks such as granites) and the younger detrital sediments derived from such rocks. Faure (1982, p. 78) hypothesizes that fluctuations in the $^{87}Sr/^{86}Sr$ ratios of the oceans through geologic time "are primarily caused by changes in inputs of strontium derived from young volcanic rocks and old sialic rocks. Increased erosion of old rocks on the continents may cause the $^{87}Sr/^{86}Sr$ ratio of the oceans to rise, whereas increased volcanic activity may lower it. The fluctuations of the $^{87}Sr/^{86}Sr$ ratio of seawater therefore reflect tectonic processes on a global scale." Orogenies and associated continental erosion rates could be associated with increases in the strontium ratios of the oceans, whereas increased volcanic activity, often associated with plate rifting, could be reflected in lowered strontium isotope ratios. Given the rapid mixing time of the oceans (on the order of a thousand years or less), fluctuations in the isotopic composition of strontium preserved in the stratigraphic record potentially could provide worldwide, relatively synchronous, markers tied to natural geological events. A major problem of the method at present, however, is the questionable precision and accuracy with which such events (fluctuations in isotopic ratios of strontium) can, or theoretically could be, distinguished and correlated; currently it appears that only relatively crude or coarse correlations can be derived from the method.

GEOCHEMICAL DIAGENETIC MARKERS

After a sediment is laid down and buried under more sediment, it may undergo various physical and chemical changes that convert the initially relatively unconsolidated sediment to more or less consolidated rock. Such diagenetic processes can include compaction, recrystallization, cementation, and replacement resulting in various textural, mineralogical, and geochemical changes in the rock material. During the diagenesis of a rock the geochemical composition of the sediment typically changes. Strontium, in particular, typically is progressively lost during the diagenesis of carbonate rocks; so it has been suggested by various researchers that the relative "strontium loss" of a particular carbonate could be used to date a sediment. Other elements that may be progressively lost, or in some circumstances gained, with age from carbonate sediments in particular include magnesium, manganese, and uranium (see review by Odin, Renard, and Grazzini, 1982).

Various workers have compiled global curves of Sr/Ca ratios in carbonate rocks plotted against age for the entire Phanerozoic and, in some cases, well back into the Precambrian (Odin, Renard, and Grazzini, 1982). However, the value and meaning of such curves are questionable; although they all exhibit approximately the same trends, the details of various curves established by differing workers are not always congruent. There is general agreement that Sr/Ca ratios in carbonates decrease relatively linearly with age throughout the Cenozoic and into the Mesozoic, but in the early Mesozoic the Sr/Ca ratio curves generally exhibit Sr/Ca values similar to those found in mid-Tertiary rocks. Sr/Ca values for pre-Mesozoic rocks vary widely.

Odin, Renard, and Grazzini (1982) suggest that a major problem inherent in such large-scale studies is that samples of different origin have been somewhat indiscriminitely mixed together to arrive at global curves. The original strontium content of carbonates appears to depend in great measure on the sedimentary environments in which they are formed; the initial Sr/Ca ratio of a rock is bound to affect subsequent Sr/Ca values obtained after varying degrees of diagenesis. Furthermore, the nature of the diagenetic processes that a particular rock is subjected to will influence the changes in the Sr/Ca ratio; for instance, very different values may be obtained, depending on the composition of interstitial waters during the diagenetic process or the depth to which a sediment was buried below the surface. Odin, Renard, and Grazzini (1982) suggest that in attempting to determine dates or chronologies of any rocks on the basis of diagenetic markers, such as strontium loss and the resulting changes in the Sr/Ca ratios with ages, one must be extremely careful to compare rocks that are truly comparable. Essentially, separate curves of variations in trace elements with diagenesis/age should be established for different types of rocks that have identical or similar initial trace element compositions and are subjected to similar diagenetic processes. As a first attempt toward this end, Odin, Renard, and Grazzini (1982) suggest distinguishing between carbonate sediments of the shallow water and sea floor neritic zone and pelagic carbonate sequences composed primarily of planktonic foraminifera and coccolithophorids. Within sediments of either class one must also take the latitudinal and climatic considerations into account. Although the application of diagenetic studies to geochronology and chronocorrelation is still in a state of infancy, given adequate care the analysis of parallel trends in the diagenesis of comparable rocks may prove extremely useful in a stratigraphic context.

SEA LEVEL CHANGES

The concept of sea level changes has been suggested to explain certain geologic phenomena since the birth of geology (see review in Fairbridge, 1961). For stratigraphic purposes, it has been proposed that worldwide changes in sea level potentially could be used in establishing long-distance correlations. Today speculations concerning global sea level (eustatic) changes remain a subject of general interest, and curves of sea level changes through time have been estimated in various ways (see Hallam, 1984a; Harland et al., 1982; Miall, 1984a). One approach has been to estimate the extent or area of continental flooding documented for different intervals of time in the past, and from such information to derive estimates of eustatic changes through time (e.g., Cogley, 1981; Hallam, 1977a; Wise, 1974). Another approach has been to estimate volumetric changes, and consequently water displacement, in spreading ocean ridges through time (e.g., Flemming and Roberts, 1973; Hays and Pitman, 1973). A third major approach to this subject is provided by the work of P. R. Vail and his colleagues (see the series of papers in Payton, 1977; Vail and Mitchum, 1977), using the concept of coastal onlap as interpreted from stratigraphic cross sections. Vail, Mitchum, and Thompson (1977a,b; see also Vail and Todd, 1982) developed a methodology, and accompanying terminology, to reconstruct and describe relative changes of sea level on a regional and global scale using reflection seismic stratigraphic sections in particular (although their basic methods also are applicable to other large-scale stratigraphic sections). The concept of relative sea level changes can provide an extremely powerful tool for certain stratigraphic studies, and the work of Vail et al. specifically warrants discussion here.

As Vail, Mitchum, and Thompson (1977a,b) define it, a relative change in sea leve is any rise or fall in the level of the sea relative to the local land surface. Relative changes in sea level can be brought about by either changes in the sea level itself, changes in the elevation of the land surface, or some combination of the two. It is important to keep in mind that curves or plots of relative changes in sea level are not eustatic (worldwide sea level) curves; taking other factors into account, such as subsidence, rates of sea floor spreading, and changes in the volumes of mid-ocean ridges through time, one can attempt to reconstruct eustatic curves from sea level curves (Vail, Mitchum, and Thompson, 1977b).

Vail, Mitchum, and Thompson (1977a, p. 63) defined a cycle (= third-order cycles, or global cycles of Vail, Mitchum, and Thompson, 1977b) of relative sea level change as "an interval of time during which a relative rise and fall of sea level takes place." Such cycles can be recognized on different geographic scales: local, regional, or global. According to these authors, a typical cycle (Fig. 7.2) of relative change of sea level is composed of a gradual relative rise in sea level, a period of stillstand (an interval of time durng which the surface of the land and the level of the sea remain stationary relative to one another), and finally a relatively rapid drop in sea level (I would question whether drops in sea level seem rapid because of erosion). Furthermore, these authors contend that on the basis of their data (much of which is proprietary material of oil companies), the cumulative rise in sea level during any one cycle consists of a series of small-scale rapid rises in sea level alternating with stillstands. Such a small-scale relative rise in sea level followed by a stillstand is termed a paracycle (Vail, Mitchum, and Thompson, 1977a).

On a larger scale than a cycle, Vail, Mitchum, and Thompson, (1977a) distinguish supercycles (= second-order cycles of Vail, Mitchum, and Thompson, 1977b). A supercycle is a set of adjacent or successive cycles of a regioinal or global scale that

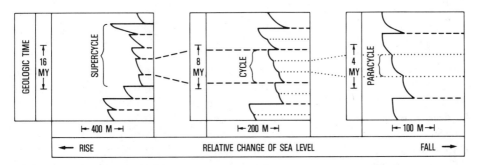

Figure 7.2. Conventions for charts of relative changes of sea level utilized by Vail, Mitchum, and Thompson (1977a, p. 64, reprinted by permission of the American Association of Petroleum Geologists): "Charts of relative changes of sea level. Cycles consist of relative rises and falls of sea level, commonly containing several paracycles, which are smaller scale pulses of relative rises to stillstands. Several cycles usually form a higher order cycle (supercycle) with [a] pattern of successive rises between major falls. Note asymmetry of gradual rises and abrupt falls at each scale."

exhibit a progressive overall tendency toward higher and higher relative sea levels. The top of a supercycle is marked by a major relative fall of sea level, perhaps to a level at or below the level at the beginning of a supercycle. Thus any supercycle normally is composed of a number of cycles, and any particular cycle in turn is composed of a number of paracycles. Finally, on the largest scale, Vail, Mitchum, and Thompson, (1977b) distinguish two first-order global cycles in Phanerozoic time, one lasting from the late Precambrian to the Early Triassic and the second having a duration from the Middle Triassic to the present. Throughout their durations each of these first-order cycles exhibits a gradual rise and subsequent fall of relative sea level.

The basic operational stratigraphic unit that Vail, Mitchum, and Thompson, (1977a) use in reconstructuring relative changes in sea level is the depositional sequence (see previous section). According to these authors, as long as there is a sufficient sediment supply, at least one depositional sequence will be deposited during a cycle of relative rise and fall of sea level. If during a particular cycle there is a continuous relative rise of the sea level, followed by stillstand, and finally an abrupt fall in sea level at the end of the cycle, then it is likely that only one depositional sequence will be produced during this interval. The sharp fall in sea level at the end of the cycle will produce an unconformity (perhaps manifested as erosional truncation, see Chapter 3) at the top of the sequence. If a cycle contains more than one paracycle, then there is a good chance that more than one sequence will be deposited during the cycle. Each paracycle may be distinguishable as a smaller-scale sequence; the boundaries between two such adjacent sequences commonly are marked by downlap of the superjacent beds lying immediately above the boundary and sometimes by toplap of the beds lying immediately under the boundary. Vail, Mitchum, and Thompson, (1977a) also point out that numerous smaller-scale sequences may form during paracycles, and one cannot necessarily expect a simple one-to-one relationship between cycles or paracycles and depositional sequences. However, when one is using primarily seismic data, many of the smaller-scale sequences (that is, those corresponding to units below the regional paracycle) may be too thin to be distinguished. Of course, in general the greater the fall in sea level at the end of a cycle, the sharper and more easily distinguishable will be the boundaries between adjacent cycles.

It should also be explicitly noted that not all depositional sequences are of equal value in reconstructing relative sea level changes. Vail, Mitchum, and Thompson, (1977a, p. 65) apply the term maritime sequence to depositional sequences that consist of "genetically related coastal and/or marine deposits." The coastal or most landward portion of a maritime sequence is controlled primarily by sea level acting as a base level for erosion and deposition. The shallow marine facies is controlled partially by sea level, but the relative sea level exerts little, if any, direct control on the deep marine facies. Therefore, the landward portions of maritime sequences (especially the coastal facies) are of utmost value in developing regional curves of relative changes of sea level. A hinterland sequence (Vail, Mitchum, and Thompson, 1977a) is a sequence that consists exclusively of nonmarine sediments and facies deposited inland to a coastal area. In hinterland sequences depositional mechanisms appear to be independent of (or at most controlled extremely indirectly by) relative changes in sea level, and thus are currently of no value in developing regional or global plots of relative changes in sea level.

Building on the work of such investigators as Curray (1964), Van Andel and Curray (1960), and Weller (1960; see also Boggs, 1987; Matthews, 1984, and many references cited in these works), Vail, Mitchum, and Thompson, (1977a) have also discussed basic concepts that are useful in attempting to determine relative changes in sea level from maritime sequences and stratigraphic cross sections.

During a relative sea level rise, if there is a sufficient supply of sediment, coastal deposits (particularly littoral deposits [those formed between the points of high and low tide] and nonmarine coastal deposits [for example, those laid down on the coastal plain above the point of high tide]) will tend to progressively migrate or onlap landward over the underlying depositional surface (Fig. 7.3); this phenomenon, termed coastal onlap (Vail, Mitchum, and Thompson, 1977a) is of prime importance in measuring relative changes in sea level using the approach of Vail, Mitchum, and Thompson. In such cases the level of the sea acts as, or approximates, the base level of deposition; and thus as the sea level rises, deposits can accumulate and migrate laterally landward.

Vail, Mitchum, and Thompson, (1977a) suggest that relative rises can be most accurately measured using littoral deposits that onlap the underlying depositional surface. Nonmarine coastal deposits are perhaps more commonly encountered in the stratigraphic record but tend to build up a few meters above sea level and thus introduce a small error (up to several tens of meters on coastal plains hundreds of kilometers wide, according to these authors) when used to estimate relative rises in sea level.

Vail, Mitchum, and Thompson, (1977a) demonstrate that a relative rise in sea level can be measured on suitable stratigraphic cross sections, although sometimes

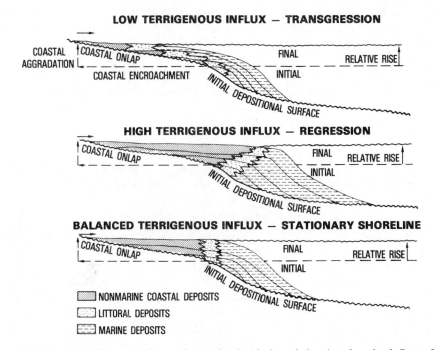

Figure 7.3. "Transgression, regression, and coastal onlap during relative rise of sea level. Rate of terrigenous influx determines whether transgression, regression, or stationary shoreline is produced during relative rise of sea level." (Figure and caption from Vail, Mitchum, and Thompson, 1977a, p. 66, reprinted by permission of the American Association of Petroleum Geologists.)

adjustments must be made. One can use either vertical or horizontal components of coastal onlap, termed coastal aggradation and coastal encroachment, respectively. In using coastal aggradation as a direct measure of a relative rise in sea level, one may have to take into account, and adjust for, basinward subsidence or perhaps compaction rates. Utilizing coastal encroachment to measure a sea level rise, one may have to allow for such factors as variations in the slope of the surface upon which the onlapping deposits were laid. In either case, one must be very wary of situations in which the onlapping coastal deposits may have been removed by erosion from a given locality or area. Vail, Mitchum, and Thompson, (1977a, p. 67) suggest that "in some cases the missing strata may be restored to a projection of the underlying initial deposition surface, sometimes with the help of isolated erosional remnants, but generally it is better to search the region for a section in which the coastal onlap is preserved."

A relative rise in sea level will not always result in coastal onlap, however. If the relative rise in sea level is more rapid than the rate of deposition, then marine onlap (the encroachment of marine strata landward) may result (Vail, Mitchum, and Thompson, 1977a). In such cases the relative rise of sea level may be much harder to determine, primarily beause "paleobathymetric control will be needed to help measure the relative rise. Assuming that structural movements are not complicated, the amount of the relative rise may be approximated by measuring the vertical component of marine onlap (marine aggradation) if the paleobathymetry remains constant. If not, the amount of rise may be estimated by determining the marine aggradation plus any amount of deepening, or minus a lesser amount of shallowing. Because paleobathymetric measurements are given in intervals of several hundreds of feet, the measurement of relative rise with marine onlap is only an approximation and should be checked against measurements of coastal onlap of the same unit in other areas" (Vail, Mitchum, and Thompson, 1977a p. 67).

It seems to be a common misconception among those who have not given thought to the matter that transgressions (landward migrations of the littoral facies) are correlated with relative rises in sea level, and regressions (seaward migrations of the littoral facies) are correlated with relative falls in sea level; but this is not the case. As is illustrated in Fig. 7.3 (see also Vail, Mitchum, and Thompson, 1977a; Weller, 1960), during a relative rise in sea level the shoreline may remain stationary, or it may experience transgression or regression. Indeed, during one cycle or paracycle of relative sea level rise a particular shoreline could experience any combination of the three conditions, depending, for instance, primarily on the relative influx of terrigenous clastic debris at any one time. During a relative rise in sea level, a relatively high terrigenous influx will result in a regression (Fig. 7.3), a relatively low terrigenous influx will result in a transgression, and a balanced terrigenous influx will result in a stationary shoreline. Note, however, that in all three cases coastal onlap is associated with a relative rise in sea level.

Just as a transgression or a regression may be associated with a relative rise in sea level, so too at any one place either a deepening of the sea bottom, a shallowing of the sea bottom, or a constant water depth may be associated with a relative rise in sea level. Again, these phenomena are dependent upon the rate of accumulation of deposits in the particular area under consideration. It is even possible along a single section roughly perpendicular to the shoreline that, during a relative rise in sea level, the littorial faces can be regressive while the coresponding marine section is deepening (for instance, due to a distribution of sediment input that feeds the littoral facies but starves the marine facies: Vail, Mitchum, and Thompson, 1977a).

To generalize, it appears that in many cases an observed transgression or deepening indeed will correlate with a relative rise in sea level. Moreover, a relative fall in sea level normally would be expected to produce a regression and shallowing (see discussion below). However, regressions or shallowing also can occur during a relative rise in sea level, as well as during a stillstand or fall in sea level. Thus, regression and shallowing per se do not indicate the nature of the sea level change, if any. Vail, Mitchum, and Thompson, (1977a, p. 68) do suggest, however, that "regression is most common during a relative rise or stillstand." In other words, even during a relative rise in sea level or a stillstand, a regressive pattern of deposition appears to be more common than a transgressive pattern or a stationary shoreline pattern.

A relative stillstand, an interval of time during which the position of sea level is constant with respect to the initial surface of deposition, often is characterized by coastal toplap (that is, newly deposited strata prograde seaward; Vail, Mitchum, and Thompson, 1977a). Given a sufficient supply of sediment, deposition cannot take place above the base level (approximated by, or determined by, the relative level of the sea); so coastal deposits cannot overlap on the initial depositional surface landward but rather build laterally seaward and lap (via toplap) landward. Vail, Mitchum, and Thompson, (1977a, p. 68) note that in cases where a stillstand occurs after a rapid rise in sea level—a rise that is more rapid than the local rate of deposition—marine onlap may occur during the stillstand. During the stillstand at any one such spot shallowing will occur at the same rate as marine aggradation (taking any structural complications into account).

Toplap is not associated solely with true stillstand conditions. Toplap features also can be produced by the rapid deposition of sediment even during a relative rise in sea level, especially locally (for example, within a deltaic complex).

During a relative fall in sea level, the resulting sedimentation pattern may exhibit an abrupt downward shift of coastal onlap (marked by an unconformity separating the two adjacent sequences: Fig. 7.4a) or a series of units successively formed at lower levels seaward (Fig. 7.4b; see Vail, Mitchum, and Thompson, 1977a; Weller, 1960). The first pattern often is associated with a relatively rapid fall in sea level, whereas the second may be indicative of a relatively gradual fall in sea level. In the seismic sections studied by Vail and his colleagues (in Payton, 1977), the first pattern prevailed. Vail, Mitchum, and Thompson (1977a) suggest a couple of possible explanations for this observation. The features used to distinguish the second pattern (lateral progradation of units at successively lower levels) often may be beyond the resolution of seismic data analysis. The areas commonly studied seismically by Vail et al. contained thick sedimentary deposits; in such areas regional subsidence in general may have proceeded at a relatively fast rate, perhaps faster than any eustatic falls. In such cases, even if sea level were actually gradually falling globally (there was a eustatic fall), regionally a relative rise of sea level might be recorded in the sediments. Of course, such a consideration perhaps casts uncertainty on the validity of the global curves of relative sea level change compiled by Vail, Mitchum, and Thompson, (1977b).

To measure the relative fall in sea level recorded between two sequences, Vail, Mitchum, and Thompson, (1977a) ideally calculate the difference in elevation between the highest coastal onlap recorded in the subjacent sequence and the lowest coastal onlap in the superjacent sequence (see Fig. 7.5). However, it is rarely so simple. As is to be expected, the underlying sequence usually suffers some erosion during a relative sea level fall; the amount of the underlying sequence that is lost must be approximated. Differential subsidence along the cross section being used

a) DOWNWARD SHIFT IN COASTAL ONLAP INDICATES RAPID FALL

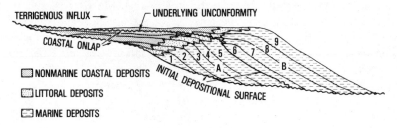

b) DOWNWARD SHIFT IN CLINOFORM PATTERN INDICATES GRADUAL FALL

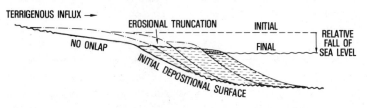

Figure 7.4. "Downward shift of coastal onlap indicates relative fall of sea level. With relative fall of base level, erosion is likely: deposition is resumed with coastal onlap during subsequent rise. a) Downward shift in coastal onlap indicates rapid fall observed in all cases studied so far. b) Downward shift in clinoform pattern (after Weller, 1960), indicates gradual fall; but has not been observed on seismic data." (Figure and caption from Vail, Mitchum, and Thompson, 1977a, p. 72, reprinted by permission of the American Association of Petroleum Geologists.)

may have occurred, which must be taken into account. Also, it is quite common to encounter marine (instead of coastal) onlap in the oldest beds of the overlying sequence, and then one must calculate paleobathymetric depths for these rocks in order to reconstruct what the elevation of the lowest coastal onlap of the overlying sequence probably was. Such marine onlap frequently is encountered in sequences immediately above unconformities corresponding to abrupt and major drops in relative sea level, because after a major drop in sea level the majority of the coastal plain and shelf (now exposed to subareal erosion) tends to be bypassed by, or cut across by, rivers and streams that deposit their sediment in the form of areally (or laterally, parallel to the shoreline) restricted fans on what was formerly the steep slope of the ocean margin. Coastal onlap is produced in the restricted areas where the rivers and streams feed into the sea, but otherwise marine onlap occurs on the landward margins of submarine fans, some of which may represent very deep water conditions. According to Vail, Mitchum, and Thompson, (1977a, p. 72): "Because a major fall is so difficult to measure accurately, it rarely can be plotted quantitatively with any degree of confidence."

Vail, Mitchum, and Thompson, (1977b) constructed charts of relative changes of sea level on a global scale down to the level of supercycles for the Cambrian to Triassic, and for the Cretaceous, and to the level of cycles for the Jurassic and Cenozoic. This work was quickly criticized (this criticism is summarized in Harland et al., 1982, and Miall, 1986a). In particular, various investigators have questioned the presumed rapidity of sea level falls, the applicabiity of the generalized stratigraphic models utilized to arrive at such curves when one is dealing with specific sections, and the interpretation of such compilations as genuinely indicative of

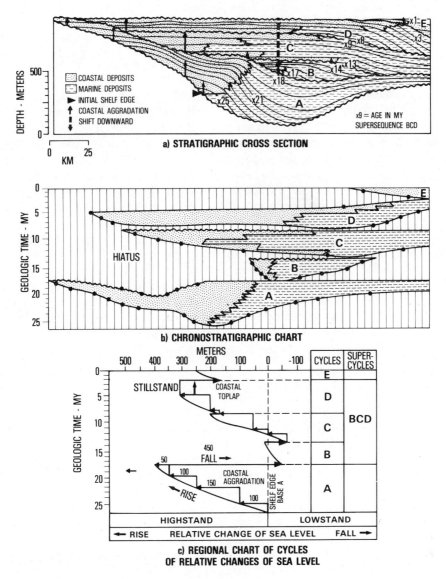

Figure 7.5. "Procedure for constructing regional chart of cycles of relative changes of sea level." (Figure and caption from Vail, Mitchum, and Thompson, 1977a, p. 78, reprinted by permission of the American Association of Petroleum Geologists.)

global changes in sea level. In partial answer to these criticisms, charts similar to the early ones derived by Vail and his colleagues are now usually considered to reflect "relative changes of coastal onlap" rather than "relative changes of sea level" per se (e.g., Haq, Hardenbol, and Vail, 1987; Vail and Todd, 1981; Vail, Hardenbol, and Todd, 1984). This is an actively developing area of research; for recent discussions and references on eustatic sea level curves interpreted from seismic stratigraphy the reader should consult Bally (1987), Ross and Haman (1987), and Schlee (1984).

CHRONOCORRELATION BASED ON POSITION IN A TRANSGRESSIVE–REGRESSIVE CYCLE

The concept of local to global transgressions and regressions, due, for instance, to worldwide eustatic changes in sea level or to local changes (such as subsidence, uplift, or changes in sediment supply), has long been known. Likewise, the principle of basing correlations on analogous points within local sections relative to a transgressive–regressive cycle that affected all of the sections under consideration has been utilized for a number of decades (most recently by Ager, 1981, and Boggs, 1987). Within any transgressive–regressive cycle (represented by an interval of time) it is suggested that there will be one point (or short interval) in time when maximum transgression will be reached. This point will be recorded in the strata of any particular locality by a local maximum in water depth (interpreted on the basis of fossil content or lithology). Ideally, a plane connecting all of the points in local sections that represent the maxima of water depths should correspond to a synchronous time plane marking maximum transgression. In any local section the rocks and biotas below this point will exhibit increasing water depths, working up section, and rock above this point will exhibit decreasing water depths with stratigraphic elevation. Potentially this method also could be used to establish time planes other than that corresponding to maximum transgression. However, to do so might involve detailed analysis of the paleogeometry of the ancient basin in which the rocks composing the local stratigraphic sections under consideration were found.

GEOLOGIC-CLIMATE UNITS

The predecessor of the current *North American Stratigraphic Code,* the *Code of Stratigraphic Nomenclature* of the American Commission on Stratigraphic Nomenclature (1961, 1970), recognized what were called "geologic-climate units" for use in the Quaternary (see Appendix 2). The fundamental units of geologic-climate classification were glaciations and interglaciations; subdivisions of glaciation were said to be stades and interstades. These designations were in keeping with the contemporary concepts of continental stratigraphy for the Quaternary, particularly as developed and used in Europe and North America, based on climatically recognized stratigraphic units (see Butzer, 1971, for example). Cold intervals were variously labeled glacials or stadials, and warm intervals were referred to as interglacials or interstadials. In Africa and other lower-latitude areas, periods of moist climate were postulated to occur during the glacial periods of the more northern latitudes. Some workers distinguished a succession of pluvial and interpluvial episodes that presumably correlated, more or less, with the glacial and interglacial episodes. However, investigators experienced great difficulty in attempting to recognize and correlate these climate-based units from one geographic area to another. Much of this classic work is now generally regarded to be of dubious value; for example, even the basic European fourfold glacial sequence of the Günz, Mindel, Riss, and Würm is now obsolete (Butzer, 1971; however, such terms persist in the literature [e.g., Day, 1986]). Because of the problems inherent in attempting to define, recognize and correlate the geologic-climate units of the *Code of Stratigraphic Nomenclature* (ACSN, 1961, 1970), they were eliminated from the latest *Code* (see NACSN, 1983, p. 849 [reprinted as Appendix 1]).

HOLOCENE CHRONOSTRATIGRAPHIC UNITS

The Holocene generally is regarded as an epoch/series (Harland et al., 1982; Palmer, 1983, Van Eysinga, 1978) although it is admittedly thus far an extremely short epoch (it has not yet been terminated), and some workers have suggested that the Holocene should be considered merely an interglacial period of the Pleistocene or Quaternary. However, as Königsson (1984) and many others have pointed out, the Holocene indeed appears to demarcate a distinct and important interval of geologic time that should be elevated to moderately high rank in the chronostratigraphic hierarchy. At approximately the beginning of the Holocene the nature of the surface of the planet began to change irrevocably. "About 10,000 years ago Man started the Neolithic Revolution, probably in many places of the Middle East simultaneously, but in Anatolia demonstrably. This was the end of the natural development of our planet. The reshaping of it, the manipulating of its ecosystems, the utilization of its biotic and abiotic wealth, and the subsequent polluting of its environment started when the first goat was captured and domesticated, and the first seeds were selected for plantation in the first cleared field" (Königsson, 1984, p. 107.).

The beginning of the Holocene usually has been dated to about 10,000 years ago with the onset of climatic conditions roughly similar to those of the present day. This "climatic event" corresponds more or less to the boundaries between European Pollen Zones III and IV, the Younger Dryas and Preboral, and the Late Glacial and Postglacial (see further discussion below). On the basis of varves, De Geer (1940; see Butzer, 1971) dates the Holocene–Pleistocene boundary to 7912 B.C. (= 9862 B.P., given that the present is defined as A.D. 1950). In 1969 the Holocene Commission of INQUA (= International Union for Quaternary Research) recommended that the Holocene–Pleistocene boundary be fixed chronostratigraphically (by a boundary stratotype) at approximately 10,000 years B.P. A boundary stratotype probably will be established in a varved lacustrine sequence in Sweden (Harland et al., 1982; Königsson, 1984).

The concept of defining and subdividing the Holocene chronostratigraphically (using the same basic principles as are applied to the remainder of the Phanerozoic) has not been universally accepted, and when accepted has been often misused. For example, in some instances climatic/environmental conditions that are time-transgressive have been applied to establish supposed chronostratigraphic correlations. Königsson (1984) attributes this situation to the fact that many Holocene workers concerned with the stratigraphy of the epoch are trained not as geologists or stratigraphers but as archeologists and anthropologists, zoologists, botanists, and so on.

Two major systems of chronostratigraphic subdivision of the Holocene have been widely used, both based on the dominant vegetational history of Europe (and Sweden in particular) and the climatic conditions inferred by the vegetation. In the late nineteenth century Rutger Sernander, building on the work of Axel Blytt, proposed what has become known as the Blytt-Sernander zone system (see Königsson, 1984, for further discussion and references). Sernander originally proposed this system in the context of a late Pleistocene to Holocene vegetational history of the island of Gotland, Sweden, and it was subsequently elaborated upon by other investigators and widely adopted even far from the area and climatic zone where the system was originally defined. The Blytt-Sernander system simply divides the latest Pleistocene and Holocene into a number of successive named time intervals (chronozones). The temporal limits of the chronozones, however, are subject to disagreement, particularly as the physical basis upon which they were originally defined (vegetational changes) may well be time-transgressive (Butzer, 1971). Relatively recently it has been suggested that the well-established names of the chronozones simply be applied to chronometrically defined intervals of the Holocene as such: the Subatlantic would correspond to the interval 0 to 2500, the Subboreal 2500 to 5000, the Atlantic 5000 to 8000, the Boreal 8000 to 9000, and the Preboreal 9000 to 10,000 years B.P. (Mangerud et al., 1974).

In the early twentieth century, standard pollen zones numbered I to VIII (oldest to youngest) were established for the latest Pleistocene and Holocene of Europe (Butzer, 1971; A. M. Davis, 1984; Godwin, 1981). The pollen zones generally have been regarded as corresponding more or less to the chronozones of the Blytt-Sernander system. The boundary between pollen zones III and IV often is taken to represent the boundary between the last glacial interval and the present interglacial or postglacial, and the boundary between the Pleistocene and Holocene. However, with the exact fixing of the Pleistocene–Holocene boundary chronostratigraphically with a boundary stratotype, and the redefining of the Blytt-Sernander chronozones chronometrically (or chronostratigraphically), there well may be discrepancies between the observed pollen peaks and zones (which form biostratigraphic horizons and units) at any one section and the chronostratigraphic units or geochronologic intervals to which they once were considered to correspond.

If the Holocene is taken as a series/epoch, then conventionally it should be subdivided into stages/ages, ideally of worldwide applicability. It could be suggested that a redefined Blytt-Sernander system could serve as the basis for division of the Holocene into stages/ages. As reported by Königsson (1984, based primarily on Mangerud, Birks, and Jäger, 1982), among specialists on Holocene stratigraphy there is no universal agreement on the feasibility of establishing even cogent regional systems of subdivision and correlation, much less a global chronostratigraphic subdivision of the Holocene. As might be expected, in general northern European workers tend to support and use the Blytt-Sernander system or systems correlated to the Nordic system, whereas workers in areas far removed geographically (for example, the Americas, Africa, Australia, New Zealand) do not. Strictly regional systems have been established in some areas.

Some investigators suggest that formal chronostratigraphic subdivision of the Holocene is unnecessary. Instead, numerical dates (especially radiocarbon dates) are widely used and sufficient for ordering and relating events.

NUMERICAL DATING

Since the beginnings of geology there have been attempts to calculate the numerical ages in years (or some other convenient unit of temporal duration) of the earth and various rock bodies. These attempts received renewed emphasis with the development of isotopic/radiometric dating methods in the twentieth century. The designation "absolute age" came into wide currency to signify ages, in years, of rocks determined by isotopic analysis. Of course, there is nothing "absolute" about such age determinations (see discussion and references cited below), and the North American Commission on Stratigraphic Nomenclature (1983, p. 854) has strongly recommended that "absolute age" be dropped. The NACSN suggests that "isotopic age" be used for an age determined by means of isotopic ratios and that "numerical age" be used to refer to an age determined by any quantifiable age-related phenomenon (including isotopic ratios). Similarly, Harland et al. (1982) suggest that a radiometrically determined age be referred to as an apparent age (versus the true age of a rock). The NACSN also recommends that the term calibration be used to refer to the dating by means of numerical ages of chronostratigraphic boundaries (and thus the geochronologic units based upon chronostratigraphic units). In stratigraphy the terms date and dating can be used to refer to a numerical age or to a chronostratigraphic/geochronologic (and implied temporal) correlation. Thus one might date a certain formation to the Early and Middle Paleocene, or to approximately 66 to 60 million years ago. Depending on the situation, it may be more accurate or precise to use a chronostratigraphic correlation or a numerical age; certainly the former is preferable if the age is based on correlations of mammalian faunas without any direct means of numerically dating the formation, whereas the latter is preferable if the formation is devoid of fossils but has been dated isotopically.

When numerical dates are given, certain conventions usually are followed. "The unit of time is the modern year as presently recognized worldwide" (NACSN, 1983, p. 854). Presumably the standard for this year may be taken as the tropical year at 1900 January 0 days 12 hours ephemeris time. The fundamental unit of time used internationally is the second, formerly the mean solar second (1/86,400 of the mean solar day), now variously defined as 1/31,556,925.9747 of the above-mentioned year or as "the duration of 9,192,631,770 periods of the radiation of the atom of Cesium 133" (see Harland et al., 1982, p. 118). Thus, for the purposes of numerical dating in stratigraphy a year as a standard interval of time is approximately 3.15569×10^7 seconds. It should always be kept in mind that this year is solely a unit of time measurement; even if a rock could be dated as having formed exactly 400 million calendar years ago, the earth may not have orbited the sun exactly 400 million times since the rock was formed (the mean interval of time that it takes the earth to orbit the sun may have changed during the course of geological time).

Numerical dates commonly are expressed in years before the present, where, by general agreement, the present is A.D. 1950. The NACSN recommends that qualifiers such as "ago" and "before the present" be omitted (they are implicit) and that the abbreviations of the multipliers of the International System of Units (SI; the "metric" system) be combined with "a" for annum; ka (kilo-annum = 10^3 years), Ma (Mega-annum = 10^6 years), Ga (Giga-annum = 10^9 years), and so on. Thus our hypothetical Early to Middle Paleocene formation might by numerically dated to 66–60 Ma. However, the duration of this formation would not be 6 Ma. According to the NACSN (1983), durations and abbreviations for numbers of years that do not refer to a date before the present are informal; the abbreviations "m.y.," "m.yr.," or simply "my" might all be used to express the concept that our hypothetical formation spans approximately 6 million years.

Other terms commonly encountered in the stratigraphic literature that refer to chronometric units of specified durations are millennium = 10^3 years, and eon [not to be confused with the geochronologic rank of eon] = aeon = gigennium = 10^9 years. It should be kept in mind that the term billion is used differently on the two sides of the Atlantic. In America a billion equals 10^9, whereas according to the British system a billion equals 10^{12}. Also commonly used, especially in the Quaternary and archaeological literature, is the abbreviation B.P. for years before the present (A.D. 1950).

The importance of numerical dating methods to the stratigrapher is primarily two-fold. One, they provide a potential means of chronocorrelation, although numerical "dates," like any other stratigraphic evidence, must be used with due caution. Chronocorrelations based on numerical dates may well be inherently less accurate in some cases than correlations based on biostratigraphy or magnetostratigraphy, for instance. Two, they provide a means of calibrating in terms of standard temporal units of equal duration (years) the historical stratigraphic record and the geochronologic time scale. In this section the natures of various means that have been used to arrive at numerical ages for rock bodies are discussed; for more detailed discussions the reader is referred to the literature cited below.

Numerical Ages Based on Sedimentation Rates

Of primarily historical interest now, in the late nineteenth and early twentieth centuries there were many attempts to date the age of the earth and calibrate the geochronologic time scale using accumulated thicknesses of sedimentary rocks and estimates as to average rates of deposition (reviewed by Eicher, 1968; see also Osborn, 1916, and Walcott, 1893). Numerous difficulties were encountered. The method itself is based on extreme uniformitarian assumptions. Deposition may be episodic and interrupted by numerous periods of nondeposition and erosion. There is no known complete sedimentary record within a single small area or section that records all of the earth's history. Depositional rates are very inadequately known and vary with rock type. Nevertheless, using composite stratigraphic columns numerous geologists attempted to date the age of the earth and calibrate the geochronologic time scale using this general method, with wide-ranging results. In general the composite stratigraphic thickness, since the origin of the earth, was (depending on the worker) calculated to range from about 100,000 to 300,000 feet, and average rates of deposition were calculated at 100 to 8000 years per foot of sediment; the age of the earth was calculated to be anywhere from 17 Ma to 1.5 Ga.

In the late twentieth century stratigraphic thicknesses and estimated rates of deposition have been used not to determine numerical ages of the geochronologic scale, but (once a geochronologic scale has been determined by other means) to estimate the completeness of stratigraphic sections. In this context completeness is defined as the portion of a given time span, in a local stratigraphic section or area, that is represented by sediment or rock (Sadler, 1981). Completeness estimates are calculated by dividing the net rate of accumulation of a sedimentary section (thickness of a section divided by the length of time over which it was deposited) by a short-term rate of sedimentation (based on modern analogues and compilations of sedimentation rates from various sedimentary environments). As an example (from Schoch, 1986b), in fluvial environments median sedimentation rates observed over 100-year intervals are approximately 600 mm (mm = millimeters) per 100 years (= 6000 m/1 m.y. [m = meters]; McKinney and Schoch, 1983; Sadler, 1981; Schindel, 1980). Let us assume that we are interested in calculating the completeness of a 500-m-thick fluvial sequence that was deposited over the course of 1 m.y. The completeness of this section at a resolution of 100 years would be 500 m/1 m.y. divided by 6000 m/1 m.y. = 8.3%; about 830 of the 10,000 100-year intervals constituting the 1 m.y. period saw some deposition. In hindsight, completeness studies demonstrate in part the difficulties of using average depositional rates and composite sedimentary thicknesses to calibrate the geochronologic time scale.

Besides using deposition rates of sediments, some early geologists also attempted to date the earth by means of rates of denudation and influx of sodium from river drainage into the oceans. The latter procedure was based on some highly speculative assumptions, such as that oceans are as old as the earth itself, that the oceans were originally fresh, that the present rate of sodium being dumped into the oceans is representative of the mean sodium influx through geologic time, and that the cycling of sodium is understood. Such a method, even if it had been successful, only would have dated the age of the earth (or oceans) and done little to directly calibrate the geochronologic time scale otherwise.

Dating on the Basis of Physical and Chemical Processes

Besides sedimentation rates, many relatively constant or repetitious physical or chemical processes potentially can be utilized in both numerical and relative dating of the geologic time scale (nuclear processes, namely, isotopic decay systems, are discussed below). Indeed, any process that lends itself to quantification over time might be applied to this purpose; a few examples will be mentioned here. For relatively recent geologic time, varves in sedimentary sequences have been considered to be deposited on an annual basis, and in some cases one can simply count back from the present (see discussion above on the dating of the Holocene–Pleistocene boundary). Recently House (1985b) has suggested that sedimentary microrhythms that are commonly found in the stratigraphic record may be the result of Milankovitch-type cycles; specifically House suggests that some sedimentary microrhythms may represent cycles that result from the obliquity of the ecliptic. The period of the cycle of the obliquity of the ecliptic is currently approximately 41,000 years. Assuming that the obliquity periodicity has changed little through time, and that one can recognize sedimentary microrhythms that correspond to this periodicity, House (1985b) suggests that one can measure the duration of selected intervals of the stratigraphic record. To give an example, House (1985b) estimates that the "Planorbis Zone" of the Lower Jurassic (Lower Lias: see Arkell, 1933) contains 22 microrhythms; if each microrhythm corresponds to 41,000 years, then the duration of the Planorbis Zone was 0.86 million years.

Physical weathering rates, rates of paleosol development, and rates of geomorphological development (landform and landscape development) have been applied to the problem of estimating numerical and relative dates, particularly for Pleistocene sequences (see, e.g., Brookes, 1985; Coates, 1984; Evans, 1985; Finkl, 1984; Vreeken, 1984). In this general category may be included obsidian hydration dating and amino acid racemization dating.

When a fresh surface of obsidian (natural volcanic glass) is exposed to external conditions at the surface of the earth, it reacts with this environment, particularly by the absorption of water. A hydration shell or ring forms on the outer surface of the obsidian as water penetrates it. By measuring the thickness of the hydration shell and knowing or estimating the rate of hydration, the length of time the obsidian surface has been exposed to the environment can be calculated (Trembour and Friedman, 1984). Calculating the actual numerical age of such an obsidian surface (as opposed to the relative dating of two or more obsidian surfaces that have been exposed to the same environmental parameters) can be difficult, as the rate of hydration is a function of both the environmental temperature and the geochemistry of the particular obsidian under consideration. The obsidian hydration technique has been utilized by both archaeologists and geologists to date events that range in age from a few hundred to several million years (see Trembour and Friedman, 1984, and references cited therein).

Amino acids that occur in living organisms as we know them are 100% L form (that is, levo or "left-handed" isomers). Various amino acids, as components of structural proteins, often are entrapped in the bony or shelly skeletal parts of organisms and thus are subject to preservation and fossilization in the stratigraphic record. At the death of the organisms, the amino acids undergo progressive racemization; that is, the amino acid distributions convert from a 100% L form into an equilibrium mixture containing both L forms and D (dextro or "right-handed") forms with time. Thus, increasing D/L ratios of amino acids indicate increasing relative ages, given that all other considerations are equal, and thus can be utilized in correlations, especially over a relatively limited geographic area. If the rate of conversion from L to D forms is known (or can be estimated), then numerical ages can be calculated using this information (see Rutter, Crawford, and Hamilton, 1985; Wehmiller, 1984). Amino acid racemization rates vary, depending on the particular amino acid under consideration, and are also a function of the thermal history of the specimen. The problem of the temperature dependence of racemization rates, and problems of contamination of the material studied, are the two major difficulties that have been encountered with this technique thus far. Amino acid racemization has been applied to various shells, bones, woods, and other organic materials spanning approximately the last million years.

Dating on the Basis of Growth Rates of Organisms

For the Pleistocene and Holocene the growth rates of organisms sometimes can be used to directly date geological materials and events. For example, knowing the growth rate of particular lichen species, investigators have dated the age of Holocene rocky substrates by measuring the sizes of lichens that colonize rock surfaces shortly after such surfaces are exposed at the surface of the earth (lichenometry: Calkin and Ellis, 1984). Many organisms bear growth rings or the equivalent. Probably the first case that comes to mind for most persons is the annual growth rings of many trees. Locally, different individual trees tend to respond to the annual climatic conditions, particularly moisture supply, and thus vary in growth from year to year. Thus, one can composite for a particular area a generalized history of the characteristics of the annual tree rings that extends well beyond the lifetime of any individual tree. Once such a chronology is established, an unknown piece of wood can be dated by fitting it to the master chronology (in many respects this is conceptually similar to correlations based upon magnetostratigraphy). Tree-ring dating (dendrochronology) has been used in the dating of historic and prehistoric structures (buildings) and artifacts, and potentially tree-ring chronologies can be composited that extend back from the present for thousands of years in the past (Parker et al., 1984).

Many corals also register growth increments, and although they cannot necessarily be used in an analogous manner to tree-ring dating, they have been used creatively in attempts to date portions of the stratigraphic record. Some modern corals bear both yearly growth ridges and finer striations that represent daily growth increments; the same apparently is true of some fossil corals. By counting the number of daily striations between the annual growth ridges of a fossil coral, one can determine the apparent number of days that made up the year when the particular coral was growing. Based on the hypothesis that the earth's rotation has been slowly decreasing because of tidal friction, but that the period of revolution of the earth around the sun has been relatively constant, one can calculate the numbers of days that should have composed the year for any point in the past. Thus in the Middle Devonian it has been calculated that there were approximately 398 days to one year (Eicher, 1968). A coral that indicated there were 398 days to a year would be attributed to the Middle Devonian, whereas a coral that recorded only 376 days per year would be considered to be of Early Cretaceous age (John W. Wells in Eicher, 1968; Runcorn, 1966). Certainly this dating method is relatively coarse, but it does corroborate the general numerical time scale that has been established on other bases (particularly through the use of isotopic dating, discussed below).

Isotopic Dating Methods

The major breakthrough needed to accurately calibrate the geochronologic scale came in the last years of the nineteenth century and the early twentieth century with the discovery of radioactivity and the development of the principles of radiometric dating (Becquerel, 1896; Boltwood, 1907; W. E. Davis, 1974; Holmes, 1946, 1947; Rutherford, 1904). As early as 1907, Bertram Borden Boltwood (1870–1927) tentatively calculated the following, remarkably accurate, numerical ages for some major periods: Carboniferous, 340 Ma; Devonian, 370 Ma; and Ordovician or Silurian, 430 Ma (Eicher, 1968).

Although the analytical techniques used in radiometric dating may require great precision, the conceptual basis of radiometric dating is not difficult and is explained in many introductory and more technical references (e.g., Faure, 1977; Mahaney, 1984; Odin, 1982a; Prothero, in press). In a typical chronometer system a particular parent radionuclide (radioactive producer) spontaneously decays, usually through a series of intermediate steps, to a daughter nuclide (radiogenic product) with a characteristic half-life (the time for one-half of the parent material to decay to daughter product). The steps in spontaneous radioactive decay commonly may take such forms as alpha decay (the nucleus of the parent radionuclide ejects an alpha particle composed of two protons and two neutrons), beta decay (an electron is emitted from the parent nucleus, and a neutron is converted to a proton), or electron capture (a proton in the parent nucleus picks up an orbital electron and is converted to a neutron). Given any particular closed system (such as a mineral), the ratio of parent to daughter will change systematically through time. The probability, or rate, of decay can be specified either in terms of the half-life ($t_{1/2}$) or by means of a decay constant (λ) where the decay constant is simply the proportion of existing radioactive parent atoms that will decay in a given unit of time. Given appropriate units, the half-life is equal to the natural log of 2 over the decay constant:

$$t_{1/2} = \frac{\ln 2}{\lambda} = \frac{0.693}{\lambda} .$$

Thus, given assumptions about the initial ratio (for instance, that initially no daughter nuclides were present), and knowing the current ratio and half-life (both determined experimentally in the laboratory), the substance can be dated. The basic equation used in calculating the age of simple radiometric systems is:

$$t = \frac{1}{\lambda} \ln \left(\frac{D}{P} + 1 \right)$$

where t is time elapsed since the system was set (closed with only the parent radio-nuclide and no radiogenic product), and D/P is the ratio of daughter to parent atoms. This equation must of course be modified for more complicated systems, such as the potassium–argon method, which involves two decay constants because ^{40}K decays both to ^{40}Ca and to ^{40}Ar (see below). Even if the system is not closed (that is, if the daughter product escapes), a date sometimes can be obtained if the amount of original parent nuclide in the system can be determined or estimated. Selected nuclide systems (radioactive isotope systems used as chronometers) commonly utilized in obtaining numerical ages of geological bodies are commented upon below.

There are many uncertainties and constraints involved in utilizing isotopic chronometers to date stratigraphic boundaries and geologic events (Obradovich, 1984; Odin, 1982b). As Obradovich (1984), among others, has pointed out, in order for a particular chronometer to be useful in stratigraphic dating, it must fulfill certain minimal criteria. The half-life must be appropriate to the particular event being dated; it must be long enough that not all of the parent material has decayed away, but short enough that a measurable amount of daughter product has formed. Chronometers with relatively longer half-lives are suited to dating relatively older events; likewise, shorter half-lives are suited to younger events. The half-life must be known with suitable accuracy. The parent radionuclide initially must be present in measurable quantities in the material to be dated. One must be able to distinguish between the daughter product of in situ decay of the parent and any similar material that may be present. The material (rock, mineral, or similar substance) must act as a closed geochemical system relative to parent and daughter isotopes, or if it does not, then appropriate corrections must be made. Metamorphism, weathering, hydrothermal alteration, and so on, as well as simply unrecognized open system behavior, may be important in this context. The time when an isotopic system closes (such as during crystallization of an igneous rock) or is reset (for example, during a metamorphic event) must bear some relationship to the event it is desired to date (perhaps a stratigraphic boundary). Often stratigraphic events, layers, or boundaries that one wishes to date can only be bracketed by isotopic dates.

Any apparent age arrived at isotopically is, in the purest sense, simply an analytical ratio of daughter to parent subsequently interpreted as an age (Odin, 1982b). Such an analytical ratio may be extremely accurate (precise and reproducible), but may not be geologically or stratigraphically accurate. "Analytical accuracy and geological accuracy need not be synonymous. That is, although an age determined on a mineral or on a series of whole rock samples is analytically correct, the age itself may have little bearing on the true age of the event being measured such as the time of eruption of a lava or ash found interbedded with fossiliferous strata" (Obradovich, 1984, p. 18).

Elaborating upon the uncertainties of isotopic dating, Odin (1982b) has identified four classes of uncertainties in the use of such geochronometers: stratigraphical uncertainties, genetic uncertainties, historical uncertainties, and analytical uncertainties.

Stratigraphical uncertainties involve problems of correlation (chronocorrelation) between the chronometer used and the events or boundaries to be dated. Usually one wishes to date a fossiliferous sequence or boundary between two biostratigraphic or chronostratigraphic units (the latter often being defined for practical purposes biostratigraphically). The best chronometers, however, may be contained in rocks of igneous origin that are not easily or directly correlatable with paleontological sequences. Studies of plutonic rocks, for example, may yield "good" dates on the rocks themselves, but there may be major disconformities between such rocks and any sedimentary or fossiliferous rocks. Volcanic flows and the like may be interlayered with fossiliferous, correlatable strata and thus bracket the ages of such strata. Sometimes datable material may occur within fossiliferous strata, such as bentonites, ash fall deposits, authigenic minerals, or biologically produced material (such as material useful for radiocarbon dating in later Pleistocene and Recent sediments). As Odin (1982b) points out, in some cases a certain fossil sequence may be well dated chronometrically, but it is impossible to correlate the particular fossil sequence accurately with standard or classical fossil sequences. "For example, the very interesting data obtained from Cenozoic volcanics in North America (Evernden *et al.,* 1964), although paleontologically quite well documented with mammals, are of little use for calibration of the time scale due to the fact that it is not possible to achieve an accurate correlation with European mammals and classical stages (Evernden and Evernden, 1970). In fact, correlations can always be attempted, but the stratigraphical uncertainties quickly become more important than the analytical ones and the result is too inaccurate for comparison with data obtained from better correlated chronometers, even when the latter are *a priori* regarded as less reliable" (Odin, 1982b, pp. 4–5, italics in the original).

Genetic uncertainties may be extremely important in evaluating the usefulness of apparent dates arrived at using various geochronometers. The time zero (blocking temperature) of an intrusive igneous rock is dependent on the cooling rate, may vary from mineral to mineral or chronometer to chronometer, and can significantly postdate the time of intrusion. A rock also may inherit unrecognized radiogenic daughter product or older minerals, or in the case of some rocks (for example, bentonites, tuff layers) the minerals in a dated rock may have crystallized before deposition in a stratigraphic sequence. Authigenic minerals in sedimentary sequences may have zero times (that is, the time of closure of the relevant chronometers) that occur after deposition and during burial and diagenesis. Bridging the concepts of genetic and historical uncertainties in the use of chronometers, one must always be aware of the possibility that alteration has taken place, perhaps during deposition, diagenesis, or much later.

Odin (1982b) considers historical uncertainties to include thermodynamic events that may reset the zero time of the chronometer (for instance, metamorphism) and superficial changes that may affect a chronometer (such as weathering). Such historical uncertainties along with genetic uncertainties may together be termed geochemical uncertainties (Odin, 1982b).

Analytical uncertainties include such factors as measurement errors in determining the amounts of daughter and parent material, interlaboratory differences in analytical techniques and calibrations, uncertainties in decay constants, and so on. Odin (1982b) suggests that within a single laboratory reproducibility of analytical ratios is commonly better than 1%, whereas the absolute calibrations from one laboratory to another may account for differences of more than 2%. Commonly isotopically determined numerical ages are given with indicated errors (for instance, 150 ± 10 m.y.); what such errors really mean, however, may vary from investigator to investigator. Often the indicated error simply refers to the precision or reproducibility of the analytical techniques and may be given in terms of plus or minus one or two sigmas (67% or 95% confidence levels, respectively). An error given with a date may represent scatter due to random variations in laboratory procedure, scatter due to variations among daughter/parent ratios of samples from a particular chronometer analyzed, or inherent error due to limitations of the machinery used (for instance, the particular mass spectrometer). The accuracy of an apparent date relative to its true (and unknown) age is not necessarily indicated by plus and minus errors per se, but may be dependent upon many other factors (such as undetected loss of parent or daughter material, and so on).

Prior to about 1978 many different decay constants and isotopic abundance ratios were utilized in isotopic dating (Harland et al., 1982), a practice that resulted in many discrepancies in calculated ages. In 1976 the IUGS Subcommission on Geochronology recommended a standardization of isotopic abundance and decay constant values utilized, such that comparisons between the results of different laboratories could be more readily made (Gale, 1982; Steiger and Jäger, 1977, 1978). These conventions have been generally adopted, and Harland et al. (1982, their Appendix 1) provide a useful table for converting most pre-1978 nonstandardized isotopic ages into standardized ages. It should be noted, however, that not all constants have been standardized; for instance, differing values for the fission rate of decay of ^{238}U are currently in use (see section below on fission-track dating). Also, logically, even if standardized conventions are established for all of the constants utilized in isotopic dating, this does not ensure that all calculated dates will be of equal accuracy (close to their true values). If the decay constant used by convention for one chronometer is extremely accurate, whereas the decay constant for a second chronometer is much less accurate, the apparent ages obtained using the two different systems may not be directly comparable.

Radiocarbon

Carbon-14, or radiocarbon, dating is one of the most frequently used methods of numerical dating for late Pleistocene and Holocene (approximately 30 to 50 ka and younger) deposits and archaeological objects. A brief discussion of the method here will point up some of the assumptions and uncertainties inherent in many forms of isotopic dating.

Radiocarbon dating was developed in the late 1940s and early 1950s and is based on the following principles (Libby, 1952; Rucklidge, 1984; Terasmae, 1984). ^{14}C is continuously produced from ^{14}N in the upper atmosphere by cosmic radiation. The ^{14}C combines with oxygen and forms radioactive carbon dioxide, which is uniformly mixed throughout the atmosphere, biosphere, and hydrosphere, and a global equilibrium state of radiocarbon results. Given that there is a state of equilibrium in the system, the rate of decay of ^{14}C to ^{14}N is balanced by the rate of production. According to Rucklidge (1984), the equilibrium concentration in living organisms is approximately 1.2×10^{-12} atoms of ^{14}C per atom of ^{12}C. When a part of the subsystem is isolated from the global system (for example, an organism dies and is buried) the radiocarbon is not replaced as it decays, and consequently the amount of ^{14}C decreases according to its characteristic half-life. Given the half-life of ^{14}C of approximately 5730 years, the amount of radiocarbon remaining in a sample is a function of the length of time since it was isolated from the global system. Conventionally, the amount of radiocarbon remaining in the sample has been determined by measuring the rate of beta decay of the remaining ^{14}C atoms. More recently, techniques have been developed to identify ^{14}C atoms directly by mass spectrometry using a particle accelerator (Rucklidge, 1984).

A radiometric date normally is reported as a mean date with an error figure. The statistics of nuclear decay approximate a Poisson distribution, which is asymmetrical around a mean (Ogden, 1977). In some cases the asymmetry is relatively insignificant, and a date is reported with a "±" one sigma error. In other cases (depending on the age of the sample dated), the asymmetry may be significant, and the date may be reported with separate "+" and "−" values. In either case "all that is implied [by the date] is that the reported radiocarbon age of the submitted sample has two chances out of three of being within the quoted limits" (Terasmae, 1984, p. 2) and approximately a 95% probability of lying within two sigmas of the quoted date. It is important to remember this and not place exaggerated emphasis on a single radiometrically calculated numerical date.

As Terasmae (1984) has pointed out, the validity of the basic methodology used in ^{14}C dating is dependent upon at least five basic assumptions: (1) ^{14}C has been produced in the atmosphere at a constant rate for at least the last 50,000 to 100,000 years. (2) The mixing of radiocarbon in the global system has been relatively rapid and uniform. (3) The decay rate of ^{14}C has been constant through time. (4) Any particular sample has not been contaminated by younger or older carbon since it was isolated from the global system. (5) Isotopic fractionation has not occurred that would alter the $^{14}C:^{13}C:^{12}C$ ratios of the sample.

The assumption of the absolute constancy of isotopic decay rates in general may be open to a slight amount of criticism (possibly, under certain conditions, some decay constants could vary by up to 4%: Emery, 1972), but it seems to be well established both experimentally and theoretically (the randomness of the process is a consequence of quantum theory; see Brush, 1983, for a general discussion of the constancy of decay rates). A more serious practical consideration is that different laboratories, and even the same laboratory at different times, may use varying half-lives for the same parent–daughter series. In the case of radiocarbon the half-life was initially determined to be 5568 ± 30 years, whereas it has more recently been calculated at 5730 ± 40 years.

It has now been demonstrated that the production, and hence amount, of ^{14}C has fluctuated through time independent of any human-induced fluctuations, involving, for example, the detonation of nuclear weapons or the burning of fossil fuels (Damon, Lerman, and Long, 1978; Suess, 1980; Terasmae, 1984). These natural fluctuations appear to be on the order of 1 or 2% over a few decades or centuries and up to 10% over the last 10 ka. The ^{14}C fluctuations are due to such factors as variations in the rate of primary production in the atmosphere (for instance, caused by variations in the cosmic-ray flux), variations in the CO_2 content of geochemical reservoirs and their exchange with the rest of the geosphere (for instance, CO_2 solubility is affected by temperature variations), and variations in the total amount of CO_2 in the geosphere (periods of intense volcanism may result in CO_2 degassing). Nuclear weapons testing of the early 1960s caused a sudden increase of nearly 100% in atmospheric ^{14}C concentration (see articles in Berger and Suess, 1979). ^{14}C fluctuations create problems in the dating method; but, as Terasmae (1984) noted, once they are sufficiently well known and understood, they can be accommodated.

It has been demonstrated that carbon isotope fractionation does occur during the growth of living organisms (Terasmae, 1984) and varies from species to species and even between different parts of a single organism. This isotopic fractionation can be corrected or compensated for, but this is not done by all laboratories; furthermore, one must know the species of plant or animal to make the correction. Samples that are composed of many species present a real problem with respect to correcting for isotopic fractionation. According to Terasmae (1984, p. 6), corrections for isotopic fractionation may be insignificant for samples on the order of 30,000 years old, but "become increasingly significant for younger samples."

Sample contamination is a problem with any method of radiometric dating. Simple carelessness in the field or laboratory may introduce contaminants. More insidious may be problems associated with the reworking of sediments, "resetting" of radiometric clocks (perhaps not a great concern in radiocarbon dating), and mixing by natural processes of substances of different ages (for example, groundwater of one age passing through a porous deposit of a different age). Some of these problems can be obviated or alleviated by dating multiple samples or in some cases using larger samples. Of course, these practices can cost extra money, which may not be available.

Another practical problem in ^{14}C dating concerns reference standards used by laboratories for calibration, checking of equipment, and calculating the final dates. It is due in part to reference standards that the dates produced by different laboratories are comparable. Various standards have been proposed and used over the last 30 years (reviewed by Terasmae, 1984), both for ^{14}C activity and for carbon isotope ratios. Such proliferation of standards easily can lead to confusion over the comparability of various dates produced by different laboratories using differing standards. Also, any physical standard runs the chance of being depleted, and even if it was artificially produced, a new batch may vary significantly from the original. As an example, the U. S. Bureau of Standards produced a 1000-pound batch of oxalic acid that provided the standard for many radiocarbon dating laboratories, but it was depleted after 20 years. A new batch was produced in 1977; however, upon analysis the new "standard" was found to differ from the old in ^{13}C abundance.

Many of the problems associated with radiocarbon dating may be ironed out by international agreement, cooperation, and standardization. Terasmae (1984) has stressed that practical difficulties and concerns that arise in using ^{14}C and many other types or radiometric ages (How reliable is a certain date? What do the numbers really represent? Is it comparable to other radiometric dates produced at different times by different laboratories?) could be greatly alleviated simply by adequate publication, not only of the date per se but also of the laboratory, the laboratory sample number, any background or qualifying information provided by the laboratory that produced the date (such as method of sample preparation, standards used, method of age calculation, any corrections made), and correct identification of the sample material (such as the species of organism).

Uranium-Lead Dating Methods

Among the first elements utilized in radioactive dating within a geologic context were ^{235}U, ^{238}U, and the related ^{232}Th (Donovan, 1966), all three of which decay to different isotopes of lead (Pb). Of the uranium on the earth, over 99% is ^{238}U, and less than 1% is ^{235}U. These two isotopes decay through complex decay schemes before reaching stable lead end products. ^{238}U decays to ^{206}Pb with a half-life of approximately 4.47 × 10^9 years (recently accepted value) or 4.51 × 10^9 years (traditionally used value), whereas ^{235}U decays to ^{207}Pb with a half-life of approximately 7.04 × 10^8 years (recently accepted value) or 7.13 × 10^8 years (traditionally accepted value) (Boggs, 1987; Eicher, 1968; Gale, 1982; Obradovich, 1984). ^{232}Th decays to ^{208}Pb with a half-life of approximately 1.4 × 10^{10} years (Gale, 1982). Common lead isotopes on the earth are the three mentioned above plus ^{204}Pb, which, as far as is known, is not the daughter product of any radioactive parent.

At first uranium–lead methods were applied in a straightforward manner to minerals that contain high concentrations of uranium, such as uranite and pitchblende. Such minerals are relatively rare, however, and with the development of more precise analytical techniques it became theoretically possible to date any of a number of minerals that commonly contain trace amounts of uranium, such as allanite, apatite, monazite, sphene, xenotime, and zircon. In practice, zircon is the mineral most often used in current uranium–lead dating techniques. Given that a particular mineral grain acted as a closed system and originally contained only the radioactive parent and no daughter product, then the age of the mineral could be calculated simply from the ratio of the daughter to parent (such as ^{206}Pb/^{238}U). If any lead was originally present in the system, the apparent age would be greater than the actual age; however, if any lead was originally present in the system, it also should contain some nonradiogenically produced lead, ^{204}Pb. Indeed, the abundance of ^{204}Pb in the system can be used to estimate the original abundances of the other isotopes of lead in the system.

Theoretically the ages determined for a single mineral (or rock) on the basis of the ^{206}Pb/^{238}U, ^{207}Pb/^{235}U, and ^{208}Pb/^{232}Th ratios should all agree, but in practice they often do not. If they agree, they are said to be concordant, and it is standard practice to plot actual ratios (corresponding to ages) relative to concordia curves. A concordia curve is a plot of concordant ratios of ^{207}Pb/^{235}U (commonly plotted on the horizontal axis) relative to ^{206}Pb/^{238}U (commonly plotted on the vertical axis) for various ages. The concordia curve is the theoretical curve that the ratios of a sample initially containing normal abundances of the uranium isotopes, but no radiogenically produced lead, should follow with time, starting at the origin at time zero (when the ratios are both equal to zero). The shape of the concordia curve, when plotted as described above on arithmetic scales, is convex up.

If lead/uranium ratios for a particular sample or set of samples do not fall on the concordia curve (that is, the ages determined using the two different lead/uranium systems are not in agreement), they are said to be discordant. Discordant results normally will plot below the concordia curve (when plotted as described above) and indicate some type of open system behavior, namely, the loss of lead by some thermal or metamorphic event, or perhaps by continuous diffusion. It has been demonstrated that in many cases a series of samples (such as individual zircon crystals) from a single rock often will define a straight line below the concordia curve. This line defined by discordant points then can be projected to intersect the concordia curve, and the upper intersection (to the right on a standard plot) can be interpreted as the age of formation of the samples, while the lower intersection may represent the age of a disturbance that disrupted the two lead–uranium systems (see Eicher, 1968; Obradovich, 1984; Wetherill, 1956). In some cases the lower interception, in particular, may not correspond to any real event, but rather the discordance may be due to lead loss spread out over a considerable period of time, or even to continuous diffusion of lead from the samples.

A problem often associated with uranium-based methods of isotopic dating is interference due to the properties of the minerals commonly used in such dating. Zircon, for instance, is a stable, resistant, and refractory mineral. Detrital zircon grains may, of course, be much older than the rocks in which they are contained. Igneous rocks may also include older xenocrysts of zircon (or other minerals); if the xenocrystic nature of such minerals is not recognized, the age of such minerals may be incorrectly interpreted as the age of the rock itself.

The ratios of isotopes of lead to one another also have been utilized as a dating technique. Here the basic concept is that the earth began with a common reservoir of lead (composed of certain percentages of the four common isotopes of lead), and that since the origin of the earth the common lead found on the planet has been evolving by the progressive addition of ^{206}Pb, ^{207}Pb, and ^{208}Pb produced by the decay of uranium and thorium. Thus the ^{206}Pb/^{204}Pb, ^{207}Pb/^{204}Pb, and ^{208}Pb/^{204}Pb ratios have been increasing with time. If at any time a sample of the common lead were removed (for instance, in a lead-rich mineral or ore), isolated from the parent isotopes that produce more daughter lead, and preserved, then the isotopic ratios of the sample would reflect its age.

There are numerous potential problems associated with common lead dating methods, however. In order to apply this technique, growth curves for the isotopic composition of the common lead of the earth must be constructed. Some investigators initially assumed that the entire earth has acted, in general, as a single lead system (which perhaps implies that uranium, thorium, and lead are distributed more or less uniformly throughout the earth); but it may be that the earth can be treated as composed of a number of independent regional subsystems, as least as far as uranium, thorium, and lead compositions are concerned. If the latter hypothesis is in fact true, then different lead growth curves must be developed and utilized for different subsystems, and one must be concerned with the possibility of spurious ages due to the mixing of lead from different subsystems. Furthermore, there is always the possibility that any particular sample of common lead isolated at a particular time may later be contaminated by radiogenically produced lead or by lead that had been isolated at a different time. Such complicating factors may be very hard to detect without independent evidence from other dating systems of the general geologic environment.

Rubidium–Strontium Methods

Approximately 28% of the rubidium found on the earth is the radioactive ^{87}Rb, which decays to ^{87}Sr with a half-life of approximately 4.88×10^{10} years (recently accepted value) or 4.7×10^{10} years (traditionally used value: Boggs, 1987; Obradovich, 1984). Both rubidium and strontium are trace elements that behave in a geochemical context similarly to potassium and calcium, respectively. Rubidium-strontium dating methods have been used on igneous, sedimentary, and metamorphic rocks; particular minerals suitable for such analysis include micas (for example, biotite and muscovite), potassium feldspars, amphibole, and glauconite, along with whole rock analyses. Given the long half-life of rubidium and the fact that it usually occurs in trace amounts, rubidium–strontium methods generally give better results for older rocks. The results of a single analysis of the $^{87}Sr/^{87}Rb$ ratio may be compromised by the fact that the mineral or rock analyzed inherited some ^{87}Sr at the time of its origin. In such instances there will also be nonradiogenically produced ^{86}Sr, and corrections may be attempted in order to compensate for the inherited ^{87}Sr on the basis of reconstructed $^{87}Sr/^{86}Sr$ curves for the earth throughout its history; but inaccuracies in such corrections may be significant relative to the amount of newly developed ^{87}Sr in the sample.

In rubidium–strontium dating it is common to deal with the problem of inherited ^{87}Sr, particularly in igneous rocks, by what has been termed the isochron method (Ehlers and Blatt, 1982; Obradovich, 1984; Prothero, in press). Given any magma, it can be hypothesized that initially the magma will be homogeneous, so any inherited strontium will be evenly distributed isotopically throughout the magma. As crystallization occurs, the $^{87}Sr/^{86}Sr$ ratio will be a constant value for all minerals initially containing strontium, regardless of how much or how little strontium they contain. Even though it may have been homogeneously distributed throughout the magma, rubidium will tend to be concentrated in certain generally potassium-rich minerals; therefore, different minerals (or different samples more generally) will have different initial $^{87}Rb/^{86}Sr$ ratios. If one plots $^{87}Rb/^{86}Sr$ values for the whole rock and various minerals (for instance, on the horizontal axis) against $^{87}Sr/^{86}Sr$ values for the same samples at time zero (when the rock first crystallized), one will have a horizontal line (because all samples have the same initial $^{87}Sr/^{86}Sr$ values). With time, if each mineral acts as a closed system, for each mineral (and the rock as a whole) the ratio of $^{87}Rb/^{86}Sr$ will decrease, and the $^{87}Sr/^{86}Sr$ ratio will increase in proportion to both the amount of ^{87}Rb originally present in the particular mineral or sample and the elapsed time since crystallization. Thus, with time what was initially a horizontal line on the plot described above will rotate counterclockwise around the intercept with the vertical axis (the value of the $^{87}Sr/^{86}Sr$ ratio for a mineral that initially had a $^{87}Rb/^{86}Sr$ ratio of zero). Such a line plotted at any one time is referred to as an isochron, and the slope of such an isochron represents the age of the system (the steeper the slope, the older the rock).

In the preceding discussion it was assumed that not only the rock as a whole but also all of the minerals within the rock under consideration acted as closed systems relative to the rubidium–strontium chronometer, but this is not always the case. To take a simple example, assume that the rock was subjected to metamorphism at some point after crystallization. Metamorphism may mobilize the atoms within the rock and equalize the $^{87}Sr/^{86}Sr$ ratios throughout the minerals in the rock once again. In other words, the rubidium–strontium clock may be reset by metamorphism. Yet, if the rock acts as a closed system as a whole, the plot of the isochron described above can be imagined as pivoting about the new (that is, new since the initial crystallization of the rock) $^{87}Rb/^{86}Sr$ value for the rock as a whole; thus although a horizontal isochron again ensues after metamorphism, the absolute value of $^{87}Sr/^{86}Sr$ will be increased and that of $^{87}Rb/^{86}Sr$ decreased for the rock as a whole. As time passes, the isochron again will rotate about the intercept on the vertical axis, and the slope of the isochron will serve to date the metamorphic event.

In the above discussion it is important to note that as long as the rock as a whole acts as a closed system, metamorphism does not affect the $^{87}Rb/^{86}Sr$ or $^{87}Sr/^{86}Sr$ ratios for the whole rock sample. This fact can be used to advantage to date both the time of initial crystallization and the time of a later metamorphic event if one has available a group of related igneous rocks that are comagmatic—in other words, "a group of igneous rocks of diverse composition that originated by fractionation of a common parent magma" (Ehlers and Blatt, 1982, p. 698). If the group of rocks crystallized at essentially the same time from a single magma with a common ^{87}Sr/^{86}Sr ratio but differing ^{87}Rb/^{86}Sr ratios for the various rocks, the whole rock compositions can be plotted as an isochron, analogously to the minerals or samples within a single rock. After a given amount of time the slope of the isochron of the rock samples will correspond to the age of crystallization of the group of rocks. Note, however, that as long as each rock of the group acts as a closed system throughout its history, the whole rock compositions will not be affected by any subsequent metamorphism. Metamorphism of any particular rock will merely mobilize and rearrange atoms within the rock. Thus one could date subsequent metamorphic events on the basis of isochrons derived from minerals or subsamples taken within a single rock although the time of crystallization of the group of comagmatic igneous rocks still would be recorded by the isochron derived from plotting the whole rock compositions of the members of the comagmatic suite of rocks.

Potassium–Argon Dating Methods

The potassium–argon dating method, based on ^{40}K, is currently one of the most widely used isotopic dating techniques in geology. The popularity of this method is due, at least in part, to the fact that potassium is one of the commoner elements in the earth's crust, and the K-Ar method is applicable to rocks ranging in age from the Precambrian to the late Pleistocene (Hall and York, 1984; McDougall, 1978). The ^{40}K isotope has an atomic abundance of approximately 0.01167% relative to total K (over 90% of all potassium is ^{39}K, with ^{41}K being the second most common potassium isotope), and ^{40}K spontaneously decays into either ^{40}Ar or ^{40}Ca. Approximately 10.5% of all ^{40}K decays into ^{40}Ar, whereas 89.5% decays into ^{40}Ca; the half-life of ^{40}K is approximately 1.250×10^9 years (Gale, 1982; Hall and York, 1984; McDougall, 1978). Utilizing the dual decay system of ^{40}K, theoretically two dating methods could be based on this isotope. However, because of the general abundance of ^{40}Ca in nature (approximately 97% of all common calcium is ^{40}Ca) and the fact that calcium generally occurs in large amounts in crustal rocks (Eicher, 1968; Prothero, in press), only the ^{40}K-^{40}Ar scheme is generally utilized in isotopic dating.

Potassium occurs in a variety of minerals, such as hornblende, micas, potassium feldspars, and clays. Because argon is an inert gas, it is not generally chemically bound into the crystal lattices of such minerals; but as ^{40}K decays, ^{40}Ar can be mechanically trapped in the crystal lattice. Volcanic and shallow intrusive igneous rocks are typically free of preexisting radiogenic argon at the time of their formation; however, radiogenic argon may be incorporated into rocks that are crystallized at high pressures (McDougall, 1978). Of course, once formed the mineral or sample must remain a closed system, or any loss or gain in potassium or argon must be taken into consideration. One must also always be concerned that the mechanically trapped radiogenic argon used in this dating method might be lost, thus rendering an isotopic age too young. It has been suggested that radiogenic argon may diffuse fairly easily from crystals, particularly under conditions of relatively low-grade metamorphism, diagenesis, or simple weathering. Another problem is that because argon occurs naturally in the atmosphere, atmospheric argon may contaminate a sample. Such atmospheric contamination generally can be recognized because atmospheric argon, although composed of about 99.6% ^{40}Ar, also universally contains ^{36}Ar.

K-Ar dates have been derived for single minerals from a particular rock, and in the case of fine-grained rocks from whole rock samples. In the latter situation the rock should have a nonporous groundmass (such as is found in some volcanic rocks) that can trap radiogenic argon as it is released, and the sample used should be unweathered.

In traditional K-Ar dating techniques the potassium and argon concentrations are determined on two separate aliquots of the sample (Hall and York, 1984; Obradovich, 1984). This procedure assumes that the sample is homogeneous with respect to potassium and argon; if it is not, errors will ensue. In part to alleviate this problem, and also to render K-Ar dating more precise, the ^{40}Ar/^{39}Ar technique was developed (Albarède, 1982; Hall and York, 1984). In this technique the sample is subjected to fast neutrons in a nuclear reactor, which converts ^{39}K into ^{39}Ar. The ^{40}Ar and ^{39}Ar contents of the sample then can be determined simultaneously by mass spectrometer analysis. The ^{39}Ar concentration is proportional to the ^{40}K concentration because the ^{39}Ar concentration is proportional to the ^{39}K concentration, which in turn is proportional to the ^{40}K concentration. Thus, using the ^{40}Ar/^{39}Ar technique, only one aliquot of the sample is required. "The proportionality constant relating ^{39}Ar with ^{40}K is determined by irradiating a standard sample of known K-Ar age along with the unknown" (Hall and York, 1984, p. 69). Of course, if the age of this monitor sample is uncertain, then this uncertainty also will be present in all K-Ar ages determined using the particular monitor. If different monitors are used by different laboratories, this will generally have the effect of decreasing the interlaboratory comparability of such ages. Furthermore, even if a single standard monitor is utilized by all laboratories in order to ensure the comparability of K-Ar ages, when one is attempting to compare K-Ar ages with ages determined by other methods (such as Rb-Sr or U-Pb methods), problems of calibration and comparability will reappear; there is as yet no established single standard that can be applied simultaneously to all of the various isotopic dating techniques currently employed (Obradovich, 1984).

A variation on the basic technique of ^{40}Ar/^{39}Ar analysis is to incrementally heat the sample (after it has been subjected to fast neutrons) and analyze the argon released at each temperature level, and to interpret each analysis as an age. "Depending on the type of release pattern produced, specific information can be obtained regarding the geochemical and geologic history of the sample. In certain cases where the sample has not been disturbed, a 'plateau age' can be obtained that relates directly to the age of cystallization" (Obradovich, 1984, p. 15; see also especially Albarède, 1982; Cassignol and Gillot, 1982; Hall and York, 1984).

A potentially extremely important aspect of K-Ar dating for stratigraphy is that this methodology, along with Rb-Sr techniques to a lesser extent, has been applied directly to the dating of authigenic glauconite found in many sedimentary sequences (see papers in Odin, 1982a; Odin, 1978). The use of glauconite as a K-Ar or Rb-Sr chronometer has been seriously questioned (see, for example, Hardenbol and Berggren, 1978; Obradovich, 1984, and references cited therein). Glauconites may be especially susceptible to argon loss and potassium gain, thus making their dates too low. Furthermore, glauconite systems do not close at the time of deposition but generally at some time after deposition, meaning that even the best glauconite ages are younger than the sediments in which the glauconite is found, and adjustments must be made to such glauconite-based ages.

Other Radioisotopic Dating Methods

The isotopic dating methods reviewed above are only some of the better known and more widely used chronometer systems. Many other decay systems also can be applied to the dating of geological materials. In particular, here might be mentioned two short-lived daughters of the ^{238}U and ^{235}U series, ^{230}Th (with a half-life of approximately 75,000 years) and ^{231}Pa (half-life of approximately 34,000 years), respectively. Radioactive thorium and protactinium often chemically separate from their parents (that is, uranium) during sedimentary processes, so these isotopes have been used in the dating of Pleistocene and Holocene deposits (Schwarcz and Blackwell, 1985; Schwarcz and Gascoyne, 1984; Stearns, 1984).

Closely related to the isotopic methods described above, which depend upon the calculation of a parent/daughter ratio, are methods that date materials on the basis of cumulative radiation damage. The latter category of techniques includes fission-track dating, thermoluminescence, and electron spin resonance (discussed below).

Fission-Track Dating

^{238}U spontaneously undergoes fission, the nucleus breaking up into two lighter nuclei, with a half-life of about 9.9×10^{15} years (Naeser and Naeser, 1984). When an atom of ^{238}U within·a mineral or piece of glass undergoes fission, the fission fragments pass through the object, producing zones of damage known as fission tracks. Therefore, theoretically if one knows the fission half-life, the amount of uranium in the mineral, and the number of fission tracks, one can calculate the age of the mineral.

The fission tracks are only a few to perhaps a hundred angstroms wide (Naeser and Naeser, 1984; Storzer and Wagner, 1982) and can be seen directly only with an electron microscope. The standard method used to make the tracks observable is to polish a surface of the specimen and then enlarge the fission tracks that intersect the polished surface with a chemical etchant such as hydrofluoric acid. After such preparation the fission tracks should be observable with an optical microscope at various magnifications in the range of $\times 200$ to $\times 2500$. By simple counting one can determine the number and density of tracks per unit area.

Natural uranium consists of approximately 99.3% ^{238}U and 0.7% ^{235}U (Storzer and Wagner, 1982), and it is generally thought that the relative abundance of these two isotopes of uranium is constant in rocks (Naeser and Naeser, 1984). Therefore, if one knows (or can calculate) the amount of ^{235}U in a sample, then one can calculate the ^{238}U content. If the sample is irradiated with thermal neutrons in a nuclear reactor, fission will be induced in the ^{235}U. The fragments of this induced fission leave fission tracks very similar to those resulting from the spontaneous fission of the ^{238}U (these fission tracks are standardly etched and counted in the same manner as the spontaneous fission tracks), and the number of such induced tracks is a function of the uranium content of the sample and the neutron dose that the sample receives. The neutron dose can be calculated by including a standard of known uranium content with the sample in the nuclear reactor, and thus by this methodology the ^{238}U content of the sample can be calculated. Finally, given all of the above information, a fission-track age in years can be calculated for the sample under consideration (see equations for this presented in Naeser and Naeser, 1984, and Storzer and Wagner, 1982).

The technique of fission-track dating, potentially a very powerful tool, has been applied in the dating of minerals ranging in age from the earliest Phanerozoic (e.g., Ross and Naeser, 1984) through the Quaternary. Like any technique, however, there are limitations to its use and reliability, some of which are outlined below (see also Naeser and Naeser, 1984; Storzer and Wagner, 1982; and references cited in these works).

In order for a sample to be datable using fission-track techniques, it must contain an adequate and appropriate uranium content relative to its age and physical characteristics; in particular, there must be enough uranium in the sample that a statistically significant number of spontaneous fission tracks can be located and recorded. In order to be identified spontaneous fission tracks must be retained over time by the sample to be be dated, and must be revealed in identifiable form in the polishing and etching process. Furthermore, the induced tracks should be identifiable at a comparable level with the spontaneous tracks. As Storzer and Wagner (1982, p. 202) in particular have pointed out, many factors may influence the "efficiency of track revealing," a few of which are cited here. In some cases the induced tracks may be recorded in an external detector rather than in the sample itself (see below), and the sample and external detector may have very different track-revealing efficiencies that must be taken into consideration. The particular crystallographic face used to identify tracks, and the orientation of the face, can affect the number of tracks observed. The particular physical and chemical conditions (for example, the etchant used and its concentration) under which a sample is etched and prepared for observation can influence the number of tracks that are recognizable. Radiation damage in minerals due to natural alpha-decay may effect the efficiency of track revealing; because of such crystal damage, the induced fission tracks may have very different characteristics from those of the spontaneous fission tracks in the mineral. Any thermal treatment of samples may change the efficiency with which tracks are etched. If the uranium is not distributed homogeneously throughout the sample (mineral grain) to be analyzed, the technique may be compromised unless appropriate corrections or compensating actions are taken. One also must be aware of subjective criteria that may affect the identification and counting of etched tracks under the microscope. Spontaneous tracks and induced tracks may not appear the same, and such features as bubbles, flaws, or microlites may after etching be mistaken for fission tracks, whereas faded tracks may not be recognized as such.

For any particular mineral or glass there is a critical temperature above which it will lose its spontaneous fission tracks—that is, it will anneal. However, this healing process of the material is not dependent solely on a critical temperature, but also upon the length of time during which the mineral is subjected to the particular temperature. Various natural glasses and some minerals, such as apatite, will anneal and slowly lose some of their spontaneous fission tracks (the tracks are said to fade) at ambient surface temperature over the course of geologic time (such as a million years or more). Temperatures of 70°C or above over a period of about 100 million years may cause spontaneous fission tracks in zircon (perhaps the most widely used mineral in fission-track dating techniques) to fade and eventually to be lost (thermal fading).

Particular techniques have been developed in attempts to deal with the various problems mentioned above. Two of the better-known techniques that address these considerations, at least in part, are the "population method" and the "external track detector method." In the population method, a large group of mineral grains or glass shards, all assumed to belong to a single population of samples having a homogeneous average uranium content and formed at a single time, is divided into two subpopulations. The grains of one subpopulation are embedded into a matrix such as epoxy, polished, and etched. The spontaneous fossil-track density then can be calculated for this sample. The second population is first irradiated, then mounted, polished, and etched; from this sample the total track density (spontaneous plus induced) can be calculated. Given that the spontaneous track density should be approximately the same for both subpopulations of the sample, the induced track density can be easily derived; by using this information, the neutron dose to which the second subpopulation was exposed in the reactor, and the values of various constants (such as the isotopic ratio of ^{235}U to ^{238}U in the sample and the decay constant for spontaneous fission of ^{238}U), an age in years for the sample can be calculated. The population technique is useful in that both subpopulations can be etched at the same time, thus eliminating differences in the efficiencies of track revelation between the irradiated and nonirradiated portions of the sample. Given large enough subpopulations, certain minor nonhomogeneities in the distribution of uranium among the crystals or glass shards, as well as difficulties due to differing crystallographic orientations of crystals, may be statistically eliminated. The population method is used principally with apatite grains and glass shards; it is not so suitable for minerals such as zircon or sphene in which the uranium may be distributed extremely nonhomogeneously.

In the external track detector method, the sample first is prepared and examined for spontaneous fission tracks. Next the face or surface containing the tabulated spontaneous fission tracks is covered with an external fission track detector, such as a muscovite plate or a piece of plastic foil, and the resulting sandwich of sample and detector is irradiated. During irradiation the fragments from the induced fission at the surface of the sample leave the sample and enter the external detector. The detector is then removed, etched, and analyzed for the induced fission tracks. The external detector method is particularly suited to the analysis of single grains, such as grains of sphene or zircon, given that they contain a high enough uranium content. The external detector method can compensate for nonhomogeneously distributed uranium content within or between grains, and it is also suitable for crystals suffering from alpha-radiation damage (in which case the original sample itself may serve as a poor detector of induced fission tracks). Drawbacks of the external track detector method include the fact that the external detector and the sample itself may have very different efficiencies with which they reveal fission tracks; if so, corrections must be made. Also, with this method the spontaneous tracks (in the sample) and the induced tracks (in the external detector) are not directly comparable; such direct comparisons, when possible, can be important in recognizing and correcting for thermal fading of spontaneous fission tracks (see below).

As noted above, fission tracks tend to fade (particularly in length, density, and diameter) with time and temperature, the exact fading characteristics depending upon the substance involved. If this fading or annealing of tracks is not taken into account, ages based on the fission-track method will be relatively low. In some cases the dimensions of etched spontaneous tracks can be compared with the dimensions of induced tracks, and "track-size" corrections can be attempted (Storzer and Wagner, 1982); correction curves have been calculated for various materials subjected to differing etching conditions.

A second approach to dealing with this problem is the "plateau annealing method": "This technique is based on the observation that the thermal energy required for further track fading increases exponentially with the degree of track fading" (Storzer and Wagner, 1982, p. 208). A sample is split into two portions, and one split is irradiated. An age is calculated for the sample on the basis of the ratio of the observed spontaneous fission tracks and the induced fission tracks. Then both the irradiated split and the nonirradiated split are heated together for a designated interval of time at a certain temperature, and the age is again determined for the sample. This process of heating is repeated numerous times at increasing temperatures, the age being determined after each interval of heating. If the sample suffered from some previous annealing or fading of the contained spontaneous fission tracks, then the calculated age of the sample will increase progressively during this process until it reaches a plateau. According to Storzer and Wagner (1982), as compared to track size corrections the plateau annealing method is generally a more precise way to correct for lowered fission-track ages.

Other factors that might potentially compromise the use of fission-track dates include variation in the isotopic composition of uranium in common rocks and minerals and the accuracy with which the spontaneous fission rate of ^{238}U is known. It is generally assumed or concluded that, except in a few very special circumstances, the $^{235}U/^{238}U$ ratio of natural uranium in terrestrial samples does not exhibit large enough variations to significantly affect the accuracy of fission-track ages; indeed, even meteorites and lunar rocks exhibit approximately the same isotopic ratio of uranium (see references cited in Storzer and Wagner, 1982). The correct decay constant of ^{238}U spontaneous fission, however, has been debated in the literature (the decay constant is equal to ln 2 [0.693] divided by the half-life). Two different decay constants, differing by about 20%, have been used to calculate fission-track ages. Further difficulties in arriving at accurate fission-track ages are due to uncertainties regarding the thermal neutron flux used to create the induced fission tracks, and to differences in calibration age standards used by different laboratories and investigators (Storzer and Wagner, 1982).

Although this is potentially an extremely valuable tool in geochronology, one must interpret and use fission-track ages (like any numerical ages) in context, always keeping in mind the inherent uncertainties and potential problems with the method. Naeser and Naeser (1984, p. 94) stress that fission-track ages "should always be considered minimum ages." Based on their review of the technique, Storzer and Wagner (1982) concluded that the overall absolute uncertainty in typical fission-track ages must fall in the range of 10 to 20% even though "for age comparison of different stratigraphic layers a much higher *precision* as good as ±3% (1 sigma) may be obtained provided the samples are dated in the same way and there is no track fading" (Storzer and Wagner, 1982, p. 219, italics in the original).

Thermoluminescence and Electron Spin Resonance

Thermoluminescence and electron spin resonance techniques are just beginning to be applied to the dating of geological materials, particularly of the Quaternary. The basic principle underlying thermoluminescence dating has been summarized as follows: " In any geological environment, natural radiation induces free electrons in minerals that can be trapped into lattice defects. They may escape upon heating and recombine with holes at luminescent centers. Energy will then be released in the form of light. By recording the thermoluminescence (TL) of a mineral, the last drainage of the traps can be dated, assuming a constant radiation level, by the following equation: AGE (years) = [EQUIVALENT DOSE (rads)]/[DOSE-RATE (rads/year)]. The equivalent dose is the dose that can produce the natural TL level and is found by irradiation from known beta or gamma sources. The dose-rate is computed from the weight of the radioactive elements in the sample to which may be added a small cosmic-ray contribution" (Lamothe et al., 1984, p. 153; see also the excellent review article by Dreimanis et al., 1985).

The radiation damage responsible for the thermoluminescence of a mineral sample may be due to naturally occurring radioactive isotopes within the mineral itself (for example, ^{238}U, ^{232}Th, and ^{40}K), or it may originate from material surrounding the mineral, or from the general environment in which the sample is located. Thermoluminescence of minerals also can be caused by pressure, friction, light, or chemical reactions (Dreimanis et al., 1985), and such sources of thermoluminescence must be taken into consideration in dating a sample.

The thermoluminescence dating method has a range of dating capability between about 10^3 and 10^5 years B.P. and has been widely applied in archaeology to the "absolute" (numerical) dating of ancient pottery. When pottery is fired, any previously existing thermoluminescence is removed (that is, the thermoluminescence clock is reset, and thus this method is well suited to the dating of fired pottery. In dating geological materials prime concerns include determining the extent of any primary thermoluminescence carried by mineral grains when they are deposited, determining the loss of thermoluminescence from samples (sunlight can bleach thermoluminescence from minerals; Lamothe et al., 1984), and determining the radiation environment that the sample has been subjected to through time. At present thermoluminescence is fraught with technical and practical difficulties, but advances are rapidly being made by workers in the field (Dreimanis et al., 1985).

Another method that is being developed to date geological materials on the basis of cumulative radiation damage is the electron spin resonance method of dating (Ikeya, 1985). Natural radiation creates lattice defects in mineral structures and also unpaired electrons and electron holes as radiation ionizes atoms and molecules. As described by Ikeya (1985), electron spin resonance (ESR) refers to the absorption of microwaves when electrons undergo transitions from low to high energy states under the influence of an external magnetic field. "The concentration of stable unpaired electrons can be determined from the ESR signal intensity. If the rate of production [of stable unpaired electrons] is known, one can determine the age from the ESR signal intensity" (Ikeya, 1985, p. 75).

Electron spin resonance dating techniques span the time range of approximately 10^3 to 10^6 or 10^7 years, and thus far have been applied with varying degrees of success to secondary carbonate deposits (for instance, travertines, stalactites, and stalagmites), secondary gypsum deposits, carbonate fossils (for example, molluscs, corals, and foraminifera), silicified wood, coal, amber, volcanic tephra and tuffs, and vertebrate teeth and bones (Hennig, Geyh, and Grün, 1985; Hennig and Grün, 1983; Hennig, Grün, and Brunnacker, 1983a,b; Hennig et al., 1981, 1983; Ikeya, 1985). ESR methodology also has been applied to the dating of faults by determining the ages of deformation of mineral grains within fault gouges.

Chronometric Calibration of the Geologic Time Scale

There have been numerous recent attempts to chronometrically calibrate proposed geochronostratigraphic time scales (as noted previously, there is not yet a universally agreed-upon geochronostratigraphic scale), in order to arrive at a numerically dated geologic time scale (cf. Cohee, Glaessner, and Hedberg, 1978; Harland et al., 1982; Obradovich, 1984; Odin, 1982a). Many of these attempts involve syntheses of previously published (and in some cases corrected) isotopic ages (e.g., Harland et al., 1982; Van Hinte, 1976a,b) and have been hailed as "a paradigm for the future" (McLaren, 1978, p. 6, referring to Van Hinte's work specifically). Harland et al. (1982) have developed what they term "chronograms" as a means of presenting the data used in estimates of stratigraphic age boundaries (such as boundaries between internationally accepted systems and stages) and depicting the potential errors, strengths, and weaknesses of the calibrations. Not all investigators, however, have been pleased with such synthetic attempts at numerical time scales, especially when such time scales are proposed by workers whose major area of expertise lies outside the realm of geochronology and geochronometry per se. In this connection a comment by Obradovich (1984, p. 20) is worth pondering:

First of all there have been time scales advanced by biostratigraphers and paleontologists who have lacked an appreciation or an understanding of the foibles of isotopic dating. Certainly they have the ultimate expertise in their particular phase of paleontology, such as planktonic foraminifera, calcareous nannoplankton, or innoceramids, but when they attempt an overall synthesis of the many biostratigraphic techniques, even their contemporaries may not agree. Given this avenue of disagreement, consider the problems that arise when they venture to evaluate data for which they have no expertise or little understanding. Furthermore, they aggravate the situation when they fail to evaluate the data but simply resort to some type of averaging of the data available (Van Hinte, 1976b; Kauffman, 1977) to produce a time scale. We can also reverse the situation and accuse the geochronologists of failing to appreciate the complexities of biostratigraphic analysis.

BIBLIOGRAPHY

Ager, D. V., 1973, *The Nature of the Stratigraphical Record,* Macmillan, London.

Ager, D. V., 1981, *The Nature of the Stratigraphical Record, 2nd edition,* Macmillan, London.

Ager, D. V., 1984, The stratigraphic code and what it implies, in *Catastrophes and Earth History: The New Uniformitarianism.* W. A. Berggren and J. A. van Couvering, eds., Princeton University Press, Princeton, New Jersey, pp. 91–100.

Ager, D. V., 1986, A reinterpretation of the basal "Littoral Lias" of the Vale of Glamorgan, *Proc. Geol. Assoc.* 97:29–35.

Aguirre, E., 1972, Utilization of proboscideans in Pleistocene stratigraphy, in *International Colloquium on the Problem "The Boundary between Neogene and Quaternary": Collection of Papers, III,* M. N. Alekseev, E. A. Vangengeim, K. V. Nikiforova, and I. M. Khoreva, eds., INQUA, Subcommission on the Pliocene–Pleistocene and Subcommission on the Neogene Stratigraphy, Moscow, pp. 3–13.

Aguirre, E., 1981, Correlation of Neogene–Quaternary boundary in continental formations, in *Abstracts, International Field Conference—Neogene/Quaternary Boundary,* Tucson, Arizona, pp. 1–12.

Aguirre, E., and G. Pasini, 1985, The Pliocene–Pleistocene boundary, *Episodes* 8:116–120.

Albarède, F., 1982, The ^{39}Ar/^{40}Ar technique of dating, in *Numerical Dating in Stratigraphy,* G. S. Odin, ed., Wiley, Chichester, England, pp. 181–197.

Albritton, C. C., Jr., ed., 1963, *The Fabric of Geology,* Addison-Wesley, Reading, Massachusetts.

Albritton, C. C., Jr., ed., 1975, *Philosophy of Geohistory: 1785-1970* (Benchmark Papers in Geology, vol. 13), Dowden, Hutchinson & Ross, Stroudsburg, Pennsylvania.

Alverez, W., and W. Lowrie, 1978, Upper Cretaceous palaeomagnetic stratigraphy at Moria (Umbrian Apennines, Italy): verification of the Gubbio section, *Geophy. Jour. Roy. Astr. Soc.* 55:1–17.

Alverez, W., and W. Lowrie, 1981, Upper Cretaceous to Eocene pelagic limestones of the Scaglia Rossa are not Miocene turbidites, *Nature* 294:246–248.

Alverez, W., M. A. Arthur, A. G. Fischer, W. Lowrie, G. Napoleone, I. Premoli Silva, and W. M. Roggenthen, 1977, Upper Cretaceous–Paleocene magnetic stratigraphy at Gubbio, Italy: V. Type section for the Late Cretaceous–Paleocene geomagnetic reversal time scale, *Geol. Soc. Amer. Bull.* 88:383–389.

American Commission on Stratigraphic Nomenclature (ACSN), 1957, Nature, usage and nomenclature of biostratigraphic units, *Amer. Assoc. Petrol. Geol. Bull.* 41:1877–1891.

American Commission on Stratigraphic Nomenclature, 1961, Code of stratigraphic nomenclature, *Amer. Assoc. Petrol. Geol. Bull.* 45:645–665.

American Commission on Stratigraphic Nomenclature, 1970, *Code of Stratigraphic Nomenclature,* American Association of Petroleum Geologists, Tulsa, Oklahoma [Originally published in 1961, *Amer. Assoc. Petrol. Geol. Bull.* 45:645–665; amendments published in the *Amer. Assoc. Petrol. Geol. Bull.,* 1962, 46:1935; 1964, 48:710–711; 1966, 50:560–561; 1967, 51:1868–1869; 1969, 53:2005–2006].

American Geological Institute, 1976, *Dictionary of Geological Terms, revised edition,* Anchor Press/Doubleday, Garden City, New York.

Anstey, N. A., 1982, *Simple Seismics,* International Human Resources Development Corporation, Boston.

Arkell, W. J., 1933, *The Jurassic System in Great Britain,* The Clarendon Press, Oxford.

Arkell, W. J., 1956a, *Jurassic Geology of the World,* Oliver and Boyd, Edinburgh.

Arkell, W. J., 1956b, Comments on stratigraphic procedure and terminology, *Amer. Jour. Sci.* 254:457–467.

Arthur, M. A., and A. G. Fischer, 1977, Upper Cretaceous–Paleocene magnetic stratigraphy at Gubbio, Italy: I. Lithostratigraphy and sedimentology, *Geol. Soc. Amer. Bull.* 88:367–371.

Asquith, G. B., and C. R. Gibson, 1982, *Basic Well Log Analysis for Geologists,* American Association of Petroleum Geologists, Tulsa, Oklahoma.

Badgley, C., 1986, Taphonomy of mammalian fossil remains from Siwalik rocks of Pakistan, *Paleobiology* 12:119–142.

Badgley, C., L. Tauxe, and F. L. Bookstein, 1986a, Estimating the error of age interpolation in sedimentary rocks, *Nature* 319:139–141.

Badgley, C., L. Tauxe, and F. L. Bookstein, 1986b, Age interpolation, *Nature* 323:471–472.

Baird, G. C., and C. E. Brett, 1986a, Submarine erosion on the dysaerobic seafloor: Middle Devonian corrasional discontinuities in the Cayuga Valley region, in *58th Annual Meeting Field Trip Guidebook,* New York State Geological Association, Manfred P. Wolff, Executive Secretary, Geology Department, Hofstra University, Hempstead, New York, pp. 23–80.

Baird, G. C., and C. E. Brett, 1986b, Erosion on an anaerobic seafloor: significance of reworked pyrite deposits from the Devonian of New York State, *Palaeogeography, Palaeoclimatogy, Palaeoecology* 57:157–193.

Bakewell, R., 1815, *Introduction to Geology, 2nd edition,* London.

Bally, A. W., ed., 1983, *Seismic Expression of Structural Styles: A Picture and Work Atlas,* AAPG Studies in Geology Series #15, vol. 1, American Association of Petroleum Geologists, Tulsa, Oklahoma.

Bally, A. W., ed., 1987, *Atlas of Seismic Stratigraphy,* AAPG Studies in Geology Series #27, vol. 1, American Association of Petroleum Geologists, Tulsa, Oklahoma.

Barber, K. E., 1981, *Peat Stratigraphy and Climatic Change,* A. A. Balkema, Rotterdam.

Barendregt, R. W., 1984, Using paleomagnetic remanence and magnetic susceptibility data for the differentiation, relative correlation and absolute dating of Quaternary sediments, in *Quaternary Dating Methods,* W. C. Mahaney, ed., Elsevier, Amsterdam, pp. 101–122.

Barendregt, R. W., 1985, Dating methods of Pleistocene deposits and their problems: VI. Paleomagnetism, in *Dating Methods of Pleistocene Deposits and their Problems,* N. W. Rutter, ed., Geoscience Canada Reprint Series 2, Geological Association of Canada, Toronto, Ontario, pp. 39–51.

Barrell, J., 1917, Rhythms and the measurement of geologic time, *Geol. Soc. Amer. Bull.* 28:745–904.

Barrientos, X., and J. Selverstone, 1987, Metamorphosed soils as stratigraphic indicators in deformed terranes: an example from the Eastern Alps, *Geology* 15:841–844.

Bassett, M. G., 1985, Towards a "Common Language" in stratigraphy, *Episodes* 8:87–92.

Baum, G. R., 1986, Sequence stratigraphic concepts as applied to the Eocene carbonates of the Carolinas, in *Southeastern United States Third Annual Midyear Meeting,* D. A. Textoris, ed., Society of Economic Paleontologists and Mineralogists, SEPM Field Guidebook, pp. 264–269.

Baumgartner, P. O., 1984, Comparison of unitary associations and probabilistic ranking and scaling as applied to Mesozoic radiolarians, *Computers and Geosciences* 10:167–183.

Bayly, B., 1968, *Introduction to Petrology,* Prentice-Hall, Englewood Cliffs, New Jersey.

Beadle, S. C., and M. E. Johnson, 1986, Palaeoecology of Silurian cyclocrinitid algae, *Palaeontology* 29:585–601.

Bebout, D. G., and R. G. Loucks, 1984, *Handbook for Logging Carbonate Rocks,* Bureau of Economic Geology, The University of Texas at Austin, Austin.

Becquerel, H., 1896, Sur les radiations invisibles émises par les sels d'uranium, *Comptes Rendus* 122:689–694.

Bell, W. C., G. E. Murray, and L. L. Sloss, eds., 1959, Symposium on concepts of stratigraphic classification and correlation, *Amer. Jour. Sci.* 257:673–785.

Bengtson, P., 1979, A bioestratigrafia esquecida—avaliacao dos métodos bioestratigráficos no Cretáceo Médio do Brasil, *An. Acad. Brasil. Cienc.* 51:535–544.

Bengston, P., 1980, Orthography of geological names derived from fossil taxa, *Geologiska Föreningens i Stockholm Förhandlingar* 102:222.

Bengston, P., 1981, Formal and informal stratigraphical names, *Geologiska Föreningens i Stockholm Förhandlingar* 103:32.

Bengston, P., 1983, The Cenomanian–Turonian ammonite succession of Sergipe, Brazil, and the question of the stage boundary, in *Abstracts: Symposium on Cretaceous Stage Foundaries,* T. Birkelund, R. Bromley, W. K. Christensen, E. Hákansson, and F. Surlyk, eds., Institute of Historical Geology and Paleontology, Copenhagen, pp. 13–16.

Benson, R. H., 1984, Perfection, continuity, and common sense in historical geology, in *Catastrophes and Earth History: The New Uniformitarianism,* W. A. Berggren and J. A. van Couvering, eds., Princeton University Press, Princeton, New Jersey, pp. 35–75.

Benson, R. H., R. E. Chapman, and L. T. Deck, 1984, Paleoceanographic events and deep-sea ostracodes, *Science* 224:1334–1336.

Berg, O. R., and D. G. Woolverton, eds., 1985, *Seismic Stratigraphy II: An Integrated Approach to Hydrocarbon Exploration,* AAPG Memoir 39, American Association of Petroleum Geologists, Tulsa, Oklahoma.

Berger, R., and H. E. Suess, eds., 1979, *Radiocarbon Dating,* University of California Press, Berkeley.

Berger, W. H., 1981, Paleoceanography: the deep-sea record, in *The Oceanic Lithosphere, The Sea,* C. Emiliani, ed., Wiley, New York, vol. 7, pp. 1437–1519.

Berger, W. H., and E. Vincent, 1981, Chemostratigraphy and biostratigraphic correlation: exercises in systemic stratigraphy, *Oceanologica Acta* 1981:115–127.

Berger, W. H., E. Vincent, and H. R. Thierstein, 1981, The deep-sea record: major steps in Cenozoic ocean evolution. *SEPM Spec. Publ.* 32:489–504.

Berggren, W. A., 1972, A Cenozoic time-scale—some implications for regional geology and paleobiogeography, *Lethaia* 5:195–215.

Berggren, W. A., and J. A. van Couvering, 1978, Biochronology, in *Contributions to the Geologic Time Scale,* G. V. Cohee, M. F. Glaessner, and H. D. Hedberg, eds., American Association of Petroleum Geologists, Tulsa, Oklahoma, pp. 39–55.

Berggren, W. A., D. V. Kent, and J. J. Flynn, 1985a, Jurassic to Paleogene: Part 2, Paleogene geochronology and chronostratigraphy, in *The Chronology of the Geological Record,* N. J. Snelling, ed., Blackwell Scientific Publications, Oxford, pp. 141–195.

Berggren, W. A., D. V. Kent, and J. J. Flynn, 1985b, The Neogene: Part 2, Neogene geochronology and chronostratigraphy, in *The Chronology of the Geological Record,* N. J. Snelling, ed., Blackwell Scientific Publications, Oxford, pp. 211–260.

Berggren, W. A., D. V. Kent, J. J. Flynn, and J. A. van Couvering, 1985, Cenozoic geochronology, *Geol. Soc. Amer. Bull.* 96:1407–1418.

Berggren, W. A., M. C. McKenna, J. Hardenbol, and J. D. Obradovich, 1978, Revised Paleogene polarity time scale, *Jour. Geol.* 86:67–81.

Berry, W. B. N., 1966, Zones and zones—with exemplification from the Ordovician, *Bull. Amer. Assoc. Petrol. Geol.* 50:1487–1500.

Berry, W. B. N., 1968, *Growth of a Prehistoric Time Scale,* W. H. Freeman, San Francisco.

Berthenet, F., S. Clauser, and M. Renard, 1986, Geochemistry of the Fuente Caldera section (Spain), in *Terminal Eocene Events,* C. Pomerol and I. Premoli-Silva, eds., Elsevier, Amsterdam, pp. 71–74.

Blatt, H., G. Middleton, and R. Murray, 1972, *Origin of Sedimentary Rocks,* Prentice-Hall, Englewood Cliffs, New Jersey.

Blatt, H., G. Middleton, and R. Murray, 1980, *Origin of Sedimentary Rocks, 2nd edition,* Prentice-Hall, Englewood Cliffs, New Jersey.

Bleil, U., 1981, Paleomagnetism of Deep Sea Drilling Project Leg 60 sediments and igneous rocks from the Mariana region, *Init. Repts. Deep Sea Drilling Project* 60:855-873.

Bleil, U., 1985, The magnetostratigraphy of northwest Pacific sediments, Deep Sea Drilling Project Leg 86, *Init. Repts. Deep Sea Drilling Project* 86:441-458.

Bleil, U., V. Spiess, and N. Weinreich, 1984, A hiatus in early Quaternary sediments documented in the magnetostratigraphic record of "Meteor" core 13519 from the eastern equatorial Atlantic, *"Meteor" Forsch.-Ergebnisse* Reihe C, No. 38:1-7.

Bloos, G., 1982, Shell beds in the Lower Lias of South Germany—facies and origin, in *Cyclic and Event Stratification,* G. Einsele and A. Seilacher, eds., Springer-Verlag, Berlin, pp. 223-239.

Bloos, G., 1984, On Lower Jurassic ammonite stratigraphy—present state and possibilities of revision, in *International Symposium on Jurassic Stratigraphy, Erlangen, September 1-8, 1984, Symposium volume I,* O. Michelsen and A. Zeiss, eds., International Subcommision on Jurassic Stratigraphy, Geological Survey of Denmark, Copenhagen, pp. 146-157.

Blow, W. H., 1979, *The Cainozoic Globigerinida,* E. J. Brill, Leiden.

Bogdanov, N. A., general secretary, 1984, *Proceedings of the 27th International Geological Congress, Moscow 4-14 August 1984, Volume 1, Stratigraphy,* VNU Science Press, Utrecht, Netherlands.

Boggs, S., Jr., 1987, *Principles of Sedimentology and Stratigraphy,* Merrill, Columbus, Ohio.

Bolli, H. M., J. B. Saunders, and K. Perch-Nielsen, eds., 1985, *Plankton Stratigraphy,* Cambridge University Press, Cambridge.

Boltwood, B. B., 1907, On the ultimate disintegration products of the radioactive elements, part II. The disintegration products of uranium, *Amer. Jour. Sci.* 23:77-80, 86-88.

Boorstin, D. J., 1983, *The Discoverers: A History of Man's Search to Know His World and Himself,* Vintage Books (Random House), New York.

Bosinski, G., K. Brunnacker, K. Krumsiek, U. Hambach, W. Tillmanns, B. Urban-Küttel, 1985, Das Frühwürm im Lössprofil von Wallertheim/Rheinhessen, *Geol. Jb. Hessen* 113:187-215.

Boucot, A. J., 1983, Does evolution take place in an ecological vacuum? *Jour. Paleontol.* 57:1-30.

Boucot, A. J., 1984a, Constraints provided by ecostratigraphic methods on correlation of strata and basin analysis, by means of fossils, *Proceedings, 27th Intern. Geol. Congress* (published by VNU Science Press, Utrecht) 1:213-218.

Boucot, A. J., 1984b, Ecostratigraphy, in *Stratigraphy Quo Vadis?* E. Seibold and J. D. Meulenkamp, eds., American Association of Petroleum Geologists, Tulsa, Oklahoma, pp. 55-60.

Bowler, P. J., 1984, *Evolution, The History of an Idea,* University of California Press, Berkeley.

Boyle, E. A., 1986, Deep ocean circulation, preformed nutrients, and atmospheric carbon dioxide: theories and evidence from oceanic sediments, *Mesozoic and Cenozoic Oceans* (Geodynamics Series, American Geophysical Union) 15:49-59.

Brandt, D. S., 1986, Preservation of event beds through time, *Palaios* 1:92-96.

Brenchley, P. J., and L. R. M. Cocks, 1982, Ecological associations in a regressive sequence: the latest Ordovician of the Oslo-Asker District, Norway, *Palaeontology* 25:783-815.

Brenchley, P. J., and G. Newall, 1975, The stratigraphy of the Upper Ordovician stage 5 in the Olso-Asker District, Norway, *Norsk Geologisk Tidsskrift* 55:243-275.

Brenchley, P. J., and G. Newall, 1980, A facies analysis of Upper Ordovician regressive sequences in the Oslo Region, Norway—a record of glacio-eustatic changes, *Palaeogeography, Palaeoclimatology, Palaeoecology* 31:1-38.

Brenner, R. L., and T. R. McHargue, 1988, *Integrative Stratigraphy: Concepts and Applications,* Prentice-Hall, Englewood Cliffs, New Jersey.

Brett, C. E., ed., 1986, *Dynamic Stratigraphy and Depositional Environments of the Hamilton Group (Middle Devonian) in New York State, Part 1,* Bull. No. 457, New York State Museum, Albany, New York.

Brett, C. E., and G. C. Baird, 1985, Carbonate-shale cycles in the Middle Devonian of New York: an evaluation of models for the origin of limestones in terrigenous shelf sequences, *Geology* 13:324-327.

Brett, C. E., and G. C. Baird, 1986a, Symmetrical and upward shallowing cycles in the Middle Devonian of New York state and their implications for the punctuated aggradational hypothesis, *Paleoceanography* 1:431-445.

Brett, C. E., and G. C. Baird, 1986b, Comparative taphonomy: A key to paleoenvironmental interpretation based on fossil preservation, *Palaios* 1:207-227.

Brewer, R., K. A. W. Crook, and J. A. Speight, 1970, Proposal for soil-stratigraphic units in the Australian stratigraphic code, *Jour. Geol. Soc. Australia* 17:103-111.

Brigham-Grette, J., 1987, Current trends in Quaternary geochronology, *Episodes* 10:43-44.

Brongniart, A., 1823, *Mémoire sur les terrains de sédiment supérieurs calcaréo-trappéens du Vicentin, et sur quelques terrains d'Italia, de France, d'Allemagne, ets.,* Paris.

Brongniart, A., 1829, *Tableau des Terrains qui composent l'Ecorce du Globe, ou Essai sur la structure de la partie connue de la Terre,* Paris and Strasbourg.

Bronn, H. G., 1831, *Italiens Tertiär-Gebilde und deren organische Einschlüsse,* Heidelberg.

Brookes, I. A., 1985, Dating methods of Pleistocene deposits and their problems: VIII. Weathering, in *Dating Methods of Pleistocene Deposits and Their Problems,* N. W. Rutter, ed., Geoscience Canada Reprint Series 2, Geological Association of Canada, Toronto, Ontario, pp. 61-71.

Brown, L. F., Jr., and W. L. Fisher, 1980, *Seismic Stratigraphic Interpretation and Petroleum Exploration,* Continuing Education Course Note Series #16, American Association of Petroleum Geologists, Tulsa, Oklahoma.

Brunhes, B., 1906, Recherches sur le direction d'aimantation des roches volcaniques, *Jour. Physique* 5:705-724.

Brunnacker, K., H. Butzke, H.-D. Dahm, H. Dahm-Arens, H.-J. Dubber, F.-D. Erkwoh, H. Mertens, E. Mückenhausen, W. Paas, J. Schalich, K. Skupin, K.-H. Will, W. Wirth, and E. von Zezschwitz, 1982, Paläoböden in Nordrhein–Westfalen [Paleosols in North Rhine–Westphalia], *Geol. Jb.* F14:165-253.

Brunnacker, K., B. Urban, and W. A. Schnitzer, 1977, Der jungpleistozöne Löss am Mittel- und Neiderrhein anhand neuer Untersuchungsmethoden [The younger Pleistocene loess in the middle and lower Rhine valley on the basis of new investigation methods], *N. Jb. Geol. Paläont. Abh.* 155:253-273.

Brush, S. G., 1983, Ghosts from the nineteenth century: creationist arguments for a young earth, in *Scientists Confront Creationism,* L. R. Godfrey, ed., Norton, New York, pp. 49-84.

Bruton, D. L., ed., 1984, *Aspects of the Ordovician System,* Palaeontological Contributions from the University of Oslo, No. 295, Universtetsforlaget, Oslo.

Buckman, S. S., 1893, The Bajocian of the Sherborne District: its relation to subjacent and superjacent strata, *Geol. Soc. London Quart. Jour.* 49:479-522.

Buckman, S. S., 1898, On the grouping of some divisions of so-called "Jurassic" time, *Geol. Soc. London Quart. Jour.* 54:422-462.

Buckman, S. S., 1902, The term "hemera," *Geol. Mag.* 9:554-557.

Buffon, G., 1797, *Buffon's Natural History volume 1,* H. D. Symonds, London.

Burckle, L. H., J. J. Morley, I. Koizumi, and U. Bleil, 1985, Assessment of diatom and radiolarian high- and low-latitude zonations in northwest Pacific sediments: comparison based upon magnetostratigraphy, *Init. Repts. Deep Sea Drilling Project* 86:781-785.

Busch, D. A., 1974, *Stratigraphic Traps in Sandstones—Exploration Techniques,* American Association of Petroleum Geologists, Tulsa, Oklahoma.

Busch, R. M., 1983, Sea level correlation of punctuated aggradational cycles (PACs) of the Manlius Formation, central New York, *Northeastern Geology* 5:82–91.

Busch, R. M., and H. B. Rollins, 1984, Correlation of Carboniferous strata using a hierarchy of transgressive–regressive units, *Geology* 12:471–474.

Busch, R. M., and R. R. West, 1987, Hierarchical genetic stratigraphy: a framework for paleoceanography, *Paleoceanography* 2:141–164.

Busch, R. M., R. R. West, F. J. Barrett, and T. R. Barrett, 1985, Cyclothems versus a hierarchy of transgressive–regressive units, in *Recent Interpretations of Late Paleozoic Cyclothems,* W. L. Watney, R. L. Kaesler, and K. D. Newell, eds., Society of Economic Paleontologists and Mineralogists, Mid-Continent Section, Lawrence, Kansas, pp. 141–153.

Butzer, K. W., 1971, *Environment and Archeology, 2nd edition,* Aldine, Chicago.

Calkin, P. E., and J. M. Ellis, 1984, Development and application of a lichenometric dating curve, Brooks Range, Alaska, in *Quaternary Dating Mehtods,* W. C. Mahaney, ed., Elsevier, Amsterdam, pp. 227–246.

Callomon, J. H., and D. T. Donovan, 1966, Stratigraphic classification and terminology, *Geol. Mag.* 103:97–99.

Cant, D. J., 1984, Subsurfacc facies analysis, in *Facies Models, Second Edition,* R. G. Walker, ed., Geoscience Canada, Reprint Series 1, Geological Association of Canada, Toronto, Ontario, pp. 287–310.

Carimati, R., A. Marini, and R. G. Potenza, 1982, The mathematical formalization of the geological relations identifying the basic structure of a geological data bank, in *Quantitative Stratigraphic Correlation,* J. M. Cubitt and R. A. Reyment, eds., Wiley, Chichester, England, pp. 13–18.

Cassignol, C., and P.-Y. Gillot, 1982, Range and effectiveness of unspiked potassium–argon dating: experimental groundwork and applications, in *Numerical Dating in Stratigraphy,* G. S. Odin, ed., Wiley, Chichester, England, pp. 159–179.

Caster, K. E., 1934, The stratigraphy and paleontology of northwestern Pennsylvania, *Bull. Amer. Paleontol.* 21:1–185.

Catacosinos, P. A., 1968, Upper Cretaceous–Lower Tertiary relations west of Raven Ridge, Uintah County, Utah, *Amer. Assoc. Petrol. Geol. Bull.* 52:343–348.

Čepek, P., A. Köthe, and W. Weiss, 1985, *Paleogeographic Evolution of the Atlantic Ocean during the Late Cretaceous,* Geologisches Jahrbuch (Reihe B, Heft 62), Hanover.

Chamberlin, T. C., 1898, The ulterior basis of time divisions and the classification of geologic history, *Jour. Geol.* 6:449–462.

Chamberlin, T. C., 1909, Diastrophism as the ultimate basis of correlation, *Jour. Geol.* 17:685–693.

Chamberlin, T. C., and R. D. Salisbury, 1909, *A College Text-Book of Geology,* Henry Holt, New York.

Chang, K. H., 1975, Unconformity-bounded stratigraphic units, *Geol. Soc. Amer. Bull.* 86:1544–1552.

Channell, J. E. T., Palaeomagnetic stratigraphy as a correlation technique, in *Numerical Dating in Stratigraphy,* G. S. Odin, ed., Wiley, Chichester, England, pp. 81–103.

Channell, J. E. T., and F. Medizza, 1981, Upper Cretaceous and Palaeogene magnetic stratigraphy and biostratigraphy from the Venetian (Southern) Alps, *Earth Planet. Sci. Let.* 55:419–432.

Channell, J. E. T., W. Lowrie, and F. Medizza, 1979, Middle and Early Cretaceous magnetic stratigraphy from the Cismon section, northern Italy, *Earth Planet. Sci. Let.* 42:153–166.

Channell, J. E. T., W. Lowrie, P. Pialli, and F. Venturi, 1984, Jurassic magnetic stratigraphy from Umbrian (Italian) land sections, *Earth Planet. Sci. Let.* 68: 309–325.

Channell, J. E. T., J. G. Ogg, and W. Lowrie, 1982, Geomagnetic polarity in the early Cretaceous and Jurassic, *Phil. Trans. Roy. Soc. London* A 306:137–146.

Childs, O. E., 1985, Correlation of stratigraphic units of North America—COSUNA, *AAPG Bull.* 69:173–180.

Cisne, J. L., and B. D. Rabe, 1978, Coenocorrelation: gradient analysis of fossil communities and its applications in stratigraphy, *Lethaia* 11:341–364.

Cisne, J. L., G. O. Chandlee, B. D. Rabe, and J. A. Cohen, 1982, Clinal variation, episodic evolution, and possible parapatric speciation: the trilobite *Flexicalymene senaria* along an Ordovician depth gradient, *Lethaia* 15:325–341.

Clain-Stefanelli, E. E., 1965, Numismatics: an ancient science, *Contr. Mus. Hist. Technol., Smithsonian Inst.,* Paper 32:1–102.

Cloetingh, S., 1986, Intraplate stresses: a new tectonic mechanism for fluctuations of relative sea level, *Geology* 14:617–620.

Cloetingh, S., H. McQueen, and K. Lambeck, 1985, On a tectonic mechanism for regional sealevel variations, *Earth Planet. Sci. Let.* 75:157–166.

Coates, D. R., 1984, Landforms and landscapes as measures of relative time, in *Quaternary Dating Methods,* W. C. Mahaney, ed., Elsevier, Amsterdam, pp. 247–267.

Cocks, L. R. M., 1985, The Ordovician–Silurian boundary, *Episodes* 8:98–100.

Cogley, N. G., 1981, Late Phanerozoic extent of dry land, *Nature* 291:56–58.

Cohee, G. V., M. F. Glaessner, and H. D. Hedberg, eds., 1978, *Contributions to the Geologic Time Scale,* Studies in Geology No. 6, American Association of Petroleum Geologists, Tulsa, Oklahoma.

Cohen, C. R., 1982, Model for a passive to active continental margin transition: implications for hydrocarbon exploration, *Amer. Assoc. Petrol. Geol. Bull.* 66:708–718.

Collinson, D. W., 1983, *Methods in Rock Magnetism and Palaeomagnetism: Techniques and Instrumentation,* Chapman and Hall, London.

Compton, R. R., 1962, *Manual of Field Geology,* Wiley, New York.

Compton, R. R., 1985, *Geology in the Field,* Wiley, New York.

Conkin, B. M., and J. E. Conkin, eds., 1984, *Stratigraphy: Foundations and Concepts,* Van Nostrand Reinhold, New York.

Conkin, J. E., 1985, Late Devonian New Albany-Ohio-Chattanooga shales and their inter-basinal correlation in Indiana, Ohio, Kentucky, and Tennessee, *1984 Eastern Shale Symposium,* Commonwealth of Kentucky, Kentucky Energy Cabinet, pp. 217–259 [reprinted as *Univ. Louisville Notes Paleont. Stratigr. C*].

Conkin, J. E., 1986, Chattanooga shale in the Tennessee valley and ridge of Hamilton and Bledsoe counties, *Univ. Louisville Notes Paleont. Stratigr. A,* 11 pp.

Conkin, J. E., and B. M. Conkin, 1973, The paracontinuity and the determination of the Devonian–Mississippian boundary in the type Lower Mississippian area of North America, *Univ. Louisville Stud. Paleont. Stratigr.,* no. 1, 36 pp.

Conkin, J. E., and B. M. Conkin, 1975, The Devonian–Mississippian and Kinderhookian-Osagean boundaries in the east-central United States are paracontinuities, *Univ. Louisville Stud. Paleont. Stratigr.,* no. 4, vii + 56 pp.

Conkin, J. E., and B. M. Conkin, eds., 1979, *Devonian–Mississippian Boundary in Southern Indiana and Northwestern Kentucky,* Field Trip 7/Ninth International Congress on Carboniferous Stratigraphy and Geology, University of Louisville and Bloomington Crushed Stone, Bloomington, Indiana [available from The American Geological Institute, Falls Church, Virginia].

Conkin, J. E., and B. M. Conkin, 1983, Paleozoic metabentonites of North America: Part 2.—Metabentonites in the Middle Ordovician Tyrone Formation at Boonesborough, Clark County, Kentucky, *Univ. Louisville Stud. Paleont. Stratigr.,* no. 17, iv + 48 pp.

Conkin, J. E., and B. M. Conkin, 1984a, Devonian and Mississippian bone beds, paracontinuities, and pyroclastics, and the Silurian–Devonian paraconformity in southern Indiana and northwestern Kentucky, in *Field Trip Guides for Geological Society of America Annual Meeting, Southeastern and North-Central Sections, April 4–6, 1984, Lexington, Kentucky,* N. Rast and H. Hay, eds., University of Kentucky and Kentucky Geological Survey, Lexington, Kentucky, pp. 25–42.

Conkin, J. E., and B. M. Conkin, 1984b, Paleozoic metabentonites of North America: Part

1.—Devonian metabentonites in the eastern United States and southern Ontario: their identities, stratigraphic positions, and correlation, *Univ. Louisville Stud. Paleont. Stratigr.*, no. 16, vi + 136 pp.

Conkin, J. E., and B. M. Conkin, 1985, Cratonic Devonian–Mississippian paracontinuous boundary in the eastern United States, in *Dixième Congress International de Stratigraphie et de Géologie du Carbonifère,* Instituto Geológico y Minero de Espanña, Madrid, pp. 489–504.

Conkin, J. E., B. M. Conkin, and L. Z. Lipchinsky, 1980, Devonian black shale in the eastern United States: Part 1—Southern Indiana, Kentucky, northern and eastern Highland Rim of Tennessee, and central Ohio, *Univ. Louisville Stud. Paleont. Stratigr.,* no. 12, iii + 63 pp.

Conkin, J. E., B. M. Conkin, M. M. Walton, and E. D. Neff, 1981, Devonian and Early Mississippian smaller foraminiferans of southern Indiana and northwestern Kentucky, *Geol. Soc. Amer., Field Trip No. 2, Annual Meeting, Cincinnati,* pp. 87–112.

Conkin, J. E., C. Layton, and B. M. Conkin, 1983, Masonry carbonate stones of Kentucky, with emphasis on the Middle Silurian Louisville limestone, in *Fourth International Congress on the Deterioration and Preservation of Stone Objects, Proceedings,* K. L. Gauri, and J. A. Gwinn, eds., The University of Louisville, Louisville, Kentucky, pp. 109–118.

Conway Morris, S., 1987, The search for the Precambrian–Cambrian boundary, *Amer. Sci.* 75:156–167.

Conybeare, C. E. B., 1979, *Lithostratigraphic Analysis of Sedimentary Basins,* Academic Press, New York.

Cotillon, P., 1984, Tentative world-wide correlation of Early Cretaceous strata by limestone-marl cyclicites in pelagic deposits, *Bull. Geol. Soc. Denmark* 33:91–102.

Cotillon, P., 1985, Les variations à différentes échelles du taux d'accumulation sédimentaire dans les séries pélagiques alternantes du Crétacé inférieur, conséquences de phénomènes globaux; essai d'évaluation, *Bull. Soc. géol. France* (8) 1:59–68.

Cotillon, P., and M. Rio, 1984, Cyclic sedimentation in the Cretaceous of Deep Sea Drilling Project Sites 535 and 540 (Gulf of Mexico), 534 (Central Atlantic), and in the Vocontian Basin (France), *Init. Repts. Deep Sea Drilling Project* 77:339–376.

Cotillon, P., S. Ferry, C. Gaillard, E. Jautée, G. Latreille, and M. Rio, 1980, Fluctuation des paramètres du milieu marin dans le domaine vocontien (France Sud-Est) au Crétacé inférieur: mise en évidence par l'étude des formations marno-calcaires alternantes, *Bull. Soc. géol. France* (7) 22:735–744.

Courtillot, V., and J. Besse, 1987, Magnetic field reversals, polar wander, and core-mantle coupling, *Science* 237:1140–1147.

Cowie, J. W., 1985a, Stratigraphy and the International Commission, *Episodes* 8:86.

Cowie, J. W., 1985b, Continuing work on the Precambrian–Cambrian boundary, *Episodes* 8:93–97.

Cowie, J. W., 1986, Guidelines for boundary stratotypes, *Episodes* 9:78–82.

Cowie, J. W., and M. R. W. Johnson, 1985, Late Precambrian and Cambrian geological time-scale, in *The Chronology of the Geological Record,* N. J. Snelling, ed., Blackwell Scientific Publications, Oxford, pp. 47–64.

Cowie, J. W., W. Ziegler, A. J. Boucot, M. G. Bassett, and J. Remane, 1986, Guidelines and statutes of the International Commission on Stratigraphy (ICS), *Cour. Forsch,-Inst. Senckenberg* 83:1–14.

Cox, A., 1969, Geomagnetic reversals, *Science* 163:237–245.

Cox, L. R., 1948, William Smith and the birth of stratigraphy, *International Geological Congress, Eighteenth Session, Great Britain,* pamphlet printed by Wightman, London, 8 pp.

Crosby, W. O., 1912, Dynamic relations and terminology of stratigraphic conformity and unconformity, *Jour. Geol.* 20:289–299.

Cubitt, J. M., and R. A. Reyment, eds., 1982, *Quantitative Stratigraphic Correlation,* Wiley, Chichester, England.

Cummins, H., E. N. Powell, H. J. Newton, R. J. Stanton, Jr., and G. Staff, 1986, Assessing transportation by the covariance of species with comments on contagious and random distributions, *Lethaia* 19:1–22.

Cummins, H., E. N. Powell, R. J. Stanton, Jr., and G. Staff, 1986, The rate of taphonomic loss in modern benthic habitats: how much of the potentially preservable community is preserved? *Palaeogeography, Palaeoclimatology, Palaeoecology* 52:291–320.

Curray, J. R., 1964, Transgressions and regressions, in *Papers in Marine Geology,* R. L. Miller, ed., Macmillan, New York, pp. 175–203.

Cuvier, G., 1813, *Essay on the Theory of the Earth,* William Blackwood, Edinburgh.

Cuvier, G., and A. Brongniart, 1808, Essai sur la géographie minéralogie des environs de Paris, *Ann. Mus. Hist. Nat. Paris* 11:293–326.

Cuvier, G., and A. Brongniart, 1822, Description géologiques des couches des environs de Paris, parmi lesquelles se trouvent les gypses a ossemens, in *Récherches sur les Ossemens Fossiles,* G. Cuvier, ed., G. Dufour and E. d'Ocagne, Paris, vol. 2, pp. 239–648.

Damon, P. E., J. C. Lerman, and A. Long, 1978, Temporal fluctuations of atmospheric ^{14}C: causal factors and implications, *Ann. Rev. Earth Planet. Sci.* 6:457–494.

Darwin, C., 1859, *On the Origin of Species by Means of Natural Selection, or the Preservation of Favored Races in the Struggle for Life,* John Murray, London.

Davaud, E., 1979, Automatisation des correlations biochronologiques: un exemple d'application de l'informatique a la resolution d'un problème naturaliste complexe, *Dept. Geol. Univ. Genève, Publ. Spec.* 1979:1–48.

Davaud, E., 1982, The automation of biochronological correlation, in *Quantitative Stratigraphic Correlation,* J. M. Cubitt and R. A. Reyment, eds., Wiley, Chichester, England, pp. 85–99.

Davies, G. H., 1981, The history of the earth sciences, in *The Cambridge Encyclopedia of Earth Sciences,* D. G. Smith, ed., Crown Publishers and Cambridge University Press, New York, pp. 12–23.

Davis, A. M., 1984, Dating with pollen: methodology, applications, limitations, in *Quaternary Dating Methods,* W. C. Mahaney, ed., Elsevier, Amsterdam, pp. 283–297.

Davis, R. A., Jr., 1983, *Depositional Systems: A Genetic Approach to Sedimentary Geology,* Prentice-Hall, Englewood Cliffs, New Jersey.

Davis, T. L., 1984, Seismic-stratigraphic models, in *Facies Models, Second Edition,* R. G. Walker, ed., Geoscience Canada, Reprint Series 1, Geological Association of Canada, Toronto, Ontario, pp. 311–317.

Davis, W. E., Jr., 1974, *Early History of the Nuclear Atom,* Kendall/Hunt Publishing Co., Dubuque, Iowa.

Day, M. H., 1986, *Guide to Fossil Man, 4th edition,* University of Chicago Press, Chicago.

De Boer, P. L., 1983, Aspects of Middle Cretaceous pelagic sedimentation in southern Europe, *Geol. Ultraiectina, Utrecht* 31:1–112.

De Boer, P. L., 1986, Changes in the organic carbon burial during the Early Cretaceous, in *North Atlantic Palaeoceanography,* C. P. Summerhayes and N. J. Shackleton, eds., Geological Society Special Publication no. 21, pp. 321–331.

De Boer, P. L., and A. A. H. Wonders, 1984, Astronomically induced rhythmic bedding in Cretaceous pelagic sediments near Moria (Italy), in *Milankovitch and Climate: Understanding the Response to Astronomical Forcing,* A. Berger, J. Imbrie, J. Hays, G. Kukla, and B. Saltzman, eds., D. Reidel, Dordrecht, Netherlands, part 1, pp. 177–190.

Debus, A. G., ed., 1968, *World Who's Who in Science,* Marquis Who's Who, Chicago.

De Geer, G., 1940, Geochronologia Suecica principles, *K. Svenska Vetensk. Handl. (ser. 3)* vol. 18, no. 6:[cited from Butzer, 1971].

De La Beche, H. T., 1839, *Report on the Geology of Cornwall, Devon and West Somerset,* Longman, and the Geological Survey of England and Wales, London.

Deshayes, G. P., 1830, Tableau comparatif des espèces de coquilles vivantes avec les espèces de coquilles fossiles des terraines tertiares de l'Europe, et des espèces de fossiles de ces terrains entr'eux, *Soc. Géol. Fr. Bull.* 1:185–187.

Desmond, A. J., 1975, The discovery of marine transgression and the explanation of fossils in antiquity, *Amer. Jour. Sci.* 275:692–707.

De Vernal, A., C. Causse, C. Hillaire-Marcel, R. J. Mott, and S. Occhietti, 1986, Palynostratigraphy and Th/U ages of upper Pleistocene interglacial and interstadial deposits on Cape Breton Island, eastern Canada, *Geology* 14:554–557.

Dienes, I., 1978, Methods of plotting temporal range charts and their application in age estimation, *Computers and Geosciences* 4:269–272.

Dienes, I., 1981, The establishment of optimal time scales and their use, *Acta Geologica Academiae Scientiarum Hungaricae* 24:395–412.

Dienes, I., 1982, Formalized Eocene stratigraphy of Dorog Basin, Transdanubia, Hungary, and related areas, in *Quantitative Stratigraphic Correlation,* J. M. Cubitt and R. A. Reyment, eds., Wiley, Chichester, England, pp. 19–42.

Dienes, I., 1983, Experience with comparison of different geological clocks, *Acta Geologica Hungarica* 26:187–195.

Dienes, I., and L. B. Kovács, 1976, Maximum transitive paths and their application to a geological problem: setting up stratigraphic units, in *Survey of Mathematical Programming* (Proceedings of the 9th International Mathematical Programming Symposium, Budapest, August 23–27, 1976), A. Prékopa, ed., Publishing House of the Hungarian Academy of Sciences, Budapest, pp. 441–454.

Dienes, I., and C. J. Mann, 1977, Mathematical formalization of stratigraphic terminology, *Math. Geol.* 9:587–603.

Dineley, D. L., 1984, *Aspects of a Stratigraphic System: The Devonian,* Wiley (Halsted), New York.

Dingus, L., and P. M. Sadler, 1982, The effects of stratigraphic completeness on estimates of evolutionary rates, *Syst. Zool.* 31:400–412.

D'Iorio, M. A., 1986, Integration of foraminiferal and dinoflagellate data sets in quantitative stratigraphy of the Grand Banks and Labrador shelf, *Bull. Canadian Petroleum Geology* 34:277–283.

Dodd, J. R., and R. J. Stanton, Jr., 1981, *Paleoecology, Concepts and Applications,* Wiley, New York.

Domenico, P. A., 1972, *Concepts and Models in Groundwater Hydrology,* McGraw-Hill, New York.

Donovan, D. T., 1966, *Stratigraphy: An Introduction to Principles,* Rand McNally, Chicago; Thomas Murby, London.

D'Orbigny, A., 1842–49, *Paléontologie Francais, Terrains Jurassiques,* Victor Masson, Paris.

D'Orbigny, A., 1849–52, *Cours élémentaire de Paléontologie et de Géologie stratigraphique,* Victor Masson, Paris.

Dott, R. H., Jr., 1983, Episodic sedimentation—How normal is average? How rare is rare? Does it matter? *Jour. Sed. Petrol.* 53:5–23.

Dott, R. H., Jr., and R. L. Batten, 1976, *Evolution of the Earth, 2nd edition,* McGraw-Hill, New York.

Doyle, J. A., S. Jardiné and A. Doerenkamp, 1982, *Afropollis,* a new genus of early angiosperm pollen, with notes on the Cretaceous palynostratigraphy and paleoenvironments of northern Gondwana, *Bull. Centres Resh. Explor.-Prod. Elf-Aquitaine* 6:39–117.

Dreimanis, A., G. Hütt, A. Raukas, and P. W. Whippey, 1985, Dating methods of Pleistocene deposits and their problems: I. Thermoluminescence dating, in *Dating Methods of Pleistocene Deposits and Their Problems,* N. W. Rutter, ed., Geoscience Canada Reprint Series 2, Geological Association of Canada, Toronto, Ontario, pp. 1–7.

Dresser Atlas, 1982, *Well Logging and Interpretation Techniques,* Dresser Atlas Publications, Houston.

Dresser Atlas, 1983, *Log Interpretation Charts,* Dresser Atlas Publications, Houston.

Drooger, C. W., 1974, The boundaries and limits of stratigraphy, *Koninkl. Nederl. Akad. Wetensch. Proc. B* 77:159-167.

Drury, S. A., 1981, The Archaean Eon and before, in *The Cambridge Encyclopedia of Earth Sciences,* D. G. Smith, ed., Crown Publishers and Cambridge University Press, New York, pp. 251-261.

Dunbar, C. O., and J. Rodgers, 1957, *Principles of Stratigraphy,* Wiley, New York.

Eder, F. W., W. Engel, W. Franke, and P. M. Sadler, 1983, Devonian and Carboniferous limestone-turbidites of the Rheinisches Schiefergebirge and their tectonic significance, in *Intracontinental Fold Belts,* H. Martin and F. W. Eder, eds., Springer-Verlag, Berlin, pp. 93-124.

Edwards, L. E., 1978, Range charts and no-space graphs, *Computers and Geoscience* 4:247-255.

Edwards, L. E., 1982a, Numerical and semi-objective biostratigraphy: review and predictions, *North American Paleontol. Conv., 3rd Proc.* 1: 147-152.

Edwards, L. E., 1982b, Quantitative biostratigraphy: the methods should suit the data, in *Quantitative Stratigraphic Correlation,* J. M. Cubitt and R. A. Reyment, eds., Wiley, Chichester, England, pp. 45-60.

Edwards, L. E., 1984, Insights on why graphic correlation (Shaw's method) works, *Jour. Geol.* 92:583-597.

Edwards, L. E., and R. J. Beaver, 1978, The use of a paired comparison model in ordering stratigraphic events, *Math. Geol.* 10:261-272.

Ehlers, E. G., and H. Blatt, 1982, *Petrology: Igneous, Sedimentary, and Metamorphic,* W. H. Freeman, San Francisco.

Ehrlich, R., and I. Lerche, 1986, Age interpolation, *Nature* 323:471.

Eicher, D. L., 1968, *Geologic Time,* Prentice-Hall, Englewood Cliffs, New Jersey.

Einsele, G., and A. Seilacher, eds., 1982a, *Cyclic and Event Stratification,* Springer-Verlag, Berlin.

Einsele, G., and A. Seilacher, 1982b, Paleogeographic significance of tempestites and periodites, in *Cyclic and Event Stratification,* G. Einsele and A. Seilacher, eds., Springer-Verlag, Berlin, pp. 531-536.

Eldredge, N., 1975, Review of *Evolution and Extinction Rate Controls,* by A. J. Boucot, *Syst. Zool.* 24:389-391.

Eldredge, N., and S. J. Gould, 1972, Punctuated equilibria: an alternative to phyletic gradualism, in *Models in Paleobiology,* T. J. M. Schopf, ed., Freeman, Cooper, San Francisco, pp. 82-115.

Eldredge, N., and S. M. Stanley, eds., 1984, *Living Fossils,* Springer-Verlag, New York.

Elliott, D. K., ed., 1986, *Dynamics of Extinction,* Wiley, New York.

Emery, G. T., 1972, Perturbations of nuclear decay rates, *Ann. Rev. Nuclear Sci.* 22:165-202.

Engel, W., W. Franke, C. Grote, K. Weber, H. Ahrendt, and F. W. Eder, 1983, Nappe tectonics in the southeastern part of the Rheinisches Schiefergebirge, in *Intracontinental Fold Belts,* H. Martin, and F. W. Eder, eds., Springer-Verlag, Berlin, pp. 267-287.

Erben, H. K., 1972, Replies to opposing statements, *Newsl. Stratigr.* 2:79-95.

Erben, H. K., 1973, Correction of a reaction, *Newsl. Stratigr.* 2:185-188.

Eskola, P., 1915, Om Sambandet mellan kemisk och mineralogisk sammansättning hos Orijärvitraktens metamorfa bergarter, *Commission Géol. Finlande Bull.* 44:1-45.

Eskola, P., 1922, The mineral facies of rocks, *Norsk Geol. Tidsskrift Bd.* 6:142-194.

Evander, R. L., 1986, Formal redefinition of the Hemingfordian-Barstovian Land Mammal Age, *Jour. Vertebrate Paleontol.* 6:374-381.

Evans, L. J., 1985, Dating methods of Pleistocene deposits and their problems: VII. Paleosols, in *Dating Methods of Pleistocene Deposits and Their Problems,* N. W. Rutter, ed., Geoscience Canada Reprint Series 2, Geological Association of Canada, Toronto, Ontario, pp. 53-59.

Evernden, J. R., and A. K. Evernden, 1970, The Cenozoic time scale, *Geol. Soc. Amer. Spec. Pap.* 124:71–90.

Evernden, J. R., D. E. Savage, G. H. Curtis, and G. T. James, 1964, Potassium–argon dates and the Cenozoic mammalian chronology of North America, *Amer. Jour. Sci.* 262:145–198.

Eyles, V. A., 1969, The extent of geological knowledge in the eighteenth century and the methods by which it was diffused, in *Toward a History of Geology,* C. J. Schneer, ed., M.I.T. Press, Cambridge, Massachusetts, pp. 159–183.

Fairbridge, R. W., 1961, Eustatic changes in sea-level, in *Physics and Chemistry of the Earth, volume 4,* L. H. Ahrens, ed., Pergamon Press, London, pp. 99–185.

Fåhraeus, L. E., 1977a, Isocommunities and correlation of the North American *Didymograptus bifidus* Zone (Ordovician), *Newsl. Stratigr.* 6:85–96.

Fåhraeus, L. E., 1977b, Correlation of the Canadian/Champlainian Series boundary and the Whiterock Stage of North America with western European conodont and graptolite zones, *Bull. Canadian Petroleum Geology* 25:981–994.

Fåhraeus, L. E., 1986, Spectres of biostratigraphic resolution and precision: rock accumulation rates, processes of speciation and paleoecological constraints, *Newsl. Stratigr:* 15:150–162.

Fåhraeus, L. E., and D. R. Hunter, 1981, Paleoecology of selected species from the Cobbs Arm Formation (middle Ordovician), New World Island, north-central Newfoundland, *Canadian Jour. Earth Sci.* 18:1653–1665.

Faure, G., 1977, *Principles of Isotope Geology,* Wiley, New York.

Faure, G., 1982, The marine-strontium geochronometer, in *Numerical Dating in Stratigraphy,* G. S. Odin, ed., Wiley, Chichester, England, pp. 73–79.

Fenner, J., 1985, Late Cretaceous to Oligocene planktic diatoms, in *Plankton Stratigraphy,* H. M. Bolli, J. B. Saunders, and K. Perch-Nielsen, eds., Cambridge University Press, Cambridge, pp. 713–762.

Finkl, C. W., Jr., 1984, Evaluation of relative pedostratigraphic dating methods, with special reference to Quaternary successions overlying weathered platform materials, in *Quaternary Dating Methods,* W. C. Mahaney, ed., Elsevier, Amsterdam, pp. 323–353.

Firth, J. V., S. Srivistava, M. Arthur, B. Clement, A. Aksu, J. Baldauf, G. Bohrmann, W. Bush, T. Cederberg, M. Cremer, K. Dady, A. Devernal, F. Hall, M. Head, R. Hiscott, R. Jarrard, M. Kaminski, D. Lazarus, A. Monjanel, O. Nielsen, R. Stein, F. Thiebault, J. Zachos, and H. Zimmerman, 1987, Paleontologic and geophysical correlations in Baffin Bay and the Labrador Sea: ODP Leg 105, in *Timing and Depositional History of Eustatic Sequences: Constraints on Seismic Stratigraphy,* C. A. Ross and D. Haman, eds., Cushman Foundation for Foraminiferal Research, Special Publication 24, Washington, D.C., pp. 1–6.

Fischer, A. G., 1980, Gilbert—bedding rhythms and geochronology, *Geol. Soc. Amer. Spec. Pap.* 183:93–104.

Fitton, W. H., 1827, Observations on some of the strata between the Chalk and the Oxford Oolite, in the southeast of England, *Geol. Soc. Lond. Trans. ser.* 2. 4:104–400.

Flemming, N. C., and D. G. Roberts, 1973, Tectono-eustatic changes in sea-level and sea-floor spreading, *Nature* 243: 19–22.

Flish, M., 1982, Potassium–argon analysis, in *Numerical Dating in Stratigraphy,* G. S. Odin, ed., Wiley, Chichester, England, pp. 151–158.

Fortey, R. A., and L. R. M. Cocks, 1986a, Fossils and tectonics, *Jour. Geol. Soc. Lond.* 143:149–150.

Fortey, R. A., and L. R. M. Cocks, 1986b, Marginal faunal belts and their structural implications, with examples from the Lower Palaeozoic, *Jour. Geol. Soc. Lond.* 143:151–160.

Francis, T. J. G., 1971, Effect of earthquakes on deep-sea sediments, *Nature* 233:98–102.

Franke, W., W. Eder, and W. Engel, 1975, Sedimentology of a Lower Carboniferous shelf-margin (Velbert Anticline, Rheinisches Schiefergebirge, W-Germany), *N. Jb. Geol. Paläont. Abh.* 150:314–353.

Franke, W., W. Eder, W. Engel, and F. Langenstrassen, 1978, Main aspects of geosynclinal sedimentation in the Rhenohercynian Zone, *Z. dt. geol. Ges.* 129:201–216.

Frebold, H., 1924, Ammonitenzonen und Sedimentationszyklen in ihrer Beziehung zueinander, *Centrb. für Mineralogie* 1924:313–320.

Fritz, W. J., and J. N. Moore, 1988, *Basics of Physical Stratigraphy and Sedimentology,* Wiley, New York.

Fuller, M. D., I. S. Williams, and K. A. Hoffman, 1979, Paleomagnetic records of geomagnetic field reversals and the morphology of the transitional fields, *Rev. Geophys. Space Phys.* 17:179–203.

Gale, N. H., 1982, The physical decay constants, in *Numerical Dating in Stratigraphy,* G. S. Odin, ed., Wiley, Chichester, England, pp. 107–122.

Galbrun, B., Magnetostratigraphy of the Berriasian stratotype section (Berrias, France), *Earth Planet. Sci. Let.* 74:130–136.

Galbrun, B., L. Rasplus, and G. Le Hégarat, 1986, Données nouvelles sur le stratotype du Berriasien: corrélations entre magnétostratigraphie et biostratigraphie, *Bull. Soc. géol. France* (8) 2:575–584.

Galloway, W. E., C. D. Henry, and G. E. Smith, 1982, *Depositional Framework, Hydrostratigraphy, and Uranium Mineralization of the Oakville Sandstone (Miocene), Texas Coastal Plain,* Report of Investigations No. 113, Bureau of Economic Geology, The University of Texas at Austin, Austin.

Geikie, A., 1897, *The Founders of Geology,* Macmillan, London.

George, T. N., W. B. Harland, D. V. Ager, H. W. Ball, W. H. Blow, R. Casey, C. H. Holland, N. F. Hughes, G. A. Kellaway, P. E. Kent, W. H. C. Ramsbottom, J. Stubblefield, and A. W. Woodland, 1967, The stratigraphical code—report of the stratigraphical code subcommittee, *Proc. Geol. Soc. London* 1638:75–87.

George, T. N., W. B. Harland, D. V. Ager, H. W. Ball, W. H. Blow, R. Casey, C. H. Holland, N. F. Hughes, G. A. Kellaway, P. E. Kent, W. H. C. Ramsbottom, J. Stubblefield, and A. W. Woodland, 1969, Recommendations on stratigraphical usage, *Proc. Geol. Soc. London* 1656:139–166.

Gignoux, M., 1955, *Stratigraphic Geology,* English translation from the fourth French edition, 1950, by G. G. Woodford, W. H. Freeman, San Francisco.

Gilbert, G. K., 1895, Sedimentary measurement of geologic time, *Jour. Geol.* 3:121–127.

Gilluly, J., 1949, The distribution of mountain-building in geologic time, *Geol. Soc. Amer. Bull.* 60:561–590.

Gilluly, J., 1967, Chronology of tectonic movements in the western United States, *Amer. Jour. Sci.* 265:306–331.

Goddard, E. N., D. D. Trask, R. K. de Ford, O. N. Rove, J. T. Singlewald, and R. M. Overback, 1951, *Rock Color Chart,* Geological Society of America, New York.

Godwin, H., 1981, *The Archives of the Peat Bogs,* Cambridge University Press, Cambridge.

Gordon, A. D., 1982a, An investigation of two sequence-comparison statistics, *Austral. J. Statist.* 24:332–342.

Gordon, A. D., 1982b, On measuring and modelling the relationship between two stratigraphically-recorded variables, in *Quantitative Stratigraphic Correlations,* J. M. Cubitt and R. A. Reyment, eds., Wiley, Chichester, England, pp. 241–248.

Gordon, A.D., and R. A. Reyment, 1979, Slotting of borehole sequences, *Math Geol.* 11: 309–327.

Gould, S. J., 1965, Is uniformitarianism necessary? *Amer. Jour. Sci.* 263:223–228.

Gould, S. J., 1984, Toward the vindication of punctuational change, in *Catastrophes and Earth History: The New Uniformitarianism,* W. A. Berggren and J. A. van Couvering, eds., Princeton University Press, Princeton, New Jersey, pp. 9–34.

Gould, S. J., 1987, *Time's Arrow, Time's Cycle: Myth and Metaphor in the Discovery of Geological Time,* Harvard University Press, Cambridge, Massachusetts.

Grabau, A. W., 1913, *Principles of Stratigraphy,* A. G. Seiler, New York.

Grabau, A. W., 1920–21, *A Textbook of Geology,* Heath, New York.

Grabau, A. W., 1936, Oscillation or pulsation, *16th Intern. Geol. Congress, Washington, Rept. 1,* pp. 539–553.

Grabau, A. W., and H. W. Shimer, 1909–10, *North American Index Fossils,* A. G. Seiler, New York.

Gradstein, F. M., F. P. Agterberg, J. C. Brower, and W. S. Schwarzacher, 1985, *Quantitative Stratigraphy,* D. Reidel, Dordrecht, Netherlands.

Greeley, R., and M. H. Carr, eds., 1976, *A Geological Basis for the Exploration of the Planets,* NASA Report SP-417, 109 pp.

Gressly, A., 1838, Observations geologiques sur le Jura Soleurois, *Nouv. Mem. Soc. Helvet. Sci. Nat. (Nauchatel)* 2:1–112.

Griffiths, C. M., 1982, A proposed geologically consistent segmentation and reassignment algorithm for petrophysical borehole logs, in *Quantitative Stratigraphic Correlation,* J. M. Cubitt and R. A. Reyment, eds., Wiley, Chichester, England, pp. 287–298.

Grün, R., and P. De Cannière, 1984, ESR-dating: problems encountered in the evaluation of the naturally accumulated dose /AD/ of secondary carbonates, *J. Radioanal. Nucl. Chem. Let.* 85/4:213–226.

Guex, J., and E. Davaud, 1984, Unitary associations method: use of graph theory and computer algorithm, *Computers and Geosciences* 10:69–96.

Hall, C. M., and D. York, 1984, The applicability of $^{40}Ar/^{39}Ar$ dating to young volcanics, in *Quaternary Dating Methods,* W. C. Mahaney, ed., Elsevier, Amsterdam, pp. 67–74.

Hall, T. S., ed., 1951, *A Source Book in Animal Biology,* McGraw-Hill, New York.

Hallam, A., 1977a, Secular changes in marine inundation of USSR and North America through the Phanerozoic, *Nature* 269:769–772.

Hallam, A., ed., 1977b, *Patterns of Evolution as Illustrated by the Fossil Record,* Elsevier, Amsterdam.

Hallam, A., 1981, *Facies Interpretation and the Stratigraphic Record,* W. H. Freeman, Oxford.

Hallam, A., 1984a, Pre-Quaternary sea-level changes, *Ann. Rev. Earth Planet. Sci.* 12:205–243.

Hallam, A., 1984b, Relations between biostratigraphy, magnetostratigraphy and event stratigraphy in the Jurassic and Cretaceous, *Proceedings, 27th Intern. Geol. Congress* (published by VNU Science Press, Utrecht) 1: 189–212.

Hancock, J. M., 1977, The historic development of concepts of biostratigraphic correlation, in *Concepts and Methods of Biostratigraphy,* E. G. Kauffman and J. E. Hazel, eds., Dowden, Hutchinson, and Ross, Stroudsburg, Pennsylvania, pp. 3–22.

Hansen, T. A., 1981, Fossil molluscan larvae: a new biostratigraphic tool, *Science* 214:915–916.

Hansen, T. A., 1984, The biostratigraphic use of larval and juvenile mollusks, *Geol. Soc. Amer. Bull.* 95:1102–1107.

Haq, B. U., 1973, Evolutionary trends in the Cenozoic coccolithopore genus *Helicopontospaera, Micropaleontology* 19:32–52.

Haq, B. U., 1980, Biogeographic history of Miocene calcareous nannoplankton and paleoceanography of the Atlantic Ocean, *Micropaleontology* 26:414–443.

Haq, B. U., and G. P. Lohman, 1976, Early Cenozoic calcareous nannoplankton biogeography of the Atlantic Ocean, *Marine Micropaleontol.* 1:119–194.

Haq, B. U., and F. W. B. van Eysinga, 1987, *Geologic Time Table, 4th revised enlarged and updated edition,* Elsevier, Amsterdam.

Haq, B. U., and T. R. Worsley, 1982, Biochronology—biological events in time resolution, their potential and limitations, in *Numerical Dating in Stratigraphy,* G. S. Odin, ed., Wiley, Chichester, England, pp. 19–35.

Haq, B. U., J. Hardenbol, and P. R. Vail, 1987, The new chronostratigraphic basic of Cenozoic and Mesozoic sea level cycles, in *Timing and Depositional History of Eustatic Sequences: Constraints on Seismic Stratigraphy,* C. A. Ross and D. Haman, eds., Cushman

Foundation for Foraminiferal Research, Special Publication 24, Washington, D.C., pp. 7–13.

Harbaugh, J. W., 1968, *Stratigraphy and Geologic Time,* Wm. C. Brown, Dubuque, Iowa.

Harbaugh, J. W., and D. F. Merriam, 1968, *Computer Applications in Stratigraphic Analysis,* Wiley, New York.

Hardenbol, J., and W. A. Berggren, 1978, A new Paleogene numerical time scale, in *Contributions to the Geologic Time Scale,* G. V. Cohee, M. F. Glaessner, and H. D. Hedberg, eds., American Association of Petroleum Geologists, Tulsa, Oklahoma, pp. 213–234.

Harington, C. R., 1984, Mammoths, bison and time in North America, in *Quaternary Dating Methods,* W. C. Mahaney, ed., Elsevier, Amsterdam, pp. 299–309.

Harland, R., 1984, Quaternary dinoflagellate cysts from holes 548 and 549A, Goban Spur (Deep Sea Drilling Project Leg 80), *Init. Repts. Deep Sea Drilling Project* 80: 761–766.

Harland, W. B., 1975, The two geological time scales, *Nature* 253: 505–507.

Harland, W. B., 1978, Geochronologic scales, in *Contributions to the Geologic Time Scale,* G. V. Cohee, M. F. Glaessner, and H. D. Hedberg, eds., American Association of Petroleum Geologists, Tulsa, Oklahoma, pp. 9–32.

Harland, W. B., A. V. Cox, P. G. Llewellyn, C. A. G. Pickton, A. G. Smith, and R. Walters, 1982, *A Geologic Timescale,* Cambridge University Press, Cambridge, England.

Harper, C. W., Jr., 1980, Relative age inference in paleontology, *Lethaia* 13:239–248.

Harper, C. W., Jr., 1981, Inferring succession of fossils in time: the need for a quantitative and statistical approach, *Jour. Paleontol.* 55:442–452.

Harper, C. W., Jr., 1984, Improved methods of facies sequence analysis, in *Facies Models, Second Edition,* R. G. Walker, ed., Geoscience Canada, Reprint Series 1, Geological Association of Canada, Toronto, Ontario, pp. 11–13.

Harris, E. C., 1979, *Principles of Archaeological Stratigraphy,* Academic Press, London.

Haug, E., 1907, *Traité de Géologie,* Armand Cohn, Paris.

Hay, W. W., 1972, Probabilistic stratigraphy, *Ecolgae geol. Helv.* 65/2:255–266.

Hays, J. D., and W. C. Pitman, 1973, Lithospheric plate motion, sea-level changes, and climatic and ecological consequences, *Nature* 246:18–22.

Hazel, J. E., 1977, Use of certain multivariate and other techniques in assemblage zonal biostratigraphy: examples utilizing Cambrian, Cretaceous, and Tertiary benthic invertebrates, in *Concepts and Methods of Biostratigraphy,* E. G. Kauffman and J. E. Hazel, eds., Dowden, Hutchingson, and Ross, Stroudsburg, Pennsylvania, pp. 187–212.

Hedberg, H. D., 1937, Stratigraphy of the Rio Querecual section of northeastern Venezuela, *Bull. Geol. Soc. Amer.* 48:1971–2024.

Hedberg, H. D., 1941, Discussion of "Technique of stratigraphic nomenclature" by C. W. Tomlinson, *Amer. Assoc. Petrol. Geol. Bull.* 25:2202–2206.

Hedberg, H. D., 1948, Time-stratigraphic classification of sedimentary rocks, *Geol. Soc. Amer. Bull.* 59:447–462.

Hedberg, H. D., 1951, Nature of time-stratigraphic units and geologic time units, *Amer. Assoc. Petrol. Geol. Bull.* 37:1077–1081.

Hedberg, H. D., 1954, Procedure and terminology in stratigraphic classification, *19th Intern. Geol. Congress* 13:205–233.

Hedberg, H. D., 1958, Stratigraphic classification and terminology, *Amer. Assoc. Petrol. Geol. Bull.* 42:1881–1896.

Hedberg, H. D., 1959, Towards harmony in stratigraphic classification, *Amer. Jour. Sci.* 257:674–683.

Hedberg, H. D., ed., 1961, Stratigraphic classification and terminology (International Subcommission on Stratigraphic Terminology), *Rept. 21st Session, Intern. Geol. Congress Norden, 1960,* part 25:1–38.

Hedberg, H. D., 1965, Chronostratigraphy and biostratigraphy, *Geol. Mag.* 102:451–461.

Hedberg, H. D., 1973a, Impressions from a discussion of the ISSC International Stratigraphic Guide, Hanover, October 18, 1972, *Newsl. Stratigr.* 2:173–180.

Hedberg, H. D., 1973b, Reaction to an attack by Professor H. F. Erben on the International Subcommission on Stratigraphic Classification and its philosophy, *Newsl. Stratigr.* 2:181–183.

Hedberg, H. D., ed, 1976, *International Stratigraphic Guide,* by the International Subcommission on Stratigraphic Classification, Wiley, New York.

Hedberg, H. D., 1978, Stratotypes and an international geochronologic scale, in *Contributions to the Geologic Time Scale,* G. V. Cohee, M. F. Glaessner, and H. D. Hedberg, eds., American Association of Petroleum Geologists, Tulsa, Oklahoma, pp. 33–38.

Heirtzler, J. R., G. O. Dickson, E. M. Herron, W. C. Pitman, and X. Le Pichon, 1968, Marine magnetic anomalies, geomagnetic field reversals, and motions of the ocean floor and continents, *Jour. Geophys. Res.* 73:2119–2136.

Heizer, R. F., and J. A. Graham, 1967, *A Guide to Field Methods in Archaeology: Approaches to the Anthropology of the Dead,* The National Press, Palo Alto.

Heller, F., 1977, Palaeomagnetism of Upper Jurassic limestones from southern Germany, *Jour. Geophys.* 42:475–488.

Heller, F., and T-S. Liu, 1982, Magnetostratigraphical dating of loess deposits in China, *Nature* 300:431–433.

Heller, F., T-S. Liu, 1984, Magnetism of Chinese loess deposits, *Geophy. Jour. Roy. Astr. Soc.* 77:125–141.

Heller, F., and T-S. Liu, 1986, Palaeoclimatic and sedimentary history from magnetic susceptibility of loess in China, *Geophy. Res. Let.* 13:1169–1172.

Heller, F., and N. Petersen, 1982, Self-reversal explanation for the Laschamp/Olby geomagnetic field excursion. *Phys. Earth Planet. Inter.* 30:358–372.

Heller, F., W. Lowrie, and J. E. T. Channell, 1984, Late Miocene magnetic stratigraphy at Deep Sea Drilling Project Hole 521A, *Init. Repts. Deep Sea Drilling Project* 73: 637–644.

Heller, P. L., Z. E. Peterman, J. R. O'Neil, and M. Shafiqullah, 1985, Isotopic provenance of sandstones from the Eocene Tyee Formation, Oregon Coast Range, *Geol. Soc. Amer. Bull.* 96:770–780.

Hempel, C. G., 1965, *Aspects of Scientific Explanation, and Other Essays in the Philosophy of Science,* Free Press, New York.

Hempel, C. G., 1966, *Philosophy of Natural Science,* Prentice-Hall, Englewood Cliffs, New Jersey.

Hennig, G. J., 1982, Notes and comments on the 230 Th/234 U dating on speleothem samples from the Grotte D'Aldène, *Bull. Mus. D'Anthrop. Prehist. de Monaco* 26:21–25.

Hennig, G. J., and R. Grün, 1983, ESR dating in Quaternary geology, *Quat. Sci. Rev.* 2:157–238.

Hennig, G. J., M. A. Geyh, and R. Grün, 1985, The first inter-laboratory ESR comparison project Phase II: Evaluation of equivalent doses (ED) of calcites, *Nucl. Tracks* 10:945–952.

Hennig, G. J., R. Grün, and K. Brunnacker, 1983a, Interlaboratory comparison project of ESR-dating, Phase 1, *Pact* 9:447–452.

Hennig, G. J., R. Grün, and K. Brunnacker, 1983b, Speleothems, travertines, and paleoclimates, *Quat. Research* 20:1–29.

Hennig, G. J., R. Grün, K. Brunnacker, and M. Pécsi, 1983, Th-230/U-234 sowie ESR-Altersbestimmungen einiger Travertine in Ungarn [Th-230/U-234 and ESR age determinations of spring deposited travertines in Hungary], *Eiszeitalter u. Gegenwart* 33:9–19.

Hennig, G. J., W. Herr, E. Weber, and N. I. Xirotiris, 1981, ESR-dating of the fossil hominid cranium from Petralona Cave, Greece, *Nature* 292:533–536.

Hickey, L. J., 1977, Stratigraphy and paleobotany of the Golden Valley Formation (Early Tertiary) of western North Dakota, *Geol. Soc. Amer. Mem.* 150:i–x, 1–183, 55 plates.

Hoffman, A., 1980, Ecostratigraphy: the limits of applicability, *Acta Geol. Polonica* 30:97–108.

Hohn, M. E., 1978, Stratigraphic correlation by principal components: effects of missing data, *Jour. Geol.* 86:524–532.

Hohn, M. E., 1982, Properties of composite sections constructed by least-squares, in *Quantitative Stratigraphic Correlation,* J. M. Cubitt and R. A. Reyment, eds., Wiley, New York, pp. 107–117.

Hohn, M. E., 1985, SAS program for quantitative stratigraphic correlation by principal components, *Computers and Geosciences* 11:471–477.

Holbrook, S. H., 1967, The growing giant, in *America's Historylands,* M. B. Grosvenor, ed., National Geographic Society, Washington, D.C., pp. 517–541.

Holland, C. H., 1978, Stratigraphical classification and all that, *Lethaia* 11:85–90.

Holland, C. H., 1984, Steps to a standard Silurian, *Proceedings, 27th Intern. Geol. Congress* (published by VNU Science Press, Utrecht) 1:127–156.

Holland, C. H., 1985, Series and stages of the Silurian System, *Episodes* 8:101–103.

Holland, C. H., 1986a, Does the golden spike still glitter? *Jour. Geol. Soc. London* 143:3–21.

Holland, C. H., 1986b, Some aspects of time, *Newsl. Stratigr.* 15:172–176.

Holmes, A., 1946, An estimate of the age of the earth. *Nature* 157:680–684.

Holmes, A., 1947, The construction of a geologic time scale, *Trans. Geol. Soc. Glasgow* 21:117–152.

Holser, W. T., 1977, Catastrophic chemical events in the history of the ocean, *Nature* 267:403–408.

Holser, W. T., 1984, Gradual and abrupt shifts in ocean chemistry during Phanerozoic time, in *Patterns of Change in Earth Evolution,* H. D. Holland and A. F. Trendall, eds., Springer-Verlag, Berlin, pp. 123–143.

Holser, W. T., M. Magaritz, and D. L. Clark, 1986, Carbon-isotope stratigraphic correlations in the Late Permian, *Amer. Jour. Sci.* 286:390–402.

Hooykaas, R., 1956, The principle of uniformity in geology, biology, and theology, *Jour. Trans. Victoria Inst.* 88:101–116.

Hooykaas, R., 1963, *The Principle of Uniformity,* Brill, Leiden.

Hooykaas, R., 1970, Catastrophism in geology, its scientific character in relation to actualism and uniformitarianism, reprinted in *Philosophy of Geohistory: 1785–1970,* 1975, C. C. Albritton, Jr., ed., Dowden, Hutchinson, and Ross, Stroudsburg, Pennsylvania, pp. 310–356.

Horner, F., and F. Heller, 1983, Lower Jurassic magnetostratigraphy at the Breggia Gorge (Ticino, Switzerland) and Alpe Turati (Como, Italy), *Geophy. Jour. Roy. Astr. Soc.* 73:705–718.

House, M. R., 1985a, Correlation of mid-Paleozoic ammonoid evolutionary events with global sedimentary pertubations, *Nature* 313:17–22.

House, M. R., 1985b, A new approach to an absolute timescale from measurements of orbital cycles and sedimentary microrhythms, *Nature* 316:721–725.

Hsü, K. J., 1984., Geochemical markers of impacts and their effects on environments, in *Patterns of Change in Earth Evolution,* H. D. Holland and A. F. Trendall, eds., Springer-Verlag, Berlin, pp. 63–74.

Hsü, K. J., H. Oberhañnsli, J. Y. Gao, S. Shu, C. Haihong, and U. Krähenbühl, 1985, "Strangelove ocean" before the Cambrian explosion, *Nature* 316:809–811.

Hubbert, M. K., 1967, Critique of the principle of uniformity, *Geol. Soc. Amer. Spec. Pap.* 89:3–33.

Hughes, N. F., D. B. Williams, J. L. Cutbill, W. B. Harland, 1967, A use of reference-points in stratigraphy, *Geol. Mag.* 104:634–635.

Hughes, N. F., D. B. Williams, J. L. Cutbill, W. B. Harland, 1968, Hierarchy in stratigraphical nomenclature, *Geol. Mag.* 105:79.

Hultberg, S. U., B. A. Malmgren, and G. Gard, 1983, Integrated microplankton stratigraphy of Upper Maastrichtian sections in Scandinavia based on census data, *Abstracts: Cretaceous Stage Boundaries, Copenhagen,* pp. 86–89.

Hunt, C. B., 1959, Dating of mining camps with tin cans and bottles, *Geotimes* 3:8–10, 34.

Hutchinson, P. J., 1986, Review of *Seismic Stratigraphy II. An Integrated Approach* edited by O. Berg and D. C. Woolverton. *Geology* 14:813.

Hutton, J., 1785, Abstract of a dissertation . . . concerning the system of the earth, its duration, and stability, reprinted in *Philosophy of Geohistory: 1785-1970,* 1975, C. C. Albritton, Jr., ed., Dowden, Hutchinson, and Ross, Stroudsburg, Pennsylvania, pp. 24–52.

Hutton, J., 1788, Theory of the Earth, *Trans. Roy. Soc. Edinburgh* 1:209–305.

Hutton, J., 1795, *Theory of the Earth, with Proofs and Illustrations,* William Creech, Edinburgh.

Huxley, T. H., 1862, The anniversary address, *Quart. Jour. Geol. Soc. London* 18:xl–liv.

Iaccarino, S., 1985, Mediterranean Miocene and Pliocene planktic foraminifera, in *Plankton Stratigraphy,* H. M. Bolli, J. B. Saunders, and K. Perch-Nielsen, eds., Cambridge University Press, Cambridge, pp. 283–314.

Ikebe, N., and R. Tsuchi, eds., 1984, *Pacific Neogene Datum Planes: Contributions to Biostratigraphy and Chronology,* University of Tokyo Press, Tokyo.

Ikeya, M., 1985, Dating methods of Pleistocene deposits and their problems: IX. Electron spin resonance, in *Dating Methods of Pleistocene Deposits and their Problems,* N. W. Rutter, ed., Geoscience Canada Reprint Series 2, Geological Association of Canada, Toronto, Ontario, pp. 73–87.

Ingram, R. L., 1954, Terminology for the thickness of stratification and parting units in sedimentary rocks, *Geol. Soc. Amer. Bull.* 65:937–938.

Interdepartmental Stratigraphic Committee, USSR, 1956, Stratigraphic classification and terminology (translated by J. Rodgers, 1959, *Intern. Geol. Rev.* 1(2):22–38).

Interdepartmental Stratigraphic Committee, USSR, 1965, Stratigraphic classification, terminology, and nomenclature (English translation 1966, *Intern. Geol. Rev.* 8(10):1–36).

International Subcommission on Stratigraphic Classification (ISSC), 1976: See Hedberg, ed., 1976.

International Subcommission on Stratigraphic Classification [(of the IUGS) and IUGS/IAGA Subcommission on a Magnetic Polarity Time Scale], 1979, Magnetostratigraphic polarity units—a supplementary chapter of the ISSC *International Stratigraphic Guide, Geology* 7:578–583.

International Subcommission on Stratigraphic Classification (Amos Salvador, Chairman), 1987a, Unconformity-bounded stratigraphic units, *Geol. Soc. Amer. Bull.* 98:232–237.

International Subcommission on Stratigraphic Classification (Amos Salvador, Chairman), 1987b, Stratigraphic classification and nomenclature of igneous and metamorphic rock bodies, *Geol. Soc. Amer. Bull.* 99:440–442.

International Subcommission on Stratigraphic Classification (Amos Salvador, Chairman), 1987c, Unconformity-bounded stratigraphic units: Reply, *Geol. Soc. Amer. Bull.* 99:444.

International Subcommission on Stratigraphic Classification, (Amos Salvador, Chairman), 1988, Unconformity-bounded stratigraphic units: Reply, *Geol. Soc. Amer. Bull.* 100:156.

International Subcommission on Stratigraphic Terminology, 1964, *Definition of Geologic Systems,* H. D. Hedberg, ed., International Geological Congress, Report of the Twenty-second Session, India, 1964, New Delhi.

Irving, E., 1964, *Paleomagnetism and Its Application to Geological and Geophysical Problems,* Wiley, New York.

Ivanovich, M., C. Vita-Finzi, and G. J. Hennig, 1983, Uranium-series dating of molluscs from uplifted Holocene beaches in the Persian Gulf, *Nature* 302:408–410.

Jeletzky, J. A., 1956, Paleontology, basis of a practical geochronology, *Bull. Amer. Assoc. Petrol. Geol.* 40:679–706.

Jeletzky, J. A., 1965, Is it possible to quantify biochronological correlation? *Jour. Paleontol.* 39:135–140.

Johnson, J. G., 1979, Intent and reality in biostratigraphic zonation, *Jour. Paleontol.* 53:931–942.

Johnson, J. G., 1982, Occurrence of phyletic gradualism and punctuated equilibria through geologic time, *Jour. Paleontol.* 56:1329–1331.

Johnson, J. G., 1987, Unconformity-bounded stratigraphic units: Discussion, *Geol. Soc. Amer. Bull.* 99:443.

Johnson, M. E., 1979, Evolutionary brachiopod lineages from the Llandovery Series of eastern Iowa, *Palaeontology* 22:549–567.

Johnson, M. E., 1986, Review of *Aspects of a Stratigraphic System: The Devonian* by D. L. Dineley, *Palaios* 1:424–426.

Johnson, M. E., 1987a, North American Paleozoic oceanography: overview of progress toward a modern synthesis, *Paleoceanography* 2:123–140.

Johnson, M. E., 1987b, Extent and bathymetry of North American platform seas in the Early Silurian, *Paleoceanography* 2:185–211.

Johnson, M. E., and V. R. Colville, 1982, Regional integration of evidence for evolution in the Silurian *Pentamerus–Pentameroides* lineage, *Lethaia* 15:41–54.

Johnson, M. E., and H. L. Lescinsky, 1986a, Depositional dynamics of cyclic carbonates from the Interlake Group (Lower Silurian) of the Williston Basin, *Palaios* 1:111–121.

Johnson, M. E., J.-Y. Rong, and X.-C. Yang, 1985, Intercontinental correlation by sea-level events in the Early Silurian of North America and China (Yangtze Platform), *Geol. Soc. Amer. Bull.* 96:1384–1397.

Jones, T. A., D. E. Hamilton, and C. R. Johnson, 1986, *Contouring Geologic Surfaces with the Computer,* Van Nostrand Reinhold, New York.

Jukes-Browne, A. J., 1903, The term "hemera," *Geol. Mag.* 10:36–38.

Kauffman, E. G., 1970, Population systematics, radiometrics, and zonation—a new biostratigraphy, *Proc. North Amer. Paleont. Conv., Pt. F,* (published by Allen Press, Lawrence, Kansas), pp. 612–666.

Kauffman, E. G., 1977, Evolutionary rates and biostratigraphy, in *Concepts and Methods of Biostratigraphy,* E. G. Kauffman and J. E. Hazel, eds., Dowden, Hutchinson and Ross, Stroudsburg, Pennsylvania, pp. 109–141.

Kauffman, E. G., in preparation, *Theoretical and Applied Biostratigraphy,* Wiley, New York.

Kauffman, E. G., and J. E. Hazel, eds., 1977, *Concepts and Methods of Biostratigraphy,* Dowden, Hutchinson and Ross, Stroudsburg, Pennsylvania.

Kay, M., 1951, North American geosynclines, *Geol. Soc. Amer. Mem.* 48:1–143.

Kelvin, Lord [William Thomson], 1899, The age of the earth as an abode fitted for life, *Science* 9:665–674, 704–711.

Keroher, G. C., 1970, Lexicon of geologic names of the United States for 1961–1970, *U. S. Geol. Surv. Bull.* 1350:1–848.

Kennett, J. P., ed., 1980, *Magnetic Stratigraphy of Sediments,* Dowden, Hutchinson and Ross, Stroudsburg, Pennsylvania.

Khramov, A. N., 1957, Paleomagnetism: the basis of a new method of correlation and subdivision of sedimentary strata, *Acad. Sci. USSR Doklady, Earth Sci. Sec. Proc.* 112:129–132.

Kidwell, S. M., 1985, Palaeobiological and sedimentological implications of fossil concentrations, *Nature* 318:457–460.

Kidwell, S. M., F. T. Fürsich, and T. Aigner, 1986, Conceptual framework for the analysis and classification of fossil concentrations, *Palaios* 1:228–238.

Kirk, J. T., 1984, *The Impecunious House Restorer: Personal Vision and Historic Accuracy,* Alfred A. Knopf, New York.

Kitts, D. B., 1963, The theory of geology, in *The Fabric of Geology,* C. C. Albritton, Jr., ed., Addison-Wesley, Reading, Massachusetts, pp. 49–68.

Kitts, D. B., 1966, Geologic time, *Jour. Geol.* 74:127–146.

Kitts, D. B., 1976, Certainty and uncertainty in geology, *Amer. Jour. Sci.* 276:29–46.

Kleinpell, R. M., 1938, *Miocene Stratigraphy of California,* American Association of Petroleum Geologists, Tulsa, Oklahoma.

Kohlberger, W., and R. M. Schoch, 1985, A consideration of stratigraphical philosophy through Arkell, *Toth-Maatien Review* 4:1950–1963.

Kohlberger, W., and R. M. Schoch, 1986, Some stratigraphical concepts and their consequences for evolutionary studies, *Toth-Maatian Review* 5:2329-2334.

Königsson, L.-K., 1984, Chronostratigraphical division of the Holocene, *Proceedings, 27th Intern. Geol. Congress* (published by VNU Science Press, Utrecht) 1:107-118.

Kottlowski, F. E., 1965, *Measuring Stratigraphic Sections,* Holt, Rinehart and Winston, New York.

Kowallis, B. J., J. S. Heaton, and K. Bringhurst, 1986, Fission-track dating of volcanically derived sedimentary rocks, *Geology* 14:19-22.

Krassilov, V., 1974, Causal biostratigraphy, *Lethaia* 7:173-179.

Kraus, M. J., and T. M. Bown, 1986, Paleosols and time resolution in alluvial stratigraphy, in *Paleosols: Their Recognition and Interpretation,* V. P. Wright, ed., Blackwell Scientific Publications, Oxford, pp. 180-207.

Krumbein, W. C., and L. L. Sloss, 1963, *Stratigraphy and Sedimentation, 2nd edition,* Freeman, San Francisco.

Krynine, P. D., 1948, The megascopic study and field classification of sedimentary rocks, *Jour. Geol.* 56:130-165.

Kummel, B., and C. Teichert, eds., 1970, *Stratigraphy Boundary Problems: Permian and Triassic of West Pakistan,* University Press of Kansas, Lawrence.

Lahee, F. H., 1923, *Field Geology, 2nd edition,* McGraw-Hill, New York.

Lamarck, J. B., 1984, *Zoological Philosophy,* University of Chicago Press, Chicago.

Lamothe, M., A. Dreimanis, M. Morency, and A. Raukas, 1984, Thermoluminescence dating of Quaternary sediments, in *Quaternary Dating Methods,* W. C. Mahaney, ed., Elsevier, Amsterdam, pp. 153-170.

Landing, E., 1985, The Levis Formation: passive margin slope process and dynamic stratigraphy in the western area, in *Field Trips Guidebook,* J. F. Riva, ed., 1985 Canadian Paleontology and Biostratigraphy Seminar, Quebec, pp. 1-11.

Landing, E., and C. E. Brett, 1987, Trace fossils and regional significance of a Middle Devonian (Givetian) disconformity in southwestern Ontario, *Jour. Paleontol.* 61:205-230.

Lane, H. R., and W. L. Manger, 1985, The basis for a Mid-Carboniferous boundary, *Episodes* 8:112-115.

Lane, H. R., and W. Ziegler, 1983, Taxonomy and phylogeny of *Scaliognathus* Branson and Mehl 1941 (Conodonta, Lower Carboniferous), *Senckenbergiana Lethaea* 64:199-225.

Langer, C., and K. Brunnacker, 1983, Schotterpetrographie des Tertiärs and Quartärs im Neuwieder Becken und am unteren Mittlerhein, *Decheniana (Bonn)* 136:100-107.

Lapo, A. V., 1982, *Traces of Bygone Biospheres,* Mir Publishers, Moscow.

Laporte, L. F., 1979, *Ancient Environments, 2nd edition,* Prentice-Hall, Englewood Cliffs, New Jersey.

Layzer, D., 1975, The arrow of time, *Sci. Amer.* 233:56-69.

Leeder, M. R., 1982, *Sedimentology: Process and Product,* Allen and Unwin, London.

Leg 112 Scientific Drilling Party, 1987, Leg 112 studies continental margin, *Geotimes* 32:10-12.

Lewis, D. W., 1984, *Practical Sedimentology,* Van Nostrand Reinhold, New York.

Libby, W. F., 1952, *Radiocarbon Dating,* University of Chicago Press, Chicago.

Lindholm, R., 1987, *A Practical Approach to Sedimentology,* Allen and Unwin, London.

Little, T. A., 1987, Stratigraphy and structure of metamorphosed upper Paleozoic rocks near Mountain City, Nevada, *Geol. Soc. Amer. Bull.* 98:1-17.

Lloyd, A. J., 1964, The Luxembourg colloquium and the revision of the stages of the Jurassic System, *Geol. Mag.* 101:249-259.

Lloyd, A. J., 1965, The Luxembourg colloquium—a reply, *Geol. Mag.* 102:88.

Lohmann, G. P., and B. A. Malmgren, 1983, Equatorward migration of *Globorotalia truncatulinoides* ecophenotypes through the Late Pleistocene: gradual evolution or ocean change? *Paleobiology* 9:414-421.

Long, D., A. Bent, R. Harland, D. M. Gregory, D. K. Graham, and A. C. Morton, 1986,

Late Quaternary palaeontology, sedimentology and geochemistry of a vibrocore from the Witch Ground Basin, central North Sea, *Marine Geol.* 73:109–123.

Lowrie, W., 1979, Geomagnetic reversals and ocean crust magnetism, in *Deep Drilling Results in the Atlantic Ocean: Ocean Crust,* M. Talwani, C. G. A. Harrison, and D. E. Hayes, eds., Maurice Ewing Series, vol. 2, AGU, Washington, D.C., pp. 135–150.

Lowrie, W., 1982, A revised magnetic polarity timescale for the Cretaceous and Cainozoic, *Phil. Trans. Roy. Soc. London* A 306:129–136.

Lowrie, W., 1986, Magnetic stratigraphy of the Eocene/Oligocene boundary, in *Terminal Eocene Events,* C. Pomerol and I. Premoli Silva, eds., Elsevier, Amsterdam, pp. 357–362.

Lowrie, W., and W. Alverez, 1976, Paleomagnetic studies of the Scaglia Rossa Limestone in Umbria, *Mem. Soc. Geol. It.* 15:41–50.

Lowrie, W., and W. Alverez, 1977a, Upper Cretaceous–Paleocene magnetic stratigraphy at Gubbio, Italy: III. Upper Cretaceous magnetic stratigraphy, *Geol. Soc. Amer. Bull.* 88:374–377.

Lowrie, W., and W. Alverez, 1977b, Late Cretaceous geomagnetic polarity sequence: detailed rock and palaeomagnetic studies of the Scaglia Rosa Limestone at Gubbio, Italy, *Geophys. Jour. Roy. Astr. Soc.* 51:561–582.

Lowrie, W., and W. Alverez, 1981, One hundred million years of geomagnetic polarity history, *Geology* 9:392–397.

Lowrie, W., and W. Alverez, 1984, Lower Cretaceous magnetic stratigraphy in Umbrian pelagic limestone sections, *Earth Planet. Sci. Let.* 71:315–328.

Lowrie, W., and J. E. T. Channell, 1983, Magnetostratigraphy of the Jurassic–Cretaceous boundary in the Maiolica limestone (Umbria, Italy), *Geology* 12:44–47.

Lowrie, W., and D. V. Kent, 1983, Geomagnetic reversal frequency since the Late Cretaceous, *Earth Planet. Sci. Let.* 62:305–313.

Lowrie, W., and J. C. Ogg, 1986, A magnetic polarity time scale for the Early Cretaceous and Late Jurassic, *Earth Planet. Sci. Let.* 76:341–349.

Lowrie, W., W. Alverez, G. Napoleone, K. Perch-Nielsen, I. Premoli Silva, and M. Toumarkine, 1982, Paleogene magnetic stratigraphy in Umbrian pelagic carbonate rocks: the Contessa sections, Gubbio, *Geol. Soc. Amer. Bull.* 93:414–432.

Lowrie, W., W. Alverez, I. Premoli Silva, and S. Monechi, 1980, Lower Cretaceous magnetic stratigraphy in Umbrian pelagic carbonate rocks, *Geophys. Jour. Roy. Astr. Soc.* 60:263–281.

Lowrie, W., J. E. T. Channell, W. Alverez, 1980, A review of magnetic stratigraphy investigations in Cretaceous pelagic carbonate rocks, *Jour. Geophys. Res.* 85:3597–3605.

Lucas, S. G., 1985, Discussion: a critique [of] chronostratigraphy, *Amer. Jour. Sci.* 285:764–766.

Lucas, S. G., R. M. Schoch, E. Manning, and C. Tsentas, 1981, The Eocene biostratigraphy of New Mexico, *Geol. Soc. Amer. Bull.* 92:951–967.

Ludvigsen, R., S. R. Westrop, B. R. Pratt, P. A. Tuffnell, and G. A. Young, 1986, Dual biostratigraphy: zones and biofacies, *Geoscience Canada* 13:139–154.

Luttrell, G. W., M. L. Hubert, and V. M. Jussen, 1986, Lexicon of new formal geologic names of the United States for 1976–1980, *U. S. Geol. Surv. Bull.* 1564:1–191.

Luttrell, G. W., M. L. Hubert, W. B. Wright, V. M. Jussen, and R. W. Swanson, 1981, Lexicon of geologic names for the United States for 1968–1975, *U. S. Geol. Surv. Bull.* 1520:1–342.

Lyell, C., 1830–33, *Principles of Geology, Being an Attempt to Explain the Former Changes of the Earth's Surface by Reference to Causes Now in Operation,* John Murray, London.

Lyell, C., 1857, *A Manual of Geology reprinted from the sixth edition,* D. Appleton, New York.

MacFadden, B. J., 1977, Magnetic polarity stratigraphy of the Chamita Formation stratotype (Mio-Pliocene) of north-central New Mexico, *Amer. Jour. Sci.* 277:769–800.

MacFadden, B. J., 1985, Drifting continents, mammals, and time scales: current developments in South America, *Jour. Vertebrate Paleontol.* 5:169–174.

MacFadden, B. J., K. C. Campbell, Jr., R. L. Cifelli, O. Siles, N. M. Johnson, C. W. Naeser, and P. Seitler, 1985, Magnetic polarity stratigraphy and mammalian fauna of the Deseadan (Late Oligocene–Early Miocene) Salla Beds of northern Bolivia, *Jour. Geol.* 93:223–250.

MacFadden, B. J., N. M. Johnson, and N. D. Opdyke, 1979, Magnetic polarity stratigraphy of the Mio-Pliocene mammal-bearing Big Sandy Formation of western Arizona, *Earth Planet. Sci. Let.* 44:349–364.

MacFadden, B. J., O. Siles, P. Zeitler, and N. M. Johnson, 1983, Magnetic polarity stratigraphy of the Middle Pleistocene (Ensenadan) Tarija Formation of southern Bolivia, *Quat. Research* 19:172–187.

Magaritz, M., 1985, The carbon isotope record of dolostones as a stratigraphic tool: a case study from the Upper Cretaceous shelf sequence, Israel, *Sedimentary Geol.* 45:115–123.

Mahaney, W. C., ed., 1984, *Quaternary Dating Methods,* Elsevier, Amsterdam.

Malmgren, B. J., 1984, Analysis of the environmental influence on the morphology of *Ammonia beccarii* (Linné) in southern European Salinas, *Geobios* 17:737–746.

Malmgren, B. A., and J. P. Kennett, 1981, Phyletic gradualism in a Late Cenozoic planktonic foraminiferal lineage; DSDP 284, southwest Pacific, *Paleobiology* 7:230–240.

Malmgren, B. A., and J. P. Kennett, 1982, The potential of morphometrically based phylozonation: application of a Late Cenozoic planktonic foraminiferal lineage, *Marine Micropaleontol.* 7:285–296.

Malmgren, B. A., W. A. Berggren, and G. P. Lohmann, 1983, Evidence for punctuated gradualism in the Late Neogene *Globorotalia tumida* lineage of planktonic Foraminifera, *Paleobiology* 9:377–389.

Malmgren, B. A., W. A. Berggren, and G. P. Lohmann, 1984, Species formation through punctuated gradualism in planktonic Foraminifera, *Science* 225:317–319.

Mangerud, J., S. T. Andersen, B. E. Berglund, and J. Donner, 1974, Quaternary stratigraphy of Norden, a proposal for terminology and classification, *Boreas* 3:109–128.

Mangerud, J., H. J. B. Birks, and K.-D. Jäger, eds., 1982, Chronostratigraphic subdivision of the Holocene, *Striae* 16:1–110.

Markevich, V. P., 1960, The concept of facies, *Int. Geol. Rev.* pp. 367–379, 498–507, 582–604.

Marquis, 1985, *Who's Who in Frontiers of Science and Technology, 2nd edition,* Marquis Who's Who, Chicago.

Martinsson, A., 1972, Editor's column: descriptive palaeontology, *Lethaia* 5:249–250.

Martinsson, A., 1973, Editor's column: ecostratigraphy, *Lethaia* 6:441–443.

Martinsson, A., ed., 1977, *The Silurian-Devonian Boundary,* International Union of Geological Sciences, Series A, Number 5, E. Schweizerbart'sche Verlagsbuchhandlung, Stuttgart.

Martinsson, A., 1978, Project ecostratigraphy, *Lethaia* 11:84.

Martinsson, A., 1980a, Ecostratigraphy: Limits of applicability, *Lethaia* 13:363.

Martinsson, A., 1980b, Phylogeny unearthed, *Lethaia* 13:364.

Martinsson, A., 1981a, Last year's texts in palaeontology, *Lethaia* 14:44.

Martinsson, A., 1981b, The palaeontological thesaurus rescrambled, *Lethaia* 14:167.

Martinsson, A., 1983, Time-scales galore, *Lethaia* 16:128.

Martinsson, A., and M. G. Bassett, 1980, International Commission on Stratigraphy, *Lethaia* 13:26.

Mason, S. F., 1962, *A History of the Sciences,* Collier Books, Macmillan, New York.

Mather, K. F., and S. L. Mason, eds., 1939, *A Source Book in Geology,* McGraw-Hill, New York.

Matthews, R. K., 1984, *Dynamic Stratigraphy: An Introduction to Sedimentation and Stratigraphy, 2nd edition,* Prentice-Hall, Englewood Cliffs, New Jersey.

Matuyama, M., 1929, On the direction of magnetization of basalt in Japan, Tyosen and Manchuria, *Japan Acad. Proc.* 5:203–205.

Mayer, L., 1979a, The origin of fine scale acoustic stratigraphy in deep-sea carbonates. *Jour. Geophys. Res.* 84:6177–6184.

Mayer, L., 1979b, Deep sea carbonates: acoustic, physical, and stratigraphic properties, *Jour. Sed. Petrol.* 49:819–836.

Mayer, L. A., T. S. Shipley, F. Theyer, R. H. Wilkens, and E. L. Winterer, 1985, Seismic modeling and paleoceanography at Deep Sea Drilling Project site 574, *Init. Repts. Deep Sea Drilling Project* 85:947–970.

Mayer, L. A., T. H. Shipley, and E. L. Winterer, 1986, Equatorial Pacific seismic reflectors as indicators of global oceanographic events, *Science* 233:761–764.

Mayr, E., 1982, *The Growth of Biological Thought: Diversity, Evolution, and Inheritance,* Harvard University Press, Cambridge, Massachusetts.

McCoy, W. D., 1987, Quaternary aminostratigraphy of the Bonneville Basin, western United States, *Geol. Soc. Amer. Bull.* 98:99–112.

McDougall, I., 1978, Potassium–argon isotopic dating method and its application to physical time-scale studies, in *Contributions to the Geologic Time Scale,* G. V. Cohee, M. F. Glaessner, and H. D. Hedberg, eds., American Association of Petroleum Geologists, Tulsa, Oklahoma, pp. 119–126.

McDougall, I., K. Saemundsson, H. Johannesson, N. D. Watkins, and L. Kristjansson, 1977, Extension of the geomagnetic polarity time scale to 6.5 m.y.: K-Ar dating, geological and paleomagnetic study of a 3,500-m lava succession in western Iceland, *Geol. Soc. Amer. Bull.* 88: 1–15.

McElhinny, M. W., 1978, The magnetic polarity time scale: prospects and possibilities in magnetostratigraphy, in *Contributions to the Geologic Time Scale,* G. V. Cohee, M. F. Glaessner, and H. D. Hedberg, eds., American Association of Petroleum Geologists, Tulsa, Oklahoma, pp. 57–65.

McIntyre, D. B., 1963, James Hutton and the philosophy of geology, in *The Fabric of Geology,* C. C. Albritton, Jr., ed., Addison-Wesley, Reading, Massachusetts, pp. 1–11.

McKee, E. D., and G. W. Weir, 1953, Terminology for stratification and cross-stratification in sedimentary rocks, *Geol. Soc. Amer. Bull.* 64:381–389.

McKinney, M. L., 1986a, Biostratigraphic gap analysis, *Geology* 14:36–38.

McKinney, M. L., 1986b, Estimating volumetric fossil abundance from cross-sections: a stereological approach, *Palaios* 1:79–84.

McKinney, M. L., 1986c, How biostratigraphic gaps form, *Jour. Geol.* 94:875–884.

McKinney, M. L., and R. M. Schoch, 1983, A composite terrestrial Paleocene section with completeness estimates, based upon magnetostratigraphy, *Amer. Jour. Sci.* 283:801–814.

McLaren, D. J., 1973, The Silurian–Devonian boundary, *Geol. Mag.* 110:302–303.

McLaren, D. J., 1977, The Silurian–Devonian boundary committee, a final report, in *The Silurian–Devonian Boundary,* A. Martinsson, ed., International Union of Geological Sciences, Series A, no. 5 (published by E. Schweizerbart'sche Verlagsbuchhandlung, Stuttgart), pp. 1–34.

McLaren, D. J., 1978, Dating and correlation, a review, in *Contributions to the Geologic Time Scale,* G. V. Cohee, M. F. Glaessner, and H. D. Hedberg, eds., American Association of Petroleum Geologists, Tulsa, Oklahoma, pp. 1–7.

McPhee, J., 1980, *Basin and Range,* Farrar, Straus, and Giroux, New York.

Meinzer, O. E., 1923, The occurrence of ground water in the United States, with a discussion of principles, *Geol. Water-Supply Pap., U. S. Geol. Surv.* 489:i–x, 1–321.

Menner, V. V., 1984, Subdivisions of the international stratigraphic scale: state and perspectives. *Proceedings, 27th Intern. Geol. Congress* (published by VNU Science Press, Utrecht) 1:1–10.

Mercanton, P. L., 1926, Inversion de l'inclinaison magnétique terrestre aux âges gèologiques, *Terrestrial Magnetism and Atmospheric Electricity* 31:187–190.

Merkel, R. H., 1979, *Well Log Formation Evaluation,* Continuing Education Course Note Series #14, American Association of Petroleum Geologists, Tulsa, Oklahoma.

Merrill, G. P., 1922, Contributions to a history of American state geological and natural history surveys, *Smithsonian Institution, U. S. Nat. Mus. Bull.* 109:1–549.

Merrill, G. P., 1924, *The First One Hundred Years of American Geology,* Yale University Press, New Haven, Connecticut.

Mertie, J. B., Jr., 1922, Graphic and mechanical computation of thickness of strata and distance to a stratum, *U. S. Geol. Soc. Prof. Paper* 129-C:39–52.

Miall, A. D., 1984a, *Principles of Sedimentary Basin Analysis,* Springer-Verlag, New York.

Miall, A. D., 1984b, Flysch and molasse: the elusive models, *Ann. Soc. Geol. Poloniae* 54-3/4:281–291.

Miall, A. D., 1984c, Deltas, in *Facies Models, Second Edition,* R. G. Walker, ed., Geoscience Canada, Reprint Series 1, Geological Association of Canada, Toronto, Ontario, pp. 105–140.

Miall, A. D., 1985a, Architectural-element analysis: a new method of facies analysis applied to fluvial deposits, *Earth-Science Reviews* 22:261–308.

Miall, A. D., 1985b, Stratigraphic and structural predictions from a plate-tectonic model of an oblique-slip orogen: the Eureka Sound Formation (Campanian–Oligocene), northeast Canadian Artic Islands, in *Strike-slip Deformation, Basin Formation, and Sedimentation,* K. T. Biddle and N. Christie-Blick, eds., Society of Economic Paleontologists and Mineralogists Special Publication 37, pp. 361–374.

Miall, A. D., 1986a, Eustatic sea level changes interpreted from seismic stratigraphy: a critique of the methodology with particular reference to the North Sea Jurassic, *Amer. Assoc. Petroleum Geol. Bull.* 70:131–137.

Miall, A. D., 1986b, The Eureka Sound Group (Upper Cretaceous–Oligocene), Canadian Artic Islands, *Bull. Canadian Petroleum Geology* 34:240–270.

Middleton, G. V., 1973, Johannes Walther's law of correlation of facies, *Geol. Soc. Amer. Bull.* 84:979–988.

Middleton, G. V., 1978, Facies, In *Encyclopedia of Sedimentology,* R. W. Fairbridge and J. Bourgeois, eds., Dowden, Hutchinson and Ross, Stroudsburg, Pennsylvania, pp. 323–325.

Milankovitch, M., 1930, Mathematische Klimalehre und astronomische Theorie der Klimaschwankungen, in *Handbuch der Klimatologie,* W. Köppen and R. Geiger, eds., Borntraeger, Berlin, vol. 1, pp. 1–176.

Miller, F. X., 1977, The graphic correlation method in biostratigraphy, in *Concepts and Methods of Biostratigraphy,* E. G. Kauffman and J. E. Hazel, eds., Dowden, Hutchinson, and Ross, Stroudsburg, Pennsylvania, pp. 165–186.

Miller, K. B., 1986a, The Paleoecologic significance of distal tempestites within a Middle Devonian muddy epeiric sea, *SEPM Annual Midyear Mtg. Abstracts* 3:77–78.

Miller, K. B., 1986b, Isochronous biofacies within the transgressive phase of a symmetric cycle from the Middle Devonian of western New York, *N. E. Section, GSA Abstracts with Programs* 18:55.

Miller, K. G., and R. G. Fairbanks, 1983, Evidence for Oligocene–Middle Miocene abyssal circulation changes in the western North Atlantic, *Nature* 306:250–253.

Miller, K. G., and R. G. Fairbanks, 1985, Oligocene to Miocene carbon isotope cycles and abyssal circulation changes, in *The Carbon Cycle and Atmospheric CO_2: Natural Variations Archean to Present,* Geophysical Monograph 32, American Geophysical Union, pp. 469–486.

Miller, K. G., and D. V. Kent, 1987, Testing Cenozoic eustatic changes: the critical role of stratigraphic resolution, in *Timing and Depositional History of Eustatic Sequences: Constraints on Seismic Stratigraphy,* C. A. Ross and D. Haman, eds., Cushman Foundation for Foraminiferal Research, Special Publication 24, Washington, D.C., pp. 51–56.

Miller, K. G., M.-P. Aubry, M. J. Khan, A. J. Melillo, D. V. Kent, and W. A. Berggren, 1985, Oligocene–Miocene biostratigraphy, magnetostratigraphy, and isotopic stratigraphy of the western North Atlantic, *Geology* 13:257–261.

Miller, K. G., W. B. Curry, and D. R. Ostermann, 1984, Late Paleogene (Eocene to Oligocene) benthic foraminiferal oceanography of the Goban Spur Region, Deep Sea Drilling Project Leg 80, *Init. Repts. Deep Sea Drilling Project* 80:505–538.

Miller, K. G., G. S. Mountain, and B. E. Tucholke, 1985, Oligocene glacio-eustacy on the margins of the North Atlantic, *Geology* 13:10–13.

Miller, T. G., 1964, The Luxembourg colloquium, *Geol. Mag.* 101:469–471.

Miller, T. G., 1965, Time in stratigraphy, *Palaeontology* 8:113–131.

Mitchum, R. M., Jr., P. R. Vail, and J. B. Sangree, 1977, Seismic stratigraphy and global changes of sea level, part 6: Stratigraphic interpretation of seismic reflection patterns in depositional sequences, in *Seismic Stratigraphy—Applications to Hydrocarbon Exploration,* C. E. Payton, ed., Memoir 26, American Association of Petroleum Geologists, Tulsa, Oklahoma, pp. 117–134.

Mitchum, R. M., Jr., P. R. Vail, and S. Thompson III, 1977, Seismic stratigraphy and global changes of sea level, part 2: The depositional sequence as a basic unit for stratigraphic analysis, in *Seismic Stratigraphy—Applications to Hydrocarbon Exploration,* C. E. Payton, ed., Memoir 26, American Association of Petroleum Geologists, Tulsa, Oklahoma, pp. 53–62.

Mojsisovics, E., 1879, *Die Dolomit-Riffe von Südtirol und Venetian,* Alfred Hölder, Vienna.

Monechi, S., U. Bleil, and J. Backman, 1985, Magnetobiochronology of late Cretaceous–Paleogene and Late Cenozoic pelagic sedimentary sequences from the northwest Pacific (Deep Sea Drilling project, Leg 86, Site 577), *Init. Repts. Deep Sea Drilling Project* 86:787–797.

Montadert, L., 1984, Problems in seismic stratigraphy, in *Stratigraphy Quo Vadis?* E. Seibold and J. D. Meulenkamp, eds., American Association of Petroleum Geologists, Tulsa, Oklahoma, pp. 3–7.

Moore, R. C., 1933, *Historical Geology,* McGraw-Hill, New York.

Moore, R. C., 1936, Stratigraphic classification of the Pennsylvanian rocks of Kansas, *Bull. Univ. Kansas* 36:1–256.

Moore, R. C., 1941, Stratigraphy, in *Geological Society of America: Geology, 1888–1938, Fiftieth Anniversary Volume,* C. P. Berkey, ed., Geological Society of America, Boulder, Colorado, pp. 177–220.

Moore, R. C., 1949, Meaning of facies, in *Sedimentary Facies in Geological History,* C. R. Longwell, ed., Geological Society of America, Memoir 39, Boulder, Colorado, pp. 1–34.

Moore, R. C., 1952, Orthography as a factor in stability of stratigraphical nomenclature, *State Geol. Surv. Kansas, Bull.* 96:363–372.

Moore, R. C., 1958, *Introduction to Historical Geology, 2nd edition,* McGraw-Hill, New York and London.

Morris, H. M., ed., 1974, *Scientific Creationism,* Creation-Life, San Diego, California.

Morris, H. M., 1977, *The Scientific Case for Creationism,* Creation-Life, San Diego, California.

Morrison, R. B., and H. E. Wright, Jr., eds., 1968, *Means of Correlation of Quaternary Successions,* University of Utah Press, Salt Lake City.

Mott, R. J., 1966, Quaternary palynological sampling techniques of the Geological Survey of Canada, *Geol. Surv. Can. Pap.* 66–41:1–24.

Mott, R. J., and D. R. Grant, 1985, Pre-Late Wisconsinian paleoenvironments in Atlantic Canada, *Géographie physique et Quaternaire* 39:239–254.

Mott, R. J., D. R. Grant, R. Stea, and S. Occhietti, 1986, Late-glacial climatic oscillation in Atlantic Canada equivalent to the Allerød/younger Dryas event, *Nature* 323:247–250.

Mountain, G., 1987, Cenozoic margin construction and destruction offshore New Jersey, in *Timing and Depositional History of Eustatic Sequences: Constraints on Seismic Stratigraphy,* C. A. Ross and D. Haman, eds., Cushman Foundation for Foraminiferal Research, Special Publication 24, Washington, D.C., pp. 57–83.

Mountain, G. S., and B. E. Tucholke, 1985, Mesozoic and Cenozoic geology of the U. S.

Atlantic continental slope and rise, in *Geologic Evolution of the United States Atlantic Margin,* C. W. Poag, ed., Van Nostrand Reinhold, New York, pp. 293–341.

Murphy, M. A., 1988, Unconformity-bounded stratigraphic units: discussion, *Geol. Soc. Amer. Bull.* 100:155.

Murray, J. W., 1981, *A Guide to Classification in Geology,* Ellis Horwood, Chichester, England.

Mutch, T. A., 1970, *Geology of the Moon: A Stratigraphic View,* Princeton University Press, Princeton, New Jersey.

Mutch, T. A., 1972, *Geology of the Moon: A Stratigraphic View, revised edition,* Princeton University Press, Princeton, New Jersey.

Naeser, N. D., and C. W. Naeser, 1984, Fission-track dating, in *Quaternary Dating Methods,* W. C. Mahaney, ed., Elsevier, Amsterdam, pp. 87–100.

Neidell, N. S., 1979, *Stratigraphic Modeling and Interpretation: Geophysical Principles and Techniques,* Education Course Note Series #13, American Association of Petroleum Geologists, Tulsa, Oklahoma.

Newell, N. D., 1967, Paraconformities, in *Essays in Paleontology and Stratigraphy: R. C. Moore Commemorative Volume,* C. Teichert and E. L. Yochelson, eds., University Press of Kansas, Lawrence, pp. 349–367.

Newton, C., and L. F. Laporte, in press, *Ancient Environments, 3rd edition,* Prentice-Hall, Englewood Cliffs, New Jersey.

North, F. K., 1985, *Petroleum Geology,* Allen and Unwin, Boston.

North American Commission on Stratigraphic Nomenclature (NACSN), 1983, North American Stratigraphic Code, *Amer. Assoc. Petrol. Geol. Bull.* 67:841–875.

Obradovich, J. D., 1984, An overview of the measurement of geologic time and the paradox of geologic time scales, *Proceedings, 27th Intern. Geol. Congress* (published by VNU Science Press, Utrecht) 1:11–30.

Odin, G. S., 1978, Results of dating Cretaceous, Paleogene sediments, Europe, in *Contributions to the Geologic Time Scale,* G. V. Cohee, M. F. Glaessner, and H. D. Hedberg, eds., American Association of Petroleum Geologists, Tulsa, Oklahoma, pp. 127–141.

Odin, G. S., ed., 1982a, *Numerical Dating in Stratigraphy,* Wiley, Chichester, England.

Odin, G. S., 1982b, Introduction: uncertainties in evaluating the numerical time scale, in *Numerical Dating in Stratigraphy,* G. S. Odin, ed., Wiley, Chichester, England, pp. 3–16.

Odin, G. S., 1984, The numerical age of system, series, and stage boundaries of the Phanerozoic column, in *Stratigraphy Quo Vadis?* E. Seibold and J. D. Meulenkamp, eds., American Association of Petroleum Geologists, Tulsa, Oklahoma, pp. 61–64.

Odin, G. S., and M. H. Dodson, 1982, Zero isotopic age of glauconies, in *Numerical Dating in Stratigraphy,* G. S. Odin, ed., Wiley, Chichester, England, pp. 277–305.

Odin, G. S., M. Renard, and C. V. Grazzini, 1982, Geochemical events as a means of correlation, in *Numerical Dating in Stratigraphy,* G. S. Odin, ed., Wiley, Chichester, England, pp. 37–71.

Ogden, J. G., III, 1977, The use and abuse of radiocarbon dating, *Annals New York Acad. Sci.* 288:167–173.

Ogg, J. G., and W. Lowrie, 1986, Magnetostratigraphy of the Jurassic/Cretaceous boundary, *Geology* 14:547–550.

Oppel, A., 1856-58, *Die Juraformation Englands, Frankreichs und des sudwestlichen Deutschlands, nach ihren einzelen gliedern eingetheilt und verglichen,* Von Ebner and Seubert, Stuttgart.

O'Rourke, J. E., 1976, Pragmatism versus materialism in stratigraphy, *Amer. Jour. Sci.* 276:47–55.

Osborn, H. F., 1916, The origin and evolution of life upon the Earth, *Sci. Monthly* 3:5–22.

Owen, D. E., 1987, Commentary: usage of stratigraphic terminology in papers, illustrations, and talks, *Jour. Sed. Petrol.* 57:363–372.

Palmer, A. R., compiler, 1983, The decade of North American geology 1983 geologic time scale, *Geology* 11:503–504.

Papp, A., 1979, Tertiary, in *Treatise on Invertebrate Paleontology,* R. A. Robinson and C. Teichert, eds., Geological Society of America, Boulder, Colorado, Part A, pp. A488–A504.

Paproth, E., and M. Streel, 1985, In search of a Devonian–Carboniferous boundary, *Episodes* 8:110–111.

Parker, M. L., L. A. Jozsa, S. G. Johnson, and P. A. Bramhall, 1984, Tree-ring dating in Canada and the northwestern U.S., in *Quaternary Dating Methods,* W. C. Mahaney, ed., Elsevier, Amsterdam, pp. 211–225.

Parker, W. C., A. J. Arnold, and W. A . Berggren, 1986, Analog morphocorrelation: a new technique with implications for high-resolution stratigraphy, *Palaios* 1:183–188.

Parkhomenko, E. I., 1967, *Electrical Properties of Rocks* (translated from the Russian and edited by G. V. Keller; also with a Supplementary Guide by G. V. Keller), Plenum Press, New York.

Parkinson, J., 1811, Observations on some of the strata in the neighborhood of London, and on the fossil remains contained in them, *Geol. Soc. London Trans.* 1:324–326.

Parsons, K., C. E. Brett, and K. B. Miller, 1986, Comparative taphonomy and sedimentary dynamics in Paleozoic marine shales, *North Amer. Paleo, Conv. IV, Abstracts with Programs,* pp. A34–A35.

Partin, R., 1969, The meaning of history: "A different approach entirely," *Historian* 33:431–435.

Paul, C. R. C., 1982, The adequacy of the fossil record, *Syst. Assoc. Spec. Vol.* 21:75–117.

Payton, C. E., ed., 1977, *Seismic Stratigraphy—Applications to Hydrocarbon Exploration,* Memoir 26, American Association of Petroleum Geologists, Tulsa, Oklahoma.

Pécsi, M., ed., 1985, *Loess and the Quaternary: Chinese and Hungarian Case Studies,* Akadémiai Kiadó, Office of Statistical Publishing House, Budapest.

Peterson, F., 1984, Fluvial sedimentation on a quivering craton: influence of slight crustal movements on fluvial processes, Upper Jurassic Formation, western Colorado Plateau, *Sed. Geol.* 38:21–49.

Peterson, F., 1986, Jurassic paleotectonics in the west-central part of the Colorado Plateau, Utah and Arizona, in *Paleotectonics and Sedimentation in the Rocky Mountain Region, United States,* American Association of Petroleum Geologists Memoir 41, Tulsa, Oklahoma, pp. 563–596.

Peterson, N., F. Heller, and W. Lowrie, 1984, Magnetostratigraphy of the Cretaceous/Tertiary geological boundary, *Init. Repts. Deep Sea Drilling Project* 73:657–661.

Phillips, J., 1829, *Illustrations of the Geology of Yorkshire; or, a description of the strata and organic remains of the Yorkshire Coast,* York.

Phillips, J., 1844, *Memoirs of William Smith, L.L.D., author of the "Map of the Strata of England and Wales,"* John Murray, London

Philological Society, 1933 [reprinted 1978], *The Oxford English Dictionary, vol. X,* Clarendon Press, Oxford.

Pirsson, L. V., and C. Schuchert, 1915, *A Text-Book of Geology,* Wiley, New York.

Playfair, J., 1802, *Illustrations of the Huttonian Theory of the Earth,* Cadell and Davies, London; and William Creech, Edinburgh.

Poag, C. W., ed., 1985, *Geologic Evolution of the United States Atlantic Margin,* Van Nostrand Reinhold, New York.

Pomerol, C., and I. Premoli-Silva, 1987, The Eocene–Oligocene boundary, *Episodes* 10:53–54.

Pompeckj, J. F., 1914, *Die Bedeutung des Schwäbischen Jura für das Erdegeschichte,* Schweizerbart, Stuttgart.

Popper, K., 1959, *The Logic of Scientific Discovery,* Basic Books, New York.

Popper, K., 1968, *The Logic of Scientific Discovery,* Harper and Row, New York.

Popper, K., 1984, Evolutionary epistemology, in *Evolutionary Theory: Paths into the Future,* J. W. Pollard, ed., Wiley, Chichester, England, pp. 239–255.

Porter, R., 1976, Charles Lyell and the principles of the history of geology, *Brit. Jour. Hist. Sci.* 9:91–103.

Powell, E. N., R. J. Stanton, Jr., H. Cummins, and G. Staff, 1982, Temporal fluctuations in bay environments—the death assemblage as a key to the past, in *Proceedings of the Symposium on Recent Benthonological Investigations in Texas and Adjacent States,* J. R. Davis, ed., Aquatic Sciences Section, Texas Academy of Science, Austin, pp. 203–232.

Premoli Silva, I., 1977, Upper Cretaceous–Paleocene magnetic stratigraphy at Gubbio, Italy: II. Biostratigraphy, *Geol. Soc. Amer. Bull.* 88:371–374.

Prothero, D. R., in press, *Interpreting the Stratigraphic Record,* W. H. Freeman, New York.

Quenstedt, F. A., 1856-58, *Der Jura,* H. Lauppschen, Tubingen.

Ramsbottom, W. H. C., 1977, Major cycles of transgression and regression (mesothems) in the Namurian, *Yorkshire Geol. Soc. Proc.* 41:261–291.

Ramsbottom, W. H. C., 1978, Namurian mesothems in South Wales and northern France, *Geol. Soc. London Jour.* 135:307–312.

Ramsbottom, W. H. C., 1979, Rates of transgression and regression in the Carboniferous of NW Europe, *Geol. Soc. London Jour.* 136:147–153.

Rapp, G., Jr., and J. A. Gifford, eds., 1985, *Archaeological Geology,* Yale University Press, New Haven, Connecticut.

Rapp, S. D., Jr., B. J. MacFadden, and J. A. Schiebout, 1983, Magnetic polarity stratigraphy of the Early Tertiary Black Peaks Formation, Big Bend National Park, Texas, *Jour. Geol.* 91:555–572.

Reading, H. G., ed., 1986, *Sedimentary Environments and Facies, 2nd edition,* Blackwell Publications, Oxford.

Renard, M., 1984, Trace element contents of carbonate sediments from the Mazagan escarpment off central Morocco (CYAMAZ, 1982), *Oceanologica Acta* 1984:153–159.

Renard, M., 1986, Pelagic carbonate chemostratigraphy (Sr, Mg, ^{18}O, ^{13}C), *Marine Micropaleont.* 10:117–164.

Renard, M., and D. Ambroise, 1983, Study of geochemical data (Eocene–Santonian samples) from Hole 516F (Leg 72) by the method of correspondence analysis, *Init. Repts. Deep Sea Drilling Project* 72:829–832.

Renard, M., F. Berthenet, S. Clauser, and G. Richebois, 1986, Geochemical events (trace elements and stable isotopes) recorded on bulk carbonates near the Eocene–Oligocene boundary. Application to the Contessa section (Gubbio, Umbria, Italia), in *Terminal Eocene Events,* C. Pomerol and I. Premoli Silva, eds., Elsevier, Amsterdam, pp. 331–348.

Retallack, G. J., 1981a, Two new approaches for reconstructing fossil vegetation with examples from the Triassic of Eastern Australia, in *Communities of the Past,* J. Gray, A. J. Boucot, and W. B. N. Berry, eds., Hutchinson Ross, Stroudsburg, Pennsylvania, pp. 271–295.

Retallack, G. J., 1981b, Preliminary observations on fossil soils in the Clarno Formation (Eocene to Early Oligocene) near Clarno, Oregon, *Oregon Geol.* 43:147–150.

Retallack, G. J., 1983, A paleopedological approach to the interpretation of terrestrial sedimentary rocks: the mid-Tertiary fossil soils of Badlands National Park, South Dakota, *Geol. Soc. Amer. Bull.* 94:823–840.

Retallack, G. J., 1984a, Trace fossils of burrowing beetles and bees in an Oligocene paleosol, Badlands National Park, South Dakota, *Jour. Paleontol.* 58:571–592.

Retallack, G. J., 1984b, Completeness of the rock and fossil record: some estimates using fossil soils, *Paleobiology* 10:59–78.

Retallack, G. J., 1986, The fossil record of soils, in *Paleosols: Their Recognition and Interpretations,* V. P. Wright, ed., Blackwell Publications, Oxford, pp. 1–57.

Retallack, G. J., and D. L. Dilcher, 1981, A coastal hypothesis for the dispersal and rise to dominance of flowering plants, in *Paleobotany, Paleoecology, and Evolution,* K. J. Niklas, ed., Praeger, New York, vol. 2, pp. 27–77.

Reyment, R. A., 1982, Correlating between electrical borehole logs in paleoecology, in

Quantitative Stratigraphic Correlation, J. M. Cubitt and R. A. Reyment, eds., Wiley, Chichester, England, pp. 233–240.

Reyment, R. A., R. E. Blackith, and N. A. Campbell, 1984, *Multivariate Morphometrics, 2nd Edition,* Academic Press, London.

Ricken, W., 1986, *Diagenetic Bedding,* Lecture Notes in Earth Sciences, vol. 6, Springer-Verlag, Berlin.

Ride, W. D. L., G. W. Sabrosky, G. Bernadi, and R. V. Melville, 1985, *International Code of Zoological Nomenclature, 3rd edition,* International Trust for Zoological Nomenclature, London.

Rider, M. H., 1986, *The Geological Interpretation of Well Logs,* Blackie, Glasgow; Halstead, Wiley, New York.

Riding, J. B., and W. A. S. Sarjeant, 1985, The role of dinoflagellate cysts in the biostratigraphical subdivision of the Jurassic System, *Newsl. Stratigr.* 14:96–109.

Rigby, J. K., and W. K. Hamblin, eds., 1972, *Recognition of Ancient Sedimentary Environments,* Special Publication No. 16, Society of Economic Paleontologists and Mineralogists, Lawrence, Kansas.

Rittmann, A., 1962, *Volcanoes and Their Activity,* Wiley, New York.

Riva, J. F., ed., 1985, *Field Trips Guidebook, 1985 Canadian Paleontology and Biostratigraphy Seminar,* Ste.-Foy, Quebec.

Roberts, J. L., 1982, *Introduction to Geological Maps and Structures,* Pergamon, Oxford.

Rodgers, J., 1959, The meaning of correlation, *Amer. Jour. Sci.* 257:684–691.

Roggenthen, W. M., and G. Napoleone, 1977, Upper Cretaceous–Paleocene magnetic stratigraphy at Gubbio, Italy: IV. Upper Maastrichtian–Paleocene magnetic stratigraphy, *Geol. Soc. Amer. Bull.* 88:378–382.

Ross, C. A., and D. Haman, eds., 1987, *Timing and Depositional History of Eustatic Sequences: Constraints on Seismic Stratigraphy,* Cushman Foundation for Foraminiferal Research, Special Publication 24, Washington, D.C.

Ross, R. J., and C. W. Naeser, 1984, The Ordovician time scale—new refinements, in *Aspects of the Ordovician System,* D. L. Bruton, ed., Paleontological Contributions of the University of Oslo, No. 295, Universitetsforlaget, Oslo, pp. 5–10.

Rowell, A. J., R. A. Robison, and D. K. Strickland, 1982, Aspects of Cambrian agnostoid phylogeny and chronocorrelation, *Jour. Paleontol.* 56:161–182.

Rucklidge, J. C., 1984, Radioisotope detection and dating with particle accelerators, in *Quaternary Dating Methods,* W. C. Mahaney, ed., Elsevier, Amsterdam, pp. 17–32.

Rudwick, M. J. S., 1976, *The Meaning of Fossils,* Neale Watson, New York.

Runcorn, S. K., 1966, Corals paleontological clocks, *Sci. Amer.* 215:26–33.

Rutherford, E., 1904, The radiation and emanation of radium, *Technics, for 1904,* pp. 11–16, 171–175.

Rutten, M. G., and H. Wensink, 1960, Paleomagnetic dating, glaciations and the chronology of the Plio-Pleistocene in Iceland, *Intern. Geol. Congress, 21st, Part IV, Proceedings,* pp. 62–70.

Rutter, N. W., ed., 1985, *Dating Methods of Pleistocene Deposits and Their Problems,* Geoscience Canada, Reprint Series 2, Geological Association of Canada, Toronto, Ontario.

Rutter, N. W., and R. J. Crawford, 1984, Utilizing wood in amino acid dating, in *Quaternary Dating Methods,* W. C. Mahaney, ed., Elsevier, Amsterdam, pp. 195–209.

Rutter, N. W., R. J. Crawford, and R. D. Hamilton, 1985, Dating methods of Pleistocene deposits and their problems: IV. Amino acid racemization dating, in *Dating Methods of Pleistocene Deposits and Their Problems,* N. W. Rutter, ed., Geoscience Canada Reprint Series 2, Geological Association of Canada, Toronto, Ontario, pp. 23–30.

Rzhevsky, V., and G. Novik, 1971, *The Physics of Rocks* (translated from the Russian by A. K. Chatterjee; translation edited by A. A. Beknazarov), Mir Publishers, Moscow.

Sadler, P. M., 1981, Sediment accumulation rates and the completeness of stratigraphic sections, *Jour. Geol.* 89:569–584.

Sadler, P. M., 1983, Is the present long enough to measure the past? *Nature* 302:752.

Salop, L. J., 1983, *Geological Evolution of the Earth During the Precambrian* (translated by V. P. Grudina), Springer-Verlag, Berlin.

Salvador, A., 1985, Chronostratigraphic and geochronometric scales in COSUNA stratigraphic correlation charts of the United States, *Amer. Assoc. Petrol. Geol. Bull.* 69:181–189.

Schadewald, R. J., 1983, The evolution of Bible-science, in *Scientists Confront Creationism*, L. R. Godfrey, ed., W. W. Norton, New York, pp. 283–299.

Schenck, H. G., and S. W. Muller, 1941, Stratigraphic terminology, *Geol. Soc. Amer. Bull.* 52:1419–1426.

Schindel, D. E., 1980, Microstratigraphic sampling and the limits of paleontologic resolution, *Paleobiology* 6:408–426.

Schindel, D. E., 1982a, The gaps in the fossil record, *Nature* 297:282–284.

Schindel, D. E., 1982b, Resolution analysis: a new approach to the gaps in the fossil record, *Paleobiology* 8:340–353.

Schindewolf, O. H., 1970a, Stratigraphical principles, *Newsl. Stratig.* 1:17–24.

Schindewolf, O. H., 1970b, Stratigraphie und Stratotypus, *Akad. Wiss. Lit. Mainz, Abh. Math.-Naturwiss,* K1:1–134.

Schlee, J. S., ed., 1984, *Interregional Unconformities and Hydrocarbon Accumulation*, AAPG Memoir 36, American Association of Petroleum Geologists, Tulsa, Oklahoma.

Schlumberger, 1972, *Log Interpretation, Volume 1—Principles,* Schlumberger Limited, New York.

Schlumberger, 1974, *Log Interpretation, Volume 2—Applications,* Schlumberger Limited, New York.

Schlumberger, 1986, *Log Interpretation Charts,* Schlumberger Well Services, Houston.

Schneer, C. J., ed., 1969, *Toward a History of Geology,* M.I.T. Press, Cambridge, Massachusetts.

Schoch, R. M., 1982, Gaps in the fossil record: fossils and stratigraphy, *Nature* 299:490.

Schoch, R. M., ed., 1984, *Vertebrate Paleontology,* Van Nostrand Reinhold, New York.

Schoch, R. M., 1986a, Systematics, functional morphology and macroevolution of the extinct mamalian order Taeniodonta, *Yale Univ. Peabody Mus. Nat. Hist. Bull.* 42:i–xii, 1–307.

Schoch, R. M., 1986b, *Phylogeny Reconstruction in Paleontology,* Van Nostrand Reinhold, New York.

Schoch, R. M., 1987, Phylogeny reconstruction for fossil versus living organisms, *Toth-Maatian Rev.* 6:2909–2911.

Schoch, R. M., 1988, A comment on the formal redefinition of the Hemingfordian–Barstovian land mammal age boundary, *Jour. Vertebrate Paleontol.* 7:472–473.

Schoch, R. M., and S. G. Lucas, 1981, A new species of *Conoryctella* (Mammalia: Taeniodonta) from the Paleocene of the San Juan Basin, New Mexico, and a revision of the genus, *Postilla* 185:1–23.

Schopf, T. J. M., 1981, Punctuated equilibrium and evolutionary stasis, *Paleobiology* 7:156–166.

Schwarcz, H. P., and B. Blackwell, 1985, Dating methods of Pleistocene deposits and their problems: II. Uranium series disequilibrium, in *Dating Methods of Pleistocene Deposits and Their Problems,* N. W. Rutter, ed., Geoscience Canada Reprint Series 2, Geological Association of Canada, Toronto, Ontario, pp. 9–17.

Schwarcz, H., and M. Gascoyne, 1984, Uranium-series dating of Quaternary deposits, in *Quaternary Dating Methods,* W. C. Mahaney, ed., Elsevier, Amsterdam, pp. 33–51.

Schwarzacher, W., 1975, *Sedimentation Models and Quantitave Stratigraphy,* Elsevier, Amsterdam.

Seibold, E., and J. D. Meulenkamp, eds., 1984, *Stratigraphy Quo Vadis?* AAPG Studies in Geology No. 16/IUGS Special Publication No. 14, American Association of Petroleum Geologists, Tulsa, Oklahoma.

Seilacher, A., 1984, Storm beds: their significance in event stratigraphy, in *Stratigraphy Quo Vadis?* E. Siebold and J. D. Meulenkamp, eds., American Association of Petroleum Geologists, Tulsa, Oklahoma, pp. 49–54.

Selley, R. C., 1976, *An Introduction to Sedimentology,* Academic Press, London.

Selley, R. C., 1978, *Concepts and Methods of Subsurface Facies Analysis,* Education Course Notes Series #9, American Association of Petroleum Geologists, Tulsa, Oklahoma.

Shackleton, N. J., M. A. Hall, and U. Bleil, 1985, Carbon-isotope stratigraphy, Site 577, *Init. Repts. Deep Sea Drilling Project* 86:503–511.

Shackleton, N. J., and N. D. Opdyke, 1976, Oxygen-isotope and paleomagnetic stratigraphy of Pacific core V28–239: Late Pliocene to latest Pleistocene, *Geol. Soc. Amer. Mem.* 145:449–464.

Shaler, N. S., J. B. Woodworth, and A. F. Foerste, 1899, *Geology of the Narragansett Basin,* U. S. Geological Survey Monograph 33, Government Printing Office, Washington, D.C.

Shaw, A. B., 1964, *Time in Stratigraphy,* McGraw-Hill, New York.

Shaw, B. R., 1982, A short note on the correlation of geologic sequences, in *Quantitative Stratigraphic Correlation,* J. M. Cubitt and R. A. Reyment, eds., Wiley, Chichester, England, pp. 7–11.

Sheriff, R. E., 1980, *Seismic Stratigraphy,* International Human Resources Development Corporation, Boston.

Sheriff, R. E., and L. P. Geldart, 1982, *Exploration Seismology, Volume 1, History, Theory, and Data Acquisition,* Cambridge University Press, Cambridge.

Sheriff, R. E., and L. P. Geldart, 1983, *Exploration Seismology, Volume 2. Data-Processing and Interpretation,* Cambridge University Press, Cambridge.

Shoemaker, E. M., and R. J. Hackman, 1962, Stratigraphic basis for a lunar time scale, in *The Moon,* Z. Kopal and Z. K. Mikhailov, eds., Academic Press, London, pp. 289–300.

Shrock, R .R., 1948, *Sequence in Layered Rocks,* McGraw-Hill, New York.

Silver, L. T., and P. H. Schultz, eds., 1982, Geological implications of impacts of large asteroids and comets on the earth, *Geol. Soc. Amer. Spec. Pap.* 190:i–xx, 1–528.

Simpson, G. G., 1963, Historical science, in *The Fabric of Geology,* C. C. Albritton, Jr., ed., Addison-Wesley, Reading, Massachusetts, pp. 24–48.

Simpson, G. G., 1970, Uniformitarianism. An inquiry into principle, theory, and method in geohistory and biohistory, in *Essays in Evolution and Genetics in Honor of Theodosius Dobzhansky,* M. K. Hecht and W. C. Steere, eds., Appleton-Century-Crofts, New York, pp. 43–96.

Sloss, L. L., 1963, Sequences in the cratonic interior of North America, *Geol. Soc. Amer. Bull.* 74:93–113.

Sloss, L. L., 1984, Comparative anatomy of cratonic unconformities, in *Interregional Unconformities and Hydrocarbon Accumulation,* J. S. Schlee, ed., American Association of Petroleum Geologists, Tulsa, Oklahoma, pp. 1–6.

Sloss, L. L., W. C. Krumbein, and E. C. Dapples, 1949, Integrated facies analysis, *Geol. Soc. Amer. Mem.* 39:91–124.

Smalley, I., 1986, INQUA's Commission on Loess, *Episodes* 9:170–172.

Smith, D. G., ed., 1981a, *The Cambridge Encyclopedia of Earth Sciences,* Crown Publishers, Cambridge University Press, New York.

Smith, D. G., 1981b, Historical geology: layers of earth history, in *The Cambridge Encyclopedia of Earth Sciences,* D. G. Smith, ed., Crown Publishers, Cambridge University Press, New York, pp. 386–409.

Smith, W., 1815, *A Memoir to the Map and Delineation of the Strata of England and Wales, with part of Scotland,* John Cary, London.

Smith, W., 1816–19, *Strata Identified by Organized Fossils, Containing Prints on Coloured Paper of the Most Characteristic Specimens in each Stratum,* London.

Smith, W., 1817, *Stratigraphical System of Organized Fossils, with reference to the specimens of the original geological collection in the British Museum: explaining their state of preservation and their use in identifying the British Strata,* E. Williams, London.

Sohl, N. F., and R. A. Christopher, 1983, The Black Creek–Peedee formational contact (Upper Cretaceous) in the Cape Fear River Region of North Carolina, *U. S. Geol. Surv. Prof. Paper* 1285:1–37.

Sorokin, V. S., 1984, Correlation of polyfacies sediments using historic-geological methods (on the unity of stratigraphy), *Proceedings, 27th Intern. Geol. Congress* (published by VNU Science Press, Utrecht) 1:219–240.

Southam, J. R., and W. W. Hay, 1978, Correlation of stratigraphic sections by continuous variables, *Computers and Geosciences* 4:257–260.

Southam, J. R., W. W. Hay, and T. R. Worsley, 1975, Quantitative formulation of reliability in stratigraphy, *Science* 188:357–359.

Staff, G. M., R. J. Stanton, Jr., E. N. Powell, and H. Cummins, 1986, Time-averaging, taphonomy, and their impact on paleocommunity reconstruction: death assemblages in Texas bays, *Geol. Soc. Amer. Bull.* 97:428–443.

Stalker, A. M., 1984, Field use of macrofeatures for correlating tills and estimating their ages: a review, in *Quaternary Dating Methods,* W. C. Mahaney, ed., Elsevier, Amsterdam, pp. 311–322.

Stanley, S. M., 1986, *Earth and Life through Time,* W. F. Freeman, New York.

Stearns, C. E., 1984, Uranium-series dating and the history of sea level, in *Quaternary Dating Methods,* W. C. Mahaney, ed., Elsevier, Amsterdam, pp. 53–66.

Steiger, R. H., and E. Jäger, 1977, Subcommission on Geochronology: convention on the use of decay constants in geo- and cosmochronology, *Earth Planet, Sci. Let.* 36:359–362.

Steiger, R. H., and E. Jäger, 1978, Subcommission on Geochronology: convention on the use of decay constants in geochronology and cosmochronology, in *Contributions to the Geologic Time Scale,* G. V. Cohee, M. F. Glaessner, and H. D. Hedberg, eds., American Association of Petroleum Geologists, Tulsa, Oklahoma, pp. 67–71.

Stein, R., and U. Bleil, 1986, Deep-water circulation in the northeast Atlantic and climatic changes during the Late Neogene (DSDP Site 141), *Marine Geol.* 70:191–209.

Steel, R., and A. P. Harvey, eds., 1979, *The Encyclopedia of Prehistoric Life,* McGraw-Hill, New York.

Steno, N., 1669, *De solido intra solidium naturaliter contento dissertationis prodromus,* The Star, Florence.

Stephens, G. C., and T. O. Wright, 1981, Stratigraphy of the Martinsburg Formation, west of Harrisburg in the Great Valley of Pennsylvania, *Amer. Jour. Sci.* 281: 1009–1020.

Stockdale, P. S., 1939, Lower Mississippian rocks of the east-central interior, *Geol. Soc. Amer. Spec. Pap.* 22:1–248.

Storey, T. S., and J. R. Patterson, 1959, Stratigraphy—traditional and modern concepts, *Amer. Jour. Sci.* 257:707–721.

Storzer, D., and G. A. Wagner, 1982, The application of fission track dating in stratigraphy: a critical review, in *Numerical Dating in Stratigraphy,* G. S. Odin, ed., Wiley, Chichester, England, pp. 199–221.

Strachey, J., 1719, Observation on the strata in the coal-mines of Mendip, in Somersetshire, *Royal Soc. Philos. Trans.* 30:968–973.

Strakhov, N. M., 1962, *Principles of Historical Geology, Part 1* (translated from Russian), Published for the National Science Foundation, Washington, D.C., by the Israel Program for Scientific Translations, Jerusalem.

Størmer, L., 1966, Concepts of stratigraphic classification and terminology, *Earth Sci. Rev.* 1:5–28.

Stupavsky, M., and C. P. Gravenor, 1984, Paleomagnetic dating of Quaternary sediments: a review, in *Quaternary Dating Methods,* W. C. Mahaney, ed., Elsevier, Amsterdam, pp. 123–140.

Suess, H. E., 1980, The radiocarbon record in tree rings of the last 8000 years, *Radiocarbon* 22:200–209.

Symons, D. T. A., M. Stupavsky, and C. P. Gravenor, 1980, Remanence resetting by shock-induced thixotropy in the Seminary Till, Scarborough, Ontario, Canada, *Geol. Soc. Amer. Bull.* 91:593–598.

Tauxe, L., and C. Badgley, 1984, Transition stratigraphy and the problem of remanence lock-in times in the Siwalik red beds, *Geophy. Res. Let.* 11:611–613.

Teichert, C., 1950, Zone concept in stratigraphy, *Amer. Assoc. Petrol. Geol. Bull.* 34:1585–1588.

Teichert, C., 1958, Concepts of facies, *Amer. Assoc. Petrol. Bull.* 42:2718–2744.

Teichert, C., and E. L. Yochelson, eds., 1967, *Essays in Paleontology and Stratigraphy: R. C. Moore Commemorative Volume,* Department of Geology, University of Kansas, Special Publication 2, University Press of Kansas, Lawrence.

Terasmae, J., 1984, Radiocarbon dating: some problems and potential developments, in *Quaternary Dating Methods,* W. C. Mahaney, ed., Elsevier, Amsterdam, pp. 1–15.

Theyer, F., L. A. Mayer, J. A. Barron, and E. Thomas, 1985, The equatorial Pacific high-productivity belt: elements for a synthesis of Deep Sea Drilling Project Leg 85 results, *Init. Repts. Deep Sea Drilling Project* 85:971–985.

Thiede, J., and W. U. Ehrmann, 1986, Late Mesozoic and Cenozoic sediment flux to the central North Atlantic Ocean, in *North Atlantic Palaeoceanography,* C. P. Summerhayes and N. J. Shackleton, eds., Geological Society Special Publication No. 21, pp. 3–15.

Thiede, J., A. Boersma, R. R. Schmidt, and E. Vincent, 1981, Reworked fossils in Mesozoic and Cenozoic pelagic central Pacific ocean sediments, Deep Sea Drilling Project Sites 463, 464, 465, and 466, Leg 62, *Init. Repts. Deep Sea Drilling Project* 62:495–512.

Thomson, W. [Lord Kelvin], 1862, On the secular cooling of the earth, *Trans. Roy. Soc. Edinburgh* 23:157–169.

Tipper, J. C., 1983, Rates of sedimentation, and stratigraphical completeness, *Nature* 302:696–698.

Tobin, R. C., 1986, An assessment of the lithostratigraphic and interpretive value of the traditional "biostratigraphy" of the type Upper Ordovician of North America, *Amer. Jour. Sci.* 286:673–701.

Tomida, Y., and R. F. Butler, 1980, Dragonian mammals and Paleocene magnetic polarity stratigraphy, North Horn Formation, Central Utah, *Amer. Jour. Sci.* 280:787–811.

Tomlinson, C. W., 1940, Technique of stratigraphic nomenclature, *Amer. Assoc. Petrol. Geol. Bull.* 24:2038–2048.

Tóth, J., 1978, Gravity-induced cross-formational flow of fluids, Red Earth Region, Alberta, Canada: analysis, patterns, and evolution, *Water Resources Research* 14:805–843.

Toulmin, G. H., 1783, *The Antiquity of the World, 2nd edition,* Cadell, London [cited from Simpson, 1970].

Toulmin, S., 1962–63, The discovery of time, *Manchester Lit. Phil. Soc. Mem. Proc.* 105:100–112.

Trembour, F., and T. Friedman, 1984, The present status of obsidian hydration dating, in *Quaternary Dating Methods,* W. C. Mahaney, ed., Elsevier, Amsterdam, pp. 141–151.

Trueman, A. E., 1923, Some theoretical aspects of correlation, *Geol. Assoc. London, Proc.* 34:193–206.

Tucholke, B. E., and G. S. Mountain, 1979, Seismic stratigraphy, lithostratigraphy and paleosedimentation patterns in the North American Basin, *Maurice Ewig Symposium (Amer. Geophy. Union), Series* 3:58–86.

Turner-Peterson, C. E., E. S. Santos, and N. S. Fishman, eds., 1986, *A Basin Analysis Case Study: The Morrison Formation Grants Uranium Region New Mexico,* AAPG Studies in Geology #22, Energy Minerals Division of The American Association of Petroleum Geologists, Tulsa, Oklahoma.

Ulrich, E. O., 1911, Revision of the Paleozoic systems, *Geol. Soc. Amer. Bull.* 22:281–680.

Ulrich, E. O., 1916, Correlation by displacements of the strandline and the function and proper use of fossils in correlation, *Geol. Soc. Amer. Bull.* 27:451–490.

Vail, P. R., and R. M. Mitchum, Jr., 1977, Seismic stratigraphy and global changes of sea level, part 1: overview, in *Seismic Stratigraphy—Applications to Hydrocarbon Exploration,* C. E. Payton, ed., Memoir 26, American Association of Petroleum Geologists, Tulsa, Oklahoma, pp. 51–52.

Vail, P. R., and R. G. Todd, 1981, North Sea Jurassic unconformities, chronostratigraphy and sea-level changes from seismic stratigraphy, *Proc. Petrol. Geol. Continental Shelf, Northwest Europe,* pp. 216–235

Vail, P. R., J. Hardenbol, and R. G. Todd, 1984, Jurassic unconformities, chronostratigraphy, and sea-level changes from seismic stratigraphy and biostratigraphy, in *Interregional Unconformities and Hydrocarbon Accumulation,* J. S. Schlee, ed., American Association of Petroleum Geologists, Tulsa, Oklahoma, pp. 129–144.

Vail, P. R., R. M. Mitchum, Jr., and S. Thompson III, 1977a, Seismic stratigraphy and global changes of sea level, part 3: relative changes of sea level from coastal onlap, in *Seismic Stratigraphy—Applications to Hydrocarbon Exploration,* C. E. Payton, ed., Memoir 26, American Association of Petroleum Geologists, Tulsa, Oklahoma, pp. 63–81.

Vail, P. R., R. M. Mitchum, Jr., and S. Thompson III, 1977b, Seismic stratigraphy and global changes of sea level, part 4: global cycles of relative changes of sea level, in *Seismic Stratigraphy—Applications to Hydrocarbon Exploration,* C. E. Payton, ed., Memoir 26, American Association of Petroleum Geologists, Tulsa, Oklahoma, pp. 83–97.

Vail, P. R., R. M. Mitchum, Jr., R. G. Todd, J. M. Widmier, S. Thompson III, J. B. Sangree, J. N. Bubb, and W. G. Hatlelid, 1977, Seismic stratigraphy and global changes of sea level [consisting of eleven parts], in *Seismic Stratigraphy—Applications to Hydrocarbon Exploration,* C. E. Payton, ed., Memoir 26, American Association of Petroleum Geologists, Tulsa, Oklahoma, pp. 49–212.

Vail, P. R., R. G. Todd, and J. B. Sangree, 1977, Seismic stratigraphy and global changes of sea level, part 5: Chronostratigraphic significance of seismic reflections, in *Seismic Stratigraphy—Applications to Hydrocarbon Exploration,* C. E. Payton, ed., Memoir 26, American Association of Petroleum Geologists, Tulsa, Oklahoma, pp. 99–116.

Valentine, J. W., 1963, Biogeographic units as biostratigraphic units, *Amer. Assoc. Petrol. Geol. Bull.* 47:457–466.

Valentine, J. W., 1977, Biogeography and biostratigraphy, in *Concepts and Methods of Biostratigraphy,* E. G. Kauffman and J. E. Hazel, eds., Dowden, Hutchinson and Ross, Stroudsburg, Pennsylvania, pp. 143–162.

Van Andel, T. H., 1981, Consider the incompleteness of the geological record, *Nature* 294:397–398.

Van Andel, T. H., 1985, *New Views of an Old Planet: Continental Drift and the History of Earth,* Cambridge University Press, Cambridge.

Van Andel, T. H., and J. R. Curray, 1960, Regional aspects of modern sedimentation in northern Gulf of Mexico and similar basins, and paleogeographic significance, in *Recent Sediments, Northwest Gulf of Mexico,* F. P. Shepard, F. B. Phleger, and T. H. Van Andel, eds., American Association of Petroleum Geologists, Tulsa, Oklahoma, pp. 345–364.

Van Devender, T. R., 1986, Pleistocene climates and endemism in the Chihuahuan desert flora, in *Invited Papers from the Second Symposium on Resources of the Chihuahuan Desert Region, United States and Mexico,* J. C. Barlow, A. M. Powell, and B. N. Timmermann, eds., Chihuahuan Desert Research Institute, Alpine, Texas, pp. 1–19.

Van Devender, T. R., J. L. Betancourt, and M. Wimberly, 1984, Biogeographic implications of a packrat midden sequence from the Sacramento Mountains, south-central New Mexico, *Quat. Research* 22:344–360.

Van Devender, T. R., P. S. Martin, R. S. Thompson, K. L. Cole, A. J. T. Jull, A. Long, L. J. Toolin, and D. J. Donahue, 1985, Fossil packrat middens and the tandem accelerator mass spectrometer, *Nature* 317:610–613.

Van Eysinga, F. W. B., 1978, *Geological Time Table, 3rd edition,* Elsevier, Amsterdam.

Van Hinte, J. E., 1976a, A Jurrasic time scale, *Amer. Assoc. Petrol. Geol. Bull.* 60:489–497.

Van Hinte, J. E., 1976b, A Cretaceous time scale, *Amer. Assoc. Petrol. Geol. Bull.* 60:498–516.

Verosub, K. L., 1975, Paleomagnetic excursions as magnetostratigraphic horizons, *Science* 190:48–50.

Verosub, K. L., 1977a, Depositional and post-depositional processes in the magnetization of sediments, *Rev. Geophys. Space Phys.* 15:145–155.

Verosub, K. L., 1977b, The absence of the Mono Lake geomagnetic excursion from the paleomagnetic record of Clear Lake, California, *Earth Planet. Sci. Let.* 36:219–230.

Verosub, K. L., 1979, Paleomagnetic evidence for the occurrence of rapid shifts in the position of the geomagnetic pole, *EOS Trans. Amer. Geophys. Union* 60:244.

Verosub, K. L., and S. K. Banerjee, 1977, Geomagnetic excursions and their paleomagnetic record, *Rev. Geophys. Space Phys.* 15:145–155.

Visher, G. S., 1984, *Exploration Stratigraphy,* PennWell, Tulsa, Oklahoma.

Vreeken, W. J., 1984, Relative dating of soils and paleosols, in *Quaternary Dating Methods,* W. C. Mahaney, ed., Elsevier, Amsterdam, pp. 269–281.

Wadia, D. N., 1953, *Geology of India, 3rd edition,* Macmillan, London.

Walcott, C. D., 1893, Geologic time, as indicated by the sedimentary rocks of North America, *Jour. Geol.* 1:639–676.

Walker, R. G., ed., 1984a, *Facies Models, Second Edition,* Geoscience Canada, Reprint Series 1, Geological Association of Canada, Toronto, Ontario.

Walker, R. G., 1984b, General introduction: facies, facies sequences and facies models, in *Facies Models, Second Edition,* R. G. Walker, ed., Geoscience Canada, Reprint Series 1, Geological Association of Canada, Toronto, Ontario, pp. 1–9.

Walther, J., 1893–94, *Einleitung in die Geologie als historische Wissenschaft,* Verlag von Gustav Fischer, Jena.

Walther, J., 1897, Ueber die Lebensweise fossiler Meeresthiere, *Deutsch, Geol. Gesell. Zeitschr.* 49:209–273.

Warner, R., 1801, *The History of Bath,* R. Cruttwell, Bath; and J. Robinson, London.

Watkins, N. D., 1972, Review of the development of the geomagnetic polarity time scale and discussion of prospects for its finer definition, *Geol. Soc. Amer. Bull.* 83:551–574.

Watson, R. A., 1966, Discussion: is geology different? A critical discussion of "The Fabric of Geology" [C. C. Albritton, Jr., ed.], *Philosophy of Science* 33:172–185.

Watson, R. A., 1969, Explanation and prediction in geology, *Jour. Geol.* 77:488–494.

Watson, R. A., 1976a, Reviews of "Recent Earth History" by C. Vita-Finzi and "The Nature of the Stratigraphical Record" by D. V. Ager, *Philosophy of Science* 43:458–459.

Watson, R. A., 1976b, Laws, systems, certainty, and particularities, *Amer. Anthropologist* 78:341–343.

Watson, R. A., 1982, Absence as evidence in geology, *Jour. Geol. Education* 30:300–301.

Watson, R. A., 1983, A critique [of] chronostratigraphy, *Amer. Jour. Sci.* 283:173–177.

Watson, R. A., 1985, Reply [to Lucas, 1985], *Amer. Jour. Sci.* 285:766–767.

Watson, R. A., and H. E. Wright, Jr., 1963, Landslides on the east flank of the Chuska Mountains, northwestern New Mexico, *Amer. Jour. Sci.* 261:525–548.

Watson, R. A., and H. E. Wright, Jr., 1980, The end of the Pleistocene: a general critique of chronostratigraphic classification, *Boreas* 9:153–163.

Weddige, K., and W. Ziegler, 1979, Evolutionary patterns in Middle Devonian conodont genera *Polygnathus* and *Icriodus, Geologica et Palaeontologica* 13:157–164.

Wehmiller, J. F., 1984, Relative and absolute dating of Quaternary mollusks with amino acid racemization: evaluation, applications and questions, in *Quaternary Dating Methods,* W. C. Mahaney, ed., Elsevier, Amsterdam, pp. 171–193.

Weimer, R. J., 1980, Recurrent movement on basement faults, a tectonic style for Colorado and adjacent areas, in *Colorado Geology,* H. C. Kent and K. W. Porter, eds., Rocky Mountain Association of Geologists Symposium, Denver, pp. 23–35.

Weimer, R. J., 1984, Relation of unconformities, tectonics, and sea-level changes, Cretaceous of Western Interior, U.S.A., in *Interregional Unconformities and Hydrocarbon Accumulation,* J. S. Schlee, ed., AAPG Memoir 36, pp. 7–35.

Weinreich, N., and F. Theyer, 1986, Magnetostratigraphy of Neogene equatorial Pacific pelagic sediments, *Jour. Geophys.* 59:183–194.

Weisburd, S., 1985, Earth's "pulses" tied to plate rates, *Science News* 127:324.

Wells, J. W., 1947, Provisional paleoecological analysis of Devonian rocks of the Columbus region, *Ohio Jour. Sci.* 47:119–126.

Wells, J. W., 1963, Early investigations of the Devonian System in New York, 1656–1836, *Geol. Soc. Amer. Spec. Pub.* 74:1–74.

Weller, J. M., 1960, *Stratigraphic Principles and Practice,* Harper, New York.

West, R. R., and R. M. Busch, 1985, Sixth-order transgressive–regressive units in the Wewoka Formation (Ana Shale Member) of Oklahoma, in *Recent Interpretations of Late Paleozoic Cyclothems,* W. L. Watney, R. L. Kaesler, and K. D. Newell, eds., Society of Economic Paleontologists and Mineralogists, Mid-Continent Section, Lawrence, Kansas, pp. 155–170.

Westgate, J. A., and N. D. Naeser, 1985, Dating methods of Pleistocene deposits and their problems: V. Tephrochronology and fission-track dating, in *Dating Methods of Pleistocene Deposits and Their Problems,* N. W. Rutter, ed., Geoscience Canada Reprint Series 2, Geological Association of Canada, Toronto, Ontario, pp. 31–38.

Wetherill, G. S., 1956, An interpretation of the Rhodesia and Witwatersrand age patterns, *Geochim. Cosmochim. Acta* 9:290–292.

Wheeler, H. E., 1959, Stratigraphic units in space and time, *Amer. Jour. Sci.* 257:692–706.

Wheeler, H. E., and E. M. Beasley, 1948, Critique of the time-stratigraphic concept, *Geol. Soc. Amer. Bull.* 59:75–86.

Wheeler, H. E., and V. S. Mallory, 1953, Designation of stratigraphic units, *Amer. Assoc. Petrol. Geol. Bull.* 37:2407–2421.

Wheeler, H. E., and V. S. Mallory, 1956, Factors in lithostratigraphy, *Amer. Assoc. Petrol. Geol. Bull.* 40:2711–2723.

[Whewell, W.], 1832, [Review of vol. 2 of Lyell, 1830–33], *Quart. Rev.* 47:103–132.

Whitcomb, J. C., and H. M. Morris, 1961, *The Genesis Flood,* Presbyterian and Reformed, Nutley, New Jersey.

White, R. S., 1983, The Little Murray Ridge, in *Seismic Expression of Structural Styles,* A. W. Bally, ed., American Association of Petroleum Geologists, Tulsa, Oklahoma, vol. 1, 1.3–19 to 1.3–23.

Wiedmann, J., 1970, Problems of stratigraphic classification and the definition of stratigraphic boundaries. *Newsl. Stratigr.* 1:35–48.

Wilhelms, D. E., 1970, Summary of lunar stratigraphy—telescopic observations, *U. S. Geol. Surv. Prof. Pap.* 599-F:F1–F47.

Wilhelms, D. E., 1987, The geologic history of the moon, *U. S. Geol. Surv. Prof. Pap.* 1348, Washington, D.C.

Williams, H. S., 1894, Dual nomenclature in geological classification, *Jour. Geol.* 2:145–160.

Williams, H. S., 1901, The discrimination of time-values in geology, *Jour. Geol.* 9:570–585.

Williams, H. S., 1903, The correlation of geological faunas: a contribution to Devonian paleontology, *U. S. Geol. Surv. Bull.* 210:1–147.

Williams, H., F. J. Turner, and C. M. Gilbert, 1954, *Petrology: An Introduction to the Study of Rocks in Thin Sections,* W. H. Freeman, San Francisco.

Williams, H. L., 1963, *Country Furniture of Early America,* A. S. Barnes, New York.

Wilmarth, M. G., 1925, The geological time classification of the United States Geological Survey compared with other classifications, accompanied by the original definitions of era, period, and epoch terms, *U. S. Geol. Surv. Bull.* 769:1–138.

Wilmarth, M. G., 1957, Lexicon of geologic names of the United States, *U. S. Geol. Surv. Bull.* 896:1–2396.

Wilson, D., G. C. Keroher, and B. E. Hansen, 1959, Index to the geologic names of North America, *U. S. Geol. Surv. Bull.* 1056-B:407–622.

Wilson, J. A., 1959a, Transfer, a synthesis of stratigraphic processes, *Bull. Amer. Assoc. Petrol. Geol.* 43:2861–2862.

Wilson, J. A., 1959b, Stratigraphic concepts in vertebrate paleontology, *Amer. Jour. Sci.* 257:770–778.

Wilson, J. A., 1971, Stratigraphy and classification, *Abh. hess. L.-Amt Bodenforsch* 60:195–202.

Wilson, J. A., 1975, Geochronology, stratigraphy, and typology, *Fieldiana (Geology)* 33:193–204.

Wilson, J. A., 1986, Stratigraphic occurrence and correlation of Early Tertiary vertebrate faunas, Trans-Pecos Texas: Agua Fria-Green Valley areas, *Jour. Vertebrate Paleontol.* 6:350–373.

Wilson, M. A., 1982, Origin of brachiopod–bryozoan assemblages in an Upper Carboniferous limestone: importance of physical and ecological controls, *Lethaia* 15:263–273.

Wilson, M. A., 1985a, Conodont biostratigraphy and paleoenvironments at the Mississippian–Pennsylvanian boundary (Carboniferous: Namurian) in the Spring Mountains of southern Nevada, *Newsl. Stratigr.* 14:69–80.

Wilson, M. A., 1985b, Disturbance and ecological succession in an Upper Ordovician cobble-dwelling hardground fauna, *Science* 228:575–577.

Wilson, M. A., 1985c, A taxonomic diversity measure for encrusting organisms, *Lethaia* 18:166.

Wise, D. U., 1974, Continental margins, freeboard and the volumes of continents and oceans through time, in *The Geology of Continental Margins,* C. A. Burke and C. L. Drake, eds., Springer-Verlag, New York, pp. 45–58.

Wobus, R. A., 1985, Changes in the nomenclature and stratigraphy of Proterozoic metamorphic rocks, Tusas Mountains, north-central New Mexico, *U. S. Geol. Surv. Bull.* 1571:1–19.

Wood, H. E., 2nd, R. W. Chaney, J. Clark, E. H. Colbert, G. L. Jepsen, J. B. Reeside, Jr., and C. Stock, 1941, Nomenclature and correlation of the North American continental Tertiary, *Bull. Geol. Soc. Amer.* 52:1–48.

Woodburne, M. O., ed., 1987, *Cenozoic Mammals of North America: Geochronology and Biostratigraphy,* University of California Press, Berkeley.

Woodford, A. O., 1963, Correlation by fossils, in *The Fabric of Geology,* C. C. Albritton, ed., Addison-Wesley, New York, pp. 75–111.

Woodward, J., 1695, *An Essay toward a Natural History of the Earth,* R. Wilkin, London.

Woolley, C. L., 1931, *Digging Up the Past,* Charles Scribner's Sons, New York.

Wright, A. A., U. Bleil, S. Monechi, H. V. Michel, N. J. Shackleton, B. R. T. Simoneit, and J. C. Zachos, 1985, Summary of Cretaceous/Tertiary boundary studies, Deep Sea Drilling Project Site 577, Shatsky Rise, *Init. Repts. Deep Sea Drilling Project* 86:799–804.

Wright, T. O., 1981, Sedimentology of the Robertson Bay Group, North Victoria Land, Antarctica, *Geol. Jb.* B 41:127–138.

Wright, T. O., 1985, Late Precambrian and early Paleozoic tectonism and associated sedimentation in northern Victoria Land, Antarctica, *Geol. Soc. Amer. Bull.* 96:1332–1339.

Wright, T. O., and R. H. Findlay, 1984, Relationships between the Robertson Bay Group and the Bowers Supergroup—new progress and complications from the Victory Mountains, North Victoria Land, *Geol. Jb.* B 60:105–116.

Wright, T. O., R. J. Ross, Jr., and J. E. Repetski, 1984, Newly discovered youngest Cambrian or oldest Ordovician fossils from the Robertson Bay terrane (formerly Precambrian), northern Victoria Land, Antarctica, *Geology* 12:301–305.

Yochelson, E. L., ed., 1970–71, *Proceedings of the North American Paleontological Convention, Field Museum of Natural History, Chicago, September 5–7, 1969,* Allen, Lawrence, Kansas.

Young, G., and J. Bird, 1822, *A Geological Survey of the Yorkshire Coast: describing the strata and fossils occurring between the Humber and the Tees, from the Germain Ocean to the Plain of York,* George Clark, Whitby, England.

Ziegler, W., 1982, Conodont age of *Pharciceras lunulicost*-Zone, *Cour. Forsch.-Inst. Senckenberg* 55:493–495.

Ziegler, W., and G. Klapper, 1982a, Subcommission on Devonian stratigraphy: decisions since 1973 and present status, *Cour. Forsch.-Inst. Senckenberg* 55:7–11.

Ziegler, W., and G. Klapper, 1982b, The *disparilis* conodont zone, the proposed level for the Middle–Upper Devonian boundary, *Cour. Forsch.-Inst. Senckenberg* 55:463–491.

Ziegler, W., and G. Klapper, 1985, Stages of the Devonian System, *Episodes* 8:104–109.

Ziegler, W., and C. A. Sandberg, 1984a, *Palmatolepis*-based revision of the upper part of standard late Devonian conodont zonation, *Geol. Soc. Amer. Spec. Paper* 196:179–194.

Ziegler, W., and C. A. Sandberg, 1984b, Important candidate sections for stratotype of conodont based Devonian-Carboniferous boundary, *Cour. Forsch.-Inst. Senckenberg* 67:231–239.

Zittel, K. von, 1901, *History of Geology and Palaeontology to the End of the Nineteenth Century,* Walter Scott, London.

APPENDIX 1

NORTH AMERICAN STRATIGRAPHIC CODE (1983)

by

The North American Commission on Stratigraphic Nomenclature

[Reprinted from the *American Association of Petroleum Geologists Bulletin* 67:841–875, 1983, with the permission of the American Association of Petroleum Geologists.]

North American Stratigraphic Code[1]

NORTH AMERICAN COMMISSION ON STRATIGRAPHIC NOMENCLATURE

FOREWORD

This code of recommended procedures for classifying and naming stratigraphic and related units has been prepared during a four-year period, by and for North American earth scientists, under the auspices of the North American Commission on Stratigraphic Nomenclature. It represents the thought and work of scores of persons, and thousands of hours of writing and editing. Opportunities to participate in and review the work have been provided throughout its development, as cited in the Preamble, to a degree unprecedented during preparation of earlier codes.

Publication of the International Stratigraphic Guide in 1976 made evident some insufficiencies of the American Stratigraphic Codes of 1961 and 1970. The Commission considered whether to discard our codes, patch them over, or rewrite them fully, and chose the last. We believe it desirable to sponsor a code of stratigraphic practice for use in North America, for we can adapt to new methods and points of view more rapidly than a worldwide body. A timely example was the recognized need to develop modes of establishing formal nonstratiform (igneous and high-grade metamorphic) rock units, an objective which is met in this Code, but not yet in the Guide.

The ways in which this Code differs from earlier American codes are evident from the Contents. Some categories have disappeared and others are new, but this Code has evolved from earlier codes and from the International Stratigraphic Guide. Some new units have not yet stood the test of long practice, and conceivably may not, but they are introduced toward meeting recognized and defined needs of the profession. Take this Code, use it, but do not condemn it because it contains something new or not of direct interest to you. Innovations that prove unacceptable to the profession will expire without damage to other concepts and procedures, just as did the geologic-climate units of the 1961 Code.

This Code is necessarily somewhat innovative because of: (1) the decision to write a new code, rather than to revise the old; (2) the open invitation to members of the geologic profession to offer suggestions and ideas, both in writing and orally; and (3) the progress in the earth sciences since completion of previous codes. This report strives to incorporate the strength and acceptance of established practice, with suggestions for meeting future needs perceived by our colleagues; its authors have attempted to bring together the good from the past, the lessons of the Guide, and carefully reasoned provisions for the immediate future.

Participants in preparation of this Code are listed in Appendix I, but many others helped with their suggestions and comments. Major contributions were made by the members, and especially the chairmen, of the named subcommittees and advisory groups under the guidance of the Code Committee, chaired by Steven S. Oriel, who also served as principal, but not sole, editor. Amidst the noteworthy contributions by many, those of James D. Aitken have been outstanding. The work was performed for and supported by the Commission, chaired by Malcolm P. Weiss from 1978 to 1982.

This Code is the product of a truly North American effort. Many former and current commissioners representing not only the ten organizational members of the North American Commission on Stratigraphic Nomenclature (Appendix II), but other institutions as well, generated the product. Endorsement by constituent organizations is anticipated, and scientific communication will be fostered if Canadian, United States, and Mexican scientists, editors, and administrators consult Code recommendations for guidance in scientific reports. The Commission will appreciate reports of formal adoption or endorsement of the Code, and asks that they be transmitted to the Chairman of the Commission (c/o American Association of Petroleum Geologists, Box 979, Tulsa, Oklahoma 74101, U.S.A.).

Any code necessarily represents but a stage in the evolution of scientific communication. Suggestions for future changes of, or additions to, the North American Stratigraphic Code are welcome. Suggested and adopted modifications will be announced to the profession, as in the past, by serial Notes and Reports published in the *Bulletin* of the American Association of Petroleum Geologists. Suggestions may be made to representatives of your association or agency who are current commissioners, or directly to the Commission itself. The Commission meets annually, during the national meetings of the Geological Society of America.

[1]Manuscript received, December 20, 1982; accepted, January 21, 1983.
Copies are available at $1.00 per copy postpaid. Order from American Association of Petroleum Geologists, Box 979, Tulsa, Oklahoma 74101.

1982 NORTH AMERICAN COMMISSION
ON STRATIGRAPHIC NOMENCLATURE

CONTENTS

North American Stratigraphic Code

North American Commission on Stratigraphic Nomenclature

North American Stratigraphic Code

North American Commission on Stratigraphic Nomenclature

North American Stratigraphic Code

North American Commission on Stratigraphic Nomenclature

PART I. PREAMBLE

BACKGROUND

PERSPECTIVE

Codes of Stratigraphic Nomenclature prepared by the American Commission on Stratigraphic Nomenclature (ACSN, 1961) and its predecessor (Committee on Stratigraphic Nomenclature, 1933) have been used widely as a basis for stratigraphic terminology. Their formulation was a response to needs recognized during the past century by government surveys (both national and local) and by editors of scientific journals for uniform standards and common procedures in defining and classifying formal rock bodies, their fossils, and the time spans represented by them. The most recent Code (ACSN, 1970) is a slightly revised version of that published in 1961, incorporating some minor amendments adopted by the Commission between 1962 and 1969. The Codes have served the profession admirably and have been drawn upon heavily for codes and guides prepared in other parts of the world (ISSC, 1976, p. 104-106). The principles embodied by any code, however, reflect the state of knowledge at the time of its preparation, and even the most recent code is now in need of revision.

New concepts and techniques developed during the past two decades have revolutionized the earth sciences. Moreover, increasingly evident have been the limitations of previous codes in meeting some needs of Precambrian and Quaternary geology and in classification of plutonic, high-grade metamorphic, volcanic, and intensely deformed rock assemblages. In addition, the important contributions of numerous international stratigraphic organizations associated with both the International Union of Geological Sciences (IUGS) and UNESCO, including working groups of the International Geological Correlation Program (IGCP), merit recognition and incorporation into a North American code.

For these and other reasons, revision of the American Code has been undertaken by committees appointed by the North American Commission on Stratigraphic Nomenclature (NACSN). The Commission, founded as the American Commission on Stratigraphic Nomenclature in 1946 (ACSN, 1947), was renamed the NACSN in 1978 (Weiss, 1979b) to emphasize that delegates from ten organizations in Canada, the United States, and Mexico represent the geological profession throughout North America (Appendix II).

Although many past and current members of the Commission helped prepare this revision of the Code, the participation of all interested geologists has been sought (for example, Weiss, 1979a). Open forums were held at the national meetings of both the Geological Society of America at San Diego in November, 1979, and the American Association of Petroleum Geologists at Denver in June, 1980, at which comments and suggestions were offered by more than 150 geologists. The resulting draft of this report was printed, through the courtesy of the Canadian Society of Petroleum Geologists, on October 1, 1981, and additional comments were invited from the profession for a period of one year before submittal of this report to the Commission for adoption. More than 50 responses were received with sufficient suggestions for improvement to prompt moderate revision of the printed draft (NACSN, 1981). We are particularly indebted to Hollis D. Hedberg and Amos Salvador for their exhaustive and perceptive reviews of early drafts of this Code, as well as to those who responded to the request for comments. Participants in the preparation and revisions of this report, and conferees, are listed in Appendix I.

Some of the expenses incurred in the course of this work were defrayed by National Science Foundation Grant EAR 7919845, for which we express appreciation. Institutions represented by the participants have been especially generous in their support.

SCOPE

The North American Stratigraphic Code seeks to describe explicit practices for classifying and naming all formally defined geologic units. *Stratigraphic procedures* and principles, although developed initially to bring order to strata and the events recorded therein, are applicable to all earth materials, not solely to strata. They promote systematic and rigorous study of the composition, geometry, sequence, history, and genesis of rocks and unconsolidated materials. They provide the framework within which time and space relations among rock bodies that constitute the Earth are ordered systematically. Stratigraphic procedures are used not only to reconstruct the history of the Earth and of extra-terrestrial bodies, but also to define the distribution and geometry of some commodities needed by society. *Stratigraphic classification* systematically arranges and partitions bodies of rock or unconsolidated materials of the Earth's crust into units based on their inherent properties or attributes.

A *stratigraphic code* or guide is a formulation of current views on stratigraphic principles and procedures designed to promote standardized classification and formal nomenclature of rock materials. It provides the basis for formalization of the language used to denote rock units and their spatial and temporal relations. To be effective, a code must be widely accepted and used; geologic organizations and journals may adopt its recommendations for nomenclatural procedure. Because any code embodies only current concepts and principles, it should have the flexibility to provide for both changes and additions to improve its relevance to new scientific problems.

Any system of nomenclature must be sufficiently explicit to enable users to distinguish objects that are embraced in a class from those that are not. This stratigraphic code makes no attempt to systematize structural, petrographic, paleontologic, or physiographic terms. Terms from these other fields that are used as part of formal stratigraphic names should be sufficiently general as to be unaffected by revisions of precise petrographic or other classifications.

The objective of a system of classification is to promote unambiguous communication in a manner not so restrictive as to inhibit scientific progress. To minimize ambiguity, a code must promote recognition of the distinction between observable features (reproducible data) and inferences or interpretations. Moreover, it should be sufficiently adaptable and flexible to promote the further development of science.

Stratigraphic classification promotes understanding of the *geometry* and *sequence* of rock bodies. The development of stratigraphy as a science required formulation of the Law of Superposition to explain sequential stratal relations. Although superposition is not applicable to many igneous, metamorphic, and tectonic rock assemblages, other criteria (such as cross-cutting relations and isotopic dating) can be used to determine sequential arrangements among rock bodies.

The term *stratigraphic unit* may be defined in several ways. Etymological emphasis requires that it be a stratum or assemblage of adjacent strata distinguished by any or several of the many properties that rocks may possess (ISSC, 1976, p. 13). The scope of stratigraphic classification and procedures, however, suggests a broader definition: a naturally occurring body of rock or rock material distinguished from adjoining rock on the basis of some stated property or properties. Commonly used properties include composition, texture, included fossils, magnetic signature, radioactivity, seismic velocity, and age. Sufficient care is required in defining the boundaries of a unit to enable others to distinguish the material body from those adjoining it. Units based on one property commonly do not coincide with those based on another and, therefore, distinctive terms are needed to identify the property used in defining each unit.

The adjective *stratigraphic* is used in two ways in the remainder of this report. In discussions of lithic (used here as synonymous with "lithologic") units, a conscious attempt is made to restrict the term to lithostratigraphic or layered rocks and sequences that obey the Law of Superposition. For nonstratiform rocks (of plutonic or tectonic origin, for example), the term *lithodemic* (see Article 27) is used. The adjective *stratigraphic* is

North American Stratigraphic Code

also used in a broader sense to refer to those procedures derived from stratigraphy which are now applied to all classes of earth materials.

An assumption made in the material that follows is that the reader has some degree of familiarity with basic principles of stratigraphy as outlined, for example, by Dunbar and Rodgers (1957), Weller (1960), Shaw (1964), Matthews (1974), or the International Stratigraphic Guide (ISSC, 1976).

RELATION OF CODES TO INTERNATIONAL GUIDE

Publication of the International Stratigraphic Guide by the International Subcommission on Stratigraphic Classification (ISSC, 1976), played a part in prompting examination of the American Stratigraphic Code and the decision to revise it.

The International Guide embodies principles and procedures that had been adopted by several national and regional stratigraphic committees and commissions. More than two decades of effort by H. D. Hedberg and other members of the Subcommission (ISSC, 1976, p. VI, 1, 3) developed the consensus required for preparation of the Guide. Although the Guide attempts to cover all kinds of rocks and the diverse ways of investigating them, it is necessarily incomplete. Mechanisms are needed to stimulate individual innovations toward promulgating new concepts, principles, and practices which subsequently may be found worthy of inclusion in later editions of the Guide. The flexibility of national and regional committees or commissions enables them to perform this function more readily than an international subcommission, even while they adopt the Guide as the international standard of stratigraphic classification.

A guiding principle in preparing this Code has been to make it as consistent as possible with the International Guide, which was endorsed by the ACSN in 1976, and at the same time to foster further innovations to meet the expanding and changing needs of earth scientists on the North American continent.

OVERVIEW

CATEGORIES RECOGNIZED

An attempt is made in this Code to strike a balance between serving the needs of those in evolving specialties and resisting the proliferation of categories of units. Consequently, more formal categories are recognized here than in previous codes or in the International Guide (ISSC, 1976). On the other hand, no special provision is made for formalizing certain kinds of units (deep oceanic, for example) which may be accommodated by available categories.

Four principal categories of units have previously been used widely in traditional stratigraphic work; these have been termed lithostratigraphic, biostratigraphic, chronostratigraphic, and geochronologic and are distinguished as follows:

1. A *lithostratigraphic unit* is a stratum or body of strata, generally but not invariably layered, generally but not invariably tabular, which conforms to the Law of Superposition and is distinguished and delimited on the basis of lithic characteristics and stratigraphic position. Example: Navajo Sandstone.

2. A *biostratigraphic unit* is a body of rock defined and characterized by its fossil content. Example: *Discoaster multiradiatus* Interval Zone.

3. A *chronostratigraphic unit* is a body of rock established to serve as the material reference for all rocks formed during the same span of time. Example: Devonian System. Each boundary of a chronostratigraphic unit is synchronous. Chronostratigraphy provides a means of organizing strata into units based on their age relations. A chronostratigraphic body also serves as the basis for defining the specific interval of geologic time, or geochronologic unit, represented by the referent.

4. A *geochronologic unit* is a division of time distinguished on the basis of the rock record preserved in a chronostratigraphic

unit. Example: Devonian Period.

The first two categories are comparable in that they consist of material units defined on the basis of content. The third category differs from the first two in that it serves primarily as the standard for recognizing and isolating materials of a specific age. The fourth, in contrast, is not a material, but rather a conceptual, unit; it is a division of time. Although a geochronologic unit is not a stratigraphic body, it is so intimately tied to chronostratigraphy that the two are discussed properly together.

Properties and procedures that may be used in distinguishing geologic units are both diverse and numerous (ISSC, 1976, p. 1, 96; Harland, 1977, p. 230), but all may be assigned to the following principal classes of categories used in stratigraphic classification (Table 1), which are discussed below:

    I. Material categories based on content, inherent attributes, or physical limits,

    II. Categories distinguished by geologic age:

        A. Material categories used to define temporal spans, and

        B. Temporal categories.

Table 1. Categories of Units Defined*

MATERIAL CATEGORIES BASED ON CONTENT
OR PHYSICAL LIMITS

      Lithostratigraphic (22)
      Lithodemic (31)**
      Magnetopolarity (44)
      Biostratigraphic (48)
      Pedostratigraphic (55)
      Allostratigraphic (58)

CATEGORIES EXPRESSING OR RELATED TO
GEOLOGIC AGE

    Material Categories Used to Define Temporal Spans
      Chronostratigraphic (66)
      Polarity-Chronostratigraphic (83)
    Temporal (Non-Material) Categories
      Geochronologic (80)
      Polarity-Chronologic (88)
      Diachronic (91)
      Geochronometric (96)

*Numbers in parentheses are the numbers of the Articles where units are defined.

**Italicized categories are those introduced or developed since publication of the previous code (ACSN, 1970).

Material Categories Based on Content or Physical Limits

The basic building blocks for most geologic work are rock bodies defined on the basis of composition and related lithic characteristics, or on their physical, chemical, or biologic content or properties. Emphasis is placed on the relative objectivity and reproducibility of data used in defining units within each category.

Foremost properties of rocks are composition, texture, fabric, structure, and color, which together are designated *lithic characteristics*. These serve as the basis for distinguishing and defining the most fundamental of all formal units. Such units based primarily on composition are divided into two categories (Henderson and others, 1980): lithostratigraphic (Article 22) and lithodemic (defined here in Article 31). A lithostratigraphic unit obeys the Law of Superposition, whereas a lithodemic unit does not. A *lithodemic unit* is a defined body of predominantly intrusive, highly metamorphosed, or intensely deformed rock that, because it is intrusive or has lost primary structure through metamorphism or tectonism, generally does not conform to the Law of Superposition.

North American Commission on Stratigraphic Nomenclature

Recognition during the past several decades that remanent magnetism in rocks records the Earth's past magnetic characteristics (Cox, Doell, and Dalrymple, 1963) provides a powerful new tool encompassed by magnetostratigraphy (McDougall, 1977; McElhinny, 1978). *Magnetostratigraphy* (Article 43) is the study of remanent magnetism in rocks; it is the record of the Earth's magnetic polarity (or field reversals), dipole-field-pole position (including apparent polar wander), the non-dipole component (secular variation), and field intensity. Polarity is of particular utility and is used to define a *magnetopolarity unit* (Article 44) as a body of rock identified by its remanent magnetic polarity (ACSN, 1976; ISSC, 1979). Empirical demonstration of uniform polarity does not necessarily have direct temporal connotations because the remanent magnetism need not be related to rock deposition or crystallization. Nevertheless, polarity is a physical attribute that may characterize a body of rock.

Biologic remains contained in, or forming, strata are uniquely important in stratigraphic practice. First, they provide the means of defining and recognizing material units based on fossil content (biostratigraphic units). Second, the irreversibility of organic evolution makes it possible to partition enclosing strata temporally. Third, biologic remains provide important data for the reconstruction of ancient environments of deposition.

Composition also is important in distinguishing pedostratigraphic units. A *pedostratigraphic unit* is a body of rock that consists of one or more pedologic horizons developed in one or more lithic units now buried by a formally defined lithostratigraphic or allostratigraphic unit or units. A pedostratigraphic unit is the part of a buried soil characterized by one or more clearly defined soil horizons containing pedogenically formed minerals and organic compounds. Pedostratigraphic terminology is discussed below and in Article 55.

Many upper Cenozoic, especially Quaternary, deposits are distinguished and delineated on the basis of content, for which lithostratigraphic classification is appropriate. However, others are delineated on the basis of criteria other than content. To facilitate the reconstruction of geologic history, some compositionally similar deposits in vertical sequence merit distinction as separate stratigraphic units because they are the products of different processes; others merit distinction because they are of demonstrably different ages. Lithostratigraphic classification of these units is impractical and a new approach, allostratigraphic classification, is introduced here and may prove applicable to older deposits as well. An *allostratigraphic unit* is a mappable stratiform body of sedimentary rock defined and identified on the basis of bounding discontinuities (Article 58 and related Remarks).

Geologic-Climate units, defined in the previous Code (ACSN, 1970, p. 31), are abandoned here because they proved to be of dubious utility. Inferences regarding climate are subjective and too tenuous a basis for the definition of formal geologic units. Such inferences commonly are based on deposits assigned more appropriately to lithostratigraphic or allostratigraphic units and may be expressed in terms of diachronic units (defined below).

Categories Expressing or Related to Geologic Age

Time is a single, irreversible continuum. Nevertheless, various categories of units are used to define intervals of geologic time, just as terms having different bases, such as Paleolithic, Renaissance, and Elizabethan, are used to designate specific periods of human history. Different temporal categories are established to express intervals of time distinguished in different ways.

Major objectives of stratigraphic classification are to provide a basis for systematic ordering of the time and space relations of rock bodies and to establish a time framework for the discussion of geologic history. For such purposes, units of geologic time traditionally have been named to represent the span of time during which a well-described sequence of rock, or a chronostratigraphic unit, was deposited ("time units based on material referents," Fig. 1). This procedure continues, to the exclusion of other possible approaches, to be standard practice in studies of Phanerozoic rocks. Despite admonitions in previous American codes and the International Stratigraphic Guide (ISSC, 1976, p. 81) that similar procedures should be applied to the Precambrian, no comparable chronostratigraphic units, or geochronologic units derived therefrom, proposed for the Precambrian have yet been accepted worldwide. Instead, the IUGS Subcommission on Precambrian Stratigraphy (Sims, 1979) and its Working Groups (Harrison and Peterman, 1980) recommend division of Precambrian time into *geochronometric units* having no material referents.

A distinction is made throughout this report between *isochronous* and *synchronous*, as urged by Cumming, Fuller, and Porter (1959, p. 730), although the terms have been used synonymously by many. *Isochronous* means of equal duration; *synchronous* means simultaneous, or occurring at the same time. Although two rock bodies of very different ages may be formed during equal durations of time, the term *isochronous* is not applied to them in the earth sciences. Rather, isochronous bodies are those bounded by synchronous surfaces and formed during the same span of time. *Isochron*, in contrast, is used for a line connecting points of equal age on a graph representing physical or chemical phenomena; the line represents the same or equal time. The adjective *diachronous* is applied either to a rock unit with one or two bounding surfaces which are not synchronous, or to a boundary which is not synchronous (which "transgresses time").

Two classes of time units based on material referents, or stratotypes, are recognized (Fig. 1). The first is that of the traditional and conceptually isochronous time units, and includes *geochronologic units*, which are based on *chronostratigraphic units*, and *polarity-geochronologic units*. These isochronous units have worldwide applicability and may be used even in areas lacking a material record of the named span of time. The second class of time units, newly defined in this Code, consists of *diachronic units* (Article 91), which are based on rock bodies known to be diachronous. In contrast to isochronous units, a diachronic term is used only where a material referent is present; a diachronic unit is coextensive with the material body or bodies on which it is based.

A *chronostratigraphic unit*, as defined above and in Article 66, is a body of rock established to serve as the material reference for all rocks formed during the same span of time; its boundaries are synchronous. It is the referent for a *geochronologic unit*, as defined above and in Article 80. Internationally accepted and traditional chronostratigraphic units were based initially on the time spans of lithostratigraphic units, biostratigraphic units, or other features of the rock record that have specific durations. In sum, they form the Standard Global Chronostratigraphic Scale (ISSC, 1976, p. 76-81; Harland, 1978), consisting of established systems and series.

A *polarity-chronostratigraphic unit* is a body of rock that contains a primary magnetopolarity record imposed when the rock was deposited or crystallized (Article 83). It serves as a material standard or referent for a part of geologic time during which the Earth's magnetic field had a characteristic polarity or sequence of polarities; that is, for a *polarity-chronologic unit* (Article 88).

A *diachronic unit* comprises the unequal spans of time represented by one or more specific diachronous rock bodies (Article 91). Such bodies may be lithostratigraphic, biostratigraphic, pedostratigraphic, allostratigraphic, or an assemblage of such units. A diachronic unit is applicable only where its material referent is present.

A *geochronometric* (or chronometric) *unit* is an isochronous direct division of geologic time expressed in years (Article 96). It has no material referent.

849

North American Stratigraphic Code

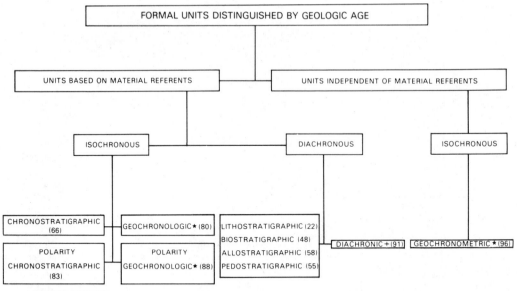

★ Applicable world-wide.
+Applicable only where material referents are present.
( )Number of article in which defined.

FIG. 1.—Relation of geologic time units to the kinds of rock-unit referents on which most are based.

Pedostratigraphic Terms

The definition and nomenclature for pedostratigraphic[2] units in this Code differ from those for soil-stratigraphic units in the previous Code (ACSN, 1970, Article 18), by being more specific with regard to content, boundaries, and the basis for determining stratigraphic position.

The term "soil" has different meanings to the geologist, the soil scientist, the engineer, and the layman, and commonly has no stratigraphic significance. The term *paleosol* is currently used in North America for any soil that formed on a landscape of the past; it may be a buried soil, a relict soil, or an exhumed soil (Ruhe, 1965; Valentine and Dalrymple, 1976).

A *pedologic soil* is composed of one or more soil horizons.[3] A *soil horizon* is a layer within a pedologic soil that (1) is approximately parallel to the soil surface, (2) has distinctive physical, chemical, biological, and morphological properties that differ from those of adjacent, genetically related, soil horizons, and (3) is distinguished from other soil horizons by objective compositional properties that can be observed or measured in the field. The physical boundaries of buried pedologic horizons are objective traceable boundaries with stratigraphic significance. A buried pedologic soil provides the material basis for definition of a stratigraphic unit in pedostratigraphic classification (Article 55), but a buried pedologic soil may be somewhat more inclusive than a pedostratigraphic unit. A pedologic soil may contain both an 0-horizon and the entire C-horizon (Fig. 6), whereas the former is excluded and the latter need not be included in a pedostratigraphic unit.

The definition and nomenclature for pedostratigraphic units

[2]From Greek, *pedon*, ground or soil.
[3]As used in a geological sense, a *horizon* is a surface or line. In pedology, however, it is a body of material, and such usage is continued here.

in this Code differ from those of soil stratigraphic units proposed by the International Union for Quaternary Research and International Society of Soil Science (Parsons, 1981). The pedostratigraphic unit, geosol, also differs from the proposed INQUA-ISSS soil-stratigraphic unit, pedoderm, in several ways, the most important of which are: (1) a geosol may be in any part of the geologic column, whereas a pedoderm is a surficial soil; (2) a geosol is a buried soil, whereas a pedoderm may be a buried, relict, or exhumed soil; (3) the boundaries and stratigraphic position of a geosol are defined and delineated by criteria that differ from those for a pedoderm; and (4) a geosol may be either all or only a part of a buried soil, whereas a pedoderm is the entire soil.

The term *geosol*, as defined by Morrison (1967, p. 3), is a laterally traceable, mappable, geologic weathering profile that has a consistent stratigraphic position. The term is adopted and redefined here as the fundamental and only unit in formal pedostratigraphic classification (Article 56).

FORMAL AND INFORMAL UNITS

Although the emphasis in this Code is necessarily on formal categories of geologic units, informal nomenclature is highly useful in stratigraphic work.

Formally named units are those that are named in accordance with an established scheme of classification; the fact of formality is conveyed by capitalization of the initial letter of the *rank* or *unit* term (for example, Morrison Formation). Informal units, whose unit terms are ordinary nouns, are not protected by the stability provided by proper formalization and recommended classification procedures. Informal terms are devised for both economic and scientific reasons. Formalization is appropriate for those units requiring stability of nomenclature, particularly those likely to be extended far beyond the locality in which they were first recognized. Informal terms are appropriate for casually mentioned, innovative, and most economic units, those

North American Commission on Stratigraphic Nomenclature

defined by unconventional criteria, and those that may be too thin to map at usual scales.

Casually mentioned geologic units not defined in accordance with this Code are informal. For many of these, there may be insufficient need or information, or perhaps an inappropriate basis, for formal designations. Informal designations as beds or lithozones (the pebbly beds, the shaly zone, third coal) are appropriate for many such units.

Most economic units, such as aquifers, oil sands, coal beds, quarry layers, and ore-bearing "reefs," are informal, even though they may be named. Some such units, however, are so significant scientifically and economically that they merit formal recognition as beds, members, or formations.

Innovative approaches in regional stratigraphic studies have resulted in the recognition and definition of units best left as informal, at least for the time being. Units bounded by major regional unconformities on the North American craton were designated "sequences" (example: Sauk sequence) by Sloss (1963). Major unconformity-bounded units also were designated "synthems" by Chang (1975), who recommended that they be treated formally. Marker-defined units that are continuous from one lithofacies to another were designated "formats" by Forgotson (1957). The term "chronosome" was proposed by Schultz (1982) for rocks of diverse facies corresponding to geographic variations in sedimentation during an interval of deposition identified on the basis of bounding stratigraphic markers. Successions of faunal zones containing evolutionarily related forms, but bounded by non-evolutionary biotic discontinuities, were termed "biomeres" (Palmer, 1965). The foregoing are only a few selected examples to demonstrate how informality provides a continuing avenue for innovation.

The terms *magnafacies* and *parvafacies*, coined by Caster (1934) to emphasize the distinction between lithostratigraphic and chronostratigraphic units in sequences displaying marked facies variation, have remained informal despite their impact on clarifying the concepts involved.

Tephrochronologic studies provide examples of informal units too thin to map at conventional scales but yet invaluable for dating important geologic events. Although some such units are named for physiographic features and places where first recognized (e.g., Guaje pumice bed, where it is not mapped as the Guaje Member of the Bandelier Tuff), others bear the same name as the volcanic vent (e.g., Huckleberry Ridge ash bed of Izett and Wilcox, 1981).

Informal geologic units are designated by ordinary nouns, adjectives or geographic terms and lithic or unit-terms that are not capitalized (chalky formation or beds, St. Francis coal).

No geologic unit should be established and defined, whether formally or informally, unless its recognition serves a clear purpose.

CORRELATION

Correlation is a procedure for demonstrating correspondence between geographically separated parts of a geologic unit. The term is a general one having diverse meanings in different disciplines. Demonstration of temporal correspondence is one of the most important objectives of stratigraphy. The term "correlation" frequently is misused to express the idea that a unit has been identified or recognized.

Correlation is used in this Code as the demonstration of correspondence between two geologic units in both some defined property and relative stratigraphic position. Because correspondence may be based on various properties, three kinds of correlation are best distinguished by more specific terms. *Lithocorrelation* links units of similar lithology and stratigraphic position (or sequential or geometric relation, for lithodemic

units). *Biocorrelation* expresses similarity of fossil content and biostratigraphic position. *Chronocorrelation* expresses correspondence in age and in chronostratigraphic position.

Other terms that have been used for the similarity of content and stratal succession are homotaxy and chronotaxy. *Homotaxy* is the similarity in separate regions of the serial arrangement or succession of strata of comparable compositions or of included fossils. The term is derived from *homotaxis*, proposed by Huxley (1862, p. xlvi) to emphasize that similarity in succession does not prove age equivalence of comparable units. The term *chronotaxy* has been applied to similar stratigraphic sequences composed of units which are of equivalent age (Henbest, 1952, p. 310).

Criteria used for ascertaining temporal and other types of correspondence are diverse (ISSC, 1976, p. 86-93) and new criteria will emerge in the future. Evolving statistical tests, as well as isotopic and paleomagnetic techniques, complement the traditional paleontologic and lithologic procedures. Boundaries defined by one set of criteria need not correspond to those defined by others.

PART II. ARTICLES

INTRODUCTION

Article 1.—**Purpose.** This Code describes explicit stratigraphic procedures for classifying and naming geologic units accorded formal status. Such procedures, if widely adopted, assure consistent and uniform usage in classification and terminology and therefore promote unambiguous communication.

Article 2.—**Categories.** Categories of formal stratigraphic u-nits, though diverse, are of three classes (Table 1). The first class is of rock-material categories based on inherent attributes or content and stratigraphic position, and includes litho-stratigraphic, lithodemic, magnetopolarity, biostratigraphic, pe-dostratigraphic, and allostratigraphic units. The second class is of material categories used as standards for defining spans of ge-ologic time, and includes chronostratigraphic and polarity-chro-nostratigraphic units. The third class is of non-material temporal categories, and includes geochronologic, polarity-chronologic, geochronometric, and diachronic units.

GENERAL PROCEDURES

DEFINITION OF FORMAL UNITS

Article 3.—**Requirements for Formally Named Geologic Units.** Naming, establishing, revising, redefining, and abandoning formal geologic units require publication in a recognized scientific medium of a comprehensive statement which includes: (i) intent to designate or modify a formal unit; (ii) designation of category and rank of unit; (iii) selection and derivation of name; (iv) specification of stratotype (where applicable); (v) description of unit; (vi) definition of boundaries; (vii) historical background; (viii) dimensions, shape, and other regional aspects; (ix) geologic age; (x) correlations; and possibly (xi) genesis (where applicable). These requirements apply to subsurface and offshore, as well as exposed, units.

Article 4.—**Publication.**[4] "Publication in a recognized scientific medium" in conformance with this Code means that a work, when first issued, must (1) be reproduced in ink on paper or by some method that assures numerous identical copies and wide distribution; (2) be issued for the purpose of scientific, public, permanent record; and (3) be readily obtainable by purchase or free distribution.

Remarks. (a) **Inadequate publication.**—The following do not constitute publication within the meaning of the Code: (1) distribution of microfilms, microcards, or matter reproduced by similar methods; (2)

---

[4]This article is modified slightly from a statement by the International Commission of Zoological Nomenclature (1964, p. 7-9).

North American Stratigraphic Code

Table 2. Categories and Ranks of Units Defined in This Code*

A. Material Units

LITHOSTRATIGRAPHIC	LITHODEMIC	MAGNETOPOLARITY	BIOSTRATIGRAPHIC	PEDOSTRATIGRAPHIC	ALLOSTRATIGRAPHIC
Supergroup	Supersuite (Complex)				
Group	Suite (Complex)	Polarity Superzone			Allogroup
Formation	*Lithodeme*	*Polarity zone*	*Biozone* (Interval, Assemblage or Abundance)	*Geosol*	*Alloformation*
Member (or Lens, or Tongue)		Polarity Subzone	Subbiozone		Allomember
Bed(s) or Flow(s)					

B. Temporal and Related Chronostratigraphic Units

CHRONO-STRATIGRAPHIC	GEOCHRONOLOGIC GEOCHRONOMETRIC	POLARITY CHRONO-STRATIGRAPHIC	POLARITY CHRONOLOGIC	DIACHRONIC
Eonothem	Eon	Polarity Superchronozone	Polarity Superchron	
Erathem (Supersystem)	Era (Superperiod)			
System (Subsystem)	*Period* (Subperiod)	*Polarity Chronozone*	*Polarity Chron*	*Episode*
Series	Epoch			Phase (Diachron)
Stage (Substage)	Age (Subage)	Polarity Subchronozone	Polarity Subchron	Span (Diachron)
Chronozone	Chron			Cline

*Fundamental units are italicized.

distribution to colleagues or students of a note, even if printed, in explanation of an accompanying illustration; (3) distribution of proof sheets; (4) open-file release; (5) theses, dissertations, and dissertation abstracts; (6) mention at a scientific or other meeting; (7) mention in an abstract, map explanation, or figure caption; (8) labeling of a rock specimen in a collection; (9) mere deposit of a document in a library; (10) anonymous publication; or (11) mention in the popular press or in a legal document.

(b). **Guidebooks.**—A guidebook with distribution limited to participants of a field excursion does not meet the test of availability. Some organizations publish and distribute widely large editions of serial guidebooks that include refereed regional papers; although these do meet the tests of scientific purpose and availability, and therefore constitute valid publication, other media are preferable.

Article 5.—**Intent and Utility.** To be valid, a new unit must serve a clear purpose and be duly proposed and duly described, and the intent to establish it must be specified. Casual mention of a unit, such as "the granite exposed near the Middleville schoolhouse," does not establish a new formal unit, nor does mere use in a table, columnar section, or map.

Remark. (a) **Demonstration of purpose served.**—The initial definition or revision of a named geologic unit constitutes, in essence, a proposal. As such, it lacks status until use by others demonstrates that a clear purpose has been served. A unit becomes established through repeated demonstration of its utility. The decision not to use a newly proposed or a newly revised term requires a full discussion of its unsuitability.

Article 6.—**Category and Rank.** The category and rank of a new or revised unit must be specified.

Remark. (a) **Need for specification.**—Many stratigraphic controversies have arisen from confusion or misinterpretation of the category of a unit (for example, lithostratigraphic vs. chronostratigraphic). Specification and unambiguous description of the category is of paramount importance. Selection and designation of an appropriate rank from the distinctive terminology developed for each category help serve this function (Table 2).

Article 7.—**Name.** The name of a formal geologic unit is compound. For most categories, the name of a unit should consist of a geographic name combined with an appropriate rank (Wasatch Formation) or descriptive term (Viola Limestone). Biostratigraphic units are designated by appropriate biologic forms (*Exus albus* Assemblage Biozone). Worldwide chronostratigraphic units bear long established and generally accepted names of diverse origins (Triassic System). The first letters of all words used in the names of formal geologic units are capitalized (except for the trivial species and subspecies terms in the name of a biostratigraphic unit).

Remarks. (a) **Appropriate geographic terms.**—Geographic names derived from permanent natural or artificial features at or near which the unit is present are preferable to those derived from impermanent features such as farms, schools, stores, churches, crossroads, and small communities. Appropriate names may be selected from those shown on topographic, state, provincial, county, forest service, hydrographic, or comparable maps, particularly those showing names approved by a national board for geographic names. The generic part of a geographic name, e.g., river, lake, village, should be omitted from new terms, unless required to distinguish between two otherwise identical names (e.g., Redstone Formation and Redstone River Formation). Two names should not be derived from the same geographic feature. A unit should not be named for the source of its components; for example, a deposit inferred to have been derived from the Keewatin glaciation center should not be designated the "Keewatin Till."

(b) **Duplication of names.**—Responsibility for avoiding duplication,

North American Commission on Stratigraphic Nomenclature

either in use of the same name for different units (homonymy) or in use of different names for the same unit (synonymy), rests with the proposer. Although the same geographic term has been applied to different categories of units (example: the lithostratigraphic Word Formation and the chronostratigraphic Wordian Stage) now entrenched in the literature, the practice is undesirable. The extensive geologic nomenclature of North America, including not only names but also nomenclatural history of formal units, is recorded in compendia maintained by the Committee on Stratigraphic Nomenclature of the Geological Survey of Canada, Ottawa, Ontario; by the Geologic Names Committee of the United States Geological Survey, Reston, Virginia; by the Instituto de Geología, Ciudad Universitaria, México, D.F.; and by many state and provincial geological surveys. These organizations respond to inquiries regarding the availability of names, and some are prepared to reserve names for units likely to be defined in the next year or two.

(c) **Priority and preservation of established names.**—Stability of nomenclature is maintained by use of the rule of priority and by preservation of well-established names. Names should not be modified without explaining the need. Priority in publication is to be respected, but priority alone does not justify displacing a well-established name by one neither well-known nor commonly used; nor should an inadequately established name be preserved merely on the basis of priority. Redefinitions in precise terms are preferable to abandonment of the names of well-established units which may have been defined imprecisely but nonetheless in conformance with older and less stringent standards.

(d) **Differences of spelling and changes in name.**—The geographic component of a well-established stratigraphic name is not changed due to differences in spelling or changes in the name of a geographic feature. The name Bennett Shale, for example, used for more than half a century, need not be altered because the town is named Bennet. Nor should the Mauch Chunk Formation be changed because the town has been renamed Jim Thorpe. Disappearance of an impermanent geographic feature, such as a town, does not affect the name of an established geologic unit.

(e) **Names in different countries and different languages.**—For geologic units that cross local and international boundaries, a single name for each is preferable to several. Spelling of a geographic name commonly conforms to the usage of the country and linguistic group involved. Although geographic names are not translated (Cuchillo is not translated to Knife), lithologic or rank terms are (Edwards Limestone, Caliza Edwards; Formación La Casita, La Casita Formation).

Article 8.—**Stratotypes.** The designation of a unit or boundary stratotype (type section or type locality) is essential in the definition of most formal geologic units. Many kinds of units are best defined by reference to an accessible and specific sequence of rock that may be examined and studied by others. A stratotype is the standard (original or subsequently designated) for a named geologic unit or boundary and constitutes the basis for definition or recognition of that unit or boundary; therefore, it must be illustrative and representative of the concept of the unit or boundary being defined.

Remarks. (a) **Unit stratotypes.**—A unit stratotype is the type section for a stratiform deposit or the type area for a nonstratiform body that serves as the standard for definition and recognition of a geologic unit. The upper and lower limits of a unit stratotype are designated points in a specific sequence or locality and serve as the standards for definition and recognition of a stratigraphic unit's boundaries.

(b) **Boundary stratotype.**—A boundary stratotype is the type locality for the boundary reference point for a stratigraphic unit. Both boundary stratotypes for any unit need not be in the same section or region. Each boundary stratotype serves as the standard for definition and recognition of the base of a stratigraphic unit. The top of a unit may be defined by the boundary stratotype of the next higher stratigraphic unit.

(c) **Type locality.**—A type locality is the specified geographic locality where the stratotype of a formal unit or unit boundary was originally defined and named. A type area is the geographic territory encompassing the type locality. Before the concept of a stratotype was developed, only type localities and areas were designated for many geologic units which are now long- and well-established. Stratotypes, though now mandatory in defining most stratiform units, are impractical in definitions of many large nonstratiform rock bodies whose diverse major components may be best displayed at several reference localities.

(d) **Composite-stratotype.**—A composite-stratotype consists of several reference sections (which may include a type section) required to demonstrate the range or totality of a stratigraphic unit.

(e) **Reference sections.**—Reference sections may serve as invaluable standards in definitions or revisions of formal geologic units. For those well-established stratigraphic units for which a type section never was specified, a principal reference section (lectostratotype of ISSC, 1976, p. 26) may be designated. A principal reference section (neostratotype of ISSC, 1976, p. 26) also may be designated for those units or boundaries whose stratotypes have been destroyed, covered, or otherwise made inaccessible. Supplementary reference sections often are designated to illustrate the diversity or heterogeneity of a defined unit or some critical feature not evident or exposed in the stratotype. Once a unit or boundary stratotype section is designated, it is never abandoned or changed; however, if a stratotype proves inadequate, it may be supplemented by a principal reference section or by several reference sections that may constitute a composite-stratotype.

(f) **Stratotype descriptions.**—Stratotypes should be described both geographically and geologically. Sufficient geographic detail must be included to enable others to find the stratotype in the field, and may consist of maps and/or aerial photographs showing location and access, as well as appropriate coordinates or bearings. Geologic information should include thickness, descriptive criteria appropriate to the recognition of the unit and its boundaries, and discussion of the relation of the unit to other geologic units of the area. A carefully measured and described section provides the best foundation for definition of stratiform units. Graphic profiles, columnar sections, structure-sections, and photographs are useful supplements to a description; a geologic map of the area including the type locality is essential.

Article 9.—**Unit Description.** A unit proposed for formal status should be described and defined so clearly that any subsequent investigator can recognize that unit unequivocally. Distinguishing features that characterize a unit may include any or several of the following: composition, texture, primary structures, structural attitudes, biologic remains, readily apparent mineral composition (e.g., calcite vs. dolomite), geochemistry, geophysical properties (including magnetic signatures), geomorphic expression, unconformable or cross-cutting relations, and age. Although all distinguishing features pertinent to the unit category should be described sufficiently to characterize the unit, those not pertinent to the category (such as age and inferred genesis for lithostratigraphic units, or lithology for biostratigraphic units) should not be made part of the definition.

Article 10.—**Boundaries.** The criteria specified for the recognition of boundaries between adjoining geologic units are of paramount importance because they provide the basis for scientific reproducibility of results. Care is required in describing the criteria, which must be appropriate to the category of unit involved.

Remarks. (a) **Boundaries between intergradational units.**—Contacts between rocks of markedly contrasting composition are appropriate boundaries of lithic units, but some rocks grade into, or intertongue with, others of different lithology. Consequently, some boundaries are necessarily arbitrary as, for example, the top of the uppermost limestone in a sequence of interbedded limestone and shale. Such arbitrary boundaries commonly are diachronous.

(b) **Overlaps and gaps.**—The problem of overlaps and gaps between long-established adjacent chronostratigraphic units is being addressed by international IUGS and IGCP working groups appointed to deal with various parts of the geologic column. The procedure recommended by the Geological Society of London (George and others, 1969; Holland and others, 1978), of defining only the basal boundaries of chronostratigraphic units, has been widely adopted (e.g., McLaren, 1977) to resolve the problem. Such boundaries are defined by a carefully selected and agreed-upon boundary-stratotype (marker-point type section or "golden spike") which becomes the standard for the base of a chronostratigraphic unit. The concept of the mutual-boundary stratotype (ISSC, 1976, p. 84-86), based on the assumption of continuous deposition in selected sequences, also has been used to define chronostratigraphic units.

North American Stratigraphic Code

Although international chronostratigraphic units of series and higher rank are being redefined by IUGS and IGCP working groups, there may be a continuing need for some provincial series. Adoption of the basal boundary-stratotype concept is urged.

Article 11.—**Historical Background.** A proposal for a new name must include a nomenclatorial history of rocks assigned to the proposed unit, describing how they were treated previously and by whom (references), as well as such matters as priorities, possible synonymy, and other pertinent considerations. Consideration of the historical background of an older unit commonly provides the basis for justifying definition of a new unit.

Article 12.—**Dimensions and Regional Relations.** A perspective on the magnitude of a unit should be provided by such information as may be available on the geographic extent of a unit; observed ranges in thickness, composition, and geomorphic expression; relations to other kinds and ranks of stratigraphic units; correlations with other nearby sequences; and the bases for recognizing and extending the unit beyond the type locality. If the unit is not known anywhere but in an area of limited extent, informal designation is recommended.

Article 13.—**Age.** For most formal material geologic units, other than chronostratigraphic and polarity-chronostratigraphic, inferences regarding geologic age play no proper role in their definition. Nevertheless, the age, as well as the basis for its assignment, are important features of the unit and should be stated. For many lithodemic units, the age of the protolith should be distinguished from that of the metamorphism or deformation. If the basis for assigning an age is tenuous, a doubt should be expressed.

Remarks. (a) **Dating.**—The geochronologic ordering of the rock record, whether in terms of radioactive-decay rates or other processes, is generally called "dating." However, the use of the noun "date" to mean "isotopic age" is not recommended. Similarly, the term "absolute age" should be suppressed in favor of "isotopic age" for an age determined on the basis of isotopic ratios. The more inclusive term "numerical age" is recommended for all ages determined from isotopic ratios, fission tracks, and other quantifiable age-related phenomena.

(b) **Calibration**—The dating of chronostratigraphic boundaries in terms of numerical ages is a special form of dating for which the word "calibration" should be used. The geochronologic time-scale now in use has been developed mainly through such calibration of chronostratigraphic sequences.

(c) **Convention and abbreviations.**—The age of a stratigraphic unit or the time of a geologic event, as commonly determined by numerical dating or by reference to a calibrated time-scale, may be expressed in years before the present. The unit of time is the modern year as presently recognized worldwide. Recommended (but not mandatory) abbreviations for such ages are SI (International System of Units) multipliers coupled with "a" for annum: ka, Ma, and Ga[5] for kilo-annum (10^3 years), Mega-annum (10^6 years), and Giga-annum (10^9 years), respectively. Use of these terms after the age value follows the convention established in the field of C-14 dating. The "present" refers to 1950 AD, and such qualifiers as "ago" or "before the present" are omitted after the value because measurement of the duration from the present to the past is implicit in the designation. In contrast, the duration of a remote interval of geologic time, as a number of years, should not be expressed by the same symbols. Abbreviations for numbers of years, without reference to the present, are informal (e.g., y or yr for years; my, m.y., or m.yr. for millions of years; and so forth, as preference dictates). For example, boundaries of the Late Cretaceous Epoch currently are calibrated at 63 Ma and 96 Ma, but the interval of time represented by this epoch is 33 m.y.

(d) **Expression of "age" of lithodemic units.**—The adjectives "early," "middle," and "late" should be used with the appropriate

geochronologic term to designate the age of lithodemic units. For example, a granite dated isotopically at 510 Ma should be referred to using the geochronologic term "Late Cambrian granite" rather than either the chronostratigraphic term "Upper Cambrian granite" or the more cumbersome designation "granite of Late Cambrian age."

Article 14.—**Correlation.** Information regarding spatial and temporal counterparts of a newly defined unit beyond the type area provides readers with an enlarged perspective. Discussions of criteria used in correlating a unit with those in other areas should make clear the distinction between data and inferences.

Article 15.—**Genesis.** Objective data are used to define and classify geologic units and to express their spatial and temporal relations. Although many of the categories defined in this Code (e.g., lithostratigraphic group, plutonic suite) have genetic connotations, inferences regarding geologic history or specific environments of formation may play no proper role in the definition of a unit. However, observations, as well as inferences, that bear on genesis are of great interest to readers and should be discussed.

Article 16.—**Subsurface and Subsea Units.** The foregoing procedures for establishing formal geologic units apply also to subsurface and offshore or subsea units. Complete lithologic and paleontologic descriptions or logs of the samples or cores are required in written or graphic form, or both. Boundaries and divisions, if any, of the unit should be indicated clearly with their depths from an established datum.

Remarks. (a) **Naming subsurface units.**—A subsurface unit may be named for the borehole (Eagle Mills Formation), oil field (Smackover Limestone), or mine which is intended to serve as the stratotype, or for a nearby geographic feature. The hole or mine should be located precisely, both with map and exact geographic coordinates, and identified fully (operator or company, farm or lease block, dates drilled or mined, surface elevation and total depth, etc).

(b) **Additional recommendations.**—Inclusion of appropriate borehole geophysical logs is urged. Moreover, rock and fossil samples and cores and all pertinent accompanying materials should be stored, and available for examination, at appropriate federal, state, provincial, university, or museum depositories. For offshore or subsea units (Clipperton Formation of Tracey and others, 1971, p. 22; Argo Salt of McIver, 1972, p. 57), the names of the project and vessel, depth of sea floor, and pertinent regional sampling and geophysical data should be added.

(c) **Seismostratigraphic units.**—High-resolution seismic methods now can delineate stratal geometry and continuity at a level of confidence not previously attainable. Accordingly, seismic surveys have come to be the principal adjunct of the drill in subsurface exploration. On the other hand, the method identifies rock types only broadly and by inference. Thus, formalization of units known only from seismic profiles is inappropriate. Once the stratigraphy is calibrated by drilling, the seismic method may provide objective well-to-well correlations.

REVISION AND ABANDONMENT OF FORMAL UNITS

Article 17.—**Requirements for Major Changes.** Formally defined and named geologic units may be redefined, revised, or abandoned, but revision and abandonment require as much justification as establishment of a new unit.

Remark. (a) **Distinction between redefinition and revision.**—Redefinition of a unit involves changing the view or emphasis on the content of the unit without changing the boundaries or rank, and differs only slightly from redescription. Neither redefinition nor redescription is considered revision. A redescription corrects an inadequate or inaccurate description, whereas a redefinition may change a descriptive (for example, lithologic) designation. Revision involves either minor changes in the definition of one or both boundaries or in the rank of a unit (normally, elevation to a higher rank). Correction of a misidentification of a unit outside its type area is neither redefinition nor revision.

---

[5]Note that the initial letters of Mega- and Giga- are capitalized, but that of kilo- is not, by SI convention.

North American Commission on Stratigraphic Nomenclature

Article 18.—Redefinition. A correction or change in the descriptive term applied to a stratigraphic or lithodemic unit is a redefinition which does not require a new geographic term.

Remarks. (a) **Change in lithic designation.**—Priority should not prevent more exact lithic designation if the original designation is not everywhere applicable; for example, the Niobrara Chalk changes gradually westward to a unit in which shale is prominent, for which the designation "Niobrara Shale" or "Formation" is more appropriate. Many carbonate formations originally designated "limestone" or "dolomite" are found to be geographically inconsistent as to prevailing rock type. The appropriate lithic term or "formation" is again preferable for such units.

(b) **Original lithic designation inappropriate.**—Restudy of some long-established lithostratigraphic units has shown that the original lithic designation was incorrect according to modern criteria; for example, some "shales" have the chemical and mineralogical composition of limestone, and some rocks described as felsic lavas now are understood to be welded tuffs. Such new knowledge is recognized by changing the lithic designation of the unit, while retaining the original geographic term. Similarly, changes in the classification of igneous rocks have resulted in recognition that rocks originally described as quartz monzonite now are more appropriately termed granite. Such lithic designations may be modernized when the new classification is widely adopted. If heterogeneous bodies of plutonic rock have been misleadingly identified with a single compositional term, such as "gabbro," the adoption of a neutral term, such as "intrusion" or "pluton," may be advisable.

Article 19.—Revision. Revision involves either minor changes in the definition of one or both boundaries of a unit, or in the unit's rank.

Remarks. (a) **Boundary change.**—Revision is justifiable if a minor change in boundary or content will make a unit more natural and useful. If revision modifies only a minor part of the content of a previously established unit, the original name may be retained.

(b) **Change in rank.**—Change in rank of a stratigraphic or temporal unit requires neither redefinition of its boundaries nor alteration of the geographic part of its name. A member may become a formation or vice versa, a formation may become a group or vice versa, and a lithodeme may become a suite or vice versa.

(c) **Examples of changes from area to area.**—The Conasauga Shale is recognized as a formation in Georgia and as a group in eastern Tennessee; the Osgood Formation, Laurel Limestone, and Waldron Shale in Indiana are classed as members of the Wayne Formation in a part of Tennessee; the Virgelle Sandstone is a formation in western Montana and a member of the Eagle Sandstone in central Montana; the Skull Creek Shale and the Newcastle Sandstone in North Dakota are members of the Ashville Formation in Manitoba.

(d) **Example of change in single area.**—The rank of a unit may be changed without changing its content. For example, the Madison Limestone of early work in Montana later became the Madison Group, containing several formations.

(e) **Retention of type section.**—When the rank of a geologic unit is changed, the original type section or type locality is retained for the newly ranked unit (see Article 22c).

(f) **Different geographic name for a unit and its parts.**—In changing the rank of a unit, the same name may not be applied both to the unit as a whole and to a part of it. For example, the Astoria Group should not contain an Astoria Sandstone, nor the Washington Formation, a Washington Sandstone Member.

(g) **Undesirable restriction.**—When a unit is divided into two or more of the same rank as the original, the original name should not be used for any of the divisions. Retention of the old name for one of the units precludes use of the name in a term of higher rank. Furthermore, in order to understand an author's meaning, a later reader would have to know about the modification and its date, and whether the author is following the original or the modified usage. For these reasons, the normal practice is to raise the rank of an established unit when units of the same rank are recognized and mapped within it.

Article 20.—Abandonment. An improperly defined or obsolete stratigraphic, lithodemic, or temporal unit may be formally abandoned, provided that (a) sufficient justification is presented to demonstrate a concern for nomenclatural stability, and (b) recommendations are made for the classification and nomenclature to be used in its place.

Remarks. (a) **Reasons for abandonment.**—A formally defined unit may be abandoned by the demonstration of synonymy or homonymy, of assignment to an improper category (for example, definition of a lithostratigraphic unit in a chronostratigraphic sense), or of other direct violations of a stratigraphic code or procedures prevailing at the time of the original definition. Disuse, or the lack of need or useful purpose for a unit, may be a basis for abandonment; so, too, may widespread misuse in diverse ways which compound confusion. A unit also may be abandoned if it proves impracticable, neither recognizable nor mappable elsewhere.

(b) **Abandoned names.**—A name for a lithostratigraphic or lithodemic unit, once applied and then abandoned, is available for some other unit only if the name was introduced casually, or if it has been published only once in the last several decades and is not in current usage, and if its reintroduction will cause no confusion. An explanation of the history of the name and of the new usage should be a part of the designation.

(c) **Obsolete names.**—Authors may refer to national and provincial records of stratigraphic names to determine whether a name is obsolete (see Article 7b).

(d) **Reference to abandoned names.**—When it is useful to refer to an obsolete or abandoned formal name, its status is made clear by some such term as "abandoned" or "obsolete," and by using a phrase such as "La Plata Sandstone of Cross (1898)". (The same phrase also is used to convey that a named unit has not yet been adopted for usage by the organization involved.)

(e) **Reinstatement.**—A name abandoned for reasons that seem valid at the time, but which subsequently are found to be erroneous, may be reinstated. Example: the Washakie Formation, defined in 1869, was abandoned in 1918 and reinstated in 1973.

CODE AMENDMENT

Article 21.—Procedure for Amendment. Additions to, or changes of, this Code may be proposed in writing to the Commission by any geoscientist at any time. If accepted for consideration by a majority vote of the Commission, they may be adopted by a two-thirds vote of the Commission at an annual meeting not less than a year after publication of the proposal.

FORMAL UNITS DISTINGUISHED BY CONTENT, PROPERTIES, OR PHYSICAL LIMITS

LITHOSTRATIGRAPHIC UNITS

Nature and Boundaries

Article 22.—Nature of Lithostratigraphic Units. A lithostratigraphic unit is a defined body of sedimentary, extrusive igneous, metasedimentary, or metavolcanic strata which is distinguished and delimited on the basis of lithic characteristics and stratigraphic position. A lithostratigraphic unit generally conforms to the Law of Superposition and commonly is stratified and tabular in form.

Remarks. (a) **Basic units.**—Lithostratigraphic units are the basic units of general geologic work and serve as the foundation for delineating strata, local and regional structure, economic resources, and geologic history in regions of stratified rocks. They are recognized and defined by observable rock characteristics; boundaries may be placed at clearly distinguished contacts or drawn arbitrarily within a zone of gradation. Lithification or cementation is not a necessary property; clay, gravel, till, and other unconsolidated deposits may constitute valid lithostratigraphic units.

(b) **Type section and locality.**—The definition of a lithostratigraphic unit should be based, if possible, on a stratotype consisting of readily accessible rocks in place, e.g., in outcrops, excavations, and mines, or of rocks accessible only to remote sampling devices, such as those in drill holes and underwater. Even where remote methods are used, definitions must be based on lithic criteria and not on the geophysical characteristics of the rocks, nor the implied age of their contained fossils. Definitions

North American Stratigraphic Code

must be based on descriptions of actual rock material. Regional validity must be demonstrated for all such units. In regions where the stratigraphy has been established through studies of surface exposures, the naming of new units in the subsurface is justified only where the subsurface section differs materially from the surface section, or where there is doubt as to the equivalence of a subsurface and a surface unit. The establishment of subsurface reference sections for units originally defined in outcrop is encouraged.

(c) **Type section never changed.**—The definition and name of a lithostratigraphic unit are established at a type section (or locality) that, once specified, must not be changed. If the type section is poorly designated or delimited, it may be redefined subsequently. If the originally specified stratotype is incomplete, poorly exposed, structurally complicated, or unrepresentative of the unit, a principal reference section or several reference sections may be designated to supplement, but not to supplant, the type section (Article 8e).

(d) **Independence from inferred geologic history.**—Inferred geologic history, depositional environment, and biological sequence have no place in the definition of a lithostratigraphic unit, which must be based on composition and other lithic characteristics; nevertheless, considerations of well-documented geologic history properly may influence the choice of vertical and lateral boundaries of a new unit. Fossils may be valuable during mapping in distinguishing between two lithologically similar, noncontiguous lithostratigraphic units. The fossil content of a lithostratigraphic unit is a legitimate lithic characteristic; for example, oyster-rich sandstone, coquina, coral reef, or graptolitic shale. Moreover, otherwise similar units, such as the Formación Mendez and Formación Velasco mudstones, may be distinguished on the basis of coarseness of contained fossils (foraminifera).

(e) **Independence from time concepts.**—The boundaries of most lithostratigraphic units may transgress time horizons, but some may be approximately synchronous. Inferred time-spans, however measured, play no part in differentiating or determining the boundaries of any lithostratigraphic unit. Either relatively short or relatively long intervals of time may be represented by a single unit. The accumulation of material assigned to a particular unit may have begun or ended earlier in some localities than in others; also, removal of rock by erosion, either within the time-span of deposition of the unit or later, may reduce the time-span represented by the unit locally. The body in some places may be entirely younger than in other places. On the other hand, the establishment of formal units that straddle known, identifiable, regional disconformities is to be avoided, if at all possible. Although concepts of time or age play no part in defining lithostratigraphic units nor in determining their boundaries, evidence of age may aid recognition of similar lithostratigraphic units at localities far removed from the type sections or areas.

(f) **Surface form.**—Erosional morphology or secondary surface form may be a factor in the recognition of a lithostratigraphic unit, but properly should play a minor part at most in the definition of such units. Because the surface expression of lithostratigraphic units is an important aid in mapping, it is commonly advisable, where other factors do not countervail, to define lithostratigraphic boundaries so as to coincide with lithic changes that are expressed in topography.

(g) **Economically exploited units.**—Aquifers, oil sands, coal beds, and quarry layers are, in general, informal units even though named. Some such units, however, may be recognized formally as beds, members, or formations because they are important in the elucidation of regional stratigraphy.

(h) **Instrumentally defined units.**—In subsurface investigations, certain bodies of rock and their boundaries are widely recognized on borehole geophysical logs showing their electrical resistivity, radioactivity, density, or other physical properties. Such bodies and their boundaries may or may not correspond to formal lithostratigraphic units and their boundaries. Where other considerations do not countervail, the boundaries of subsurface units should be defined so as to correspond to useful geophysical markers; nevertheless, units defined exclusively on the basis of remotely sensed physical properties, although commonly useful in stratigraphic analysis, stand completely apart from the hierarchy of formal lithostratigraphic units and are considered informal.

(i) **Zone.**—As applied to the designation of lithostratigraphic units, the term "zone" is informal. Examples are "producing zone," "mineralized zone," "metamorphic zone," and "heavy-mineral zone." A zone may include all or parts of a bed, a member, a formation, or even a group.

(j) **Cyclothems.**—Cyclic or rhythmic sequences of sedimentary rocks, whose repetitive divisions have been named cyclothems, have been recognized in sedimentary basins around the world. Some cyclothems have

been identified by geographic names, but such names are considered informal. A clear distinction must be maintained between the division of a stratigraphic column into cyclothems and its division into groups, formations, and members. Where a cyclothem is identified by a geographic name, the word *cyclothem* should be part of the name, and the geographic term should not be the same as that of any formal unit embraced by the cyclothem.

(k) **Soils and paleosols.**—Soils and paleosols are layers composed of the in-situ products of weathering of older rocks which may be of diverse composition and age. Soils and paleosols differ in several respects from lithostratigraphic units, and should not be treated as such (see "Pedostratigraphic Units," Articles 55 et seq).

(l) **Depositional facies.**—Depositional facies are informal units, whether objective (conglomeratic, black shale, graptolitic) or genetic and environmental (platform, turbiditic, fluvial), even when a geographic term has been applied, e.g., Lantz Mills facies. Descriptive designations convey more information than geographic terms and are preferable.

Article 23.—**Boundaries.** Boundaries of lithostratigraphic units are placed at positions of lithic change. Boundaries are placed at distinct contacts or may be fixed arbitrarily within zones of gradation (Fig. 2a). Both vertical and lateral boundaries are based on the lithic criteria that provide the greatest unity and utility.

Remarks. (a) **Boundary in a vertically gradational sequence.**—A named lithostratigraphic unit is preferably bounded by a single lower and a single upper surface so that the name does not recur in a normal stratigraphic succession (see Remark b). Where a rock unit passes vertically into another by intergrading or interfingering of two or more kinds of rock, unless the gradational strata are sufficiently thick to warrant designation of a third, independent unit, the boundary is necessarily arbitrary and should be selected on the basis of practicality (Fig. 2b). For example, where a shale unit overlies a unit of interbedded limestone and shale, the boundary commonly is placed at the top of the highest readily traceable limestone bed. Where a sandstone unit grades upward into shale, the boundary may be so gradational as to be difficult to place even arbitrarily; ideally it should be drawn at the level where the rock is composed of one-half of each component. Because of creep in outcrops and caving in boreholes, it is generally best to define such arbitrary boundaries by the highest occurrence of a particular rock type, rather than the lowest.

(b) **Boundaries in lateral lithologic change.**—Where a unit changes laterally through abrupt gradation into, or intertongues with, a markedly different kind of rock, a new unit should be proposed for the different rock type. An arbitrary lateral boundary may be placed between the two equivalent units. Where the area of lateral intergradation or intertonguing is sufficiently extensive, a transitional interval of interbedded rocks may constitute a third independent unit (Fig. 2c). Where tongues (Article 25b) of formations are mapped separately or otherwise set apart without being formally named, the unmodified formation name should not be repeated in a normal stratigraphic sequence, although the modified name may be repeated in such phrases as "lower tongue of Mancos Shale" and "upper tongue of Mancos Shale." To show the order of superposition on maps and cross sections, the unnamed tongues may be distinguished informally (Fig. 2d) by number, letter, or other means. Such relationships may also be dealt with informally through the recognition of depositional facies (Article 22-1).

(c) **Key beds used for boundaries.**—Key beds (Article 26b) may be used as boundaries for a formal lithostratigraphic unit where the internal lithic characteristics of the unit remain relatively constant. Even though bounding key beds may be traceable beyond the area of the diagnostic overall rock type, geographic extension of the lithostratigraphic unit bounded thereby is not necessarily justified. Where the rock between key beds becomes drastically different from that of the type locality, a new name should be applied (Fig. 2e), even though the key beds are continuous (Article 26b). Stratigraphic and sedimentologic studies of stratigraphic units (usually informal) bounded by key beds may be very informative and useful, especially in subsurface work where the key beds may be recognized by their geophysical signatures. Such units, however, may be a kind of chronostratigraphic, rather than lithostratigraphic, unit (Article 75, 75c), although others are diachronous because one, or both, of the key beds are also diachronous.

(d) **Unconformities as boundaries.**—Unconformities, where recognizable objectively on lithic criteria, are ideal boundaries for lithostratigraphic units. However, a sequence of similar rocks may include an

North American Commission on Stratigraphic Nomenclature

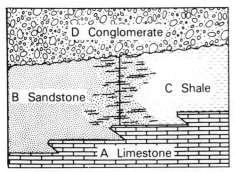

A.--Boundaries at sharp lithologic contacts and in laterally gradational sequence.

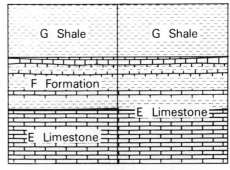

B.--Alternative boundaries in a vertically gradational or interlayered sequence.

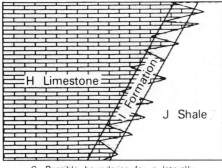

C.--Possible boundaries for a laterally intertonguing sequence.

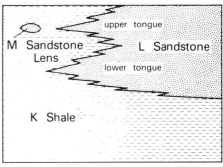

D.--Possible classification of parts of an intertonguing sequence.

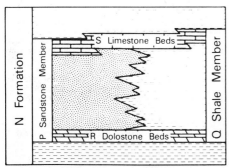

E.--Key beds, here designated the R Dolostone Beds and the S Limestone Beds, are used as boundaries to distinguish the Q Shale Member from the other parts of the N Formation. A lateral change in composition between the key beds requires that another name, P Sandstone Member, be applied. The key beds are part of each member.

EXPLANATION

- Conglomerate
- Sandstone
- Siltstone
- Mudstone, Shale
- Limestone
- Dolostone(dolomite)

FIG. 2.—Diagrammatic examples of lithostratigraphic boundaries and classification.

North American Stratigraphic Code

obscure unconformity so that separation into two units may be desirable but impracticable. If no lithic distinction adequate to define a widely recognizable boundary can be made, only one unit should be recognized, even though it may include rock that accumulated in different epochs, periods, or eras.

(e) **Correspondence with genetic units.**—The boundaries of lithostratigraphic units should be chosen on the basis of lithic changes and, where feasible, to correspond with the boundaries of genetic units, so that subsequent studies of genesis will not have to deal with units that straddle formal boundaries.

Ranks of Lithostratigraphic Units

Article 24.—**Formation.** The formation is the fundamental unit in lithostratigraphic classification. A formation is a body of rock identified by lithic characteristics and stratigraphic position; it is prevailingly but not necessarily tabular and is mappable at the Earth's surface or traceable in the subsurface.

Remarks. (a) **Fundamental unit.**—Formations are the basic lithostratigraphic units used in describing and interpreting the geology of a region. The limits of a formation normally are those surfaces of lithic change that give it the greatest practicable unity of constitution. A formation may represent a long or short time interval, may be composed of materials from one or several sources, and may include breaks in deposition (see Article 23d).

(b) **Content.**—A formation should possess some degree of internal lithic homogeneity or distinctive lithic features. It may contain between its upper and lower limits (i) rock of one lithic type, (ii) repetitions of two or more lithic types, or (iii) extreme lithic heterogeneity which in itself may constitute a form of unity when compared to the adjacent rock units.

(c) **Lithic characteristics.**—Distinctive lithic characteristics include chemical and mineralogical composition, texture, and such supplementary features as color, primary sedimentary or volcanic structures, fossils (viewed as rock-forming particles), or other organic content (coal, oil-shale). A unit distinguishable only by the taxonomy of its fossils is not a lithostratigraphic but a biostratigraphic unit (Article 48). Rock type may be distinctively represented by electrical, radioactive, seismic, or other properties (Article 22h), but these properties by themselves do not describe adequately the lithic character of the unit.

(d) **Mappability and thickness.**—The proposal of a new formation must be based on tested mappability. Well-established formations commonly are divisible into several widely recognizable lithostratigraphic units; where formal recognition of these smaller units serves a useful purpose, they may be established as members and beds, for which the requirement of mappability is not mandatory. A unit formally recognized as a formation in one area may be treated elsewhere as a group, or as a member of another formation, without change of name. Example: the Niobrara is mapped at different places as a member of the Mancos Shale, of the Cody Shale, or of the Colorado Shale, and also as the Niobrara Formation, as the Niobrara Limestone, and as the Niobrara Shale.

Thickness is not a determining parameter in dividing a rock succession into formations; the thickness of a formation may range from a feather edge at its depositional or erosional limit to thousands of meters elsewhere. No formation is considered valid that cannot be delineated at the scale of geologic mapping practiced in the region when the formation is proposed. Although representation of a formation on maps and cross sections by a labeled line may be justified, proliferation of such exceptionally thin units is undesirable. The methods of subsurface mapping permit delineation of units much thinner than those usually practicable for surface studies; before such thin units are formalized, consideration should be given to the effect on subsequent surface and subsurface studies.

(e) **Organic reefs and carbonate mounds.**—Organic reefs and carbonate mounds ("buildups") may be distinguished formally, if desirable, as formations distinct from their surrounding, thinner, temporal equivalents. For the requirements of formalization, see Article 30f.

(f) **Interbedded volcanic and sedimentary rock.**—Sedimentary rock and volcanic rock that are interbedded may be assembled into a formation under one name which should indicate the predominant or distinguishing lithology, such as Mindego Basalt.

(g) **Volcanic rock.**—Mappable distinguishable sequences of stratified volcanic rock should be treated as formations or lithostratigraphic units of higher or lower rank. A small intrusive component of a dominantly stratiform volcanic assemblage may be treated informally.

(h) **Metamorphic rock.**—Formations composed of low-grade metamorphic rock (defined for this purpose as rock in which primary structures are clearly recognizable) are, like sedimentary formations, distinguished mainly by lithic characteristics. The mineral facies may differ from place to place, but these variations do not require definition of a new formation. High-grade metamorphic rocks whose relation to established formations is uncertain are treated as lithodemic units (see Articles 31 et seq).

Article 25.—**Member.** A member is the formal lithostratigraphic unit next in rank below a formation and is always a part of some formation. It is recognized as a named entity within a formation because it possesses characteristics distinguishing it from adjacent parts of the formation. A formation need not be divided into members unless a useful purpose is served by doing so. Some formations may be divided completely into members; others may have only certain parts designated as members; still others may have no members. A member may extend laterally from one formation to another.

Remarks. (a) **Mapping of members.**—A member is established when it is advantageous to recognize a particular part of a heterogeneous formation. A member, whether formally or informally designated, need not be mappable at the scale required for formations. Even if all members of a formation are locally mappable, it does not follow that they should be raised to formational rank, because proliferation of formation names may obscure rather than clarify relations with other areas.

(b) **Lens and tongue.**—A geographically restricted member that terminates on all sides within a formation may be called a lens (lentil). A wedging member that extends outward beyond a formation or wedges ("pinches") out within another formation may be called a tongue.

(c) **Organic reefs and carbonate mounds.**—Organic reefs and carbonate mounds may be distinguished formally, if desirable, as members within a formation. For the requirements of formalization, see Article 30f.

(d) **Division of members.**—A formally or informally recognized division of a member is called a bed or beds, except for volcanic flow-rocks, for which the smallest formal unit is a flow. Members may contain beds or flows, but may never contain other members.

(e) **Laterally equivalent members.**—Although members normally are in vertical sequence, laterally equivalent parts of a formation that differ recognizably may also be considered members.

Article 26.—**Bed(s).** A bed, or beds, is the smallest formal lithostratigraphic unit of sedimentary rocks.

Remarks. (a) **Limitations.**—The designation of a bed or a unit of beds as a formally named lithostratigraphic unit generally should be limited to certain distinctive beds whose recognition is particularly useful. Coal beds, oil sands, and other beds of economic importance commonly are named, but such units and their names usually are not a part of formal stratigraphic nomenclature (Articles 22g and 30g).

(b) **Key or marker beds.**—A key or marker bed is a thin bed of distinctive rock that is widely distributed. Such beds may be named, but usually are considered informal units. Individual key beds may be traced beyond the lateral limits of a particular formal unit (Article 23c).

Article 27.—**Flow.** A flow is the smallest formal lithostratigraphic unit of volcanic flow rocks. A flow is a discrete, extrusive, volcanic body distinguishable by texture, composition, order of superposition, paleomagnetism, or other objective criteria. It is part of a member and thus is equivalent in rank to a bed or beds of sedimentary-rock classification. Many flows are informal units. The designation and naming of flows as formal rock-stratigraphic units should be limited to those that are distinctive and widespread.

Article 28.—**Group.** A group is the lithostratigraphic unit next higher in rank to formation; a group may consist entirely of named formations, or alternatively, need not be composed entirely of named formations.

Remarks. (a) **Use and content.**—Groups are defined to express the natural relationships of associated formations. They are useful in small-

North American Commission on Stratigraphic Nomenclature

scale mapping and regional stratigraphic analysis. In some reconnaissance work, the term "group" has been applied to lithostratigraphic units that appear to be divisible into formations, but have not yet been so divided. In such cases, formations may be erected subsequently for one or all of the practical divisions of the group.

(b) **Change in component formations.**—The formations making up a group need not necessarily be everywhere the same. The Rundle Group, for example, is widespread in western Canada and undergoes several changes in formational content. In southwestern Alberta, it comprises the Livingstone, Mount Head, and Etherington Formations in the Front Ranges, whereas in the foothills and subsurface of the adjacent plains, it comprises the Pekisko, Shunda, Turner Valley, and Mount Head Formations. However, a formation or its parts may not be assigned to two vertically adjacent groups.

(c) **Change in rank.**—The wedge-out of a component formation or formations may justify the reduction of a group to formation rank, retaining the same name. When a group is extended laterally beyond where it is divided into formations, it becomes in effect a formation, even if it is still called a group. When a previously established formation is divided into two or more component units that are given formal formation rank, the old formation, with its old geographic name, should be raised to group status. Raising the rank of the unit is preferable to restricting the old name to a part of its former content, because a change in rank leaves the sense of a well-established unit unchanged (Articles 19b, 19g).

Article 29.—Supergroup. A supergroup is a formal assemblage of related or superposed groups, or of groups and formations. Such units have proved useful in regional and provincial syntheses. Supergroups should be named only where their recognition serves a clear purpose.

Remark. (a) **Misuse of "series" for group or supergroup.**—Although "series" is a useful general term, it is applied formally only to a chronostratigraphic unit and should not be used for a lithostratigraphic unit. The term "series" should no longer be employed for an assemblage of formations or an assemblage of formations and groups, as it has been, especially in studies of the Precambrian. These assemblages are groups or supergroups.

Lithostratigraphic Nomenclature

Article 30.—Compound Character. The formal name of a lithostratigraphic unit is compound. It consists of a geographic name combined with a descriptive lithic term or with the appropriate rank term, or both. Initial letters of all words used in forming the names of formal rock-stratigraphic units are capitalized.

Remarks. (a) **Omission of part of a name.**—Where frequent repetition would be cumbersome, the geographic name, the lithic term, or the rank term may be used alone, once the full name has been introduced; as "the Burlington," "the limestone," or "the formation," for the Burlington Limestone.

(b) **Use of simple lithic terms.**—The lithic part of the name should indicate the predominant or diagnostic lithology, even if subordinate lithologies are included. Where a lithic term is used in the name of a lithostratigraphic unit, the simplest generally acceptable term is recommended (for example, limestone, sandstone, shale, tuff, quartzite). Compound terms (for example, clay shale) and terms that are not in common usage (for example, calcirudite, orthoquartzite) should be avoided. Combined terms, such as "sand and clay," should not be used for the lithic part of the names of lithostratigraphic units, nor should an adjective be used between the geographic and the lithic terms, as "Chattanooga Black Shale" and "Biwabik Iron-Bearing Formation."

(c) **Group names.**—A group name combines a geographic name with the term "group," and no lithic designation is included; for example, San Rafael Group.

(d) **Formation names.**—A formation name consists of a geographic name followed by a lithic designation or by the word "formation." Examples: Dakota Sandstone, Mitchell Mesa Rhyolite, Monmouth Formation, Halton Till.

(e) **Member names.**—All member names include a geographic term and the word "member;" some have an intervening lithic designation, if useful; for example, Wedington Sandstone Member of the Fayetteville Shale. Members designated solely by lithic character (for example, siliceous shale member), by position (upper, lower), or by letter or number, are informal.

(f) **Names of reefs.**—Organic reefs identified as formations or members are formal units only where the name combines a geographic name with the appropriate rank term, e.g., Leduc Formation (a name applied to the several reefs enveloped by the Ireton Formation), Rainbow Reef Member.

(g) **Bed and flow names.**—The names of beds or flows combine a geographic term, a lithic term, and the term "bed" or "flow;" for example, Knee Hills Tuff Bed, Ardmore Bentonite Beds, Negus Variolitic Flows.

(h) **Informal units.**—When geographic names are applied to such informal units as oil sands, coal beds, mineralized zones, and informal members (see Articles 22g and 26a), the unit term should not be capitalized. A name is not necessarily formal because it is capitalized, nor does failure to capitalize a name render it informal. Geographic names should be combined with the terms "formation" or "group" only in formal nomenclature.

(i) **Informal usage of identical geographic names.**—The application of identical geographic names to several minor units in one vertical sequence is considered informal nomenclature (lower Mount Savage coal, Mount Savage fireclay, upper Mount Savage coal, Mount Savage rider coal, and Mount Savage sandstone). The application of identical geographic names to the several lithologic units constituting a cyclothem likewise is considered informal.

(j) **Metamorphic rock.**—Metamorphic rock recognized as a normal stratified sequence, commonly low-grade metavolcanic or metasedimentary rocks, should be assigned to named groups, formations, and members, such as the Deception Rhyolite, a formation of the Ash Creek Group, or the Bonner Quartzite, a formation of the Missoula Group. High-grade metamorphic and metasomatic rocks are treated as lithodemes and suites (see Articles 31, 33, 35).

(k) **Misuse of well-known name.**—A name that suggests some well-known locality, region, or political division should not be applied to a unit typically developed in another less well-known locality of the same name. For example, it would be inadvisable to use the name "Chicago Formation" for a unit in California.

LITHODEMIC UNITS

Nature and Boundaries

Article 31.—Nature of Lithodemic Units. A lithodemic[6] unit is a defined body of predominantly intrusive, highly deformed, and/or highly metamorphosed rock, distinguished and delimited on the basis of rock characteristics. In contrast to lithostratigraphic units, a lithodemic unit generally does not conform to the Law of Superposition. Its contacts with other rock units may be sedimentary, extrusive, intrusive, tectonic, or metamorphic (Fig. 3).

Remarks. (a) **Recognition and definition.**—Lithodemic units are defined and recognized by observable rock characteristics. They are the practical units of general geological work in terranes in which rocks generally lack primary stratification; in such terranes they serve as the foundation for studying, describing, and delineating lithology, local and regional structure, economic resources, and geologic history.

(b) **Type and reference localities.**—The definition of a lithodemic unit should be based on as full a knowledge as possible of its lateral and vertical variations and its contact relationships. For purposes of nomenclatural stability, a type locality and, wherever appropriate, reference localities should be designated.

(c) **Independence from inferred geologic history.**—Concepts based on inferred geologic history properly play no part in the definition of a lithodemic unit. Nevertheless, where two rock masses are lithically similar but display objective structural relations that preclude the possibility of their being even broadly of the same age, they should be assigned to different lithodemic units.

(d) **Use of "zone."**—As applied to the designation of lithodemic units, the term "zone" is informal. Examples are: "mineralized zone," "contact zone," and "pegmatitic zone."

---

[6]From the Greek *demas, -os:* "living body, frame".

North American Stratigraphic Code

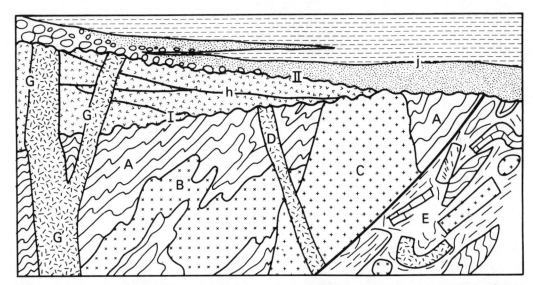

FIG. 3.—Lithodemic (upper case) and lithostratigraphic (lower case) units. A *lithodeme* of *gneiss* (A) contains an *intrusion* of diorite (B) that was deformed with the gneiss. A and B may be treated jointly as a *complex*. A younger *granite* (C) is cut by a dike of *syenite* (D), that is cut in turn by unconformity I. All the foregoing are in fault contact with a *structural complex* (E). A *volcanic complex* (G) is built upon unconformity I, and its feeder dikes cut the unconformity. Laterally equivalent volcanic strata in orderly, mappable succession (h) are treated as lithostratigraphic units. A *gabbro* feeder (G′), to the volcanic complex, where surrounded by gneiss is readily distinguished as a separate lithodeme and named as a *gabbro* or an *intrusion*. All the foregoing are overlain, at unconformity II, by sedimentary rocks (j) divided into formations and members.

Article 32.—**Boundaries.** Boundaries of lithodemic units are placed at positions of lithic change. They may be placed at clearly distinguished contacts or within zones of gradation. Boundaries, both vertical and lateral, are based on the lithic criteria that provide the greatest unity and practical utility. Contacts with other lithodemic and lithostratigraphic units may be depositional, intrusive, metamorphic, or tectonic.

Remark. (a) **Boundaries within gradational zones.**—Where a lithodemic unit changes through gradation into, or intertongues with, a rockmass with markedly different characteristics, it is usually desirable to propose a new unit. It may be necessary to draw an arbitrary boundary within the zone of gradation. Where the area of intergradation or intertonguing is sufficiently extensive, the rocks of mixed character may constitute a third unit.

Ranks of Lithodemic Units

Article 33.—**Lithodeme.** The lithodeme is the fundamental unit in lithodemic classification. A lithodeme is a body of intrusive, pervasively deformed, or highly metamorphosed rock, generally non-tabular and lacking primary depositional structures, and characterized by lithic homogeneity. It is mappable at the Earth's surface and traceable in the subsurface. For cartographic and hierarchical purposes, it is comparable to a formation (see Table 2).

Remarks. (a) **Content.**—A lithodeme should possess distinctive lithic features and some degree of internal lithic homogeneity. It may consist of (i) rock of one type, (ii) a mixture of rocks of two or more types, or (iii) extreme heterogeneity of composition, which may constitute in itself a form of unity when compared to adjoining rock-masses (see also "complex," Article 37).

(b) **Lithic characteristics.**—Distinctive lithic characteristics may include mineralogy, textural features such as grain size, and structural features such as schistose or gneissic structure. A unit distinguishable

from its neighbors only by means of chemical analysis is informal.

(c) **Mappability.**—Practicability of surface or subsurface mapping is an essential characteristic of a lithodeme (see Article 24d).

Article 34.—**Division of Lithodemes.** Units below the rank of lithodeme are informal.

Article 35.—**Suite.** A *suite* (metamorphic suite, intrusive suite, plutonic suite) is the lithodemic unit next higher in rank to lithodeme. It comprises two or more associated lithodemes of the same class (e.g., plutonic, metamorphic). For cartographic and hierarchical purposes, suite is comparable to group (see Table 2).

Remarks. (a) **Purpose.**—Suites are recognized for the purpose of expressing the natural relations of associated lithodemes having significant lithic features in common, and of depicting geology at compilation scales too small to allow delineation of individual lithodemes. Ideally, a suite consists entirely of named lithodemes, but may contain both named and unnamed units.

(b) **Change in component units.**—The named and unnamed units constituting a suite may change from place to place, so long as the original sense of natural relations and of common lithic features is not violated.

(c) **Change in rank.**—Traced laterally, a suite may lose all of its formally named divisions but remain a recognizable, mappable entity. Under such circumstances, it may be treated as a lithodeme but retain the same name. Conversely, when a previously established lithodeme is divided into two or more mappable divisions, it may be desirable to raise its rank to suite, retaining the original geographic component of the name. To avoid confusion, the original name should not be retained for one of the divisions of the original unit (see Article 19g).

Article 36.—**Supersuite.** A supersuite is the unit next higher in rank to a suite. It comprises two or more suites or complexes having a degree of natural relationship to one another, either in the vertical or the lateral sense. For cartographic and hierarchical purposes, supersuite is similar in rank to supergroup.

North American Commission on Stratigraphic Nomenclature

Article 37.—Complex. An assemblage or mixture of rocks of *two or more genetic classes*, i.e., igneous, sedimentary, or metamorphic, with or without highly complicated structure, may be named a *complex*. The term "complex" takes the place of the lithic or rank term (for example, Boil Mountain Complex, Franciscan Complex) and, although unranked, commonly is comparable to suite or supersuite and is named in the same manner (Articles 41, 42).

Remarks (a) **Use of "complex."**—Identification of an assemblage of diverse rocks as a complex is useful where the mapping of each separate lithic component is impractical at ordinary mapping scales. "Complex" is unranked but commonly comparable to suite or supersuite; therefore, the term may be retained if subsequent, detailed mapping distinguishes some or all of the component lithodemes or lithostratigraphic units.

(b) **Volcanic complex.**—Sites of persistent volcanic activity commonly are characterized by a diverse assemblage of extrusive volcanic rocks, related intrusions, and their weathering products. Such an assemblage may be designated a *volcanic complex*.

(c) **Structural complex.**—In some terranes, tectonic processes (e.g., shearing, faulting) have produced heterogeneous mixtures or disrupted bodies of rock in which some individual components are too small to be mapped. *Where there is no doubt that the mixing or disruption is due to tectonic processes*, such a mixture may be designated as a structural complex, whether it consists of two or more classes of rock, or a single class only. A simpler solution for some mapping purposes is to indicate intense deformation by an overprinted pattern.

(d) **Misuse of "complex".**—Where the rock assemblage to be united under a single, formal name consists of diverse types of a *single class* of rock, as in many terranes that expose a variety of either intrusive igneous or high-grade metamorphic rocks, the term "intrusive suite," "plutonic suite," or "metamorphic suite" should be used, rather than the unmodified term "complex." Exceptions to this rule are the terms *structural complex* and *volcanic complex* (see Remarks c and b, above).

Article 38.—Misuse of "Series" for Suite, Complex, or Supersuite. The term "series" has been employed for an assemblage of lithodemes or an assemblage of lithodemes and suites, especially in studies of the Precambrian. This practice now is regarded as improper; these assemblages are suites, complexes, or supersuites. The term "series" also has been applied to a sequence of rocks resulting from a succession of eruptions or intrusions. In these cases a different term should be used; "group" should replace "series" for volcanic and low-grade metamorphic rocks, and "intrusive suite" or "plutonic suite" should replace "series" for intrusive rocks of group rank.

Lithodemic Nomenclature

Article 39.—General Provisions. The formal name of a lithodemic unit is compound. It consists of a geographic name combined with a descriptive or appropriate rank term. The principles for the selection of the geographic term, concerning suitability, availability, priority, etc, follow those established in Article 7, where the rules for capitalization are also specified.

Article 40.—Lithodeme Names. The name of a lithodeme combines a geographic term with a lithic or descriptive term, e.g., Killarney Granite, Adamant Pluton, Manhattan Schist, Skaergaard Intrusion, Duluth Gabbro. The term *formation* should not be used.

Remarks. (a) **Lithic term.**—The lithic term should be a common and familiar term, such as schist, gneiss, gabbro. Specialized terms and terms not widely used, such as websterite and jacupirangite, and compound terms, such as graphitic schist and augen gneiss, should be avoided.

(b) **Intrusive and plutonic rocks.**—Because many bodies of intrusive rock range in composition from place to place and are difficult to characterize with a single lithic term, and because many bodies of plutonic rock

are considered not to be intrusions, latitude is allowed in the choice of a lithic or descriptive term. Thus, the descriptive term should preferably be compositional (e.g., gabbro, granodiorite), but may, if necessary, denote form (e.g., dike, sill), or be neutral (e.g., intrusion, pluton[7]). In any event, specialized compositional terms not widely used are to be avoided, as are form terms that are not widely used, such as bysmalith and chonolith. Terms implying genesis should be avoided as much as possible, because interpretations of genesis may change.

Article 41.—Suite Names. The name of a suite combines a geographic term, the term "suite," and an adjective denoting the fundamental character of the suite; for example, Idaho Springs Metamorphic Suite, Tuolumne Intrusive Suite, Cassiar Plutonic Suite. The geographic name of a suite may not be the same as that of a component lithodeme (see Article 19f). Intrusive assemblages, however, may share the same geographic name if an intrusive lithodeme is representative of the suite.

Article 42.—Supersuite Names. The name of a supersuite combines a geographic term with the term "supersuite."

MAGNETOSTRATIGRAPHIC UNITS

Nature and Boundaries

Article 43.—Nature of Magnetostratigraphic Units. A magnetostratigraphic unit is a body of rock unified by specified remanent-magnetic properties and is distinct from underlying and overlying magnetostratigraphic units having different magnetic properties.

Remarks. (a) **Definition.**—Magnetostratigraphy is defined here as all aspects of stratigraphy based on remanent magnetism (paleomagnetic signatures). Four basic paleomagnetic phenomena can be determined or inferred from remanent magnetism: polarity, dipole-field-pole position (including apparent polar wander), the non-dipole component (secular variation), and field intensity.

(b) **Contemporaneity of rock and remanent magnetism.**—Many paleomagnetic signatures reflect earth magnetism at the time the rock formed. Nevertheless, some rocks have been subjected subsequently to physical and/or chemical processes which altered the magnetic properties. For example, a body of rock may be heated above the blocking temperature or Curie point for one or more minerals, or a ferromagnetic mineral may be produced by low-temperature alteration long after the enclosing rock formed, thus acquiring a component of remanent magnetism reflecting the field at the time of alteration, rather than the time of original rock deposition or crystallization.

(c) **Designations and scope.**—The prefix *magneto* is used with an appropriate term to designate the aspect of remanent magnetism used to define a unit. The terms "magnetointensity" or "magnetosecularvariation" are possible examples. This Code considers only polarity reversals, which now are recognized widely as a stratigraphic tool. However, apparent-polar-wander paths offer increasing promise for correlations within Precambrian rocks.

Article 44.—Definition of Magnetopolarity Unit. A magnetopolarity unit is a body of rock unified by its remanent magnetic polarity and distinguished from adjacent rock that has different polarity.

Remarks. (a) **Nature.**—Magnetopolarity is the record in rocks of the polarity history of the Earth's magnetic-dipole field. Frequent past reversals of the polarity of the Earth's magnetic field provide a basis for magnetopolarity stratigraphy.

(b) **Stratotype.**—A stratotype for a magnetopolarity unit should be designated and the boundaries defined in terms of recognized lithostratigraphic and/or biostratigraphic units in the stratotype. The formal definition of a magnetopolarity unit should meet the applicable specific requirements of Articles 3 to 16.

(c) **Independence from inferred history.**—Definition of a magnetopolarity unit does not require knowledge of the time at which the unit acquired its remanent magnetism; its magnetism may be primary or secondary. Nevertheless, the unit's present polarity is a property that may be

---

[7]Pluton—a mappable body of plutonic rock.

North American Stratigraphic Code

ascertained and confirmed by others.

(d) **Relation to lithostratigraphic and biostratigraphic units.**—Magnetopolarity units resemble lithostratigraphic and biostratigraphic units in that they are defined on the basis of an objective recognizable property, but differ fundamentally in that most magnetopolarity unit boundaries are thought not to be time transgressive. Their boundaries may coincide with those of lithostratigraphic or biostratigraphic units, or be parallel to but displaced from those of such units, or be crossed by them.

(e) **Relation of magnetopolarity units to chronostratigraphic units.**—Although transitions between polarity reversals are of global extent, a magnetopolarity unit does not contain within itself evidence that the polarity is primary, or criteria that permit its unequivocal recognition in chronocorrelative strata of other areas. Other criteria, such as paleontologic or numerical age, are required for both correlation and dating. Although polarity reversals are useful in recognizing chronostratigraphic units, magnetopolarity alone is insufficient for their definition.

Article 45.—**Boundaries.** The upper and lower limits of a magnetopolarity unit are defined by boundaries marking a change of polarity. Such boundaries may represent either a depositional discontinuity or a magnetic-field transition. The boundaries are either polarity-reversal horizons or polarity transition-zones, respectively.

Remark. (a) **Polarity-reversal horizons and transition-zones.**—A polarity-reversal horizon is either a single, clearly definable surface or a thin body of strata constituting a transitional interval across which a change in magnetic polarity is recorded. Polarity-reversal horizons describe transitional intervals of 1 m or less; where the change in polarity takes place over a stratigraphic interval greater than 1 m, the term "polarity transition-zone" should be used. Polarity-reversal horizons and polarity transition-zones provide the boundaries for polarity zones, although they may also be contained within a polarity zone where they mark an internal change subsidiary in rank to those at its boundaries.

Ranks of Magnetopolarity Units

Article 46.—**Fundamental Unit.** A polarity zone is the fundamental unit of magnetopolarity classification. A polarity zone is a unit of rock characterized by the polarity of its magnetic signature. Magnetopolarity zone, rather than polarity zone, should be used where there is risk of confusion with other kinds of polarity.

Remarks. (a) **Content.**—A polarity zone should possess some degree of internal homogeneity. It may contain rocks of (1) entirely or predominantly one polarity, or (2) mixed polarity.

(b) **Thickness and duration.**—The thickness of rock of a polarity zone or the amount of time represented should play no part in the definition of the zone. The polarity signature is the essential property for definition.

(c) **Ranks.**—When continued work at the stratotype for a polarity zone, or new work in correlative rocks elsewhere, reveals smaller polarity units, these may be recognized formally as polarity subzones. If it should prove necessary or desirable to group polarity zones, these should be termed polarity superzones. The rank of a polarity unit may be changed when deemed appropriate.

Magnetopolarity Nomenclature

Article 47.—**Compound Name.** The formal name of a magnetopolarity zone should consist of a geographic name and the term *Polarity Zone*. The term may be modified by *Normal*, *Reversed*, or *Mixed* (example: Deer Park Reversed Polarity Zone). In naming or revising magnetopolarity units, appropriate parts of Articles 7 and 19 apply. The use of informal designations, e.g., numbers or letters, is not precluded.

BIOSTRATIGRAPHIC UNITS

Nature and Boundaries

Article 48.—**Nature of Biostratigraphic Units.** A biostratigraphic unit is a body of rock defined or characterized by its fossil content. The basic unit in biostratigraphic classification is the biozone, of which there are several kinds.

Remarks. (a) **Enclosing strata.**—Fossils that define or characterize a biostratigraphic unit commonly are contemporaneous with the body of rock that contains them. Some biostratigraphic units, however, may be represented only by their fossils, preserved in normal stratigraphic succession (e.g., on hardgrounds, in lag deposits, in certain types of remanié accumulations), which alone represent the rock of the biostratigraphic unit. In addition, some strata contain fossils derived from older or younger rocks or from essentially coeval materials of different facies; such fossils should not be used to define a biostratigraphic unit.

(b) **Independence from lithostratigraphic units.**—Biostratigraphic units are based on criteria which differ fundamentally from those for lithostratigraphic units. Their boundaries may or may not coincide with the boundaries of lithostratigraphic units, but they bear no inherent relation to them.

(c) **Independence from chronostratigraphic units.**—The boundaries of most biostratigraphic units, unlike the boundaries of chronostratigraphic units, are both characteristically and conceptually diachronous. An exception is an abundance biozone boundary that reflects a mass-mortality event. The vertical and lateral limits of the rock body that constitutes the biostratigraphic unit represent the limits in distribution of the defining biotic elements. The lateral limits never represent, and the vertical limits rarely represent, regionally synchronous events. Nevertheless, biostratigraphic units are effective for interpreting chronostratigraphic relations.

Article 49.—**Kinds of Biostratigraphic Units.** Three principal kinds of biostratigraphic units are recognized: *interval, assemblage*, and *abundance* biozones.

Remark: (a) **Boundary definitions.**—Boundaries of interval zones are defined by lowest and/or highest occurrences of single taxa; boundaries of some kinds of assemblage zones (Oppel or concurrent range zones) are defined by lowest and/or highest occurrences of more than one taxon; and boundaries of abundance zones are defined by marked changes in relative abundances of preserved taxa.

Article 50.—**Definition of Interval Zone.** An interval zone (or subzone) is the body of strata between two specified, documented lowest and/or highest occurrences of single taxa.

Remarks. (a) **Interval zone types.**—Three basic types of interval zones are recognized (Fig. 4). These include the range zones and interval zones of the International Stratigraphic Guide (ISSC, 1976, p. 53, 60) and are:

1. The interval between the documented lowest and highest occurrences of a single taxon (Fig. 4A). This is the *taxon range zone* of ISSC (1976, p. 53).

2. The interval included between the documented lowest occurrence of one taxon and the documented highest occurrence of another taxon (Fig. 4B). When such occurrences result in stratigraphic overlap of the taxa (Fig. 4B-1), the interval zone is the *concurrent range zone* of ISSC (1976, p. 55), that involves only two taxa. When such occurrences do not result in stratigraphic overlap (Fig. 4B-2), and are used to partition the range of a third taxon, the interval is the *partial range zone* of George and others (1969).

3. The interval between documented successive lowest occurrences or successive highest occurrences of two taxa (Fig. 4C). When the interval is between successive documented lowest occurrences within an evolutionary lineage (Fig. 4C-1), it is the *lineage zone* of ISSC (1976, p. 58). When the interval is between successive lowest occurrences of unrelated taxa or between successive highest occurrences of either related or unrelated taxa (Fig. 4C-2), it is a kind of *interval zone* of ISSC (1976, p. 60).

(b) **Unfossiliferous intervals.**—Unfossiliferous intervals between or within biozones are the *barren interzones* and *intrazones* of ISSC (1976, p. 49).

Article 51.—**Definition of Assemblage Zone.** An assemblage zone is a biozone characterized by the association of three or more taxa. It may be based on all kinds of fossils present, or restricted to only certain kinds of fossils.

North American Commission on Stratigraphic Nomenclature

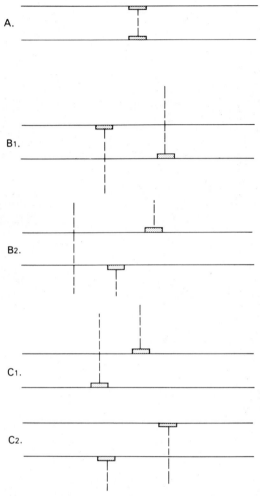

A.

B₁.

B₂.

C₁.

C₂.

FIG. 4.—Examples of biostratigraphic interval zones.
Vertical broken lines indicate ranges of taxa; bars indicate lowest or highest documented occurrences.

Remarks. (a) **Assemblage zone contents.**—An assemblage zone may consist of a geographically or stratigraphically restricted assemblage, or may incorporate two or more contemporaneous assemblages with shared characterizing taxa (*composite assemblage zones* of Kauffman, 1969) (Fig. 5c).

(b) **Assemblage zone types.**—In practice, two assemblage zone concepts are used:

1. The *assemblage zone* (or cenozone) of ISSC (1976, p. 50), which is characterized by taxa without regard to their range limits (Fig. 5a). Recognition of this type of assemblage zone can be aided by using techniques of multivariate analysis. Careful designation of the characterizing taxa is especially important.

2. The *Oppel zone*, or the *concurrent range zone* of ISSC (1976, p. 55, 57), a type of zone characterized by more than two taxa and having boundaries based on two or more documented first and/or last occurrences of the included characterizing taxa (Fig. 5b).

Article 52.—**Definition of Abundance Zone.** An abundance

zone is a biozone characterized by quantitatively distinctive maxima of relative abundance of one or more taxa. This is the *acme zone* of ISSC (1976, p. 59).

Remark. (a) **Ecologic controls.**—The distribution of biotic assemblages used to characterize some assemblage and abundance biozones may reflect strong local ecological control. Biozones based on such assemblages are included within the concept of ecozones (Vella, 1964), and are informal.

Ranks of Biostratigraphic Units

Article 53.—**Fundamental Unit.** The fundamental unit of biostratigraphic classification is a biozone.

Remarks. (a) **Scope.**—A single body of rock may be divided into various kinds and scales of biozones or subzones, as discussed in the International Stratigraphic Guide (ISSC, 1976, p. 62). Such usage is recommended if it will promote clarity, but only the unmodified term *biozone* is accorded formal status.

(b) **Divisions.**—A biozone may be completely or partly divided into formally designated sub-biozones (subzones), if such divisions serve a useful purpose.

Biostratigraphic Nomenclature

Article 54.—**Establishing Formal Units.** Formal establishment of a biozone or subzone must meet the requirements of Article 3 and requires a unique name, a description of its content and its boundaries, reference to a stratigraphic sequence in which the zone is characteristically developed, and a discussion of its spatial extent.

Remarks. (a) **Name.**—The name, which is compound and designates the kind of biozone, may be based on:

1. One or two characteristic and common taxa that are restricted to the biozone, reach peak relative abundance within the biozone, or have their total stratigraphic overlap within the biozone. These names most commonly are those of genera or subgenera, binomial designations of species, or trinomial designations of subspecies. If names of the nominate taxa change, names of the zones should be changed accordingly. Generic or subgeneric names may be abbreviated. Trivial species or subspecies names should not be used alone because they may not be unique.

2. Combinations of letters derived from taxa which characterize the biozone. However, alpha-numeric code designations (e.g., N1, N2, N3...) are informal and not recommended because they do not lend themselves readily to subsequent insertions, combinations, or eliminations. Biozonal systems based *only* on simple progressions of letters or numbers (e.g., A, B, C, or 1, 2, 3) are also not recommended.

(b) **Revision.**—Biozones and subzones are established empirically and may be modified on the basis of new evidence. Positions of established biozone or subzone boundaries may be stratigraphically refined, new characterizing taxa may be recognized, or original characterizing taxa may be superseded. If the concept of a particular biozone or subzone is substantially modified, a new unique designation is required to avoid ambiguity in subsequent citations.

(c) **Specifying kind of zone.**—Initial designation of a formally proposed biozone or subzone as an abundance zone, or as one of the types of interval zones, or assemblage zones (Articles 49-52), is strongly recommended. Once the type of biozone is clearly identified, the designation may be dropped in the remainder of a text (e.g., *Exus albus* taxon range zone to *Exus albus* biozone).

(d) **Defining taxa.**—Initial description or subsequent emendation of a biozone or subzone requires designation of the defining and characteristic taxa, and/or the documented first and last occurrences which mark the biozone or subzone boundaries.

(e) **Stratotypes.**—The geographic and stratigraphic position and boundaries of a formally proposed biozone or subzone should be defined precisely or characterized in one or more designated reference sections. Designation of a stratotype for each new biostratigraphic unit and of reference sections for emended biostratigraphic units is required.

North American Stratigraphic Code

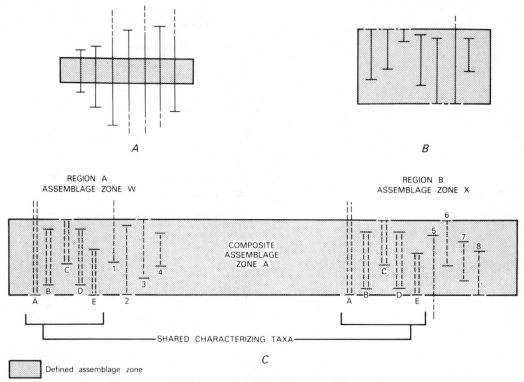

FIG. 5.—Examples of assemblage zone concepts.

PEDOSTRATIGRAPHIC UNITS

Nature and Boundaries

Article 55.—**Nature of Pedostratigraphic Units.** A pedostratigraphic unit is a body of rock that consists of one or more pedologic horizons developed in one or more lithostratigraphic, allostratigraphic, or lithodemic units (Fig. 6) and is overlain by one or more formally defined lithostratigraphic or allostratigraphic units.

Remarks. (a) **Definition.**—A pedostratigraphic[8] unit is a buried, traceable, three-dimensional body of rock that consists of one or more differentiated pedologic horizons.

(b) **Recognition.**—The distinguishing property of a pedostratigraphic unit is the presence of one or more distinct, differentiated, pedologic horizons. Pedologic horizons are products of soil development (pedogenesis) which occurred subsequent to formation of the lithostratigraphic, allostratigraphic, or lithodemic unit or units on which the buried soil was formed; these units are the parent materials in which pedogenesis occurred. Pedologic horizons are recognized in the field by diagnostic features such as color, soil structure, organic-matter accumulation, texture, clay coatings, stains, or concretions. Micromorphology, particle size, clay mineralogy, and other properties determined in the laboratory also may be used to identify and distinguish pedostratigraphic units.

(c) **Boundaries and stratigraphic position.**—The upper boundary of a pedostratigraphic unit is the top of the uppermost pedologic horizon

[8]Terminology related to pedostratigraphic classification is summarized on page 850.

formed by pedogenesis in a buried soil profile. The lower boundary of a pedostratigraphic unit is the lowest *definite* physical boundary of a pedologic horizon within a buried soil profile. The stratigraphic position of a pedostratigraphic unit is determined by its relation to overlying and underlying stratigraphic units (see Remark d).

(d) **Traceability.**—Practicability of subsurface tracing of the upper boundary of a buried soil is essential in establishing a pedostratigraphic unit because (1) few buried soils are exposed continuously for great distances, (2) the physical and chemical properties of a specific pedostratigraphic unit may vary greatly, both vertically and laterally, from place to place, and (3) pedostratigraphic units of different stratigraphic significance in the same region generally do not have unique identifying physical and chemical characteristics. Consequently, extension of a pedostratigraphic unit is accomplished by lateral tracing of the contact between a buried soil and an overlying, formally defined lithostratigraphic or allostratigraphic unit, or between a soil and two or more demonstrably correlative stratigraphic units.

(e) **Distinction from pedologic soils.**—Pedologic soils may include organic deposits (e.g., litter zones, peat deposits, or swamp deposits) that overlie or grade laterally into differentiated buried soils. The organic deposits are not products of pedogenesis, and O horizons are not included in a pedostratigraphic unit (Fig. 6); they may be classified as biostratigraphic or lithostratigraphic units. Pedologic soils also include the entire C horizon of a soil. The C horizon in pedology is not rigidly defined; it is merely the part of a soil profile that underlies the B horizon. The base of the C horizon in many soil profiles is gradational or unidentifiable; commonly it is placed arbitrarily. The need for clearly defined and easily recognized physical boundaries for a stratigraphic unit requires that the lower boundary of a pedostratigraphic unit be defined as the lowest *definite* physical boundary of a pedologic horizon in a buried soil profile, and part or all of the C horizon may be excluded from a pedostratigraphic unit.

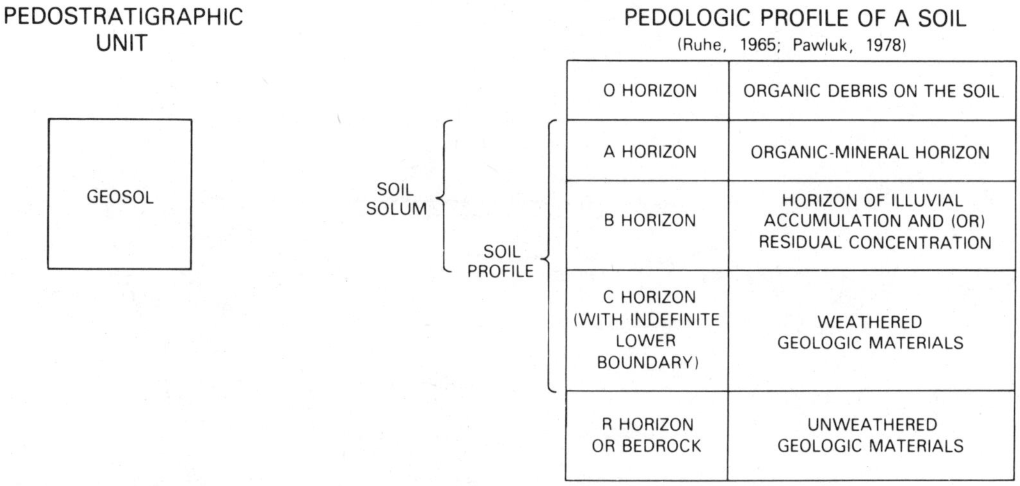

North American Commission on Stratigraphic Nomenclature

PEDOSTRATIGRAPHIC UNIT

PEDOLOGIC PROFILE OF A SOIL
(Ruhe, 1965; Pawluk, 1978)

GEOSOL	SOIL SOLUM / SOIL PROFILE		

O HORIZON	ORGANIC DEBRIS ON THE SOIL
A HORIZON	ORGANIC-MINERAL HORIZON
B HORIZON	HORIZON OF ILLUVIAL ACCUMULATION AND (OR) RESIDUAL CONCENTRATION
C HORIZON (WITH INDEFINITE LOWER BOUNDARY)	WEATHERED GEOLOGIC MATERIALS
R HORIZON OR BEDROCK	UNWEATHERED GEOLOGIC MATERIALS

FIG. 6.—Relationship between pedostratigraphic units and pedologic profiles.
The base of a geosol is the lowest clearly defined physical boundary of a pedologic horizon in a buried soil profile. In this example it is the lower boundary of the B horizon because the base of the C horizon is not a clearly defined physical boundary. In other profiles the base may be the lower boundary of a C horizon.

(f) **Relation to saprolite and other weathered materials.**—A material derived by in situ weathering of lithostratigraphic, allostratigraphic, and(or) lithodemic units (e.g., saprolite, bauxite, residuum) may be the parent material in which pedologic horizons form, but is not a pedologic soil. A pedostratigraphic unit may be based on the pedologic horizons of a buried soil developed in the product of in-situ weathering, such as saprolite. The parents of such a pedostratigraphic unit are both the saprolite and, indirectly, the rock from which it formed.

(g) **Distinction from other stratigraphic units.**—A pedostratigraphic unit differs from other stratigraphic units in that (1) it is a product of surface alteration of one or more older material units by specific processes (pedogenesis), (2) its lithology and other properties differ markedly from those of the parent material(s), and (3) a single pedostratigraphic unit may be formed in situ in parent material units of diverse compositions and ages.

(h) **Independence from time concepts.**—The boundaries of a pedostratigraphic unit are time-transgressive. Concepts of time spans, however measured, play no part in defining the boundaries of a pedostratigraphic unit. Nonetheless, evidence of age, whether based on fossils, numerical ages, or geometrical or other relationships, may play an important role in distinguishing and identifying non-contiguous pedostratigraphic units at localities away from the type areas. The name of a pedostratigraphic unit should be chosen from a geographic feature in the type area, and not from a time span.

Pedostratigraphic Nomenclature and Unit

Article 56.—**Fundamental Unit.** The fundamental and only unit in pedostratigraphic classification is a geosol.

Article 57.—**Nomenclature.**—The formal name of a pedostratigraphic unit consists of a geographic name combined with the term "geosol." Capitalization of the initial letter in each word serves to identify formal usage. The geographic name should be selected in accordance with recommendations in Article 7 and should not duplicate the name of another formal geologic unit. Names based on subjacent and superjacent rock units, for example the super-Wilcox–sub-Claiborne soil, are informal, as are

those with time connotations (post-Wilcox–pre-Claiborne soil).

Remarks. (a) **Composite geosols.**—Where the horizons of two or more merged or "welded" buried soils can be distinguished, formal names of pedostratigraphic units based on the horizon boundaries can be retained. Where the horizon boundaries of the respective merged or "welded" soils cannot be distinguished, formal pedostratigraphic classification is abandoned and a combined name such as Hallettville-Jamesville geosol may be used informally.

(b) **Characterization.**—The physical and chemical properties of a pedostratigraphic unit commonly vary vertically and laterally throughout the geographic extent of the unit. A pedostratigraphic unit is characterized by the *range* of physical and chemical properties of the unit in the type area, rather than by "typical" properties exhibited in a type section. Consequently, a pedostratigraphic unit is characterized on the basis of a composite stratotype (Article 8d).

(c) **Procedures for establishing formal pedostratigraphic units.**—A formal pedostratigraphic unit may be established in accordance with the applicable requirements of Article 3, and additionally by describing major soil horizons in each soil facies.

ALLOSTRATIGRAPHIC UNITS

Nature and Boundaries

Article 58.—**Nature of Allostratigraphic Units.** An allostratigraphic[9] unit is a mappable stratiform body of sedimentary rock that is defined and identified on the basis of its bounding discontinuities.

Remarks. (a) **Purpose.**—Formal allostratigraphic units may be defined to distinguish between different (1) superposed discontinuity-bounded deposits of similar lithology (Figs. 7, 9), (2) contiguous discontinuity-bounded deposits of similar lithology (Fig. 8), or (3) geographically separated discontinuity-bounded units of similar lithology (Fig. 9), or to distinguish as single units discontinuity-bounded deposits characterized by lithic heterogeneity (Fig. 8).

(b) **Internal characteristics.**—Internal characteristics (physical, chemical, and paleontological) may vary laterally and vertically throughout the unit.

(c) **Boundaries.**—Boundaries of allostratigraphic units are laterally traceable discontinuities (Figs. 7, 8, and 9).

(d) **Mappability.**—A formal allostratigraphic unit must be mappable

[9]From the Greek *allo*: "other, different."

North American Stratigraphic Code

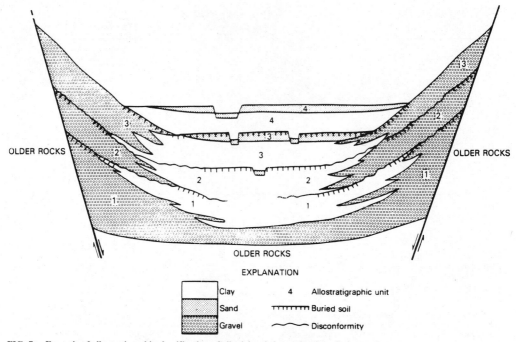

EXPLANATION

▢ Clay	4	Allostratigraphic unit
▢ Sand	⊤⊤⊤⊤⊤	Buried soil
▢ Gravel	∿∿∿	Disconformity

FIG. 7.—Example of allostratigraphic classification of alluvial and lacustrine deposits in a graben.

The alluvial and lacustrine deposits may be included in a single formation, or may be separated laterally into formations distinguished on the basis of contrasting texture (gravel, clay). Textural changes are abrupt and sharp, both vertically and laterally. The gravel deposits and clay deposits, respectively, are lithologically similar and thus cannot be distinguished as members of a formation. Four allostratigraphic units, each including two or three textural facies, may be defined on the basis of laterally traceable discontinuities (buried soils and disconformities).

at the scale practiced in the region where the unit is defined.

(e) **Type locality and extent.**—A type locality and type area must be designated; a composite stratotype or a type section and several reference sections are desirable. An allostratigraphic unit may be laterally contiguous with a formally defined lithostratigraphic unit; a vertical cut-off between such units is placed where the units meet.

(f) **Relation to genesis.**—Genetic interpretation is an inappropriate basis for defining an allostratigraphic unit. However, genetic interpretation may influence the choice of its boundaries.

(g) **Relation to geomorphic surfaces.**—A geomorphic surface may be used as a boundary of an allostratigraphic unit, but the unit should not be given the geographic name of the surface.

(h) **Relation to soils and paleosols.**—Soils and paleosols are composed of products of weathering and pedogenesis and differ in many respects from allostratigraphic units, which are depositional units (see "Pedostratigraphic Units," Article 55). The upper boundary of a surface or buried soil may be used as a boundary of an allostratigraphic unit.

(i) **Relation to inferred geologic history.**—Inferred geologic history is not used to define an allostratigraphic unit. However, well-documented geologic history may influence the choice of the unit's boundaries.

(j) **Relation to time concepts.**—Inferred time spans, however measured, are not used to define an allostratigraphic unit. However, age relationships may influence the choice of the unit's boundaries.

(k) **Extension of allostratigraphic units.**—An allostratigraphic unit is extended from its type area by tracing the boundary discontinuities or by tracing or matching the deposits between the discontinuities.

Ranks of Allostratigraphic Units

Article 59.—**Hierarchy.** The hierarchy of allostratigraphic units, in order of decreasing rank, is allogroup, alloformation, and allomember.

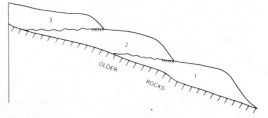

FIG. 8.—Example of allostratigraphic classification of contiguous deposits of similar lithology.

Allostratigraphic units 1, 2, and 3 are physical records of three glaciations. They are lithologically similar, reflecting derivation from the same bedrock, and constitute a single lithostratigraphic unit.

Remarks. (a) **Alloformation.**—The alloformation is the fundamental unit in allostratigraphic classification. An alloformation may be completely or only partly divided into allomembers, if some useful purpose is served, or it may have no allomembers.

(b) **Allomember.**—An allomember is the formal allostratigraphic unit next in rank below an alloformation.

(c) **Allogroup.**—An allogroup is the allostratigraphic unit next in rank above an alloformation. An allogroup is established only if a unit of that rank is essential to elucidation of geologic history. An allogroup may consist entirely of named alloformations or, alternatively, may contain one or more named alloformations which jointly do not comprise the entire allogroup.

North American Commission on Stratigraphic Nomenclature

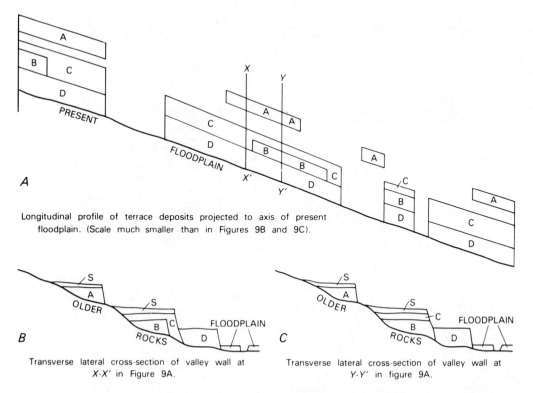

Longitudinal profile of terrace deposits projected to axis of present floodplain. (Scale much smaller than in Figures 9B and 9C).

Transverse lateral cross-section of valley wall at X-X' in Figure 9A.

Transverse lateral cross-section of valley wall at Y-Y' in figure 9A.

FIG. 9.—Example of allostratigraphic classification of lithologically similar, discontinuous terrace deposits.
A, B, C, and D are terrace gravel units of similar lithology at different topographic positions on a valley wall. The deposits may be defined as separate formal allostratigraphic units if such units are useful and if bounding discontinuities can be traced laterally. Terrace gravels of the same age commonly are separated geographically by exposures of older rocks. Where the bounding discontinuities cannot be traced continuously, they may be extended geographically on the basis of objective correlation of internal properites of the deposits other than lithology (e.g., fossil content, included tephras), topographic position, numerical ages, or relative-age criteria (e.g., soils or other weathering phenomena). The criteria for such extension should be documented. Slope deposits and eolian deposits (S) that mantle terrace surfaces may be of diverse ages and are not included in a terrace-gravel allostratigraphic unit. A single terrace surface may be underlain by more than one allostratigraphic unit (units B and C in sections b and c).

(d) **Changes in rank.**—The principles and procedures for elevation and reduction in rank of formal allostratigraphic units are the same as those in Articles 19b, 19g, and 28.

Allostratigraphic Nomenclature

Article 60.—**Nomenclature.** The principles and procedures for naming allostratigraphic units are the same as those for naming of lithostratigraphic units (see Articles 7, 30).

Remark. (a) **Revision.**—Allostratigraphic units may be revised or otherwise modified in accordance with the recommendations in Articles 17 to 20.

FORMAL UNITS DISTINGUISHED BY AGE

GEOLOGIC-TIME UNITS

Nature and Types

Article 61.—**Types.** Geologic-time units are conceptual, rather than material, in nature. Two types are recognized: those based

on material standards or referents (specific rock sequences or bodies), and those independent of material referents (Fig. 1).

Units Based on Material Referents

Article 62.—**Types Based on Referents.** Two types of formal geologic-time units based on material referents are recognized: they are isochronous and diachronous units.

Article 63.—**Isochronous Categories.** Isochronous time units and the material bodies from which they are derived are twofold: geochronologic units (Article 80), which are based on corresponding material chronostratigraphic units (Article 66), and polarity-geochronologic units (Article 88), based on corresponding material polarity-chronostratigraphic units (Article 83).

Remark. (a) **Extent.**—Isochronous units are applicable worldwide; they may be referred to even in areas lacking a material record of the named span of time. The duration of the time may be represented by a unit-stratotype referent. The beginning and end of the time are represented by point-boundary-stratotypes either in a single stratigraphic sequence or in separate stratotype sections (Articles 8b, 10b).

North American Stratigraphic Code

Article 64.—Diachronous Categories. Diachronic units (Article 91) are time units corresponding to diachronous material allostratigraphic units (Article 58), pedostratigraphic units (Article 55), and most lithostratigraphic (Article 22) and biostratigraphic (Article 48) units.

Remarks. (a) **Diachroneity.**—Some lithostratigraphic and biostratigraphic units are clearly diachronous, whereas others have boundaries which are not demonstrably diachronous within the resolving power of available dating methods. The latter commonly are treated as isochronous and are used for purposes of chronocorrelation (see biochronozone, Article 75). However, the assumption of isochroneity must be tested continually.

(b) **Extent.**—Diachronic units are coextensive with the diachronous material stratigraphic units on which they are based and are not used beyond the extent of their material referents.

Units Independent of Material Referents

Article 65.—Numerical Divisions of Time. Isochronous geologic-time units based on numerical divisions of time in years are geochronometric units (Article 96) and have no material referents.

CHRONOSTRATIGRAPHIC UNITS

Nature and Boundaries

Article 66.—Definition. A chronostratigraphic unit is a body of rock established to serve as the material reference for all rocks formed during the same span of time. Each of its boundaries is synchronous. The body also serves as the basis for defining the specific interval of time, or geochronologic unit (Article 80), represented by the referent.

Remarks. (a) **Purposes.**—Chronostratigraphic classification provides a means of establishing the temporally sequential order of rock bodies. Principal purposes are to provide a framework for (1) temporal correlation of the rocks in one area with those in another, (2) placing the rocks of the Earth's crust in a systematic sequence and indicating their relative position and age with respect to earth history as a whole, and (3) constructing an internationally recognized Standard Global Chronostratigraphic Scale.

(b) **Nature.**—A chronostratigraphic unit is a material unit and consists of a body of strata formed during a specific time span. Such a unit represents all rocks, and only those rocks, formed during that time span.

(c) **Content.**—A chronostratigraphic unit may be based upon the time span of a biostratigraphic unit, a lithic unit, a magnetopolarity unit, or any other feature of the rock record that has a time range. Or it may be any arbitrary but specified sequence of rocks, provided it has properties allowing chronocorrelation with rock sequences elsewhere.

Article 67.—Boundaries. Boundaries of chronostratigraphic units should be defined in a designated stratotype on the basis of observable paleontological or physical features of the rocks.

Remark. (a) **Emphasis on lower boundaries of chronostratigraphic units.**—Designation of point boundaries for both base and top of chronostratigraphic units is not recommended, because subsequent information on relations between successive units may identify overlaps or gaps. One means of minimizing or eliminating problems of duplication or gaps in chronostratigraphic successions is to define formally as a point-boundary stratotype only the base of the unit. Thus, a chronostratigraphic unit with its base defined at one locality, will have its top defined by the base of an overlying unit at the same, but more commonly another, locality (Article 8b).

Article 68.—Correlation. Demonstration of time equivalence is required for geographic extension of a chronostratigraphic unit from its type section or area. Boundaries of chronostratigraphic units can be extended only within the limits of resolution of available means of chronocorrelation, which currently include paleontology, numerical dating, remanent magnetism, thermoluminescence, relative-age criteria (examples are superposition and cross-cutting relations), and such indirect and inferential physical criteria as climatic changes, degree of weathering, and relations to unconformities. Ideally, the boundaries of chronostratigraphic units are independent of lithology, fossil content, or other material bases of stratigraphic division, but, in practice, the correlation or geographic extension of these boundaries relies at least in part on such features. Boundaries of chronostratigraphic units commonly are intersected by boundaries of most other kinds of material units.

Ranks of Chronostratigraphic Units

Article 69.—Hierarchy. The hierarchy of chronostratigraphic units, in order of decreasing rank, is eonothem, erathem, system, series, and stage. Of these, system is the primary unit of worldwide major rank; its primacy derives from the history of development of stratigraphic classification. All systems and units of higher rank are divided completely into units of the next lower rank. Chronozones are non-hierarchical and commonly lower-rank chronostratigraphic units. Stages and chronozones in sum do not necessarily equal the units of next higher rank and need not be contiguous. The rank and magnitude of chronostratigraphic units are related to the time interval represented by the units, rather than to the thickness or areal extent of the rocks on which the units are based.

Article 70.—Eonothem. The unit highest in rank is eonothem. The Phanerozoic Eonothem encompasses the Paleozoic, Mesozoic, and Cenozoic Erathems. Although older rocks have been assigned heretofore to the Precambrian Eonothem, they also have been assigned recently to other (Archean and Proterozoic) eonothems by the IUGS Precambrian Subcommission. The span of time corresponding to an eonothem is an *eon*.

Article 71.—Erathem. An erathem is the formal chronostratigraphic unit of rank next lower to eonothem and consists of several adjacent systems. The span of time corresponding to an erathem is an *era*.

Remark. (a) **Names.**—Names given to traditional Phanerozoic erathems were based upon major stages in the development of life on Earth: Paleozoic (old), Mesozoic (intermediate), and Cenozoic (recent) life. Although somewhat comparable terms have been applied to Precambrian units, the names and ranks of Precambrian divisions are not yet universally agreed upon and are under consideration by the IUGS Subcommission on Precambrian Stratigraphy.

Article 72.—System. The unit of rank next lower to erathem is the system. Rocks encompassed by a system represent a time-span and an episode of Earth history sufficiently great to serve as a worldwide chronostratigraphic reference unit. The temporal equivalent of a system is a *period*.

Remark. (a) **Subsystem and supersystem.**—Some systems initially established in Europe later were divided or grouped elsewhere into units ranked as systems. *Subsystems* (Mississippian Subsystem of the Carboniferous System) and *supersystems* (Karoo Supersystem) are more appropriate.

Article 73.—Series. Series is a conventional chronostratigraphic unit that ranks below a system and always is a division of a system. A series commonly constitutes a major unit of chronostratigraphic correlation within a province, between provinces, or between continents. Although many European series are being adopted increasingly for dividing systems on other continents, provincial series of regional scope continue to be useful. The temporal equivalent of a series is an *epoch*.

Article 74.—Stage. A stage is a chronostratigraphic unit of smaller scope and rank than a series. It is most commonly of greatest use in intra-continental classification and correlation, although it has the potential for worldwide recognition. The geochronologic equivalent of stage is *age*.

North American Commission on Stratigraphic Nomenclature

Remark. (a) **Substage.**—Stages may be, but need not be, divided completely into substages.

Article 75.—**Chronozone.** A chronozone is a non-hierarchical, but commonly small, formal chronostratigraphic unit, and its boundaries may be independent of those of ranked units. Although a chronozone is an isochronous unit, it may be based on a biostratigraphic unit (example: *Cardioceras cordatum* Biochronozone), a lithostratigraphic unit (Woodbend Lithochronozone), or a magnetopolarity unit (Gilbert Reversed-Polarity Chronozone). Modifiers (litho-, bio-, polarity) used in formal names of the units need not be repeated in general discussions where the meaning is evident from the context, e.g., *Exus albus* Chronozone.

Remarks. (a) **Boundaries of chronozones.**—The base and top of a *chronozone* correspond in the unit's stratotype to the observed, defining, physical and paleontological features, but they are extended to other areas by any means available for recognition of synchroneity. The temporal equivalent of a chronozone is a chron.

(b) **Scope.**—The scope of the non-hierarchical chronozone may range markedly, depending upon the purpose for which it is defined either formally or informally. The informal "biochronozone of the ammonites," for example, represents a duration of time which is enormous and exceeds that of a system. In contrast, a biochronozone defined by a species of limited range, such as the *Exus albus* Chronozone, may represent a duration equal to or briefer than that of a stage.

(c) **Practical utility.**—Chronozones, especially thin and informal biochronozones and lithochronozones bounded by key beds or other "markers," are the units used most commonly in industry investigations of selected parts of the stratigraphy of economically favorable basins. Such units are useful to define geographic distributions of lithofacies or biofacies, which provide a basis for genetic interpretations and the selection of targets to drill.

Chronostratigraphic Nomenclature

Article 76.—**Requirements.** Requirements for establishing a formal chronostratigraphic unit include: (i) statement of intention to designate such a unit; (ii) selection of name; (iii) statement of kind and rank of unit; (iv) statement of general concept of unit including historical background, synonymy, previous treatment, and reasons for proposed establishment; (v) description of characterizing physical and/or biological features; (vi) designation and description of boundary type sections, stratotypes, or other kinds of units on which it is based; (vii) correlation and age relations; and (viii) publication in a recognized scientific medium as specified in Article 4.

Article 77.—**Nomenclature.** A formal chronostratigraphic unit is given a compound name, and the initial letter of all words, except for trivial taxonomic terms, is capitalized. Except for chronozones (Article 75), names proposed for new chronostratigraphic units should not duplicate those for other stratigraphic units. For example, naming a new chronostratigraphic unit simply by adding "-an" or "-ian" to the name of a lithostratigraphic unit is improper.

Remarks. (a) **Systems and units of higher rank.**—Names that are generally accepted for systems and units of higher rank have diverse origins, and they also have different kinds of endings (Paleozoic, Cambrian, Cretaceous, Jurassic, Quaternary).

(b) **Series and units of lower rank.**—Series and units of lower rank are commonly known either by geographic names (Virgilian Series, Ochoan Series) or by names of their encompassing units modified by the capitalized adjectives Upper, Middle, and Lower (Lower Ordovician). Names of chronozones are derived from the unit on which they are based (Article 75). For series and stage, a geographic name is preferable because it may be related to a type area. For geographic names, the adjectival endings -an or -ian are recommended (Cincinnatian Series), but it is permissible to use the geographic name without any special ending, if more euphonious. Many series and stage names already in use have been based on lithic units (groups, formations, and members) and bear the names of these units

(Wolfcampian Series, Claibornian Stage). Nevertheless, a stage preferably should have a geographic name not previously used in stratigraphic nomenclature. Use of internationally accepted (mainly European) stage names is preferable to the proliferation of others.

Article 78.—**Stratotypes.** An ideal stratotype for a chronostratigraphic unit is a completely exposed unbroken and continuous sequence of fossiliferous stratified rocks extending from a well-defined lower boundary to the base of the next higher unit. Unfortunately, few available sequences are sufficiently complete to define stages and units of higher rank, which therefore are best defined by boundary-stratotypes (Article 8b).

Boundary-stratotypes for major chronostratigraphic units ideally should be based on complete sequences of either fossiliferous monofacial marine strata or rocks with other criteria for chronocorrelation to permit widespread tracing of synchronous horizons. Extension of synchronous surfaces should be based on as many indicators of age as possible.

Article 79.—**Revision of units.** Revision of a chronostratigraphic unit without changing its name is allowable but requires as much justification as the establishment of a new unit (Articles 17, 19, and 76). Revision or redefinition of a unit of system or higher rank requires international agreement. If the definition of a chronostratigraphic unit is inadequate, it may be clarified by establishment of boundary stratotypes in a principal reference section.

GEOCHRONOLOGIC UNITS

Nature and Boundaries

Article 80.—**Definition and Basis.** Geochronologic units are divisions of time traditionally distinguished on the basis of the rock record as expressed by chronostratigraphic units. A geochronologic unit is not a stratigraphic unit (i.e., it is not a material unit), but it corresponds to the time span of an established chronostratigraphic unit (Articles 65 and 66), and its beginning and ending corresponds to the base and top of the referent.

Ranks and Nomenclature of Geochronologic Units

Article 81.—**Hierarchy.** The hierarchy of geochronologic units in order of decreasing rank is *eon, era, period, epoch,* and *age.* Chron is a non-hierarchical, but commonly brief, geochronologic unit. Ages in sum do not necessarily equal epochs and need not form a continuum. An eon is the time represented by the rocks constituting an eonothem; era by an erathem; period by a system; epoch by a series; age by a stage; and chron by a chronozone.

Article 82.—**Nomenclature.** Names for periods and units of lower rank are identical with those of the corresponding chronostratigraphic units; the names of some eras and eons are independently formed. Rules of capitalization for chronostratigraphic units (Article 77) apply to geochronologic units. The adjectives Early, Middle, and Late are used for the geochronologic epochs equivalent to the corresponding chronostratigraphic Lower, Middle, and Upper series, where these are formally established.

POLARITY-CHRONOSTRATIGRAPHIC UNITS

Nature and Boundaries

Article 83.—**Definition.** A polarity-chronostratigraphic unit is a body of rock that contains the primary magnetic-polarity record imposed when the rock was deposited, or crystallized, during a specific interval of geologic time.

Remarks. (a) **Nature.**—Polarity-chronostratigraphic units depend fundamentally for definition on actual sections or sequences, or measurements on individual rock units, and without these standards they are meaningless. They are based on material units, the polarity zones of magnetopolarity classification. Each polarity-chronostratigraphic unit is the record of the time during which the rock formed and the Earth's magnetic field had a designated polarity. Care should be taken to define polarity-chronologic units in terms of polarity-chronostratigraphic units, and not vice versa.

(b) **Principal purposes.**—Two principal purposes are served by polarity-chronostratigraphic classification: (1) correlation of rocks at one place with those of the same age and polarity at other places; and (2) delineation of the polarity history of the Earth's magnetic field.

(c) **Recognition.**—A polarity-chronostratigraphic unit may be extended geographically from its type locality only with the support of physical and/or paleontologic criteria used to confirm its age.

Article 84.—**Boundaries.** The boundaries of a polarity chronozone are placed at polarity-reversal horizons or polarity transition-zones (see Article 45).

Ranks and Nomenclature of Polarity-Chronostratigraphic Units

Article 85.—**Fundamental Unit.** The polarity chronozone consists of rocks of a specified primary polarity and is the fundamental unit of worldwide polarity-chronostratigraphic classification.

Remarks. (a) **Meaning of term.**—A polarity chronozone is the worldwide body of rock strata that is collectively defined as a polarity-chronostratigraphic unit.

(b) **Scope.**—Individual polarity zones are the basic building blocks of polarity chronozones. Recognition and definition of polarity chronozones may thus involve step-by-step assembly of carefully dated or correlated individual polarity zones, especially in work with rocks older than the oldest ocean-floor magnetic anomalies. This procedure is the method by which the Brunhes, Matuyama, Gauss, and Gilbert Chronozones were recognized (Cox, Doell, and Dalrymple, 1963) and defined originally (Cox, Doell, and Dalrymple, 1964).

(c) **Ranks.**—Divisions of polarity chronozones are designated polarity subchronozones. Assemblages of polarity chronozones may be termed polarity superchronozones.

Article 86.—**Establishing Formal Units.** Requirements for establishing a polarity-chronostratigraphic unit include those specified in Articles 3 and 4, and also (1) definition of boundaries of the unit, with specific references to designated sections and data; (2) distinguishing polarity characteristics, lithologic descriptions, and included fossils; and (3) correlation and age relations.

Article 87.—**Name.** A formal polarity-chronostratigraphic unit is given a compound name beginning with that for a named geographic feature; the second component indicates the normal, reversed, or mixed polarity of the unit, and the third component is *chronozone.* The initial letter of each term is capitalized. If the same geographic name is used for both a magnetopolarity zone and a polarity-chronostratigraphic unit, the latter should be distinguished by an -an or -ian ending. Example: Tetonian Reversed-Polarity Chronozone.

Remarks: (a) **Preservation of established name.**—A particularly well-established name should not be displaced, either on the basis of priority, as described in Article 7c, or because it was not taken from a geographic feature. Continued use of Brunhes, Matuyama, Gauss, and Gilbert, for example, is endorsed so long as they remain valid units.

(b) **Expression of doubt.**—Doubt in the assignment of polarity zones to polarity-chronostratigraphic units should be made explicit if criteria of time equivalence are inconclusive.

POLARITY-CHRONOLOGIC UNITS

Nature and Boundaries

Article 88.—**Definition.** Polarity-chronologic units are divi-

sions of geologic time distinguished on the basis of the record of magnetopolarity as embodied in polarity-chronostratigraphic units. No special kind of magnetic time is implied; the designations used are meant to convey the parts of geologic time during which the Earth's magnetic field had a characteristic polarity or sequence of polarities. These units correspond to the time spans represented by polarity chronozones, e.g., Gauss Normal Polarity Chronozone. They are not material units.

Ranks and Nomenclature of Polarity-Chronologic Units

Article 89.—**Fundamental Unit.** The polarity chron is the fundamental unit of geologic time designating the time span of a polarity chronozone.

Remark. (a) **Hierarchy.**—Polarity-chronologic units of decreasing hierarchical ranks are polarity superchron, polarity chron, and polarity subchron.

Article 90.—**Nomenclature.** Names for polarity chronologic units are identical with those of corresponding polarity-chronostratigraphic units, except that the term chron (or superchron, etc) is substituted for chronozone (or superchronozone, etc).

DIACHRONIC UNITS

Nature and Boundaries

Article 91.—**Definition.** A diachronic unit comprises the unequal spans of time represented either by a specific lithostratigraphic, allostratigraphic, biostratigraphic, or pedostratigraphic unit, or by an assemblage of such units.

Remarks. (a) **Purposes.**—Diachronic classification provides (1) a means of comparing the spans of time represented by stratigraphic units with diachronous boundaries at different localities, (2) a basis for establishing in time the beginning and ending of deposition of diachronous stratigraphic units at different sites, (3) a basis for inferring the rate of change in areal extent of depositional processes, (4) a means of determining and comparing rates and durations of deposition at different localities, and (5) a means of comparing temporal and spatial relations of diachronous stratigraphic units (Watson and Wright, 1980).

(b) **Scope.**—The scope of a diachronic unit is related to (1) the relative magnitude of the transgressive division of time represented by the stratigraphic unit or units on which it is based and (2) the areal extent of those units. A diachronic unit is not extended beyond the geographic limits of the stratigraphic unit or units on which it is based.

(c) **Basis.**—The basis for a diachronic unit is the diachronous referent.

(d) **Duration.**—A diachronic unit may be of equal duration at different places despite differences in the times at which it began and ended at those places.

Article 92.—**Boundaries.** The boundaries of a diachronic unit are the times recorded by the beginning and end of deposition of the material referent at the point under consideration (Figs. 10, 11).

Remark. (a) **Temporal relations.**—One or both of the boundaries of a diachronic unit are demonstrably time-transgressive. The varying time significance of the boundaries is defined by a series of boundary reference sections (Article 8b, 8e). The duration and age of a diachronic unit differ from place to place (Figs. 10, 11).

Ranks and Nomenclature of Diachronic Units

Article 93.—**Ranks.** A diachron is the fundamental and non-hierarchical diachronic unit. If a hierarchy of diachronic units is needed, the terms episode, phase, span, and cline, in order of decreasing rank, are recommended. The rank of a hierarchical

North American Commission on Stratigraphic Nomenclature

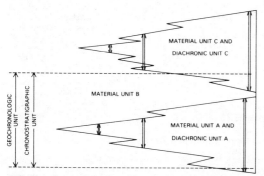

FIG. 10.—Comparison of geochronologic, chronostratigraphic, and diachronic units.

unit is determined by the scope of the unit (Article 91 b), and not by the time span represented by the unit at a particular place.

Remarks. (a) **Diachron.**—Diachrons may differ greatly in magnitude because they are the spans of time represented by individual or grouped lithostratigraphic, allostratigraphic, biostratigraphic, and(or) pedostratigraphic units.

(b) **Hierarchical ordering permissible.**—A hierarchy of diachronic units may be defined if the resolution of spatial and temporal relations of diachronous stratigraphic units is sufficiently precise to make the hierarchy useful (Watson and Wright, 1980). Although all hierarchical units of rank lower than episode are part of a unit next higher in rank, not all parts of an episode, phase, or span need be represented by a unit of lower rank.

(c) **Episode.**—An episode is the unit of highest rank and greatest scope in hierarchical classification. If the "Wisconsinan Age" were to be redefined as a diachronic unit, it would have the rank of episode.

Article 94.—**Name.** The name for a diachronic unit should be compound, consisting of a geographic name followed by the term diachron or a hierarchical rank term. Both parts of the compound name are capitalized to indicate formal status. If the diachronic unit is defined by a single stratigraphic unit, the geographic name of the unit may be applied to the diachronic unit. Otherwise, the geographic name of the diachronic unit should not duplicate that of another formal stratigraphic unit. Genetic terms (e.g., alluvial, marine) or climatic terms (e.g., gla-

cial, interglacial) are not included in the names of diachronic units.

Remarks. (a) **Formal designation of units.**—Diachronic units should be formally defined and named only if such definition is useful.

(b) **Inter-regional extension of geographic names.**—The geographic name of a diachronic unit may be extended from one region to another if the stratigraphic units on which the diachronic unit is based extend across the regions. If different diachronic units in contiguous regions eventually prove to be based on laterally continuous stratigraphic units, one name should be applied to the unit in both regions. If two names have been applied, one name should be abandoned and the other formally extended. Rules of priority (Article 7d) apply. Priority in publication is to be respected, but priority alone does not justify displacing a well-established name by one not well-known or commonly used.

(c) **Change from geochronologic to diachronic classification.**—Lithostratigraphic units have served as the material basis for widely accepted chronostratigraphic and geochronologic classifications of Quaternary nonmarine deposits, such as the classifications of Frye et al (1968), Willman and Frye (1970), and Dreimanis and Karrow (1972). In practice, time-parallel horizons have been extended from the stratotypes on the basis of markedly time-transgressive lithostratigraphic and pedostratigraphic unit boundaries. The time ("geochronologic") units, defined on the basis of the stratotype sections but extended on the basis of diachronous stratigraphic boundaries, are diachronic units. Geographic names established for such "geochronologic" units may be used in diachronic classification if (1) the chronostratigraphic and geochronologic classifications are formally abandoned and diachronic classifications are proposed to replace the former "geochronologic" classifications, and (2) the units are redefined as formal diachronic units. Preservation of well-established names in these specific circumstances retains the intent and purpose of the names and the units, retains the practical significance of the units, enhances communication, and avoids proliferation of nomenclature.

Article 95.—**Establishing Formal Units.** Requirements for establishing a formal diachronic unit, in addition to those in Article 3, include (1) specification of the nature, stratigraphic relations, and geographic or areal relations of the stratigraphic unit or units that serve as a basis for definition of the unit, and (2) specific designation and description of multiple reference sections that illustrate the temporal and spatial relations of the defining stratigraphic unit or units and the boundaries of the unit or units.

Remark. (a) **Revision or abandonment.**—Revision or abandonment of the stratigraphic unit or units that serve as the material basis for defini-

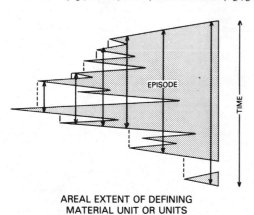

AREAL EXTENT OF DEFINING
MATERIAL UNIT OR UNITS

AREAL EXTENT OF DEFINING
MATERIAL UNIT OR UNITS

FIG. 11.—Schematic relation of phases to an episode.
Parts of a phase similarly may be divided into spans, and spans into clines. Formal definition of spans and clines is unnecessary in most diachronic unit hierarchies.

North American Stratigraphic Code

tion of a diachronic unit may require revision or abandonment of the diachronic unit. Procedure for revision must follow the requirements for establishing a new diachronic unit.

GEOCHRONOMETRIC UNITS

Nature and Boundaries

Article 96.—**Definition.** Geochronometric units are units established through the direct division of geologic time, expressed in years. Like geochronologic units (Article 80), geochronometric units are abstractions, i.e., they are not material units. Unlike geochronologic units, geochronometric units are not based on the time span of designated chronostratigraphic units (stratotypes), but are simply time divisions of convenient magnitude for the purpose for which they are established, such as the development of a time scale for the Precambrian. Their boundaries are arbitrarily chosen or agreed-upon ages in years.

Ranks and Nomenclature of Geochronometric Units

Article 97.—**Nomenclature.** Geochronologic rank terms (eon, era, period, epoch, age, and chron) may be used for geochronometric units when such terms are formalized. For example, Archean Eon and Proterozoic Eon, as recognized by the IUGS Subcommission on Precambrian Stratigraphy, are formal geochronometric units in the sense of Article 96, distinguished on the basis of an arbitrarily chosen boundary at 2.5 Ga. Geochronometric units are not defined by, but may have, corresponding chronostratigraphic units (eonothem, erathem, system, series, stage, and chronozone).

PART III: ADDENDA

REFERENCES[10]

American Commission on Stratigraphic Nomenclature, 1947, Note 1—Organization and objectives of the Stratigraphic Commission: American Association of Petroleum Geologists Bulletin, v. 31, no. 3, p. 513-518.

———, 1961, Code of Stratigraphic Nomenclature: American Association of Petroleum Geologists Bulletin, v. 45, no. 5, p. 645-665.

———, 1970, Code of Stratigraphic Nomenclature (2d ed.): American Association of Petroleum Geologists, Tulsa, Okla., 45 p.

———, 1976, Note 44—Application for addition to code concerning magnetostratigraphic units: American Association of Petroleum Geologists Bulletin, v. 60, no. 2, p. 273-277.

Caster, K. E., 1934, The stratigraphy and paleontology of northwestern Pennsylvania, Part 1, Stratigraphy: Bulletins of American Paleontology, v. 21, 185 p.

Chang, K. H., 1975, Unconformity-bounded stratigraphic units: Geological Society of America Bulletin, v. 86, no. 11, p. 1544-1552.

Committee on Stratigraphic Nomenclature, 1933, Classification and nomenclature of rock units: Geological Society of America Bulletin, v. 44, no. 2, p. 423-459, and American Association of Petroleum Geologists Bulletin, v. 17, no. 7, p. 843-868.

Cox, A. V., R. R. Doell, and G. B. Dalrymple, 1963, Geomagnetic polarity epochs and Pleistocene geochronometry: Nature, v. 198, p. 1049-1051.

———, 1964, Reversals of the Earth's magnetic field: Science, v. 144, no. 3626, p. 1537-1543.

Cross, C. W., 1898, Geology of the Telluride area: U.S. Geological Survey 18th Annual Report, pt. 3, p. 759.

Cumming, A. D., J. G. C. M. Fuller, and J. W. Porter, 1959, Separation of strata: Paleozoic limestones of the Williston basin: American Journal of Science, v. 257, no. 10, p. 722-733.

Dreimanis, Aleksis, and P. F. Karrow, 1972, Glacial history of the Great Lakes–St. Lawrence region, the classification of the Wisconsin(an) Stage, and its correlatives: International Geologic Congress, 24th Session, Montreal, 1972, Section 12, Quaternary Geology, p. 5-15.

Dunbar, C. O., and John Rodgers, 1957, Principles of stratigraphy: Wiley, New York, 356 p.

Forgotson, J. M., Jr., 1957, Nature, usage and definition of marker-defined vertically segregated rock units: American Association of Petroleum Geologists Bulletin, v. 41, no. 9, p. 2108-2113.

Frye, J. C., H. B. Willman, Meyer Rubin, and R. F. Black, 1968, Definition of Wisconsinan Stage: U.S. Geological Survey Bulletin 1274-E, 22 p.

George, T. N., and others, 1969, Recommendations on stratigraphical usage: Geological Society of London, Proceedings no. 1656, p. 139-166.

Harland, W. B., 1977, Essay review [of] International Stratigraphic Guide, 1976: Geology Magazine, v. 114, no. 3, p. 229-235.

———, 1978, Geochronologic scales, in G. V. Cohee et al, eds., Contributions to the Geologic Time Scale: American Association of Petroleum Geologists, Studies in Geology, no. 6, p. 9-32.

Harrison, J. E., and Z. E. Peterman, 1980, North American Commission on Stratigraphic Nomenclature Note 52—A preliminary proposal for a chronometric time scale for the Precambrian of the United States and Mexico: Geological Society of America Bulletin, v. 91, no. 6, p. 377-380.

Henbest, L. G., 1952, Significance of evolutionary explosions for diastrophic division of Earth history: Journal of Paleontology, v. 26, p. 299-318.

Henderson, J. B., W. G. E. Caldwell, and J. E. Harrison, 1980, North American Commission on Stratigraphic Nomenclature, Report 8—Amendment of code concerning terminology for igneous and high-grade metamorphic rocks: Geological Society of America Bulletin, v. 91, no. 6, p. 374-376.

Holland, C. H., and others, 1978, A guide to stratigraphical procedure: Geological Society of London, Special Report 10, p. 1-18.

Huxley, T. H., 1862, The anniversary address: Geological Society of London, Quarterly Journal, v. 18, p. xl-liv.

International Commission on Zoological Nomenclature, 1964: International Code of Zoological Nomenclature adopted by the XV International Congress of Zoology: International Trust for Zoological Nomenclature, London, 176 p.

International Subcommission on Stratigraphic Classification (ISSC), 1976, International Stratigraphic Guide (H. D. Hedberg, ed.): John Wiley and Sons, New York, 200 p.

International Subcommission on Stratigraphic Classification, 1979, Magnetostratigraphy polarity units—a supplementary chapter of the ISSC International Stratigraphic Guide: Geology, v. 7, p. 578-583.

Izett, G. A., and R. E. Wilcox, 1981, Map showing the distribution of the Huckleberry Ridge, Mesa Falls, and Lava Creek volcanic ash beds (Pearlette family ash beds) of Pliocene and Pleistocene age in the western United States and southern Canada: U. S. Geological Survey Miscellaneous Geological Investigations Map I-1325.

Kauffman, E. G., 1969, Cretaceous marine cycles of the Western Interior: Mountain Geologist: Rocky Mountain Association of Geologists, v. 6, no. 4, p. 227-245.

Matthews, R. K., 1974, Dynamic stratigraphy—an introduction to sedimentation and stratigraphy: Prentice-Hall, New Jersey, 370 p.

McDougall, Ian, 1977, The present status of the geomagnetic polarity time scale: Research School of Earth Sciences, Australian National University, Publication no. 1288, 34 p.

McElhinny, M. W., 1978, The magnetic polarity time scale; prospects and possibilities in magnetostratigraphy, in G. V. Cohee et al, eds., Contributions to the Geologic Time Scale, American Association of Petroleum Geologists, Studies in Geology, no. 6, p. 57-65.

McIver, N. L., 1972, Cenozoic and Mesozoic stratigraphy of the Nova Scotia shelf: Canadian Journal of Earth Science, v. 9, p. 54-70.

McLaren, D. J., 1977, The Silurian-Devonian Boundary Committee. A final report, in A. Martinsson, ed., The Silurian-Devonian boundary: IUGS Series A, no. 5, p. 1-34.

Morrison, R. B., 1967, Principles of Quaternary soil stratigraphy, in R. B. Morrison and H. E. Wright, Jr., eds., Quaternary soils: Reno, Nevada, Center for Water Resources Research, Desert Research Institute, Univ. Nevada, p. 1-69.

North American Commission on Stratigraphic Nomenclature, 1981, Draft North American Stratigraphic Code: Canadian Society of

---

[10]Readers are reminded of the extensive and noteworthy bibliography of contributions to stratigraphic principles, classification, and terminology cited by the International Stratigraphic Guide (ISSC, 1976, p. 111-187).

North American Commission on Stratigraphic Nomenclature

Petroleum Geologists, Calgary, 63 p.

Palmer, A. R., 1965, Biomere-a new kind of biostratigraphic unit: Journal of Paleontology, v. 39, no. 1, p. 149-153.

Parsons, R. B., 1981, Proposed soil-stratigraphic guide, in International Union for Quaternary Research and International Society of Soil Science: INQUA Commission 6 and ISSS Commission 5 Working Group, Pedology, Report, p. 6-12.

Pawluk, S., 1978, The pedogenic profile in the stratigraphic section, in W. C. Mahaney, ed., Quaternary soils: Norwich, England, GeoAbstracts, Ltd., p. 61-75.

Ruhe, R. V., 1965, Quaternary paleopedology, in H. E. Wright, Jr., and D. G. Frey, eds., The Quaternary of the United States: Princeton, N.J., Princeton University Press, p. 755-764.

Schultz, E. H., 1982, The chronosome and supersome--terms proposed for low-rank chronostratigraphic units: Canadian Petroleum Geology, v. 30, no. 1, p. 29-33.

Shaw, A. B., 1964, Time in stratigraphy: McGraw-Hill, New York, 365 p.

Sims, P. K., 1979, Precambrian subdivided: Geotimes, v. 24, no. 12, p. 15.

Sloss, L. L., 1963, Sequences in the cratonic interior of North America: Geological Society of America Bulletin, v. 74, no. 2, p. 94-114.

Tracey, J. I., Jr., and others, 1971, Initial reports of the Deep Sea Drilling Project, v. 8: U.S. Government Printing Office, Washington, 1037 p.

Valentine, K. W. G., and J. B. Dalrymple, 1976, Quaternary buried paleosols: A critical review: Quaternary Research, v. 6, p. 209-222.

Vella, P., 1964, Biostratigraphic units: New Zealand Journal of Geology and Geophysics, v. 7, no. 3, p. 615-625.

Watson, R. A., and H. E. Wright, Jr., 1980, The end of the Pleistocene: A general critique of chronostratigraphic classification: Boreas, v. 9, p. 153-163.

Weiss, M. P., 1979a, Comments and suggestions invited for revision of American Stratigraphic Code: Geological Society of America, News and Information, v. 1, no. 7, p. 97-99.

——— ,1979b, Stratigraphic Commission Note 50--Proposal to change name of Commission: American Association of Petroleum Geologists Bulletin, v. 63, no. 10, p. 1986.

Weller, J. M., 1960, Stratigraphic principles and practice: Harper and Brothers, New York, 725 p.

Willman, H. B., and J. C. Frye, 1970, Pleistocene stratigraphy of Illinois: Illinois State Geological Survey Bulletin 94, 204 p.

APPENDIX I: PARTICIPANTS AND CONFEREES IN CODE REVISION

Code Committee

Steven S. Oriel (U.S. Geological Survey), chairman, Hubert Gabrielse (Geological Survey of Canada), William W. Hay (Joint Oceanographic Institutions), Frank E. Kottlowski (New Mexico Bureau of Mines), John B. Patton (Indiana Geological Survey).

Lithostratigraphic Subcommittee

James D. Aitken (Geological Survey of Canada), chairman, Monti Lerand (Gulf Canada Resources, Ltd.), Mitchell W. Reynolds (U.S. Geological Survey), Robert J. Weimer (Colorado School of Mines), Malcolm P. Weiss (Northern Illinois University).

Biostratigraphic Subcommittee

Allison R. (Pete) Palmer (Geological Society of America), chairman, Ismael Ferrusquia (University of Mexico), Joseph E. Hazel (U.S. Geological Survey), Erle G. Kauffman (University of Colorado), Colin McGregor (Geological Survey of Canada), Michael A. Murphy (University of California, Riverside), Walter C. Sweet (Ohio State University).

Chronostratigraphic Subcommittee

Zell E. Peterman (U.S. Geological Survey), chairman, Zoltan de Cserna (Sociedad Geológica Mexicana), Edward H. Schultz (Suncor, Inc., Calgary), Norman F. Sohl (U.S. Geological Survey), John A. Van Couvering (American Museum of Natural History).

Plutonic-Metamorphic Advisory Group

Jack E. Harrison (U.S. Geological Survey), chairman, John B. Henderson (Geological Survey of Canada), Harold L. James (retired), Leon T. Silver (California Institute of Technology), Paul C. Bateman (U.S. Geological Survey).

Magnetostratigraphic Advisory Group

Roger W. Macqueen (University of Waterloo), chairman, G. Brent Dalrymple (U.S. Geological Survey), Walter F. Fahrig (Geological Survey of Canada), J. M. Hall (Dalhousie University).

Volcanic Advisory Group

Richard V. Fisher (University of California, Santa Barbara), chairman, Thomas A. Steven (U.S. Geological Survey), Donald A. Swanson (U.S. Geological Survey).

Tectonostratigraphic Advisory Group

Darrel S. Cowan (University of Washington), chairman, Thomas W. Donnelly (State University of New York at Binghamton), Michael W. Higgins and David L. Jones (U.S. Geological Survey), Harold Williams (Memorial University, Newfoundland).

Quaternary Advisory Group

Norman P. Lasca (University of Wisconsin-Milwaukee), chairman, Mark M. Fenton (Alberta Research Council), David S. Fullerton (U.S. Geological Survey), Robert J. Fulton (Geological Survey of Canada), W. Hilton Johnson (University of Illinois), Paul F. Karrow (University of Waterloo), Gerald M. Richmond (U.S. Geological Survey).

Conferees

W. G. E. Caldwell (University of Saskatchewan), Lucy E. Edwards (U.S. Geological Survey), Henry H. Gray (Indiana Geological Survey), Hollis D. Hedberg (Princeton University), Lewis H. King (Geological Survey of Canada), Rudolph W. Kopf (U.S. Geological Survey), Jerry A. Lineback (Robertson Research U.S.), Marjorie E. MacLachlan (U.S. Geological Survey), Amos Salvador (University of Texas, Austin), Brian R. Shaw (Samson Resources, Inc.), Ogden Tweto (U.S. Geological Survey).

APPENDIX II: 1977-1982 COMPOSITION OF THE NORTH AMERICAN COMMISSION ON STRATIGRAPHIC NOMENCLATURE

Each Commissioner is appointed, with few exceptions, to serve a 3-year term (shown by such numerals as 80-82 for 1980-1982) and a few are reappointed.

American Association of Petroleum Geologists

Timothy A. Anderson (Gulf Oil Co.) 77-83, Orlo E. Childs (Texas Tech University) 76-79, Kenneth J. Englund (U.S. Geological Survey) 74-77, Susan Longacre (Getty Oil Co.) 78-84, Donald E. Owen (Cities Service Co.) 79-82, Grant Steele (Gulf Oil Co.) 75-78.

Association of American State Geologists

Larry D. Fellows (Arizona Bureau of Geology) 81-82, Lee C. Gerhard (North Dakota Geological Survey) 79-81, Donald C. Haney (Kentucky Geological Survey) 80-83, Wallace B. Howe (Missouri Division of Geology) 74-77, Robert R. Jordan (Delaware Geological Survey) 78-84, vice-chairman, Frank E. Kottlowski (New Mexico Bureau of Mines) 76-79, Meredith E. Ostrom (Wisconsin Geological Survey) 77-80, John B. Patton (Indiana Geological Survey) 75-78.

North American Stratigraphic Code

Geological Society of America

Clarence A. Hall, Jr. (University of California, Los Angeles) 78-81, Jack E. Harrison (U.S. Geological Survey) 74-77, William W. Hay (University of Miami) 75-78, Robert S. Houston (University of Wyoming) 77-80, Michael A. Murphy (University of California, Riverside) 81-84, Allison R. Palmer (Geological Society of America) 80-83, Malcolm P. Weiss (Northern Illinois University) 76-82, chairman.

United States Geological Survey

Earl E. Brabb (Menlo Park) 78-82, David S. Fullerton (Denver) 78-84, E. Dale Jackson (Menlo Park) 76-78, Kenneth L. Pierce (Denver) 75-78, Norman F. Sohl (Washington) 74-83.

Geological Survey of Canada

James D. Aitken (Calgary) 75-78, Kenneth D. Card (Kanata) 80-83, Donald G. Cook (Calgary) 78-81, Robert J. Fulton (Ottawa) 81-84, John B. Henderson (Ottawa) 74-77, Lewis H. King (Dartmouth) 79-82, Maurice B. Lambert (Ottawa) 77-80, Christopher J. Yorath (Sydney) 76-79.

Canadian Society of Petroleum Geologists

Roland F. deCaen (Union Oil Co. of Canada) 79-82, J. Ross McWhae (Petro Canada Exploration) 77-80, Edward H. Schultz (Suncor, Inc.) 74-77, 80-83, Ulrich Wissner (Union Oil Co. of Canada) 76-79.

Geological Association of Canada

W. G. E. Caldwell (University of Saskatchewan) 76-79, R. K. Jull (University of Windsor) 78-79, Paul S. Karrow (University of Waterloo) 81-84, Alfred C. Lenz (University of Western Ontario) 79-81, David E. Pearson (British Columbia Mines and Petroleum Resources) 79-81, Paul E. Schenk (Dalhousie University) 75-78.

Asociación Mexicana de Geólogos Petróleros

Jose Carillo Bravo (Petróleos Mexicanos) 78-81, Baldomerro Carrasco V., 75-78.

Sociedad Geológica Mexicana

Zoltan de Cserna (Universidad Nacional Autónoma de México) 76-82.

Instituto de Geología de la Universidad Nacional Autónoma de México

Ismael Ferrusquia Villafranca (Universidad Nacional Autónoma de México) 76-81, Fernando Ortega Gutiérrez (Universidad Nacional Autónoma de México) 81-84.

APPENDIX III: REPORTS AND NOTES OF THE AMERICAN COMMISSION ON STRATIGRAPHIC NOMENCLATURE

Reports (formal declarations, opinions, and recommendations)
1. Moore, Raymond C., Declaration on naming of subsurface stratigraphic units: AAPG Bulletin, v. 33, no. 7, p. 1280-1282, 1949.
2. Hedberg, Hollis D., Nature, usage, and nomenclature of time-stratigraphic and geologic-time units: AAPG Bulletin, v. 36, no. 8, p. 1627-1638, 1952.
3. Harrison, J. M., Nature, usage, and nomenclature of time-stratigraphic and geologic-time units as applied to the Precambrian: AAPG Bulletin, v. 39, no. 9, p. 1859-1861, 1955.
4. Cohee, George V., and others, Nature, usage, and nomenclature of rock-stratigraphic units: AAPG Bulletin, v. 40, no. 8, p. 2003-2014, 1956.
5. McKee, Edwin D., Nature, usage and nomenclature of biostratigraphic units: AAPG Bulletin, v. 41, no. 8, p. 1877-1889, 1957.
6. Richmond, Gerald M., Application of stratigraphic classification and

nomenclature to the Quaternary: AAPG Bulletin, v. 43, no. 3, pt. I, p. 663-675, 1959.
7. Lohman, Kenneth E., Function and jurisdictional scope of the American Commission on Stratigraphic Nomenclature: AAPG Bulletin, v. 47, no. 5, p. 853-855, 1963.
8. Henderson, John B., W. G. E. Caldwell, and Jack E. Harrison, Amendment of code concerning terminology for igneous and high-grade metamorphic rocks: GSA Bulletin, pt. I, v. 91, no. 6, p. 374-376, 1980.
9. Harrison, Jack E., and Zell E. Peterman, Adoption of geochronometric units for divisions of Precambrian time: AAPG Bulletin, v. 66, no. 6, p. 801-802, 1982.

Notes (informal statements, discussions, and outlines of problems)

1. Organization and objectives of the Stratigraphic Commission: AAPG Bulletin, v. 31, no. 3, p. 513-518, 1947.
2. Nature and classes of stratigraphic units: AAPG Bulletin, v. 31, no. 3, p. 519-528, 1947.
3. Moore, Raymond C., Rules of geologic nomenclature of the Geological Survey of Canada: AAPG Bulletin, v. 32, no. 3, p. 366-367, 1948.
4. Jones, Wayne V., and Raymond C. Moore, Naming of subsurface stratigraphic units: AAPG Bulletin, v. 32, no. 3, p. 367-371, 1948.
5. Flint, Richard Foster, and Raymond C. Moore, Definition and adoption of the terms stage and age: AAPG Bulletin, v. 32, no. 3, p. 372-376, 1948.
6. Moore, Raymond C., Discussion of nature and classes of stratigraphic units: AAPG Bulletin, v. 21, no. 3, p. 376-381, 1948.
7. Records of the Stratigraphic Commission for 1947-1948: AAPG Bulletin, v. 33, no. 7, p. 1271-1273, 1949.
8. Australian Code of Stratigraphical Nomenclature: AAPG Bulletin, v. 33, no. 7, p. 1273-1276, 1949.
9. The Pliocene-Pleistocene boundary: AAPG Bulletin, v. 33, no. 7, p. 1276-1280, 1949.
10. Moore, Raymond C., Should additional categories of stratigraphic units be recognized?: AAPG Bulletin, v. 34, no. 12, p. 2360-2361, 1950.
11. Moore, Raymond C., Records of the Stratigraphic Commission for 1949-1950: AAPG Bulletin, v. 35, no. 5, p. 1074-1076, 1951.
12. Moore, Raymond C., Divisions of rocks and time: AAPG Bulletin, v. 35, no. 5, p. 1076, 1951.
13. Williams, James Steele, and Aureal T. Cross, Third Congress of Carboniferous Stratigraphy and Geology: AAPG Bulletin, v. 36, no. 1, p. 169-172, 1952.
14. Official report of round table conference on stratigraphic nomenclature at Third Congress of Carboniferous Stratigraphy and Geology, Heerlen, Netherlands, June 26-28, 1951: AAPG Bulletin, v. 36, no. 10, p. 2044-2048, 1952.
15. Records of the Stratigraphic Commission for 1951-1952: AAPG Bulletin, v. 37, no. 5, p. 1078-1080, 1953.
16. Records of the Stratigraphic Commission for 1953-1954: AAPG Bulletin, v. 39, no. 9, p. 1861-1863, 1955.
17. Suppression of homonymous and obsolete stratigraphic names: AAPG Bulletin, v. 40, no. 12, p. 2953-2954, 1956.
18. Gilluly, James, Records of the Stratigraphic Commission for 1955-1956: AAPG Bulletin, v. 41, no. 1, p. 130-133, 1957.
19. Richmond, Gerald M., and John C. Frye, Status of soils in stratigraphic nomenclature: AAPG Bulletin, v. 31, no. 4, p. 758-763, 1957.
20. Frye, John C., and Gerald M. Richmond, Problems in applying standard stratigraphic practice in nonmarine Quaternary deposits: AAPG Bulletin, v. 42, no. 8, p. 1979-1983, 1958.
21. Frye, John C., Preparation of new stratigraphic code by American Commission on Stratigraphic Nomenclature: AAPG Bulletin, v. 42, no. 8, p. 1984-1986, 1958.
22. Records of the Stratigraphic Commission for 1957-1958: AAPG Bulletin, v. 43, no. 8, p. 1967-1971, 1959.
23. Rodgers, John, and Richard B. McConnell, Need for rock-stratigraphic units larger than group: AAPG Bulletin, v. 43, no. 8, p. 1971-1975, 1959.
24. Wheeler, Harry E., Unconformity-bounded units in stratigraphy: AAPG Bulletin, v. 43, no. 8, p. 1975-1977, 1959.
25. Bell, W. Charles, and others, Geochronologic and chronostratigraphic units: AAPG Bulletin, v. 45, no. 5, p. 666-670, 1961.
26. Records of the Stratigraphic Commission for 1959-1960: AAPG Bul-

North American Commission on Stratigraphic Nomenclature

letin, v. 45, no. 5, p. 670-673, 1961.

27. Frye, John C., and H. B. Willman, Morphostratigraphic units in Pleistocene stratigraphy: AAPG Bulletin, v. 46, no. 1, p. 112-113, 1962.

28. Shaver, Robert H., Application to American Commission on Stratigraphic Nomenclature for an amendment of Article 4f of the Code of Stratigraphic Nomenclature on informal status of named aquifers, oil sands, coal beds, and quarry layers: AAPG Bulletin, v. 46, no. 10, p. 1935, 1962.

29. Patton, John B., Records of the Stratigraphic Commission for 1961-1962: AAPG Bulletin, v. 47, no. 11, p. 1987-1991, 1963.

30. Richmond, Gerald M., and John G. Fyles, Application to American Commission on Stratigraphic Nomenclature for an amendment of Article 31, Remark (b) of the Code of Stratigraphic Nomenclature on misuse of the term "stage": AAPG Bulletin, v. 48, no. 5, p. 710-711, 1964.

31. Cohee, George V., Records of the Stratigraphic Commission for 1963-1964: AAPG Bulletin, v. 49, no. 3, pt. I of II, p. 296-300, 1965.

32. International Subcommission on Stratigraphic Terminology, Hollis D. Hedberg, ed., Definition of geologic systems: AAPG Bulletin, v. 49, no. 10, p. 1694-1703, 1965.

33. Hedberg, Hollis D., Application to American Commission on Stratigraphic Nomenclature for amendments to Articles 29, 31, and 37 to provide for recognition of erathem, substage, and chronozone as time-stratigraphic terms in the Code of Stratigraphic Nomenclature: AAPG Bulletin, v. 50, no. 3, p. 560-561, 1966.

34. Harker, Peter, Records of the Stratigraphic Commission for 1964-1966: AAPG Bulletin, v. 51, no. 9, p. 1862-1869, 1967.

35. DeFord, Ronald K., John A. Wilson, and Frederick M. Swain, Application to American Commission on Stratigraphic Nomenclature for an amendment of Article 3 and Article 13, Remarks (c) and (e), of the Code of Stratigraphic Nomenclature to disallow recognition of new stratigraphic names that appear only in abstracts, guidebooks, microfilms, newspapers, or in commercial or trade journals: AAPG Bulletin, v. 51, no. 9, p. 1868-1869, 1967.

36. Cohee, George V., Ronald K. DeFord, and H. B. Willman, Amendment of Article 5, Remarks (a) and (e) of the Code of Stratigraphic Nomenclature for treatment of geologic names in a gradational or interfingering relationship of rock-stratigraphic units: AAPG Bulletin, v. 53, no. 9, p. 2005-2006, 1969.

37. Kottlowski, Frank E., Records of the Stratigraphic Commission for 1966-1968: AAPG Bulletin, v. 53, no. 10, p. 2179-2186, 1969.

38. Andrews, J., and K. Jinghwa Hsü, A recommendation to the American Commission on Stratigraphic Nomenclature concerning nomenclatural problems of submarine formations: AAPG Bulletin, v. 54, no. 9, p. 1746-1747, 1970.

39. Wilson, John Andrew, Records of the Stratigraphic Commission for 1968-1970: AAPG Bulletin, v. 55, no. 10, p. 1866-1872, 1971.

40. James, Harold L., Subdivision of Precambrian: An interim scheme to be used by U.S. Geological Survey: AAPG Bulletin, v. 56, no. 6, p. 1128-1133, 1972.

41. Oriel, Steven S., Application for amendment of Article 8 of code, concerning smallest formal rock-stratigraphic unit: AAPG Bulletin, v. 59, no. 1, p. 134-135, 1975.

42. Oriel, Steven S., Records of Stratigraphic Commission for 1970-1972: AAPG Bulletin, v. 59, no. 1, p. 135-139, 1975.

43. Oriel, Steven S., and Virgil E. Barnes, Records of Stratigraphic Commission for 1972-1974: AAPG Bulletin, v. 59, no. 10, p. 2031-2036, 1975.

44. Oriel, Steven S., Roger W. Macqueen, John A. Wilson, and G. Brent Dalrymple, Application for addition to code concerning magnetostratigraphic units: AAPG Bulletin, v. 60, no. 2, p. 273-277, 1976.

45. Sohl, Norman F., Application for amendment concerning terminology for igneous and high-grade metamorphic rocks: AAPG Bulletin, v. 61, no. 2, p. 248-251, 1977.

46. Sohl, Norman F., Application for amendment of Articles 8 and 10 of code, concerning smallest formal rock-stratigraphic unit: AAPG Bulletin, v. 61, no. 2, p. 252, 1977.

47. Macqueen, Roger W., and Steven S. Oriel, Application for amendment of Articles 27 and 34 of stratigraphic code to introduce point-boundary stratotype concept: AAPG Bulletin, v. 61, no. 7, p. 1083-1085, 1977.

48. Sohl, Norman F., Application for amendment of Code of Stratigraphic Nomenclature to provide guidelines concerning formal terminology for oceanic rocks: AAPG Bulletin, v. 62, no. 7, p. 1185-1186, 1978.

49. Caldwell, W.G.E., and N. F. Sohl, Records of Stratigraphic Commission for 1974-1976: AAPG Bulletin, v. 62, no. 7, p. 1187-1192, 1978.

50. Weiss, Malcolm P., Proposal to change name of commission: AAPG Bulletin, v. 63, no. 10, p. 1986, 1979.

51. Weiss, Malcolm P., and James D. Aitken, Records of Stratigraphic Commission, 1976-1978: AAPG Bulletin, v. 64, no. 1, p. 136-137, 1980.

52. Harrison, Jack E., and Zell E. Peterman, A preliminary proposal for a chronometric time scale for the Precambrian of the United States and Mexico: GSA Bulletin, pt. I, v. 91, no. 6, p. 377-380, 1980.

APPENDIX 2

THE DEFINITION AND CONCEPT OF GEOLOGIC-CLIMATE UNITS

[Excerpts from the *Code of Stratigraphic Nomenclature,* American Commission on Stratigraphic Nomenclature, 1970, pp. 16–17. Note that geologic-climate units are no longer recognized by the *North American Stratigraphic Code* (see Appendix 1).]

. . . A geologic-climate unit is an inferred widespread climatic episode defined from a subdivision of Quaternary rocks.

Definition.—A geologic-climate unit is defined from its records, which are bodies of rock, soil, and organic material. At any single place the time boundaries of the geologic-climate unit are defined by the boundaries of some kind of stratigraphic unit. These local boundaries may be isochronous surfaces, but the different stratigraphic boundaries that define the limits of the geologic-climate unit in different latitudes are not likely to be isochronous. In this respect geologic-climate units differ from geologic-time units, which are based on time-stratigraphic units. The locality where the geologic-climate unit is first defined is its type locality.

Principal purpose.—Geologic-climate units are used (i) in correlating episodes of deposition of Quaternary rocks in different areas, and (ii) in determining the historical sequence of events in the Quaternary Period.

Extent.—Geologic-climate units may be extended geographically as far as the record of the geologic climate can be identified, regardless of changes of facies of rocks, soils, or other materials that constitute the record.

. . . Glaciation and interglaciation are fundamental units of geologic-climate classification; stade and interstade are subdivisions of a glaciation.

Definition.—(i) A glaciation was a climatic episode during which extensive glaciers developed, attained a maximum extent, and receded. (ii) An interglaciation was an episode during which the climate was incompatible with the wide extent of glaciers that characterized a glaciation. (iii) A stade was a climatic episode within a glaciation during which a secondary advance of glaciers took place. (iv) An interstade was a climatic episode within a glaciation during which a secondary recession or a standstill of glaciers took place.

Nomenclature.—Formal names of geologic-climate units should be chosen in accordance with the rules . . . that govern the naming of rock-stratigraphic units. A geologic-climate unit may be named after a rock-stratigraphic unit, a soil-stratigraphic unit, or some other geographically named stratigraphic unit. In the type locality of the geologic-climate unit the record of its major climatic characteristics should be plain, and the evidence of climatic change at the lower and upper limits should be manifest.

APPENDIX 3

THE DECADE OF
NORTH AMERICAN GEOLOGY
1983 GEOLOGIC TIME SCALE
[FOR THE EARTH]
COMPILED BY A. R. PALMER

[Reprinted from *Geology* 11:503–504, 1983, with the permission of A. R. Palmer and the Geological Society of America.]

Note: This geologic time scale is reprinted here only as an example of a current, widely used time scale. The present author does not endorse this particular time scale over any others; the interested reader should critically compare Palmer's (1983) compilation with the time scales of other workers, such as Harland et al. (1982) and Haq and van Eysinga (1987).

The Decade of North American Geology
Geologic Time Scale
Compiled 1983

Compiled by
Allison R. Palmer
Centennial Science Program Coordinator
Geological Society of America, P.O. Box 9140, Boulder, Colorado 80301

Preparation of the 27 synthesis volumes of *The Geology of North America* for the Decade of North American Geology (DNAG) is now in progress. In order to encourage uniformity among DNAG authors in the citation of numerical ages for chronostratigraphic units of the geologic time scale, an ad hoc Time Scale Advisory Committee was established by the DNAG Steering Committee in 1982. This advisory committee, consisting of Z. E. Peterman (Chairman) and J. E. Harrison, U.S. Geological Survey; R. L. Armstrong, University of British Columbia; and W. A. Berggren, Woods Hole Oceanographic Institution, was asked to evaluate numerical dating schemes that were either recently published or in press and to provide recommendations for the best numbers to use in preparation of a DNAG time scale. The chart on the opposite side of this page was developed from the recommendations of the Time Scale Advisory Committee.

Geochronometric ages (Ma, Ga) assigned to chronostratigraphic boundaries are subject to several uncertainties in addition to those introduced by the numerical dating methods themselves; boundary stratotypes for many units are not yet chosen, so disagreement exists about exact biostratigraphic placement and correlation of a boundary; and many materials that can be numerically dated are not known in good context with biostratigraphic data, so extrapolation to a chronostratigraphic boundary is commonly required. Furthermore, with respect to the late Mesozoic and the Cenozoic, differing numerical age calibrations of the magnetic polarity-reversal scale based on differing choices of scattered isotopically dated tie points, differing interpretations of the positions of biostratigraphic boundaries with respect to the polarity-reversal scale, and uncertainties in the meaning of isotopic ages derived from glauconies lead to disagreement about ages assigned to some chronostratigraphic boundaries.

With these caveats, the numerical ages given in this chart represent interpretations acceptable to the DNAG Time Scale Advisory Committee. The uncertainty bars for Paleozoic and Mesozoic ages are from data in Harland and others (1982). Uncertainty bars for the Cenozoic are not available.

Sources for the numerical ages and for the chronostratigraphic nomenclature are given below.

CENOZOIC
Berggren, W. A., Kent, D. V., and Van Couvering, J. A., 1984, Neogene geochronology and chronostratigraphy; *in* Geochronology and the geologic record: Geological Society of London (in press).

Berggren, W. A., Kent, D. V., and Flynn, J. J., 1984, Paleogene geochronology and chronostratigraphy, *in* Geochronology and the geologic record: Geological Society of London (in press).

MESOZOIC
Base of Campanian to end of Cretaceous
Berggren, W. A., Kent, D. V., and Flynn, J. J., 1984, Appendix, *in* Geochronology and the geologic record: Geological Society of London (in press).

Base of Aptian to base of Santonian
Harland, W. B., Cox, A. V., Llewellyn, P. G., Picton, C.A.G., Smith, A. G., and Walters, R., 1982, A geological time scale: Cambridge, Cambridge University Press, 128 p.

Base of Hettangian to base of Barremian (dating and chronostratigraphic correlation of the "M" series)
Kent, D. V., and Gradstein, F. M., 1984, A Jurassic to Recent chronology, *in* Tucholke, B. E., and Vogt, P. R., eds., The Western Atlantic region, Volume M of The geology of North America: Boulder, Colorado, Geological Society of America (in press).

Note: Rhaetian has been eliminated from the Late Triassic chronostratigraphic scale following Tozer, E. T., 1979, Latest Triassic ammonoid faunas and biochronology, western Canada: Geological Survey of Canada Paper 79-1B, p. 127–135.

Base of Ladinian to base of Norian
Armstrong, R. L., 1982, Late Triassic–Early Jurassic time scale calibration in British Columbia, Canada, *in* Odin, G. S., ed., Numerical dating in stratigraphy: New York, John Wiley & Sons, p. 509–513.

Base of Scythian to base of Anisian
Webb, J. A., 1982, Triassic radiometric dates from eastern Australia: *in* Odin, G. S., ed., Numerical dating in stratigraphy: New York, John Wiley & Sons, p. 515–521.

PALEOZOIC
All numerical ages except those for the upper and lower boundaries of the Paleozoic are derived from Harland and others (see above, 1982, p. 52–55). Late Carboniferous numbers are for continentally based ages (N = "Namurian"; W = Westphalian; S = Stephanian). The marine-based ages are from Harland and others (1982, Fig. 5.6). The earlier estimate for the base of the Cambrian at 570 Ma is retained.

PRECAMBRIAN
Harrison, J. E., and Peterman, Z. E., 1982, North American Commission on Stratigraphic Nomenclature, Report 9, Adoption of geochronometric units for divisions of Precambrian time: American Association of Petroleum Geologists Bulletin, v. 66, p. 801–802.

DECADE OF NORTH AMERICAN GEOLOGY
GEOLOGIC TIME SCALE

DNAG

GEOLOGICAL SOCIETY OF AMERICA

CENOZOIC

MAGNETIC POLARITY	AGE (Ma)	PERIOD	EPOCH	AGE	PICKS (Ma)
		QUATERNARY	HOLOCENE / PLEISTOCENE	CALABRIAN	0.01 / 1.6
		NEOGENE	PLIOCENE (L)	PIACENZIAN	3.4
			PLIOCENE (E)	ZANCLEAN	
			MIOCENE (L)	MESSINIAN	5.3 / 6.5
				TORTONIAN	11.2
			MIOCENE (M)	SERRAVALLIAN	15.1
				LANGHIAN	16.6
			MIOCENE (E)	BURDIGALIAN	21.8
				AQUITANIAN	23.7
		PALEOGENE	OLIGOCENE (L)	CHATTIAN	30.0
			OLIGOCENE (E)	RUPELIAN	36.6
			EOCENE (L)	PRIABONIAN	40.0
				BARTONIAN	43.6
			EOCENE (M)	LUTETIAN	52.0
			EOCENE (E)	YPRESIAN	57.8
			PALEOCENE (L)	THANETIAN	60.6
				SELANDIAN / UNNAMED	63.6
			PALEOCENE (E)	DANIAN	66.4

TERTIARY

MESOZOIC

MAGNETIC POLARITY	AGE (Ma)	PERIOD	EPOCH	AGE	PICKS (Ma)	UNCERT (m.y.)
		CRETACEOUS	LATE	MAASTRICHTIAN	66.4	
				CAMPANIAN	74.5	4
				SANTONIAN	84.0	4.5
				CONIACIAN	87.5	
				TURONIAN	88.5	2.5
				CENOMANIAN	91	
			EARLY (NEOCOMIAN)	ALBIAN	97.5	2.5
				APTIAN	113	4
				BARREMIAN	119	9
				HAUTERIVIAN	124	9
				VALANGINIAN	131	8
				BERRIASIAN	138	5
					144	5
		JURASSIC	LATE	TITHONIAN	152	12
				KIMMERIDGIAN	156	6
				OXFORDIAN	163	15
			MIDDLE	CALLOVIAN	169	15
				BATHONIAN	176	34
				BAJOCIAN	183	34
				AALENIAN	187	34
			EARLY	TOARCIAN	193	28
				PLIENSBACHIAN	198	32
				SINEMURIAN	204	18
				HETTANGIAN	208	18
		TRIASSIC	LATE	NORIAN	225	8
				CARNIAN	230	22
			MIDDLE	LADINIAN	235	10
				ANISIAN	240	22
			EARLY	SCYTHIAN	245	20

RAPID POLARITY CHANGES

PALEOZOIC

AGE (Ma)	PERIOD	EPOCH	AGE	PICKS (Ma)	UNCERT (m.y.)
	PERMIAN	LATE	TATARIAN	245	20
			KAZANIAN	253	20
			UFIMIAN / KUNGURIAN	258	24
			ARTINSKIAN	263	22
		EARLY	SAKMARIAN	268	12
			ASSELIAN		
	CARBONIFEROUS (PENNSYLVANIAN)	LATE	GZELIAN (S.)	286	12
			KASIMOVIAN (W.)	296	10
			MOSCOVIAN		
		EARLY	BASHKIRIAN (N.)	315	20
	(MISSISSIPPIAN)	LATE	SERPUKHOVIAN	320	
		EARLY	VISEAN	333	22
			TOURNAISIAN	352	8
	DEVONIAN	LATE	FAMENNIAN	360	10
			FRASNIAN	367	12
		MIDDLE	GIVETIAN	374	18
			EIFELIAN	380	18
		EARLY	EMSIAN	387	28
			SIEGENIAN	394	22
			GEDINNIAN	401	18
	SILURIAN	LATE	PRIDOLIAN	408	12
			LUDLOVIAN	414	12
			WENLOCKIAN	421	12
		EARLY	LLANDOVERIAN	428	8
	ORDOVICIAN	LATE	ASHGILLIAN	438	12
			CARADOCIAN	448	12
		MIDDLE	LLANDEILAN	458	16
			LLANVIRNIAN	468	16
		EARLY	ARENIGIAN	478	16
			TREMADOCIAN	488	20
	CAMBRIAN	LATE	TREMPEALEAUAN	505	32
			FRANCONIAN / DRESBACHIAN	523	36
		MIDDLE		540	28
		EARLY		570	

PRECAMBRIAN

EON	ERA	BDY. AGES (Ma)	AGE (Ma)
PROTEROZOIC	LATE	570	750 / 1000
		900	1250
	MIDDLE	1600	1500 / 1750
	EARLY	2500	2000 / 2250 / 2500
ARCHEAN	LATE	3000	2750 / 3000
	MIDDLE	3400	3250 / 3500
	EARLY	3800?	3750

Compiled 1983

Published by The Geological Society of America, Inc.
3300 Penrose Place, P. O. Box 9140
Boulder, Colorado 80301

MAP AND CHART SERIES MC-50

APPENDIX 4

GENERALIZED LUNAR STRATIGRAPHY AND GEOLOGIC TIME SCALE FOR THE MOON

[Based on Wilhelms, 1987]

"MAJOR SEQUENCES"	SYSTEMS/PERIODS	SERIES/EPOCHS	APPROXIMATE DATING OF BOUNDARIES
	Copernican		
			1.1 billion yrs ago
IV	Eratosthenian		
			3.2 billion yrs ago
		Upper/Late	
III	Imbrian		3.8 billion yrs ago
		Lower/Early	
			3.85 billion yrs ago
II	Nectarian		
			3.92 billion yrs ago
I	pre-Nectarian		
			4.55 billion yrs ago

INDEX

Abandonment, of a stratigraphic unit, 141–142, 334–335
Absolute age, 258
Absolute dating, in archeology, 177
Abundance, of species, 203–204
Abundance zone, 193
Acme, 186, 187, 188, 194–197
Acme-zone, 192
Acoustic impedance, 55, 56
Acoustic log, 54–55
Acrotem, 151
Acrozone, 190
Actualism, 70, 72, 73
Aeon, 259
Age, 7, 22, 148, 186, 188, 216, 219, 222
Ager, D., quoted, vi, 75–76, 136, 234, 235, 237
A horizon, 167
Allocorrelation, 123
Alloformation, 171
Allogroup, 171
Allomember, 171
Allostratigraphic units, 168–173, 345–347
  defined, 171
Alpha decay, 263
Alternate point system, 217
American Commission on Stratigraphic Nomenclature, opinions of, 115, 168, 215, 357
American school, of stratigraphy, v
Amino acid racemization, 262
Ancestor-descendant relationships, 192, 204
Annealing temperature, 275
Anonymous publication, 138
Antiquities, 177
Apparent age, 259, 264
Aquifer, 173, 174
  defined, 173
Aquifer system, 173, 174
Archean, 22–23, 361
Archeological stratigraphy, 175–179
Archeology, 1, 257, 278
Archeozoic, 22–23
Arduino, G., 12, 21, 25
Argon-argon dating, 273
Arkell, W., quoted, 184, 185, 186, 188, 195–196
Artifacts, human, 176, 177, 178
Assemblages, natural, 200
Assemblage-zone, 190, 193, 200
Assise, 164, 186

Asteroids, 181
Atlantic, 258
Attribute-defined units, 163–164
Aufgeschwemmptgebirge, 13, 14
Auxiliary subdivisions, 150
Auxiliary Stratotype Point, 134–135
Azoic, 23

Barendregt, R., quoted, 210
Barrell, J., quoted, 94
Baselap, 85
Baselevel, 93, 94, 251, 253
Basins, on Moon, 180–181
Batholith, 38–39
Bed, 74, 151, 155, 156, 162
  thickness of, 74
Bedding. *See* Stratification
Bedding plane, 233
Benson, R., quoted, 68
Bentonite bed, 217
Berggren, W., and J. Van Couvering, quoted, 92–93, 202, 222
Beta decay, 263
B horizon, 167
Billion, 259
Biochron, 188, 196, 198–199
Biochronology, 93, 200, 222
Biocorrelation (Biostratigraphic correlation), 112, 116, 123, 194–205
  as chronocorrelation, 194, 197
Biofacies (Biologic facies), 98, 103–104, 106, 193
Biohorizon, 189
Biointerval-zone, 192
Biosome, 105, 106
Biostratigraphic units, 6, 131, 147, 189–193, 342–343
  defined, 189
  spatial, 193–194
Biostratigraphy, 18, 19–20, 182–205, 226–227, 230–232
  defined, 193
  development of, 182–188
  dual, 193–194
Biostrome, 105, 106
Biotope, 104, 106
Bioturbation, 202, 203
Biozone, 187, 188, 196
Bird, J., 97